Computational Modeling of Inorganic Nanomaterials

Series in Materials Science and Engineering

Other books in the series:

2D Materials for Nanoelectronics
M Houssa, A Dimoulas, A Molle (Eds)

Automotive Engineering: Lightweight, Functional, and Novel Materials
B Cantor, P Grant, C Johnston

Strained-Si Heterostructure Field Effect Devices
C K Maiti, S Chattopadhyay, L K Bera

Spintronic Materials and Technology
Y B Xu, S M Thompson (Eds)

Fundamentals of Fibre Reinforced Composite Materials
A R Bunsell, J Renard

Novel Nanocrystalline Alloys and Magnetic Nanomaterials
B Cantor (Ed)

3-D Nanoelectronic Computer Architecture and Implementation
D Crawley, K Nikolic, M Forshaw (Eds)

Computer Modelling of Heat and Fluid Flow in Materials Processing
C P Hong

High-K Gate Dielectrics
M Houssa (Ed)

Metal and Ceramic Matrix Composites
B Cantor, F P E Dunne, I C Stone (Eds)

High Pressure Surface Science and Engineering
Y Gogotsi, V Domnich (Eds)

Physical Methods for Materials Characterisation, Second Edition
P E J Flewitt, R K Wild

Topics in the Theory of Solid Materials
J M Vail

Solidification and Casting
B Cantor, K O'Reilly (Eds)

Fundamentals of Ceramics
M W Barsoum

Aerospace Materials
B Cantor, H Assender, P Grant (Eds)

Series in Materials Science and Engineering

Computational Modeling of Inorganic Nanomaterials

Edited by

Stefan T. Bromley
University of Barcelona
Spain

Martijn A. Zwijnenburg
University College London
UK

CRC Press
Taylor & Francis Group
Boca Raton London New York

CRC Press is an imprint of the
Taylor & Francis Group, an **informa** business

CRC Press
Taylor & Francis Group
6000 Broken Sound Parkway NW, Suite 300
Boca Raton, FL 33487-2742

First issued in paperback 2020

© 2016 by Taylor & Francis Group, LLC
CRC Press is an imprint of Taylor & Francis Group, an Informa business

No claim to original U.S. Government works

ISBN-13: 978-1-4665-7641-4 (hbk)
ISBN-13: 978-0-367-78304-4 (pbk)

Library of Congress Cataloging-in-Publication Data

Names: Bromley, Stefan T., 1971- editor. | Zwijnenburg, Martijn A., 1976- editor.
Title: Computational modeling of inorganic nanomaterials / edited by Stefan T. Bromley and Martijn A. Zwijnenburg.
Other titles: Series in materials science and engineering.
Description: Boca Raton, FL : CRC Press, Taylor & Francis Group, [2016] | ©2016 | Series: Series in materials science and engineering
Identifiers: LCCN 2016003208| ISBN 9781466576414 (alk. paper) | ISBN 1466576413 (alk. paper)
Subjects: LCSH: Nanostructured materials--Mathematical models. | Inorganic compounds--Mathematical models.
Classification: LCC TA418.9.N35 C68246 2016 | DDC 620.1/150113--dc23
LC record available at http://lccn.loc.gov/2016003208

Visit the Taylor & Francis Web site at
http://www.taylorandfrancis.com

and the CRC Press Web site at
http://www.crcpress.com

Contents

Foreword. vii

Preface ix

Editors. xi

Contributors xiii

SECTION I Structure and Dimensionality

1 Nanoclusters and Nanoparticles 3
   Scott M. Woodley

2 One-Dimensional Nanosystems 47
   John Buckeridge and Alexey A. Sokol

3 Two-Dimensional Nanosystems 83
   Benjamin J. Morgan

4 Nanocluster-Assembled Materials 113
   Elisa Jimenez-Izal, Jesus M. Ugalde, and Jon M. Matxain

SECTION II Properties

5 Melting and Phase Transitions 151
   Florent Calvo

6 Nanoparticles and Crystallization 181
   Gareth A. Tribello

7 Mechanical Properties of Inorganic Nanostructures. 213
   Eduardo R. Hernández, Yukihiro Takada, and Takahiro Yamamoto

8 Thermal Properties of Inorganic Nanostructures 247
   Yukihiro Takada, Eduardo R. Hernández, and Takahiro Yamamoto

9 Modeling Optical and Excited-State Properties 269
   Enrico Berardo and Martijn A. Zwijnenburg

SECTION III Case Studies

10 Interfaces in Nanocrystalline Oxide Materials: From Powders
    Toward Ceramics 291
    Oliver Diwald, Keith P. McKenna, and Alexander L. Shluger

11 Heterogeneous Catalysis: Vanadia-Supported Catalysts for
    Selective Oxidation Reactions. 313
    Monica Calatayud

12 Metal-Supported Oxide Nanofilms. 335
    Marek Sierka

13 Cosmic and Atmospheric Nanosilicates 369
    Stefan T. Bromley and John M. C. Plane

Index 413

Foreword

This book brings together two of the most exciting and rapidly developing areas of current science: nanoscience and computational modeling. Nanoscience is now established as a distinct area of contemporary research, which interacts with almost all fields within physical and, increasingly, bio-sciences engineering. Within the science of inorganic materials—the focus of this book—the reach is again considerable: For example, nanoscience is central to much of contemporary research in catalysis, electronic, and energy materials, and its understanding is crucial in developing models of crystal growth and nucleation—an enduring challenge in chemistry and chemical engineering. The field is full of challenges, especially as both the atomic and electronic properties of inorganic matter at the nanoscale can be radically different from those of the bulk materials, and the differences need to be understood in detail if they are to be exploited.

Computational modeling again contributes to almost all areas of physical and bio-sciences engineering and provides a range of especially powerful tools in nanoscience. Modeling methods can, with growing accuracy, predict the structures, properties, and reactivities of matter at the atomic level—an ability that is crucially important in nanoscience, particularly in view of the frequently found difficulties in the experimental characterization of nanostructures. Modeling tools are increasingly used in conjunction with experiment in modeling nanostructures and in predicting their electronic properties. Indeed they are helping to unravel the fascinating new chemistry and physics of nanoscale inorganic matter.

The chapters in this book cover a range of topics in this growing field. Their authors are leading practitioners of the techniques discussed, and this book gives an overview of the limitations and challenges and the impressive achievements in the field. It will be an invaluable guide both to those already working in the field of inorganic nanoscience and to those entering this important area of contemporary science.

Prof. Richard Catlow FRS
Department of Chemistry
University College London
London, United Kingdom

Preface

The overall aim of this book is to provide an accessible overview of the flourishing field of computational modeling of inorganic nanomaterials. Therefore, our intention was to be particularly inclusive to inorganic materials (e.g., oxides, sulfides), and thus help to fill the voids left by other volumes that have tended to primarily concentrate on carbon or metallic nanostructures. Although some chapters also touch upon these important systems, our materials coverage is intentionally more general. Inorganic nanomaterials are increasingly being employed in technological applications and industrial processes (e.g., TiO_2, CdS, GaN) and are thus of real-world significance. More fundamentally, there is a need to have a general understanding of how the structure and properties of materials with different chemical compositions are affected by reduced sizes and dimensions. We have also attempted to cover various possible interpretations of the term "nanomaterial" with chapters that cover nanoclusters and nanoparticles (quasi-zero-dimensional), nanowires and nanotubes (quasi-one-dimensional), and nanofilms (quasi-two-dimensional).

Inorganic nanomaterials, largely due to their extremely reduced dimensions, are extremely difficult to characterize accurately in experiment. Such small-scale materials systems are typically more tractable for computational modeling, which can often provide unrivaled, detailed insights to complement and guide experimental research. As such, computational modeling is, and will increasingly be, critical for understanding inorganic nanomaterials and their future development. Our book brings together experts who have provided chapters on a range of modern computational modeling methods used to study the structure and properties of a varied spectra of inorganic nanomaterial systems.

Although this book is a collection of different contributed chapters, the authors were requested to respect the intention to make the final book a readable text that should be accessible to newcomers to the field and useful to the seasoned researcher. We take this opportunity to sincerely thank all authors whose considerable efforts have made this volume possible. This book is logically organized into three sections (I–III) according to the well-defined aspects of modeling inorganic nanomaterials, with each containing individual chapters that address particular types of modeling approaches and/or systems. Section I contains four chapters that cover different types of inorganic nanosystems with increasing dimensionality. Section II introduces various properties and phenomena associated with inorganic nanomaterials and how they may be described computationally. Specifically, modeling of melting and phase transitions, crystallization, and thermal, mechanical, optical, and excited state properties are discussed. Section III highlights a diverse range of important recent case studies of systems where modeling the properties and structures of inorganic nanomaterials is fundamental to their understanding.

Overall, we hope that this book provides the reader with

1. An introduction to key concepts and a general background to key computational modeling methodologies with pointers to more detailed literature where necessary
2. A guide to the most appropriate choices of models and methods for describing particular classes of inorganic nanosystems and their structural/chemical/physical properties
3. Pertinent examples directly from the state-of-the-art research of expert authors and the wider literature, helping to give practical guidance for embarking on new modeling investigations of this type
4. An awareness of the strengths and limitations of various approaches to modeling inorganic nanomaterials and of current outstanding challenges in the field

While by no means exhaustive in its coverage, we hope that this book will serve as a unified introduction to the field of computational modeling of inorganic nanomaterials for advanced undergraduate/graduate student readers. Due to its detailed wide-ranging content, we also envisage that this book will be a useful resource for academic researchers and industry professionals in materials science, solid state physics, computational chemistry, materials chemistry, surface science, nanoscience, and nanotechnology.

Stefan T. Bromley
Martijn A. Zwijnenburg
Barcelona, Spain

Editors

Stefan T. Bromley, PhD, is a Research Professor with the Catalan Institution for Research and Advanced Studies (Institució Catalana de Recerca i Estudis Avançats—ICREA) based in the Department of Physical Chemistry (University of Barcelona, Barcelona, Spain) and leads the Nanoclusters and Nanostructured Materials group. After earning his PhD in computational physics (University of Southampton, Southampton, United Kingdom), he held research posts in the United Kingdom (postdoctoral fellow, Royal Institution), the Netherlands (associate professor, Delft University of Technology, Delft, the Netherlands), and Spain (Ramon y Cajal fellowship, University of Barcelona, Barcelona, Spain). Dr. Bromley has published more than 120 refereed articles (including three reviews on modeling inorganic nanomaterials), which in total have accrued more than 3000 citations (h-index: 30). He has also authored seven book chapters and has given about 40 invited talks. His main general research interest is in the structure and properties of inorganic nanoclusters and their use as novel materials building blocks. In recent years, he has become increasingly involved in the study of inorganic nanoclusters with respect to their astronomical importance.

Martijn A. Zwijnenburg, PhD, is a lecturer and a UK Engineering and Physical Sciences Research Council (EPRSC) Career Acceleration fellow at University College London, London, United Kingdom, where he leads the Computational Photochemistry of Materials research group. He earned his PhD in chemical engineering (Delft University of Technology [TUD], Delft, the Netherlands) in 2004 and has held research posts in the Netherlands (postdoctoral research assistant, TUD), the United Kingdom (Marie Curie fellow and Netherlands Organisation for Scientific Research [NWO] Talent fellow at the Royal Institution of Great Britain, London, United Kingdom), and Spain (Juan de la Cierva fellow at the University of Barcelona, Barcelona, Spain). Dr. Zwijnenburg has published more than 50 refereed articles, which in total have accrued more than 600 citations (h-index: 15), and has given several invited talks. His main research interests are the prediction of the excited state properties of inorganic nanostructured, polymeric, and molecular materials; the link between these excited state properties and the application of materials in photovoltaics and photocatalysis; and the generation of realistic structural models of materials.

Contributors

Enrico Berardo
Department of Chemistry
Imperial College London
and
Department of Chemistry
University College London
London, United Kingdom

Stefan T. Bromley
Departament de Quimica Fisica
Universitat de Barcelona
and
Institució Catalana de Recerca i
    Estudis Avançats (ICREA)
Barcelona, Spain

John Buckeridge
Department of Chemistry
University College London
London, United Kingdom

Monica Calatayud
Laboratoire de Chimie Théorique
University Pierre et Marie Curie
Paris, France

Florent Calvo
Laboratoire Interdisciplinaire de
    Physique
Université Joseph Fourier
Grenoble, France

and

Institut Lumière-Matière
University of Lyon
Lyon, France

Oliver Diwald
Department of Materials Science and
    Physics
University of Salzburg
Salzburg, Austria

Eduardo R. Hernández
Instituto de Ciencia de Materiales de
    Madrid
Consejo Superior de Investigaciones
    Científicas
Madrid, Spain

Elisa Jimenez-Izal
Department of Quantum Chemistry
University of the Basque Country
Donostia, Spain

and

Kimika Fakultatea
Euskal Herriko Unibertsitatea
Euskadi, Spain

Jon M. Matxain
Department of Quantum Chemistry
University of the Basque Country
Donostia, Spain

and

Kimika Fakultatea
Euskal Herriko Unibertsitatea
Euskadi, Spain

Keith P. McKenna
Department of Physics
University of York
York, United Kingdom

Benjamin J. Morgan
Centre for Sustainable Chemical
  Technologies
and
Department of Chemistry
University of Bath
Bath, United Kingdom

John M. C. Plane
School of Chemistry
University of Leeds
Leeds, United Kingdom

Alexander L. Shluger
Department of Physics and
  Astronomy
University College London
London, United Kingdom

Marek Sierka
Otto-Schott-Institut für
  Materialforschung
Friedrich-Schiller-Universität Jena
Jena, Germany

Alexey A. Sokol
Department of Chemistry
University College London
London, United Kingdom

Yukihiro Takada
Faculty of Engineering
and
Department of Electrical Engineering
Tokyo University of Science
Tokyo, Japan

Gareth A. Tribello
Atomistic Simulation Centre
Queen's University Belfast
Belfast, United Kingdom

Jesus M. Ugalde
Department of Quantum Chemistry
University of the Basque Country
Donostia, Spain

and

Kimika Fakultatea
Euskal Herriko Unibertsitatea
Euskadi, Spain

Scott M. Woodley
Department of Chemistry
University College London
London, United Kingdom

Takahiro Yamamoto
Faculty of Engineering
and
Department of Electrical Engineering
Tokyo University of Science
Tokyo, Japan

Martijn A. Zwijnenburg
Department of Chemistry
University College London
London, United Kingdom

Structure and Dimensionality

1. Nanoclusters and Nanoparticles

Scott M. Woodley*

1.1   Introduction 3

1.2   Modeling Landscapes. 4

1.3   Challenges 6

1.4   Energy Functions 10

1.5   Global Optimization. 16

1.6   Global Optimization Algorithms: A Single Monte Carlo Walker. 17

1.7   Global Optimization Algorithms: A Single Walker Who Utilizes
      Local Gradient Information 22

1.8   Global Optimization Algorithms: Interacting Multiple Walkers,
      or Population Based 27

1.9   Atomic Structures of Nanoclusters 30

1.10 Summary 40

Acknowledgments. 42

References 42

1.1   Introduction

Interest in the synthesis and characterization of nanoparticles has been driven in part by their increasing use in a wide range of applications in electronics, energy conversion, and catalysis, where their large surface-area-to-volume ratio is beneficial [1]. Three interesting examples are vanadia, titania, and silver-based particles. The vanadium dioxide and titanium dioxide–based particles are tuned to absorb light within a certain range of frequencies, which are used in smart windows and sunscreen lotions, respectively. Silver nanoparticles, which have antibacterial properties, are useful in combating infections once the particles are incorporated in materials used in medical devices and hospital equipment; and they are exploited in our everyday life: in socks and shoes to help prevent odors. Moreover, interest in nanoparticles is driven by the need to understand and gain insight into the atomic mechanism of crystal nucleation and the early stages of crystal growth; see, for example, the perspective article on modeling nanoclusters and

*   To my children's grandparents, Valerie and Ronald White, and Lorraine and Ronald Woodley.

Chapter 1

nucleation by Catlow et al. [2] and the review on experimental and computational studies of ZnS nanostructures by Hamad et al. [3].

By definition, nanoparticles are particles between 1 and 100 nm in size. Reducing the size of the particle—by the removal of atoms as opposed to varying applied pressure—can lead to a structural phase change and, in some cases, to an atomic structure that differs dramatically from that of the bulk phase(s). Nanoparticles at the bottom of this range or below, composed of less than ~100 atoms, are referred to as nanoclusters. The structural diversity at the nanoscale is well demonstrated by zinc sulfide, a wide bandgap semiconductor that is used in optoelectronics and as a photocatalyst. In bulk, ZnS can adopt either the sphalerite (cubic) or wurtzite (hexagonal) structure. The two phases coexist in nature, but sphalerite is the most stable bulk form under ambient conditions, whereas for smaller nanoparticles, the wurtzite phase becomes more thermodynamically stable [4]; moreover, nanoparticles with mixed cubic and hexagonal stacking have been synthesized [5]. For the smallest particles, or nanoclusters of ZnS, the atomic structures no longer resemble a cut taken from any bulk phase; see later. Such nanoclusters are readily created by either laser ablation of the bulk structure or nucleation in solution [6–8].

Critically, for particles with sizes below 5 and 10 nm, no clear diffraction patterns are obtainable, and accurate structure determination of these small particles using standard x-ray diffraction techniques starts to fail. Diffraction techniques require a target of sufficient size, a large single crystal or a powder formed of smaller crystals (ideally all with the same crystalline phase). Structure determination of nanoclusters currently relies on computational techniques to predict the atomic structure for each size as the structure is dependent upon the number of atoms. Matching observed and predicted properties gives confidence in the current structural predictions.

1.2 Modeling Landscapes

What are the *properties* of nanoparticles? In materials science, a computational modeler may first want to ask, what are the *atomic structures* of these nanoparticles? In fact, only approximate atomic coordinates of the system are required as there are standard computational techniques for refining structures, after which physical and electronic properties can be calculated.

But hang on one second; I have already made some assumptions in the opening paragraph. The more thorough scientist would have wanted answers to many other questions. At what *temperature* and *pressure* are the nanoparticles under? Are the nanoparticles in *vacuum* or in a *medium*, like water, for example? Nanoparticles are also commonly *capped* by organic molecules (chemically passivated). The structure of the nanoparticles will depend on all these factors.

Perhaps, the best place to start is to ask the question of what occurs in nature and what do we mean by the statement *this is the structure of the nanoparticle*. One key parameter that has not been mentioned yet is that of *time*. As the precise atomic coordinates will fluctuate with time, average coordinates are more relevant and the concept of *locally ergodic regions* is required [9–12], or should at least be discussed. Assuming that the atomic structure can be measured and that it takes t_m seconds to make this measurement (t_m is often called the observation time), then over this period of time it is important that the measured time average of the coordinates does not change too much. In particular, we would demand that further measurements yield essentially the same average atom coordinates. This range of accessible coordinates is referred to as the locally ergodic region of space for the system. In a crystal, for example, each atom vibrates about one of the lattice sites. At any instantaneous moment in time, the atoms are likely to be slightly displaced from their respective lattice sites in different random directions, the maximum magnitude of which will be dependent on a number of factors including the atoms' local environment, temperature, and applied pressure. It certainly makes more sense to measure the average or equilibrium position of each atom. The accuracy of this measurement will depend on how well the instantaneous positions are sampled, that is, the time of measurement must be much greater than the time, t_{eq}^{R}, for each atom to explore its respective part of the locally ergodic region R (to equilibrate). In turn, the time to make this measurement must be shorter than the time required for the system to escape the current locally ergodic region R, the escape time, t_{es}^{R}. That is, the following inequality must be satisfied [9–12]:

$$t_{eq}^{R} \ll t_{m} \ll t_{es}^{R}. \tag{1.1}$$

One familiar analogy often used in the field of structure prediction is that between the state of the system (a particular atomic structure) and the location on a map of a person who is out exploring the countryside, the so-called *walker*. The altitude of the walker, or height of the landscape, is analogous to the (potential) energy of the system. The *energy landscape* can be generated by mapping out the energy of the system, $E(\mathbf{r})$, as a function of atomic coordinates (x_i, y_i, z_i), where the subscript i is used to distinguish each of the N atoms and $\mathbf{r}$ is the $3N$-component vector, formed by concatenating all atomic coordinates. In general, as the number of variables, or atomic coordinates, will typically be greater than two, we often speak of an *energy hypersurface*.

At this point it is convenient to assume that the hypersurfaces of interest are both smooth and continuous—there are no discontinuities in the energy or its gradient. The landscape will typically contain hills and valleys for the walker to explore. Let us, for the moment, assume our walker is wearing roller skates. The stability of the walker, and the corresponding atomic structure, will depend on the local environment of the walker. The speed of the walker will increase (decrease) when travelling downhill (uphill). In our analogous system, the speed of the walker is related to the kinetic energy of the atoms (we assume no friction is present). If the local landscape is flat—the $3N$-dimensional gradient in all directions is zero—then the speed of the walker will remain constant. Choosing to employ Cartesian coordinates, this can be expressed as

$$\mathbf{F}_i = -\left(\frac{\partial E}{\partial x_i}, \frac{\partial E}{\partial y_i}, \frac{\partial E}{\partial z_i} \right) = \mathbf{0} \quad \forall i, \tag{1.2}$$

where $\mathbf{F}_i$ is the force on atom i and, by definition, $\mathbf{r}_i = (x_i, y_i, z_i)$. Thus, if the initial speed of this walker is zero, then the walker will remain there indefinitely. Such points are referred to as *stationary points*; on an undulating landscape, however, it is important to ascertain whether the particular stationary point represents a stable or unstable point.

A local maximum on the landscape, or hilltop, represents an unstable stationary point as a small displacement of the walker will result in the walker moving away from this region (the walker experiences a *gravitational force* that pulls the walker down the hillside). An inflexion stationary point (often called a saddle point) is also an unstable point of the landscape. Consider the 1D landscape $E(x) = x^3$, which has an inflexion stationary point at $x = 0$; even if the direction of the initial speed of the walker is toward the local uphill (toward $x > 0$), the walker will eventually return to the stationary point but with a velocity in the direction of the local downhill (toward $x < 0$). A minimum on the landscape represents a stable stationary point as a small displacement of the walker will always result in the walker experiencing a gravitational force that pulls the walker back down the hill toward the minimum regardless of its initial displacement; essentially the walker oscillates about the local minimum (LM) and thus remains in this locally ergodic region, or energy basin. The *global minimum* (GM) on the energy landscape is *the stable* point. If the *local minimum* is not the lowest point on the landscape, then it is a *metastable* stationary point; given enough time the walker may escape this locally ergodic region and find a more stable configuration (even at zero temperature, quantum tunneling may occur for the individual nanocluster). For a 1D landscape, maxima are transition points between two regions. Assuming the walker is at an LM and ignoring quantum effects, the energy difference between this LM and a neighboring maximum is the energy barrier that the walker needs to overcome in order to change state (average atomic configuration) of the system; switch locally ergodic regions. If the walker can successfully pass over this barrier (within the observation time of interest), then the locally ergodic region includes both basins. Note that since escape times usually decrease with increasing temperature, the locally ergodic regions for the time scales of interest tend to be larger at elevated temperatures [12,13].

For multidimensional, or N-dimensional landscapes where $N > 1$, the stationary point can also be an order-n saddle point, where n is the number of orthogonal directions along which the stationary point is a 1D maximum. A mountain pass is a classical example of a region about a saddle point on a 2D landscape; see Figure 1.1a. The highest-energy point along the lowest-energy pathway between two LMs is a saddle point. Low-order saddle points are now the transition points between energy basins, or locally ergodic regions, that represent different atomic configurations. The likelihood of escape from an energy basin is not only dependent upon the height of the energy barrier but also the width of the pass, how many passes there are, and how easy it is to find and navigate through at least one of them (and the probability to return [14]). Finally, locally ergodic regions need not be one or several connected energy basins, or a nested sequence of marginally ergodic regions [12] (the latter representing, e.g., a glassy material), but could also be a region that is enclosed by entropic barriers [15]: the path of escape is via a very narrow pass, with no energy barrier, but that is hard for the system to find. Schematic examples of locally ergodic regions for both a glassy nanoparticle and an entropic barrier are shown in panels b and c of Figure 1.1. For further discussions, please see References 9 through 13.

1.3   Challenges

To predict the atomic structure of a nanocluster or nanoparticle, one should overcome several challenges.

The first challenge is to find a suitable analytical expression (or numerical representation) for a cost function, $\Xi(\mathbf{r})$, which will describe the system of interest in a *manageable* manner.

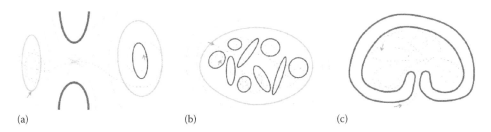

(a)   (b)   (c)

FIGURE 1.1 Example energy landscapes containing locally ergodic regions: (a) two local basins separated by a mountain range, (b) nine shallow local basins within a superbasin, and (c) an enclosed area containing no local minimum. Red, green, and dark green lines are the highest, low, and lowest contours; the broken blue line a trajectory, or path taken by a walker, and the star marks a saddle point (passing through the mountain range). In (a), the time-averaged atomic coordinates of the walker within the higher-energy basin represents a metastable structure that is predicted to be observed provided Equation 1.1 is satisfied before the walker escapes via the mountain pass to the lower basin representing the stable structure. In (b), if Equation 1.1 is satisfied for the path within the green contour, then an average of the structures from each local basin is observed. Equation 1.1 can also be satisfied with an entropic barrier, as shown in (c). Note that the paths within each locally ergodic region would be much longer and, for a random walker, the pathway more jagged as the momentum of the walker is not considered.

In particular, the cost function should be robust and cheap to evaluate in a numerical (computer-aided) procedure. An energy-based cost function is typically chosen for structural solutions within or near a locally ergodic region. It is convenient to define the region of the energy landscape, Ω, that corresponds to the thermally accessible atomic coordinates. Points outside of Ω would include, for example, structures containing two or more atoms at the same location. As we are only interested in solutions within Ω, the cost function need not be accurately defined outside of Ω and, therefore, can be redefined as a function based, for example, only on geometrical parameters without recourse to the actual physical stability. Points outside of Ω are therefore much cheaper to assess, and a larger proportion of available computer resources can be spent on assessing more feasible structures—ideally any search should be prevented from leaving Ω, which is not always easy or straightforward to enforce. An important requirement for such a surrogate cost function is that a higher value is returned for points outside Ω than within Ω. Moreover, the gradient (uphill) of the cost function outside of Ω should, ideally, point away from Ω. Another advantage of a surrogate cost function, which is discussed in more detail later, is that the energy for points outside of Ω is sometimes problematic to calculate. Even within Ω, the accuracy does not need to be very high as solutions can be refined in a subsequent stage. Importantly, the cost function is an approximation of the "true", or a more accurate landscape. For each LM of interest on the true landscape there should be an equivalent LM on the landscape defined by the cost function, and its approximate coordinates should lie within the equivalent basin on the more accurate energy landscape.

Successfully completing the first challenge does not immediately yield the required solutions as, going back to our analogous walker on a landscape, we should imagine that the entire landscape is blanketed by a fog. Regions only become visible after the walker has visited that spot, that is, $E(\mathbf{r})$ has been computed at each $\mathbf{r}$. Finding the locally ergodic regions on this energy hypersurface is *the second challenge*. In the first instance,

Chapter 1

it is common practice to search this hypersurface for LMs. Although not yet routinely performed, the locally ergodic region can be explored about each LM as a separate stage. At this point, it is important to note that as the size of the particle, or number of atoms, increases, so does the dimensionality of $E(\mathbf{r})$, the number of LMs, and hence the difficulty of finding the lowest-energy minima. Optimally, during the search we do not want to revisit the parts of the landscape, which have already been unveiled previously, or stumble in any wilderness devoid of energy basins. Global optimization techniques have, therefore, been developed and applied in this field.

Further challenges, which are not explored in detail in this chapter, include finding the local saddle points between the LMs [16], probability flows between LMs [14,17], and, finally, life (residence) times of structural solutions of a particular type [18]. The saddle points correspond to energy barriers between local basins, which can be conveniently mapped as dendritic (tree) graphs [19–21], from which the complexity of the landscape can be gleaned. For example, the task of locating the GM is much harder if the landscape is characterized by a map resembling a weeping willow (see Figure 1.2c) or where the GM resides outside a superbasin containing all other LMs (Figure 1.2b) rather than a hanging pine cone (Figure 1.2a). A plethora of tree graphs and their analysis can be found in the comprehensive book by David Wales [22].

In Sections 1.5 through 1.8, global optimization techniques employed to search for the atomic configuration with the lowest energy (GM) are described. Such approaches are readily applied to nanoclusters, whereas different techniques and assumptions are used for modeling nanoparticles as the number of possible configurations to generate (let alone evaluate) becomes intractable. In Section 1.9, it is demonstrated that $(XY)_n$ GM configurations are bulk-like for large n. It may be natural, therefore, to assume that

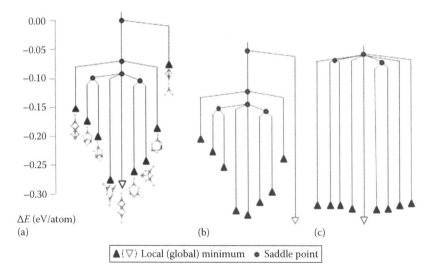

FIGURE 1.2   A dendritic tree map of (a) the B3LYP energy landscape for $(MgF_2)_3$ nanoclusters, where energies are taken from Neelamraju et al. [17], and two fictitious dendritic tree maps based on (a), but modified to demonstrate the landscape for which the global minimum (GM) is harder to locate because (b) funneling leads away from the GM, which is outside the superbasin containing all other local minima, and (c) there is no funneling. Note that only the highest-energy saddle point along the lowest-energy path between two local energy minima is marked on these maps.

nanoparticles are essentially the bulk cut that has the lowest surface energy (see discussion later on morphology). Ordering the GM configurations by size, starting from $n = 1$, the smallest stoichiometric nanocluster, one might observe (predict) a number of transition points: for example, planar to nonplanar configurations, non-bulk-like to bulk-like, or a change in the bulk phase that the bulk-like configurations adopt. Each transition can occur over a range of sizes, for example, after a series of non-bulk-like GMs, there is a sequence of non-bulk-like and bulk-like GMs, before there is only bulk-like GM. With enough atoms, it is also easy to imagine a polycrystalline nanoparticle, that is, a bulk-like configuration that contains grain boundaries. An example of a grain boundary in a nanoparticle is shown in Figure 1.3, a snapshot from a molecular dynamics simulation of a nanoparticle growing from one or more nucleation points. In nature, there may of course be many defects; for example, Sayle et al. [23] have also modeled microtwinning within a MnO_2 nanoparticle formed during a molecular dynamics simulation with a temperature schedule designed to first melt a bulk-like cut and then encourage recrystallization of a low-energy configuration (see discussion on simulated annealing in Section 1.6).

For large pure-phase single crystals, the morphology (crystal shape) can be determined by the requirement for the overall surface energies to attain a minimum. Typically, the equilibrium morphology is determined using the Wulff construction, where the distance from the center of the crystal to the center of each surface is proportional to its surface energy. Thus, low-energy surfaces are realized and dominate the morphology (see Figure 1.4). Computationally, the surface energy of each candidate face (chosen plane along which the bulk phase is cleaved) is readily determined using infinite surface models (a single slab with 2D periodic boundary conditions and the number of atomic layers parallel to the surface increased until the surface energy has converged or an infinite stack of slabs separated by vacuum with 3D periodicity and convergence is in respect to both atomic layers and the distance between slabs). For nanoclusters, the assumption that each surface ends a semi-infinite number of atomic layers is clearly wrong, although energy minima clusters can resemble cuts from the bulk phase (see cuboid examples in Section 1.9 which are cuts from the NaCl rock salt structure) but not always so. Hence, global optimization techniques are applied to search $E(\mathbf{r}_i)$ for all atoms, i. As the number of atoms of the cluster increases, there must come a point where the interior atoms of the

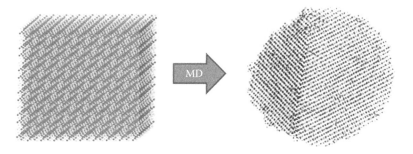

FIGURE 1.3 Structural model of a Li_2MnO_3 nanoparticle containing a grain boundary (between the gray and yellow highlighted halves), which was obtained using molecular dynamic simulations that was initialized from a cutout of the bulk phase [24]. Li, Mn, and O are represented by green, blue, and red spheres.

Chapter 1

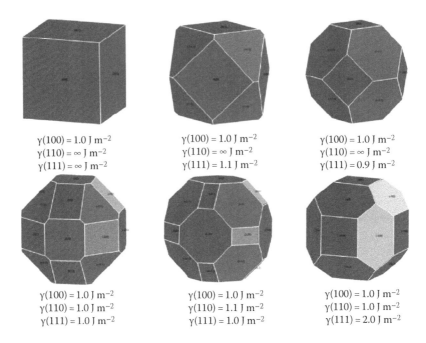

$\gamma(100) = 1.0$ J m^{-2}
$\gamma(110) = \infty$ J m^{-2}
$\gamma(111) = \infty$ J m^{-2}

$\gamma(100) = 1.0$ J m^{-2}
$\gamma(110) = \infty$ J m^{-2}
$\gamma(111) = 1.1$ J m^{-2}

$\gamma(100) = 1.0$ J m^{-2}
$\gamma(110) = \infty$ J m^{-2}
$\gamma(111) = 0.9$ J m^{-2}

$\gamma(100) = 1.0$ J m^{-2}
$\gamma(110) = 1.0$ J m^{-2}
$\gamma(111) = 1.0$ J m^{-2}

$\gamma(100) = 1.0$ J m^{-2}
$\gamma(110) = 1.1$ J m^{-2}
$\gamma(111) = 1.0$ J m^{-2}

$\gamma(100) = 1.0$ J m^{-2}
$\gamma(110) = 1.0$ J m^{-2}
$\gamma(111) = 2.0$ J m^{-2}

FIGURE 1.4 Equilibrium crystal (space group $Fm\bar{3}m$) morphology predicted using Wulff construction with the surface energies stated below each crystal. As the relative surface energy increases, the corresponding exposed surface decreases.

energy minimum structure resemble the bulk phase. Hence, the number of variables, and the space to search over, can be dramatically reduced by fixing or initializing them to bulk-like sites and then applying Monte Carlo (MC)-based global optimization or molecular dynamics to find the location of the outer atoms.

In the following section, the most commonly used cost functions, including energy definitions, will be outlined, which is aimed at addressing the first challenge. The second challenge will be addressed next under the assumptions that only an LM is sought (rather than the whole locally ergodic region) for nanoclusters in vacuum in the athermal limit and that all atomic coordinates are (initially) unknown variables. However, the methods described can be readily extended to include regions of atoms that are fixed.

1.4 Energy Functions

To assess the quality of a candidate structure, a cost function is evaluated; lower values imply a better fit to target requirements. For nanoclusters, an energy-based cost function is typically employed and the target requirements may also include m physical properties, for example, observed infrared frequencies [25,26] or, perhaps, desired energy difference between the lowest unoccupied and highest occupied electronic states [27]:

$$\Xi(\mathbf{r}) = E(\mathbf{r}) + \sum_m w_m \left| \lambda_m^{target} - \lambda_m^{calc.}(\mathbf{r}) \right|. \tag{1.3}$$

In this expression, w_m is the relative weight, with respect to the cluster's energy, given to the deviation of the calculated value(s) of property m from its target. In the field of structure prediction, the atomic coordinates, $\mathbf{r}$, are typically sought that only minimize the energy term, and thus discussions below are restricted to energy-based cost functions. Optimization algorithms presented in later sections are, however, also applicable to the more general expression in Equation 1.3, although the derivatives with respect to $\mathbf{r}$ for the physical properties are more demanding to compute.

As in other areas of molecular simulation, energies of nanoclusters can be calculated either by using methods based on interatomic potentials (IPs) or by employing electronic structure techniques. Calculating the energy using electronic structure techniques is in itself a challenge, particularly when the nanocluster does not resemble a *reasonable* chemical structure, or is a candidate structure that is outside of Ω (as defined in Section 1.3). For such candidate structures, we may encounter problems of nearly degenerate electronic terms, which are difficult to converge in the electronic space, and the computational cost of structural relaxation is typically greater than that for candidate structures within Ω. Moreover, the resulting configuration is not necessarily relevant as it could be a high-energy LM or even become fragmented. Thus, evaluation of such candidates using electronic structure approaches should be avoided as it is not just a significant computational cost but often does not aid the search for the global or other low-energy LM structures. Exploration of energy landscapes based on electronic structure methods has been successfully applied to generating low-energy minimum, atomic structures of nanoclusters; see, for example, References 17, 25, 28 through 33. Although sometimes not emphasized, some form of prescreening or filtering [33] using a surrogate cost function is employed to avoid a detailed assessment of non-Ω candidate structures. Note that electronic structure approaches will be introduced in Chapter 2, whereas we now describe atomistic approaches.

Energies of nanoclusters calculated using methods based on IPs are far less computationally expensive than electronic structure-based approaches. Therefore, larger numbers and sizes of clusters can readily be explored. These methods do, however, suffer from a number of limitations—IPs depend on coordination of the atoms, oxidation state of each atom is required to be predefined, and in some cases, key electronic effects (not modeled) are important to both atomic structure and energy ranking of minima—and consequently may in some cases give inaccurate results. Some limitations can also be advantageous; for example, if oxide solutions are required, then predefining all oxide anions with a formal oxidation state of −2 removes unwanted minima structures containing peroxides or fragmented clusters with isolated oxygen atoms (moreover, an isolated anion is more likely to have a stronger attraction to a cationic cluster).

A fruitful approach is to combine the two, with the potential-based methods being used in a first stage during the exploration of a range of cluster sizes and structures, while the electronic structure methods are used in a second stage to refine the energy ordering and structures of a selected subset of clusters (e.g., those calculated to be low in energy in the first stage). In the second stage, electronic structure methods can also be employed to test the reliability of the first-stage calculations, which are likely to be sensitive to the parameterization of the IP used. Typically the potential parameters are tuned to reproduce bulk structures and properties as these are more readily available. However, potential parameters have been refined to reproduce atomic structures of

Chapter 1

nanoclusters that have been optimized using electronic structure methods [34], especially when the local atomic structure of the nanocluster differs significantly from that in the bulk phases.

The total energy of the system, based on IPs, is written as a function of the atomic coordinates. For strongly ionic systems, such as alkali halides and alkaline earth metal oxides, the Born model provides a good basis, allowing the energy to be written as a sum of Coulomb and short-range terms:

$$E = \sum_{i=1, j<i}^{N} \left(\frac{q_i q_j}{r_{ij}} + E_{ij} \right), \tag{1.4}$$

where

the cluster is composed of N atoms

r_{ij} and E_{ij} are the distance and short-range pair-wise interaction between atoms i and j, respectively

q_k is the charge on the atom k

Note that zero energy is chosen to represent the system when all interatomic distances are infinite; Equation 1.4 gives the energy to form the cluster from its constituent ions; electron transfer between atoms is not simulated. Moreover, single-body terms are ignored as typically we want to compare clusters containing the same constituent ions and their contribution to energy differences is cancelled (the atomization energy—the difference between E and the sum of the energies of each individual atom—is more generally computed in electronic structure approaches; *cf.* the cohesive energy for bulk systems). The short-range interactions account for (simulate) Pauli repulsion between ions and Van der Waals interactions, or dispersion (London–London forces), and are typically modeled using one of the following three potentials:

$$E_{ij} = A_{ij} \exp\left(\frac{-r_{ij}}{\rho_{ij}} \right) + \frac{C_{ij}}{r_{ij}^6}, \tag{1.5}$$

the Buckingham potential [35];

$$E_{ij} = \frac{B_{ij}}{r_{ij}^{12}} + \frac{C_{ij}}{r_{ij}^6}, \tag{1.6}$$

the Lennard–Jones potential [36]; or

$$E_{ij} = D_{ij} \left\{ (1 - \exp(-\rho_{ij}(r_{ij} - a_{ij})))^2 - 1 \right\}, \tag{1.7}$$

the Morse potential [37,38]. The potential parameters, A_{ij}, B_{ij}, C_{ij}, D_{ij}, ρ_{ij}, and a_{ij}, are species dependent (hence the use of *ij* subscripts). When refining these potential parameters, it is important to remember what each term in E_{ij} or parameter represents, for example, C_{ij} controls the strength of the dispersion between atoms i and j, whereas a_{ij} is the equilibrium bond distance for a diatomic molecule composed of atoms i and j.

The shell model [39,40] is employed to describe the polarizability of individual ions; each ion, i, is represented by two point charges, a core with mass, M_i, and charge, X_i, and a massless shell with charge, Y_i, which are coupled together via a harmonic spring (with spring constant k_i). Note that the short-range potentials given earlier are typically applied between shells, that $X_i + Y_i$ is the charge of ion, and that the polarizability of the ion is proportional to Y_i^2/k_i. Using the shell model, the high-frequency dielectric constants can also be computed assuming only the positions of the shells have a chance to respond to a changing electric field.

In principle, Equation 1.4 can also include higher-order terms:

$$E = \sum_{i=1,j<i}^{N} \left(\frac{q_i q_j}{r_{ij}} + E_{ij} \right) + \sum_{i=1,j<i,k<j}^{N} E_{ijk} + \cdots. \tag{1.8}$$

For systems such as silica and silicates, where it is generally considered that there is an appreciable degree of covalence in the bonding, pair potential models still prove to be useful, but it is desirable to include some representation of the angular dependence of the bonding by incorporating simple bond angle dependence:

$$E_{ijk} = \frac{1}{2} k_{ijk} (\theta_{ijk} - \theta_0)^2, \tag{1.9}$$

where θ_0 and θ_{ijk} are the observed, in the bulk phase, and calculated, in the simulated cluster, bond angle $i\hat{j}k$.

For open-shell transition metal cations, where the atomic structure undergoes Jahn–Teller-driven distortions, and metallic clusters, a many-body potential is required [38,41,42].

Upon comparing total energies of nanoclusters, even in the athermal limit (note that various aspects of thermodynamics of inorganic nanomaterials can be found in Chapters 5 and 6), contributions to the energy from zero-point vibrations could also be included:

$$\Xi(\mathbf{r}) = E(\mathbf{r}) + \Xi^{vib}(\mathbf{r}), \tag{1.10}$$

where

$$\Xi^{vib} = \frac{1}{2} \sum_m \hbar \omega_m. \tag{1.11}$$

For the IP-based models, determining the harmonic phonon frequencies, ω_m, is straightforward, although slightly more expensive to compute than Equation 1.4. Typically, these contributions will make small changes in the energy ranking of LM clusters, so added in a second stage (after the search is completed).

Chapter 1

For nanoclusters of $(AB)_n$, $(ABC)_n$, $(ABCD)_n$,..., the stability typically improves with cluster size, n, and the total energy converges, as $n \rightarrow \infty$, to that of the bulk phase for AB, ABC, ABCD,..., respectively. Different energy functions have been employed in the quest to analyze the relative stability of clusters of different size. It makes sense to find the lowest-energy, GM, configuration for each size n and to analyze (typically plot as a function of n) binding energies with respect to the bulk phase:

$$E_n^b = E_n - E_\infty. \tag{1.12}$$

E_n and E_∞ are total energies per formula for the nanocluster of size n and the bulk phase, respectively. A different convention, where the binding energy per formula unit is defined with the opposite sign, and alternative definitions are often used including

$$\tilde{E}_n^b = E_\infty - E_n, \tag{1.13}$$

$$\check{E}_n^b = E_n - E_1, \tag{1.14}$$

and

$$\hat{E}_n^b = n(E_n - E_1). \tag{1.15}$$

These all give rise to negative, rather than positive, binding energies. Equations 1.14 and 1.15 give per formula and total binding energies, respectively, relative to the GM configuration for the smallest size, $n = 1$, and represent the stability of the nanocluster in a bath of $n = 1$ molecules. The binding energy of nanoclusters is often parameterized using simple analytical functions including inverse linear and cubic root dependencies.

To determine the stability toward fragmentation into two smaller nanoclusters of sizes m and $n - m$, the function

$$E_{n,m}^f = E_n - (E_m + E_{n-m}) \tag{1.16}$$

is evaluated for different values of m; the $E_{n,m}^f$ minima could be used to assess the fragmentation path. Choosing $m = 1$, we obtain a fragmentation energy that will influence the rate of nucleation from or removal of the gas phase monomers; a lower value would imply that the GM nanocluster of size n is more readily formed from the $n - 1$ sized nanocluster.

Alternatively, the stability of each GM cluster relative to both its smaller- and larger-sized GM neighbors can be evaluated using the second difference:

$$E_n^s = E_{n+1} + E_{n-1} - 2E_n. \tag{1.17}$$

A more positive value of E_n^s indicates a greater stability, whereas a negative value indicates that an exchange of a monomer is favorable between like pairs and leads to the depletion of nanoclusters of size n. Equation 1.17 can be further generalized to allow for the exchange of larger fragments and for an increase in the number of initial nanoclusters participating in the exchange.

If there is more than one cation (or anion) type, then ignoring changes in interatomic distances due to change in atom sizes, the difference between two nanoclusters may be simply a difference in the ordering of the different cations (or anions). Each unique configuration, upon ignoring atom types, is referred to as a structural type, and in the same manner as one would model solid solutions in a bulk phase, the stable structural type, as a function of temperature, will depend on its relative energy to other structural types and on the number of possible cation (or anion) arrangements.

Nanoclusters can also be grouped by the number of formula units they are composed from, or size, n. The mass spectra obtained from nanoclusters, which have been synthesized either through nucleation from solution or through laser ablation of a surface, sometimes indicate that particular sizes are more readily produced; see, for example, the mass spectra for $(ZnS)_n^+$, $(ZnSe)_n^+$, $(CdS)_n^+$, and $(CdSe)_n^+$ in References 43 and 44, which all show a significant increased peak for the size $n = 13$. These sizes are referred to in the literature as magic numbers [6]. For alkali halides, magic numbers occur for sizes where it is possible to construct a perfect cuboid, which are typically more stable than noncuboid cuts from the rock salt phase [7,45]. Thus, there appears to be a link between stability of the GMs and magic numbers of nanoclusters. Which configurations are synthesized may also depend on kinetics and configurational entropic contributions [46]. To include the effects of the latter will be important if the number of LMs with a similar energy as the GM differs greatly for neighboring sized nanoclusters. Comparing free energies that include configurational entropic contributions may provide a more successful route for predicting magic numbers.

For each size (or structural type), the free energy, Ξ_i^{conf}, which includes configurational entropic effects, can be written as

$$\Xi_i^{conf} = E_i^{GM} - k_B T \ln Z_i,\qquad(1.18)$$

where

   GM indicates that the first term is the energy for the GM nanocluster for a particular size (or structural type), labeled as i
   k_B is the Boltzmann constant
   T is temperature
   Z_i is the partition function:

$$Z_i = \sum_j m_i^j \exp\left\{-\frac{\left(E_i^j - E_i^{GM}\right)}{k_B T}\right\}.\qquad(1.19)$$

Note that the summation is over all unique configurations for a particular composition (and structural type), m_i^j is the degeneracy of each unique configuration, and E_i^j is the total energy of the individual nanocluster.

Chapter 1

Note that each energy term, E_i, in the cost functions (1.12) through (1.17), can be replaced by Ξ_i^{conf}. Free energies for subgroups of nanoclusters can be computed and compared. For example, one might want to either group pairs of enantiomers (left- and right-hand versions of essentially the same LM configuration), group LM configurations of the same structural type (configuration where atom type is ignored), or group LM configurations of the same size (fixed composition).

More generally, each energy term, E_i, in the cost functions (1.12) through (1.17), can be replaced by $\Xi_i^{conf} + \Xi_i^{vib}$, from Equations 1.11 and 1.18, so that configurational entropic contributions and contributions from zero-point vibrations are accounted for. Moreover, translational, rotational, and vibrational thermodynamic effects can be included, although typically cost functions include only the energy terms or energy terms and zero-point vibrations.

1.5   Global Optimization

Once a suitable cost function has been chosen—challenge one completed—attention is turned toward finding the best atomic structure as determined by exploring the landscape of this cost function. Of course during the exploration we may discover that the cost function is in fact not suitable or requires refining. In this chapter, we assume the cost function is one of the energy definitions given earlier and we are seeking the GM on the energy hypersurface for a particular composition, as well as all other atomic structures that correspond to LMs that have a slightly higher energy than the GM. These non-GM LMs may be required: as it is expected that the ranking in LM will differ slightly to that observed in nature; because more than the GM is realized in nature (LM within thermal energy of the GM); or you either want to ascertain how stable the GM is relative to other LMs or explain why a GM of a particular size of cluster is more difficult to find than anticipated. These LMs may also be discovered as a by-product during the primary search for the GM or the secondary search after the GM is found, but more sampling is required to increase our confidence that the tentative GM is the true GM. Examples in the literature include finding the lower energy configurations for clusters of $(NaCl)_n Cl^-$, which contain n Na^+ cations and $n + 1$ Cl^- anions [47]; for clusters of $(LaF_3)_n$, which contain n La^{3+} cations and $3n$ F^- anions [48]; and for clusters of $(Al_2O_3)_n$, which has $2n$ Al^{3+} cations and $3n$ O^{2-} anions [49]. Here, higher-energy LMs are targeted: in order to create disconnectivity graphs (which also requires finding the saddle points between LMs), which are then used to explain the difficulty of finding the GM [22]; as the energy ranking of structures is expected to change when a more accurate cost function is employed using the refinement of the predicted structures [34,48]; and in order to predict properties as a function of temperature (Boltzmann weighting contributions from the different configurations) [49]. As explained earlier, the computational expense will depend on how many atomic configurations are evaluated, that is, how many calculated energy points on the energy landscape, and how expensive it is to compute the energy of each atomic configuration or energy point on the landscape.

In this section, a number of global optimization algorithms and how they are applied to find the low-energy LMs on energy landscapes are described. First, algorithms based on a single walker are introduced; rather than running one long simulation (for each composition), such algorithms are typically tuned to locating LM near the initial starting point and therefore applied many times in order to help ensure that all important regions of the landscape have been explored as well as to simultaneously increase the confidence that all low-energy minima have been found. The simulation of many *independent* walkers is achieved by either running in parallel or sequentially. Note that each run is initialized using a different random seed, which determines the exact sequence of pseudorandom numbers used during any MC algorithm, typically leading to a different starting configuration (a different point on the energy landscape) and a different pathway across the energy surface. Afterward, algorithms based on interacting, multiple walkers [50] or a population of trial solutions (candidate structures) [51] will be discussed.

1.6 Global Optimization Algorithms: A Single Monte Carlo Walker

Perhaps the simplest approach to exploring a landscape is to evaluate it at a number of sample points that are chosen at random or on an evenly spaced predefined grid. The density of the sample points will determine both how thorough your search is (which in turn yields a confidence level in respect to finding LM and, in particular, the GM) and the accuracy of your LM atomic structures. Provided points near the desired LM are found, then the accuracy can always be improved by applying a standard local optimization scheme, which we will discuss later. If the sample points are evenly spaced across the entire landscape, then an improved level of confidence of whether the search has been successful can be achieved by increasing the density of sample points. The number of sample points to evaluate can quickly become too large, and bookkeeping is required to ensure that the same configuration is not evaluated more than once. Clearly it is advantageous if the region of the landscape to be sampled could be reduced without erroneously discarding the region containing the GM (e.g., regions corresponding to configurations composed of overlapping atom could be avoided).

For a nanocluster composed of two atoms, there is essentially one variable; the interatomic distance, r_{12}, where the subscript labels the atoms 1 and 2. Assuming fragmented nanoclusters are less stable, it seems sensible to restrict the search to distances near a typical bond length, for example, sample points can be taken as either a sequence of random numbers between 0.5 and 3.0 Å or a set of equally spaced points between these limits.

For a nanocluster composed of three atoms, there are three variables: the interatomic distances between the first atom and the other two, r_{12} and r_{13}, as well as the angle, θ_{23}, made by these lengths or the distance between the latter two atoms, r_{23}. Again we can restrict the search, $0.5 \text{ Å} \leq r_{ij} \leq 6.0 \text{ Å}$, where the maximum interatomic distance has been doubled in order to allow for a linear arrangement, for example, $\theta_{23}=0°$ or 180°. For nanoclusters composed of identical atoms, the search can be restricted further; for example, if atoms 2 and 3 are identical, then switching atoms 2 and 3 would yield the same configuration; many of such duplicates can be avoided by applying the constraint $r_{12} \leq r_{13}$. For ionic nanoclusters, cutoffs can be adjusted to account for shorter interatomic distances between cation–anion pairs than between either anion–anion and cation–cation pairs.

Chapter 1

For a nanocluster containing four atoms, there are six variables and now it is much easier to develop and implement an algorithm based on randomly placing N atoms within a defined region, Ω, an ellipsoid of volume that is proportional to N, under the constraint that $r_{ij} \geq 1.8r_C$ for cation–cation pairs, $r_{ij} \geq 1.8r_A$ for anion–anion pairs, and $r_{ij} \geq 0.8(r_C + r_A)$ otherwise, where r_A and r_C are the ionic radii of the anion and cation, respectively. These constraints can either be applied as each atom is randomly placed (if atom is placed too close to another already placed atom, then try another random location until okay) or at the end of defining all atomic positions (if constraints are not satisfied, then reject, i.e., generate another configuration rather than evaluate the cost function for this configuration). Generating and checking simple interatomic-based constraints are computationally far cheaper than evaluating the cost function or energy. The size of Ω will determine the ease with which the constraints on the inter-atomic distances can be satisfied when randomly placing the ions. The relative efficiency of the former approach—checking constraints upon placing each atom, rather than only after assigning atomic coordinates—increases as the chosen volume of Ω is decreased (i.e., when targeting more densely packed nanoclusters). This random approach to exploring the landscape is analogous to paratroopers exiting a plane that flies in the dark or above the clouds. Upon landing, each paratrooper calculates their altitude, unless they land on a mountain (analogous to the region of the cost function where the atomic configuration has one or more atoms too close to each other), which they then transmit back. This is a standard MC approach to finding LM, where the sequence of landing points of the paratroopers map out a random path across the landscape. The idea of a *random walker* makes more sense when the distance between consecutive points is small compared to the features of the landscape (randomly displace atoms from the current configuration by a maximum distance of ~0.1 Å in order to generate the next random configuration).

It is natural to ask how likely a random walker will find an LM, particularly when the walker is near an LM. Apart from the one-variable system, the random walker is, unfortunately, more likely to walk away from the nearby LM because there are more uphill directions than downhill. Consider a two-variable system. A step in the direction of the tangent to any contour around an LM will typically be uphill. If the number of variables increases, then the ratio of uphill to downhill directions near an LM will also increase. This extends to order M saddle points. At the saddle point, the energy is a maximum along M and a minimum along $(N - M)$ orthogonal directions, where N is the number of variables. Thus, if $N > 2M$ (a condition satisfied by saddle points that are typically nearer LM), then a walker at an order M saddle point is more likely to randomly select an uphill direction. This unfortunate result that a random walker is likely to move away from any nearby LM (an analogous result holds also for locating local maxima) can also be explained for nanoclusters in terms of atomic vibrations of the LM configuration. In general, an LM nanocluster composed of N atoms has $(3N - 6)$ normal modes of vibration. The atomic displacements, Δr_i, where i labels all the atoms in the nanocluster, required to go from this LM configuration and a random configuration near this LM can be expressed as a linear combination of the normal modes of vibration. It is also possible to generate $(3N - 5)$ orthogonal modes of vibration that are all orthogonal to this displacement. Thus, the walker at the random nearby point needs to choose one particular vibration and not any combination of the other $(3N - 5)$ orthogonal modes in order to transform its configuration to that of the LM.

Clearly, the random walker needs help to find the LM, particularly for larger systems (nanocluster composed of many atoms). Remember a *random walker* maps a random path across the landscape with each consecutive step corresponding to small random displacement of atoms. One way to drive the random walker toward LM is to only accept downhill steps, that is, any newly generated configuration that results in an uphill step (increase in energy) is discarded. Thus, all new configurations are based on small random structural changes to the lowest-energy configuration so far found (the tentative GM). The so-called successful steps form the pseudorandom path connecting the tentative GM. This procedure of finding LM is termed stochastic quenching. Starting from the same initial configuration that has an energy that is outside an energy basin, stochastic quenching may find different LMs as a random downhill path will not necessarily end up in the same energy basin. Of course, once within an energy basin, the eventual result will be the LM at the bottom of that basin (assuming the step size is smaller than any barrier width), so initializing several runs from the same point within an energy basin will result in the same LM.

The hillside leading to the GM may be sawtooth-like; a sequence of decreasing LM with each contained within a basin that has its lowest barrier in the direction of the GM—see left hand side of Figure 1.5a. Clearly, it would be advantageous to allow a certain number of uphill steps to prevent the walker from getting stuck in one of the LM on this or similar hillside. One implementation of a random walker that can successfully

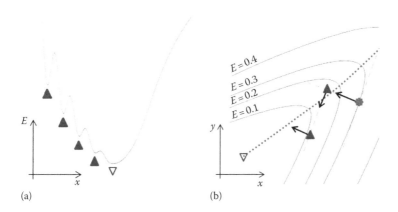

(a)

(b)

FIGURE 1.5   Energy landscapes as a function of (a) one and (b) two variables. The open and filled triangles mark the coordinates of the global minimum (GM) and the 1D local minimum (LM); the filled star indicates the starting point of a walker; arrows indicate the direction of steepest descent; dot dash lines are tangents to the contour at which the walker reaches an LM; and the dotted line is the lowest energy path along the valley floor. In (a), a walker who is reluctant to travel uphill will become stuck in an LM if starting on the left-hand hillside and steps are smaller than approximately half the distance between LMs. In (b), if the walker decides to proceed downhill along a sequence of straight lines that start in the direction of steepest descent (i.e., perpendicular to the contour) and ends at an LM (along the current direction of walking), then typically the walker will zigzag down to the GM as the direction of steepest descent is not perpendicular to the valley floor.

Chapter 1

search for the GM of superbasins (the definition of which will be provided later) employs the Metropolis acceptance criterion:

$$R_n \leq \exp\left(-\frac{\Delta E}{k_B T}\right), \tag{1.20}$$

where
R_n is a random number between 0 and 1
ΔE is the change in energy between the last accepted configuration and the new candidate
k_B is the Boltzmann constant
T is temperature

Downhill steps are always accepted, since the right-hand side would be greater or equal to one. Care is required in choosing a fixed temperature; set to zero and uphill steps will be rejected and the algorithm reverts back to a stochastic quench, whereas too high a temperature will remove any pressure on the walker toward LM. Returning to the example landscape in Figure 1.5a, the number of steps required to navigate from the top of the left-hand hillside down to the GM will depend on (1) how often a step to the right is attempted, (2) the chosen temperature, and (3) the step size. A higher temperature run would be advantageous if a larger sequence of successful uphill steps is required to escape an LM, although too high a temperature (or step size) and the walker might climb up into the next highest LM and begin an escape from the superbasin. For a higher-dimensional landscape, the efficiency will also depend on how the last accepted configuration can be modified—how the *moveclass operators* are defined (see later).

The temperature need not be fixed to a single value throughout the simulation. Switching between a high and low temperature would simulate a walker who explores the high ground for new superbasins and then, for a period of time (number of steps), is pressurized to seek deep down into the current superbasin the walker happens to be in (variations of this method include the bang-bang approach [52], thermal cycling [53], or the so-called multiquench [10, 11]). Starting the walker at a different initial point has the same effect and certainly quicker than running a high-temperature simulation; however, one is in danger of not fully exploring a superbasin. The user also has control over where individual walkers start, but this requires care to ensure the set of starting points span the key area of the landscape (include all important superbasins). For two variables, one can imagine an evenly spaced Cartesian grid of starting points, but for a nanocluster composed of more than three atoms, this is not so straightforward and, although quicker, the user may not want to influence the search for fear of missing a crucial superbasin.

One popular approach to predicting atomic structures is based on simulating the process of annealing, the so-called simulated annealing [54,55]. The simulation is run at different fixed temperatures, each lower than the last, and typically the fall in temperature is also reduced, for example,

$$T_{\text{new}} = 0.99 T_{\text{old}}. \tag{1.21}$$

Initializing the simulation at a high temperature ensures the walker is able to access all important regions of the landscape. It is important that the walker has enough time (steps) to sample the accessible regions and that the temperature is reduced slowly enough so that the walker does not become stuck in a high-energy basin. As temperature falls, the walker needs enough time to escape any such high-energy basins; once achieved the escaped walker is likely to spend more time away from this high-energy basin as the relative height of the barrier, or saddle point of escape (now a point of entrance), can quickly become higher from the side of the deeper basin (or more generally a lower region of the energy landscape). As the temperature is reduced further the walker is even less likely to be able to climb back up into the original high energy basin. In fact it is more crucial that a greater number of MC steps are performed at temperatures where barriers *become* impassable [56–58]. For small nanoclusters (clusters composed of few atoms), this technique is heavy handed as it requires *tuning* of a schedule to ensure that the walker does not get trapped in a higher LM (a minimum number of temperatures and MC steps at each temperature), that is, the number of MC steps taken to do this ends up greater than that required by a successful application of stochastic quenching. However, for larger-sized nanocluster, the fortunes of the two approaches reverse. Note that molecular dynamics could also be employed, instead of an MC walk, to probe configurations at each temperature [59].

It is desirable to gain some insight into how a simulation is progressing, but the low-energy solutions from a simulated annealing run are typically only obtained at the end of the simulation, when the simulated temperature is low. One solution is to create so-called holding points and investigate what is below the walker's current location [60,61]. At the holding point, a number of stochastic quenches are performed: if the holding point of the walker is in an energy basin, then all stochastic quenches will lead to the same LM. The analogy would be the walker resting while this children check out the neighborhood by playing roly-poly down the hillside in random directions. Before setting off again the children report back how far down they managed to roll—the altitude of the LM they found. Plotting the best energy visited by the walker and the LM the children found as a function of temperature helps clarify whether the walker is progressing well (the average and spread of LM found at each temperature gradually decreases [62]) or as has gone stuck (and a different schedule is required). Only one of the simulated annealing runs shown in Figure 1.6 (green data and longest temperature schedule) found the GM (double 3-ring); interestingly, the other runs either found the 2D ring LM (red and orange data) or the 2D edge-sharing 3-2-3-ringed LM (both blue data sets) that reside in the largest and second largest pseudo-basin (catchment area that includes the local basin and surrounding hypersurface from which the LM would be found after relaxation), respectively.

Returning to the roly-poly analogy, once the children (or indeed any walker from a stochastic quench) is within an energy basin, the eventual result will be the LM at the bottom of the basin (assuming the step size is smaller than any barrier width), but how quickly this will be found will also depend on the number of unsuccessful steps. Given the final destination is no longer determined by chance, it makes more sense to switch from a MC to a deterministic algorithm. It is typically much more efficient to relax the structure using a standard local optimization scheme, particularly if analytical gradients are available. In fact, the crosses in Figure 1.6 were obtained using such an approach; note, at lower temperatures, although both blue data sets correspond to a walker who is stuck in the basin containing the 2D edge-sharing 3-2-3-ringed LM, one of these walkers, due to the use of a larger

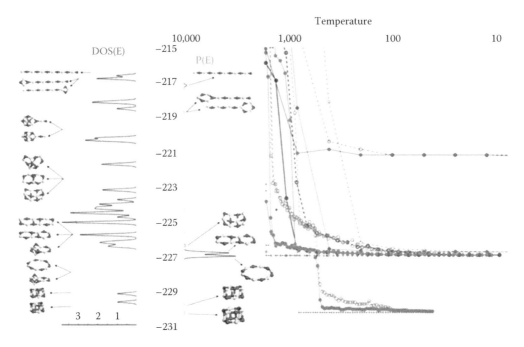

FIGURE 1.6 Global optimization schemes implemented with KLMC [48,63] found 42 local minima (LMs) on the energy landscape for $(MgO)_6$ as modeled using interatomic potentials applied to Mg^{2+} and O^{2-} rigid ions. The energies of these LMs are shown as a density of states curve, DOS(E), where ball-and-stick models of the LM configurations are also shown for 8 of the 19 peaks. From relaxing 50,000 random configurations, the probability of finding a particular LM from a random configuration was estimated and used to rescale the DOS intensities (red spectrum), resulting in the blue spectrum, P(E). The configurations for each noticeable peak in P(E) are also shown. Several Monte Carlo (MC)-based simulated annealing runs were employed to search the energy landscape for the GM: 10,000 MC steps per temperature T; a maximum MC step size of either 0.1 (highest set), 0.05 (lowest set), or 0.02 Å (otherwise) per atom; and, starting at $T = 1600$, a temperature schedule $T_{new} = \alpha T_{old}$, where $\alpha = 0.5$ (blue), 0.8 (orange), 0.9 (red), and 0.95 (green). The best (worse) accepted energy point for each temperature is plotted using closed (open) circles connected by a solid (broken) line. At the end of each temperature run, 10 independent stochastic quenches (1000 MC steps at $T = 0$) followed by local relaxation of the cluster, with a standard optimization routine, were conducted; the LMs found are shown as crosses (smallest for largest value of α). The energy scale is in eV and is applicable to DOS(E), P(E), and data from MC simulations.

MC step size, is unable to progress very far down into the basin (reduce energy difference between the closed blue circles and the blue crosses) until the standard local optimization scheme (that employs a smaller tolerance on the minimum step size) is applied.

1.7 Global Optimization Algorithms: A Single Walker Who Utilizes Local Gradient Information

A number of popular structure prediction approaches, particularly in the field of nanoclusters, combine global and standard local optimization routines to help in the quest of finding low-energy LM [64]. It is now, therefore, a convenient point to introduce local optimization schemes based on energy gradients (resultant forces on atoms). The method of conjugate gradients and quasi-Newtonian approaches are two popular local optimization routines that employ a sequence of linear minimizations. For conjugate gradients, the direction of each linear search is dependent upon the resultant atomic forces (gradient of energy with respect to small atom displacements) and the directions of the previous line searches. This approach is designed to avoid problems resulting from repeated overshooting as sometimes happens in the approach of steepest descent (on a 2D landscape, the first line minimization causes the walker to pass the valley floor as its approach to the floor of the valley is not along a contour and so can continue downhill; thus, as the next starting point is not on the valley floor, the second line is also not along the valley floor to the LM but back toward the valley floor that causes the walker to over-shoot again; see the zigzag path shown in Figure 1.5b). In a Newtonian approach, invert-ing the matrix of second derivatives (Hessian) provides both the direction and distance to the LM assuming the local landscape between the current point and the LM can be successfully approximated as a Taylor expansion to the second order. The more reason-able this approximation, the closer the calculated step will take the walker to the LM, from which a better accuracy of the Taylor expansion is likely (as the walker is nearer the LM) leading to a more accurate second step to the LM. The success of this itera-tive approach relies on a good starting point, which is typically not the case for initial structures that are not based on experimental data (e.g., extracted from x-ray diffraction studies). A more robust procedure, so-called quasi-Newtonian step, is to perform a line search for a minimum along the direction determined by the Newtonian step.

If there is not an analytical expression for the derivatives of the cost function, then, for N atoms, the $3N$ components of the forces and $9N^2$ components of the second deriva-tives can be computed numerically, for example, the force on atom i in the x-direction is

$$F_i = -\frac{\partial E(x_i)}{\partial x_i} \approx \frac{E(x_i - \delta x_i) - E(x_i + \delta x_i)}{2\delta x_i} \approx \frac{E(x_i) - E(x_i + \delta x_i)}{\delta x_i}, \tag{1.22}$$

where δx_i is a small displacement of atom i in the x-direction. Hence, to obtain the first derivatives, the energy is computed at a further $6N$ or $3N$ neighboring points on the landscape, whereas for the second derivatives

$$\frac{\partial^2 E(x_i)}{\partial x_i^2} \approx \frac{E(x_i + \delta x_i) - 2E(x_i) + E(x_i - \delta x_i)}{\delta x_i \delta x_i} \tag{1.23}$$

and

$$\frac{\partial^2 E(x_i, y_i)}{\partial x_i \partial y_i} \approx \frac{\begin{bmatrix} E(x_i + \delta x, y_i + \delta y) - E(x_i + \delta x_i, y_i - \delta y_i) \\ -E(x_i - \delta x_i, y_i + \delta y_i) + E(x_i - \delta x_i, y_i - \delta y_i) \end{bmatrix}}{4\delta x_i \delta y_i} \tag{1.24}$$

or, using energy points already calculated for energy, the first derivative and other sec-ond derivative components (e.g., Equations 1.22 and 1.23),

$$\approx \frac{\begin{bmatrix} E(x_i + \delta x, y_i + \delta y) - E(x_i + \delta x_i, y_i) - E(x_i - \delta x_i, y_i) + 2E(x_i, y_i) \\ -E(x_i, y_i - \delta y_i) - E(x_i, y_i + \delta y_i) + E(x_i - \delta x_i, y_i - \delta y_i) \end{bmatrix}}{2\delta x_i \delta y_i}$$

Chapter 1

so that a further $3N \times 3N$ energy evaluations are required. As well as computing the Hessian, the Newtonian approach still requires the Hessian matrix to be inverted, which scales poorly with the number of unknown atomic coordinates.

Rather than compute and invert the Hessian after each step, update schemes can be employed that provide an estimate to the Hessian or inverted Hessian. There is also a choice of line minimizer, but such matters can be readily found elsewhere [65]. The important take-home messages are as follows: (1) During each line search, at least three evaluations of the cost function are required (just to bracket the minimum in this direction) and the chosen step size used in each linear search will determine whether or not an energy barrier can be stepped over; (2) using derivatives comes at a computational cost that may, or may not, reduce the number of required line searches and therefore the overall cost of finding the LM; and (3) conjugate gradients are typically employed instead of a quasi-Newtonian approach when the current point is not within an energy basin and when employing electronic structure techniques that require numerical derivatives (each subject to costly self-consistent procedures).

Let us return to the topic of searching the energy landscape associated with the structure of a nanocluster and the analogy of an airplane of paratroopers. We now imagine each paratrooper to have skis and upon landing each alpine paratrooper immediately skis downhill (as determined by the local optimizer) and reports the altitude (and position) of the minimum they find. Although local optimization routines are more computationally demanding than single-point evaluations of the cost function, the alpine paratroopers do not attempt any uphill steps in order to explore a region of landscape for the LM, and the energy ranking of the sample points is more intuitive (a sample point on the wall of the basin containing the GM can be higher than a sample point in another basin). If very fortunate, one paratrooper will land in the basin of each LM that is sought. As discussed earlier, we maintain the assumption that a sufficiently small step size is employed during the local optimization to avoid steps that cross barriers. Given the number of LM found for $(LaF)_n$ [48], this would require around at least 1, 3, 11, 64, 180, 1451 alpine troopers for $n = 1, 2, 3, 4, 5,$ and 6, respectively. Note the rapid increase in the rate of change of the number of LM with n. Alpine troopers landing somewhere outside a basin may still end up at an LM; the *catchment area* of each LM is typically larger than the area of the basin. Importantly though, this catchment area is also dependent upon the chosen route the paratrooper skis along, that is, the LM found is dependent upon the choice and parameters of the local optimizer that is employed. Gradient-based quenching of random configurations is a particularly efficient approach when there are a small number of unknown variables [66]. Although one can imagine a path connecting the series of LM, each LM is not dependent on any details of the previously generated LM and the statistics of how often individual LMs are found can provide a measure of the catchment area for each LM [48]. This approach was used to estimate the probability of generating a certain configuration of $(MgO)_6$, labeled P(E) in Figure 1.6.

Combining gradient-based quenching with a global optimizer has proved popular and successful in the field of predicting low-energy atomic structures of nanoclusters [36,64,67]. Here we consider the MC walker that decides whether or not to accept an MC step (small random changes to the last accepted configuration) using the Metropolis criterion, Equation 1.20. The MC step now includes a structural relaxation, and the relaxed

FIGURE 1.7 Two example energy landscapes: the upper curve represents a typical continuous and smooth landscape, whereas the lower landscape is a set of plateaux generated by replacing each point on the upper curve with the local minimum value that would be obtained on the upper curve by rolling directly downhill from that point. Superbasins on the upper landscape are marked using a shaded background and are associated with basins on the lower landscape.

energy of the new candidate is compared to the energy of the last accepted configuration. If the step is successful, then the relaxed energy becomes the next successful energy of the walker and either (1) the LM structure or (2) the initial nonrelaxed configuration becomes the current configuration for future candidates to be based on.

For approach (2), the walker is effectively exploring a landscape composed of flat plateaux; see example shown in Figure 1.7. The transformed landscape has fewer barriers and basins and therefore presumably easier for a walker to locate the GM. The region of the original landscape that maps onto a basin on the transformed landscape is defined as a superbasin. This approach could be used to find the GM within a randomly selected superbasin with temperature set to zero and repeated enough times in order to sample all superbasins. Given the discrete nature of the transformed landscape, when a new higher-energy LM is found (walker steps onto a higher plateau), the energy difference is likely to be greater than that from a tradition MC step, so a higher temperature is required if not performing a quench (only downhill steps accepted).

For approach (1), which also benefits from the reduction in the number of LM and barriers, there is an increase in the likelihood of a walker becoming stuck; all new candidate structures relax back to the LM structure they were derived from. One solution is to simply increase/decrease the MC step size (maximum atomic displacement increased/decreased) to maintain the acceptance probability at a preset level. As the walker effectively hops from one LM to the next, the algorithm is coined Monte Carlo basin hopping (MCBH). (Many variations and implementations of this procedure exist in the literature; compare, for example, algorithms described within References 36, 61, 68, and 69.) The sequence of LM configurations found when applying such an approach, but with temperature fixed at zero (only downhill steps accepted), is shown in Figure 1.8 for three example systems. Interestingly, a similar number of attempted steps were required for the two binary systems, whereas a similar effort was required for a smaller-sized ternary system. The latter has an increased number of possible cation permutations to search through. Note, too large a MC step would remove the influence of the walker's experience and the algorithm reverts back to a set of independent random gradient-based quenches.

Chapter 1

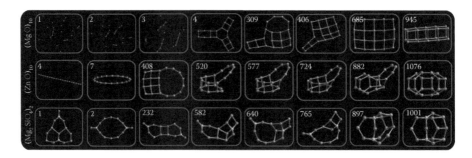

FIGURE 1.8 LM for $(MgO)_{10}$, $(ZnO)_{10}$, and $(Mg_2SiO_4)_2$ [35] obtained during Monte Carlo basin hopping (MCBH) quench of a random configuration in the upper, middle, and lower row, respectively. The minimum number of MCBH steps to obtain the shown nanocluster in the particular run is shown. The number of line searches during a relaxation was not unlimited: the fragmented configurations are less stable and unlikely to actually be LM.

Given the assumption that the lower landscape in Figure 1.7 is more readily searchable than the original upper landscape, then it seems reasonable that the lower landscape can also be transformed to a smaller set of wider plateaux that have the values of the superbasins. An algorithm for superbasin hopping has been developed by Cheng et al. [70] and applied to predicting Lennard–Jones clusters.

Only two moveclass operators (algorithm that generates a new configuration) have been discussed: a complete randomization of atomic coordinates or a set of small, random, atomic displacements that creates an MC step on the energy landscape. A sequence of the latter was visualized as the path of a walker travelling or hopping across a landscape. The success or failure of a global optimization approach can also be influenced by the choice of the moveclass operator, which can be tailored specifically for atomic structure prediction of nanoclusters. Examples include: changing the order of the atoms within the current configuration (clearly not useful for elemental nanoclusters, but, for example, extremely useful for searching homotops of bimetallic clusters [38]); displacing atoms that have a lower coordination than expected (internal atoms could even be fixed to bulk-like sites or based on a GM configuration of smaller sized nanoclusters); or displacing the least symmetrically located atom(s) to higher symmetry sites (useful if the GM nanocluster is expected to have high symmetry) [71]. As well as imposing restrictions on new candidate structures, the landscape itself can be modified; addition of penalty functions, for example, to push the walker away from regions already explored. Finally, crossover and mutation are two more examples of commonly employed moveclass operators in population-based algorithms for global optimization and will be discussed in a moment (Section 1.8).

Before leaving ideas of gradient-based algorithms and single walkers, is it worth mentioning that saddle points can also be targeted with such methods (energy lid [17,21] or threshold [72,73] algorithms). For example, starting at an LM, an MC walker can be used to explore the local region, but with the constraint that the walker remains below a certain altitude (energy lid). Holding points are used, where once again several children pay roly-poly and report back which LM they find. The energy lid is increased, and from the holding point, the walker continues. At a certain value of the lid, the children will discover a new LM, and an estimate of the barrier to escape the basin of the original LM is simply the energy difference between the current lid and the initial LM. This process can be repeated with smaller energy steps of the lid near this value to improve the accuracy of this estimate of the saddle point height. Alternatively, a grid of single-point energy calculations between the two LM can be performed and then analyzed to estimate the coordinates of the saddle point. For high-dimensional landscapes, the analysis is not straightforward, and methods based on the analogy of pouring water on the LM until water floods over the saddle point are used; see, for example, the algorithm implemented within the BUBBLE code [74]. Of course there are many gradient-based algorithms for finding saddle points: if two LMs are known, then the nudge elastic band approach [75] is popular, where points between the two LMs are kept equally spaced apart by the introduction of classical harmonic springs (elastic bands) and these stretch as, all going well, the points move toward the minimum energy path. Another example is that of *step and slide* [76] where we imagine two walkers at that same altitude who walk along their respective contour until the distance between them is minimized (slide) and then take a *step* toward each other (typically uphill) and slide again. Eventually, they should meet at the saddle point. Finally, using just one walker starting at an LM, the walker could progress uphill along the least (or greatest) incline until a saddle point is found (eigenvector following) [77]. Note that near a saddle point, local optimization can proceed based on keeping one imaginary phonon.

1.8 Global Optimization Algorithms: Interacting Multiple Walkers, or Population Based

Particle swarm [78,79], ant colony [80], taboo [81,82], and genetic algorithms [51,83,84] are population-based approaches that have been applied to atomic structure prediction. For example, in order to prevent simultaneous searching of the same location, an algorithm has been developed where multiple MC walkers effectively repel each other during their respective exploration on the locally minimized potential energy surface [50]. Of the population-based algorithms, genetic or, more generally, evolutionary algorithms are the most widely employed, and thus they are the topic of this section.

Genetic algorithms are based on mimicking the process of evolution; competition within a population to survive and procreate—creating offspring resembling parents—that drives the population toward a best fit to the environment. For nanoclusters, the lowest-energy LM, or GM, configuration is typically considered the best fit to the environment, although of course there is no reason why other properties cannot be included and a Pareto front [85] employed. Thus, the environment is typically the definition of the energy, or cost function, and the composition of the nanocluster.

The population is simply a set of candidate structures—each candidate a complete set of atomic coordinates. The population either evolves iteratively (whereby a new population replaces the last and the user can define the maximum number of iterations) or continuously (at any one moment an individual candidate or small set of candidates are replaced). The latter has become more popular as it lends itself better to parallelization; a core or node can select a subgroup from the population and when finished can replace them, if necessary, with their offspring without waiting idle for other cores or nodes to complete their tasks (as might be the case if a set number of offspring are required before the new population replaces the current).

Chapter 1

To simulate the process of procreation, a moveclass is required that can generate new candidate solutions that contain features of their parents. Two moveclass operators are in fact employed: crossover and mutation. For nanoclusters, the features are structural (atomic coordinates). Crossover typically generates two new candidates (offspring) by mixing atomic coordinates of two candidate structures taken from the current population. Traditionally, atomic coordinates are converted to binary numbers (this requires a discrete space, atoms either restricted to sites of a grid or displaced to the nearest grid point after relaxation) that are concatenated to form a sequence of 0s and 1s, the analogous DNA of an atomic configuration [86,87]. Swapping a random section of 0s and 1s between the two parents results in two new sequences corresponding to two new atomic configurations (nanoclusters), children that will have structural features of both parents. Clearly if both parents happen to be the same nanocluster, then the result of applying crossover will be two children that are identical to their parents—copies of the original nanocluster—and no evolution occurs. If the entire population has only one unique nanocluster, then crossover will never generate new candidate structures. The strength of an evolutionary algorithm, as compared to the single walker approaches described earlier, is its ability to span the search space, which is only achieved if a high diversity of the population is maintained. Diversity for nanoclusters is a measure of the number of different structural features contained within the population. The mutation operator is employed to potentially insert new structural features and help maintain a good diversity. Traditionally, this is achieved by randomly changing a small fraction of the 0s and 1s of the newly created offspring.

There is no requirement for the crossover and mutation operators to be designed so that only children that are composed of the better features are produced, as the competition to survive and then procreate will be biased toward killing off candidate nanoclusters containing bad structural features. However, avoiding the repeated chance of creating nonphysical nanoclusters can surely only be beneficial to the efficiency of the evolutionary algorithm, and many research groups employ phenotype versions of crossover and mutation [51]. The traditional versions of these moveclass operators work on an analogous gene (abstract series of 0s and 1s) and are referred to as genotype operators. From homogeneous parents, genotype operators can easily produce inhomogeneous nanoclusters, with denser regions containing nonphysical (too short) bond lengths. Working explicitly with atomic coordinates, phenotype operators have been designed to overcome, or at least reduce, the problem of sampling poor regions of the landscape. The most commonly employed phenotype crossover for nanoclusters proceeds as follows: apply a random rotation to each parent nanocluster, cut each into two fragments and create new nanoclusters by combining fragments of both parents [88]; see Figure 1.9. Of course there are some finer details missing here, for example, how to ensure the composition of the offspring matches the parents, which is easily managed; see References 38 and 51. For phenotype mutations, a standard MC step is sometimes applied, as well as the other moveclasses described earlier that target atom ordering or symmetry.

There are two commonly employed approaches to simulate competition between candidates that drives the evolution toward a better fit: the roulette wheel and direct tournaments. In the former, the sizes of the slots around the wheel are proportional to the relative fit of the individual candidates with respect to the current population (or subgroup if applying a continuous evolution). Thus, a random point on the wheel will more likely select a better candidate. Selection of n_c candidates that will survive or become

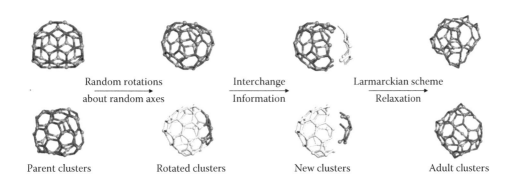

Random rotations | Interchange | Larmarckian scheme
about random axes | Information | Relaxation

Parent clusters | Rotated clusters | New clusters | Adult clusters

FIGURE 1.9 Two new configurations based on two current configurations: phenotype crossover for nanocluster. Note that although a random cut is applied to one configuration, the compositions of the exchanged fragments are identical. Without the last step, the crossover scheme is suitable for a Darwinian-based evolutionary algorithm.

parents can thus be achieved by spinning the roulette (generating a random number between 0 and the length of the circumference) n_c times. The application of the roulette is prone to reducing diversity of the population, which should be avoided, particularly in the earlier iterations. One solution is to start by having equally sized slots and gradually increase the disparity of slot size between better and worse candidates as iterations progress. Tournaments are typically simulated between two randomly chosen candidates; the best or the worst candidate is just as likely to be chosen as any other. Then a random number between 0 and 1 is generated, and if less than P_t, then the better candidate is selected. This procedure is easily generalized to tournaments between three or more candidates. The value of P_t will determine the strength and direction of the applied pressure on the population, either favoring evolution of better or worse candidates if greater or less than 0.5, respectively. A different fixed value is sometimes used for tournaments determining parenthood and that determining survival (or replacement for continuous evolution) and will be dependent upon the overall evolutionary algorithm. Alternatively, the value of P_t can depend on time—initiated at 0.5 and increased as the simulation progresses in order to help search more widely in the early stages of evolution (maintaining diversity) and gradually increasing the pressure to survive and/or procreate.

The number of ways the various components of an evolutionary algorithm, which is designed for targeting low-energy minima structures for nanoclusters, are strung together is probably greater than the number of research groups developing their own software for this task. Variations include the following: for procreation only simulate competition to become a father (equal chance for each candidate to become a mother) rather than for both parents; for survival allow candidates to be selected more than once, replace population with only new candidates, or allow elitism (best candidate always survives); allow (or not) all newly generated candidates at least one free pass into the next population; and offspring are either created by crossover or mutation rather than crossover and mutation. Ignoring the choice of moveclass operators, perhaps the most significant choice to make is whether to employ Darwinian or Lamarckian evolution. In the latter, offspring are immediately relaxed using a standard local optimization scheme (analogous to growing up). The population of adults only contains LM structures and the cost function is based on the relaxed structures. Future offspring will be composed of fragments from a relaxed structure and not

Chapter 1

the atomic coordinates passed to the parent from its parent (hence Lamarckian, rather than Darwinian). Advantages are similar to that already discussed for single walkers: as with simulated annealing, to get a good understanding on how well the Darwinian approach is proceeding, it also requires holding points whereby all (or at least the better) candidates in the current population undergo local optimization (as with stochastic quenching, the Darwinian algorithm should not be employed to find atomic coordinates to high precision); and the Lamarckian algorithm effectively searches the plateau landscape that has fewer barriers to navigate over. For the Lamarckian algorithms, each candidate structure is more expensive to evaluate and so the population is often much smaller and more prone to low diversity. The latter is solved by killing off (removing) duplicates of any candidate structure, but this in itself presents the problem of determining efficiently whether two candidate structures are the same; increasing the required accuracy during local optimization comes at a computational cost. Nevertheless, Lamarckian evolutionary algorithms are the most popular when employed to find low-energy minima for nanoclusters.

1.9　Atomic Structures of Nanoclusters

In this section, example atomic structures of nanoclusters are presented, starting with example LM configurations predicted for $n = 1$ nanoclusters of binary compounds, $(X_N Y_M)_n$, for a range of stoichiometry. Typically the LM configurations are ranked with respect to their relative energies. It is convenient to use this rank in the labeling of configurations; QM LM# (IP LM#) is ranked #th if the LM configurations are ordered with respect to the energy obtained using the QM functional in a DFT approach (IPs). Note that LM1 is therefore the GM and that the + symbol is used in Figure 1.13 to indicate that the ranking only includes nanoclusters shown in the figure. Ball-and-stick models are used in Figure 1.6, and in Figures 1.8 through 1.19 to display the atomic arrangements of the nanoclusters. In these two-dimensional figures I have chosen a rotation of each cluster that offers, in my opinion, the best view for an interested person to correctly visualize the three-dimensional configuration. Even carefully chosen rotations may lead an interested person unsure of the atomic configuration as atoms in the foreground hide other atoms further back. Hence, particularly for configurations of low symmetry and/or larger configurations that is only shown from one angle (whether in this book or elsewhere in the literature) I recommend that the atomic coordinates are obtained and uploaded into a graphical package is used where the nanoclusters can be rotated. One such freely available package forms part of the WASP@N database of published nanoclusters [89], and all real examples in this chapter are already uploaded. Moreover, nanoclusters (composition and atomic coordinates) of your own creation (via prediction software or by construction) can also be uploaded and searched against all entries in the database to reveal whether or not your structure, based on connectivity, is a *new discovery* (as far as the database is concerned). If the structure is new, then hopefully it is also lower in energy (more stable) than nanoclusters with the same composition. In the case where your nanocluster matches one or more in the database, then the relevant references are provided.

Example LM atomic structures for $(X_N Y_M)_1$ nanoclusters for a range of stoichiometry (different $N{:}M$ ratios) are shown in Figure 1.10. As these are the smallest configurations, there are not many LMs for each stoichiometry and the use of global optimization schemes is most

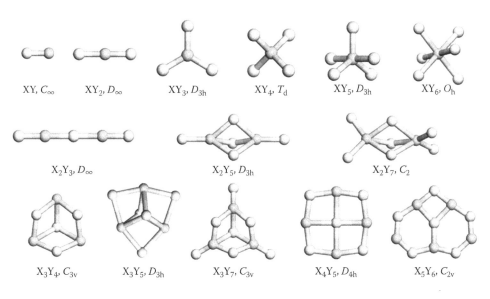

XY, C_∞ XY_2, D_∞ XY_3, D_{3h} XY_4, T_d XY_5, D_{3h} XY_6, O_h

X_2Y_3, D_∞ X_2Y_5, D_{3h} X_2Y_7, C_2

X_3Y_4, C_{3v} X_3Y_5, D_{3h} X_3Y_7, C_{3v} X_4Y_5, D_{4h} X_5Y_6, C_{2v}

FIGURE 1.10 Example local minimum atomic structures, $(X_N Y_M)_1$, for one unit of a number of compounds, $X_N Y_M$. Dark and light gray balls represent X and Y atoms, respectively.

definitely heavy handed. All the examples shown in Figure 1.10, employing any of the methods described earlier, are readily found for a rigid ion model (simple pair-wise IPs), which neglect, for example, any lone pair effects. Given no length scale is provided in Figure 1.10, the nanoclusters are representative GM configurations for many compounds; for example, the smallest binary, $(XY)_1$, corresponds to the GM for pretty much all 1–1 binary compounds (LiF, LiCl, ..., NaF, ..., CsI, ..., MgO, CaO, ..., ZnO, ZnS, ...) [25,28,46,59,84,90–95]. Examples for the other $n = 1$ nanoclusters can also be found in the literature; the next smallest composition, $(XY_2)_1$, has a higher symmetry than that of $(XY)_1$—an additional mirror plane—and is reported, for example, as the GM for $(SiO_2)_1$ [2,34] and $(MgF_2)_1$ [17].

The nanoclusters in Figure 1.10 have two common features: high symmetry and atoms with lower coordination numbers, as compared to that seen for these atoms in bulk phases. Due to the latter, these configurations are not always that easy to model using electronic structure methods as severe undercoordination can result in difficult convergence of the electronic state. Classic examples are shown in Figure 1.11, where degeneracy is lifted and symmetry is lowered, resulting in an increase in dimensionality of the configuration. The C_{2v} configuration for $(XY_2)_1$ is adopted by titania, zirconia, and hafnia [96–98], whereas the $n = 1$ configuration for lanthanum fluoride [48] and india [49,99] relaxes to the C_{3v} and C_{2v} configurations for $(XY_3)_1$ and $(X_2Y_3)_1$, respectively. As seen in molecules, like water, excess electrons act like additional atoms and reduce the symmetry; the H–O–H bond angle is slightly less than the ideal tetrahedral angle, rather than 180°. The configurations in Figure 1.11 are also problematic when a rigid ion model is employed as they are not stable, for example, the C_{2v} configuration for titania reverts back to the 1D D_∞ configuration. Polarization is extremely important for such small configurations, which requires at least the employment of a shell model (see earlier discussions). For larger-sized $(TiO_2)_n$, $(LaF_3)_n$, and $(In_2O_3)_n$ nanoclusters (and similar ionic compounds, like $(ZrO_2)_n$, $(HfO_2)_n$, $(AlF_3)_n$, $(Al_2O_3)_n$, and $(Ga_2O_3)_n$), there are less problems in converging the electronic structures, as undercoordination is

Chapter 1

XY$_2$, C_{2v} XY$_3$, C_{3v} X$_2$Y$_3$, C_{2v}

FIGURE 1.11   Polarized example local minimum atomic structures, $(X_NY_M)_1$, for one unit of a number of compounds, X_NY_M. Dark and light gray balls represent X and Y atoms, respectively.

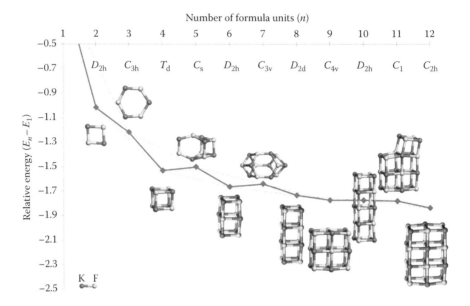

FIGURE 1.12   Ball-and-stick models of tentative global minimum (GM) (PBEsol) configurations for $(KF)_n$ nanoclusters [33], and their respective point group symmetry, overlaid on the graph based on their respective total PBEsol energies, E_n, in eV per formula unit, as a function of the number of formula units, n. Ten feint broken lines connect second nearest neighbor energy points.

minimized, and a better match of the DFT LMs are found with those obtained on the landscape defined using IPs, which is not surprising as the IP parameters are typically refined to reproduce the structure and properties of the bulk phases.

The properties of nanoparticles depend on the atomic structure, which, particularly for nanoclusters, will depend on its exact composition. Figure 1.12 contains tentative—someone may eventually discover lower-energy structures for these compositions—GM (PBEsol) configurations for nanoclusters of potassium fluoride [33]. In this series, as the stoichiometry is fixed, the composition is only dependent upon its size, n. Ignoring entropic factors, the stability (as measured by the relative PBEsol energies; Equations 1.12 through 1.15) of the tentative GM typically improves with the number of formula units and the energy per atom or formula unit will converge to the value for the bulk phase. This is a quite general result for inorganic nanoclusters (more examples will be shown later).

A comparison of the second differences, E_n^s (as defined in Equation 1.17), or the stability of each tentative GM nanocluster for $(KF)_n$ relative to both its smaller- and larger-sized GM neighbors, reveals an interesting result: an increased stability of perfect cuboids. For clarity, all configurations are shown below (above) the energy curve in Figure 1.12, if E_n^s has a positive (negative) value, that is, E_n is less (greater) than the average of its neighbors E_{n-1} and

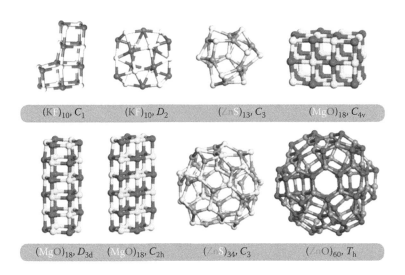

FIGURE 1.13 Ball-and-stick models, and their respective point group symmetry, of tentative PBEsol LM2 and LM4 configurations for $(KF)_{10}$ nanoclusters [33]; PBE/DNP GM configurations for $(ZnS)_{13}$ [46]; PBEsol GM, LM1+, and LM2+ configurations for $(MgO)_{18}$; IP GM configurations for $(ZnS)_{34}$; and IP GM configuration for $(ZnO)_{60}$ [100].

E_{n+1} (midpoint along the $(n-1)^{th}$ feint broken line in Figure 1.12). It is perhaps no surprise that the more stable set can be cut from the bulk phase—common rock salt structure—and of those there is an increase in stability for cuboids that are more cubic-like; $(KF)_{10}$ has a higher energy than a weighted average of energies for $(KF)_9$ and $(KF)_{12}$. For cuts that resemble a cuboid with a chunk missing, like that of $(KF)_{11}$ and the next LM for $(KF)_{10}$ (see Figure 1.13), E_n is slightly above the respective feint broken line. The other *less stable* GM configurations contain at least one hexagonal ring. Although each hexagonal ring can be transformed into two tetragonal rings, this is not favorable. For example, the resulting ladder for $(KF)_3$ is less stable than a hexagonal ring, even though the average coordination in the ladder is higher and despite the fact that the ladder and not the hexagonal ring can be cut from the KF bulk phase. At least three faces of the basic cuboid, $(KF)_4$, are required for stability; such structure features are within both $(KF)_5$ and $(KF)_7$. Simply increasing the average coordination number of the atoms does not necessarily result in a lower energy. To achieve a higher coordination, an increased in strain may be required, the balance of which may determine whether or not it is a favorable transformation. For example, the average coordination number for atoms in LM1 and LM2 for $(KF)_{10}$ is 3.6, which can be increased to 4 by transforming either of these to a chiral D_2 configuration that is composed of just tetragonal rings (shown in Figure 1.13); however, its stability, as measured by its rank, would worsen. The relative stability of cuboid cuts has been indirectly observed in time-of-flight mass spectrometry for $M(MX)_n^+$ and $X(MX)_n^-$ nanoclusters of alkali halides that were produced using laser vaporization of their respective bulk [7]; an increased abundance of a particular size matches those sizes where a perfect cuboid is possible.

Switching compounds, but keeping the same stoichiometry (1–1) and choosing a compound that adopts the same bulk structure, one might expect to find similar GM configurations and LM configurations for a particular size with similar ranking.

Chapter 1

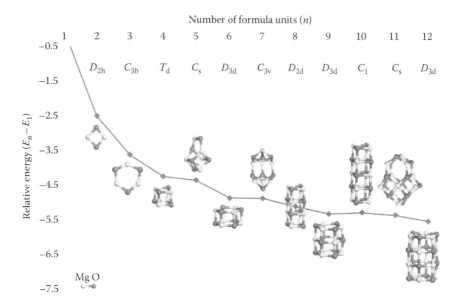

FIGURE 1.14   Ball-and-stick models of tentative global minimum (GM) (PBEsol) configurations for $(MgO)_n$ nanoclusters [33], and their respective point group symmetry, overlaid on the graph based on their respective total PBEsol energies, E_n, in eV per formula unit, as a function of the number of formula units, n. Ten feint broken lines connect second nearest neighbor energy points.

Indeed, this is what is typically found; compare $(KF)_n$ and $(MgO)_n$ examples in Figures 1.12 and 1.14, respectively. Nevertheless, there are important differences: appearance of more hexagonal rings for $(MgO)_n$ and a larger-sized nanocluster are required for $(MgO)_n$ before tentative GM configurations are typically cuboid cuts of the bulk phase. The latter is possibly the result of MgO having larger formal charges [84], ±2 as opposed to ±1, than KF. For nanocluster sizes shown in Figure 1.14, barrel-like hexagonal prisms are more stable than cuboids; note the change in the tentative GM configurations between $(KF)_n$ and $(MgO)_n$ for $n = 6$, 9, and 12. Barrel-like configurations are possible for 1–1 nanoclusters of size n when 3 is a factor of n. The barrel configuration for $(MgO)_{15}$ is also more stable (PBEsol energy) than a cuboid configuration; however, for larger sizes, this trend reverses and the GM configurations are bulk-like cuts; the most (least) cubic-like cuboids for $(MgO)_{18}$ and $(MgO)_{21}$ are more (less) stable than the barrel. The barrel and cuboid configurations for $(MgO)_{18}$ are shown in Figure 1.13.

It is natural to compare the LMs found with those already predicted by others; disagreement can be caused by the employment of a different definition of energy (including functional, basis set, IP form, and parameters) and/or missing LMs from one or both sets (a more exhaustive search is required of one or both energy landscapes). Generally, a mismatch is more likely for larger-sized nanoclusters as there is typically more LMs; however, there are examples where the GM for a slightly larger size is easier to find as the search is also dependent upon the nature of the landscape and not just the number of LMs: funneling for a particular size nanocluster can steer a search away from the GM and thus make the search for the GM more difficult (see, for example, the double-funnel energy landscape of the 38-atom

Lennard–Jones cluster [101]). The PBEsol LM $(MgO)_n$ atomic configurations (tentative GM shown in Figure 1.14) generally match, in respect of rank and structural type (motif) or connectivity (there will typically be a slight change in interatomic distances and angles if different models—energy definitions—are employed) those reported for: PBE LMs found by data mining the PBEsol LMs [33]; rigid ion (RI) LMs found using a genetic algorithm [84] or found from relaxing structures data mined from NaCl as well as random perturbations from them [102]; Coulomb–Hartree–Fock LMs found from relaxing structures data mined from results of IP calculations on MgO and alkali-halide clusters [92]; GGA LMs found from relaxing IP LMs that were obtained using a genetic algorithm [103]; and B3LYP LMs obtained using a genetic algorithm [25]. Details of the match or mismatches can be found in Reference 33, for example, nanoclusters with the same structural motifs are reported for both PBEsol and B3LYP LMs; however, a change in the respective rankings is found such that the GMs do not match for $n = 8$, 10, 11, and 12. From a comparison of the nanocluster configurations obtained in the different studies, one can conclude that successful applications of global optimization schemes produce similar sets of low-energy LMs for each composition. Crucially, a match between experimental and simulated infrared spectra has been achieved in order to validate one of the sets of predicted $(MgO)_n$ atomic structures [25] and, therefore, validating similar results from the other studies. The match of spectra was sometimes achieved by adding contributions to each spectrum from one or more of the lowest B3LYP energy LM.

Switching to a compound that adopts a different bulk structure than potassium fluoride and magnesium oxide, but keeping a 1–1 stoichiometry, one might expect to find different GM configurations. Moreover, one may expect the structural motif of the GM to match a higher-energy LM we have already seen. Zinc oxide adopts the wurtzite phase; atoms are four, rather than six, coordinated. For the smallest size, all atoms are extremely undercoordinated and the same GM motif is found for $(ZnO)_n$ as seen earlier—see Figure 1.15. A ring structural motif, with all atoms two-coordinated, is reported as the PBEsol GM for $n = 2$, 3, 4, 5, 6, and 7 [33]. The increased stability of this motif is due to the stronger polarization of the component atoms: the cation–anion–cation bond angles for $n = 3$ are smallest for $(ZnO)_3$ and largest for $(KF)_3$, and the polarization stabilizes $(ZnO)_7$ by transforming $(D_{7h} \rightarrow C_1)$ the planar ring into a crown (crenulated ring). For larger $(ZnO)_n$ nanoclusters, the GM configurations adopt a bubble-like motif of mainly hexagonal faces; six (or more if octagonal faces are present) tetragonal faces are seen as defects in a hexagonal sheet and add curvature. Although wurtzite can be visualized as corrugated hexagonal sheets, the bubbles are not bulk-like cuts and coordination of all atoms is three rather than four. A bubble structural motif is found for much larger LMs, after which other non-bulk-like motifs are also predicted to be LMs; see, for example, the double bubble example in Figure 1.13 for $n = 60$, where the $n = 12$ sodalite cage is within the $n = 48$ T_h cage. Using the IP data shown in Figure 1.16, it can be estimated that bulk-like cuts do not appear until after $n = 71$.

So far we have considered 1–1 compounds that typically adopt either the cubic or the hexagonal-based structure. In the former, we have considered phases synonymous for NaCl and CsCl, whereas in the latter we have considered the wurtzite phase. With a different stacking sequence to that of wurtzite, the zinc blende structure is perhaps the obvious missing phase. As with wurtzite, all atoms have a

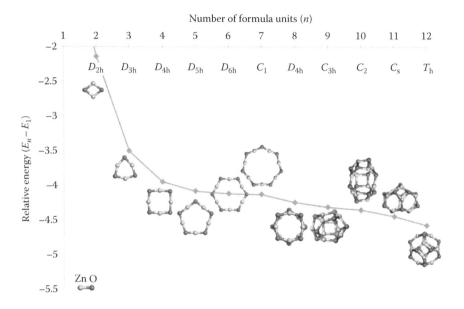

FIGURE 1.15 Ball-and-stick models of tentative global minimum (GM) (PBEsol) configurations for $(ZnO)_n$ nanoclusters [33], and their respective point group symmetry, overlaid on the graph based on their respective total PBEsol energies, E_n, in eV per formula unit, as a function of the number of formula units, n. Ten feint broken lines connect second nearest neighbor energy points.

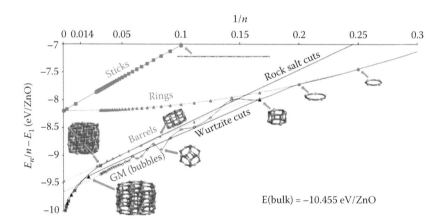

FIGURE 1.16 Relative energies of $(ZnO)_n$ local minimum (LM): 1D configuration (pink squares), 2D rings (orange triangles), barrels formed from $(ZnO)_3$ rings (purple triangles), tentative GMs that typically have a bubble motif (dark blue open diamonds), cuts from rock salt (red highlighted black triangles), and cuts from wurtzite (green highlighted black triangles). Examples of each are shown as ball-and-tick models and the broken line is shown to help ascertain where the tentative GM and bulk cuts cross.

coordination of four. The tentative GM nanoclusters for zinc sulfide, which adopts the zinc blende phase, are similar to zinc oxide, with the exception of the appearance of more filled bubble motifs, which occur for $n = 13$ and 34—see Figure 1.13. The same relative rotation of the inner bubble with respect to the outer can be chosen for any 1–1 compound; however, the ideal bond distance between the two bubble layers (or atom and outer layer in the case of $n = 13$) does depend on the relative sizes of the cation and anion [100]. The $n = 13$ and 34 examples were first suggested as atomic structures for nanoclusters of ZnS, ZnSe, CdS, and CdSe [8,44,104], and the fact that it is not always possible to perfectly match two layers is believed to be the cause of the so-called magic sizes for these compounds (see the earlier discussion in Section 1.4).

From just four example 1–1 compounds, I have already shown the rich variety of possible LM nanoclusters. One could be more systematic; for example, one could investigate the effect of increasing the size of one of the atoms on the ranking of the lowest-energy LM configurations; see Figure 1.17 where I could have easily given a 1–1 example, say, $(AO)_9$ LM nanoclusters for A^{2+} cations from group two of the periodic table. If two compounds have different GM, then one might like to know what happens when these are mixed. For example, the PBE GM for $(MgO)_8$ is a perfect bubble, whereas it is an octagonal drum for $(ZnO)_8$. As each is doped, different cation arrangements of each structural type should be considered, as well as other low-energy structural types. PBE solution energies for $(Zn_{1-x}Mg_xO)_8$ nanoclusters are plotted in Figure 1.18 for a range of structural types. As expected from earlier discussions, the ring motif becomes less stable as x increases from 0 to 1. Interestingly, the lowest-energy configuration for $x = 0.5$ has a different structural motif than those adopted by the binary end members. However, compared to the drum and bubble motifs, there is a larger spread in energies for different cation arrangements of the inflated structural motif, as there is a large penalty for magnesium atoms to occupy the two- rather than the three-coordinated sites. Comparing the free energies of solution, which includes a configurational entropic

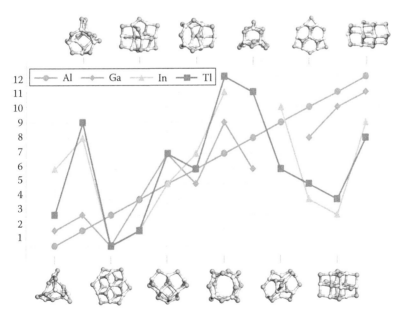

FIGURE 1.17 Ball-and-stick models of structural types for $(X_2O_3)_4$ local minimum (LM) nanoclusters and their rank based on PBEsol relaxed geometries and PBEsol0 total energies. Dark balls are oxygen anions, whereas lighter balls are X^{3+} cations, that is, either aluminum, gallium, indium, or thallium cations. The structural types are ordered with respect to the ranking of alumina nanoclusters; the greatest deviation from the linear line is found for the larger cations.

Chapter 1

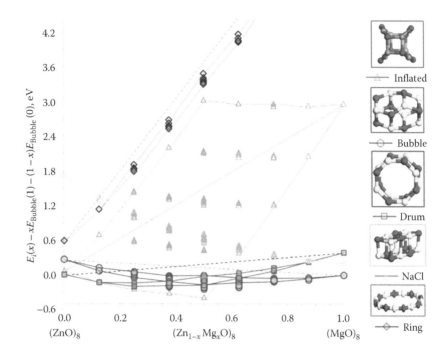

FIGURE 1.18 PBE solution energies for $(Zn_{1-x}Mg_xO)_8$ nanoclusters [105], where $E_i(x)$ is the total PBE energy of configuration type i. The solid lines connect the highest and lowest values for each structural type (shown on the right as ball-and-stick models for just one value of x); and the broken lines represent the arithmetic mean energy of mixing end members (solution energy of zero for that structural type). Data for the less stable ring nanoclusters are not shown; $E_{Ring}(1) = 6.78$ eV.

term, it is predicted that the inflated structure is only more stable at extremely low temperatures [105]. Also, as there is a significant change in the atomic structure as n is incremented, there is typically a significant change in the electronic structure and the optical properties of the GM nanoclusters as a function of n; see figures within Reference 105. This study highlights the typical result that when modeling nanoclusters, *every atom counts.*

If the stoichiometry is changed, then a completely new set of LMs needs to be found—see Figure 1.19. At what size will the atomic structure, electronic structure, and physical properties of the GM nanoclusters for each compound resemble bulk-like cuts? As with the 1–1 examples, a change in compound will again change the GM configurations as steric and polarization effects compete. Again a mix of two compounds with the same stoichiometry and that have the same (or different) structural types for their low-energy LMs can be investigated. Moreover, binary compounds with different stoichiometry can be mixed, for example, MgO and SiO_2, which will produce even more structural types.

Rather than investigating all ratios of two binary compounds, one could concentrate on a particular ratio, that is, predict the LMs for ternary or quad–ternary compounds. Nanocluster examples for four ternary oxides [35] are shown in Figure 1.20; the magnesium silicate nanoclusters are examples of the smallest grains of star dust and will be discussed in Chapter 13.

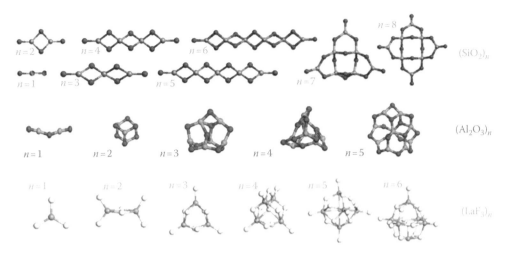

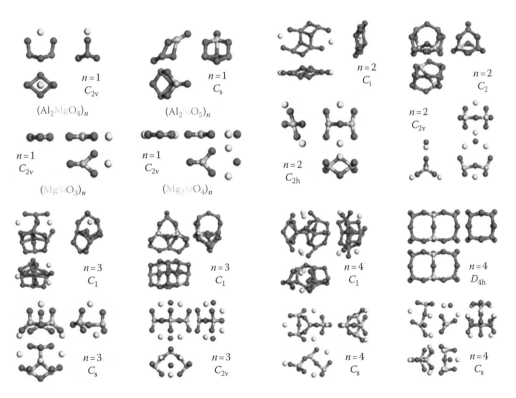

FIGURE 1.19 Example GM configurations for 1–2 [2], 2–3 [49], and 1–3 [48] nanoclusters composed of 25 atoms or less.

FIGURE 1.20 Ball-and-stick models of ternary oxide IP GM nanoclusters [35]. Front, side, and plan views are shown for each nanocluster, as well as their point group symmetry. A different background color is used for each compound.

1.10 Summary

Approximate atomic coordinates are typically the minimum requirement for solid-state or materials modeling if the phenomenon under investigation or property of interest depends on the atomic and/or electronic structure. Unlike crystalline bulk phases, approximate atomic coordinates for small nanoparticles are not readily obtainable from current experimental techniques. This chapter introduced the concepts for predicting the atomic structures of nanoparticles, as well as the challenges that need to be tackled if one is to be successful. Ideas of energy landscapes and a locally ergodic region were touched upon in order to answer the question, what structures are we actually trying to model? The first challenge is choosing the right landscape that describes your system well enough that the essential physics and/or chemistry is included while minimizing the cost of evaluating a point on this landscape. The second challenge is how efficiently can the important features of the landscape be found; typically this includes the lowest LMs and connecting saddle points. The feasibility of locating these desired features will be dependent on the size and complexity of the landscape, which generally increases rapidly with the number of variables (unknown atomic coordinates), and if gradient-based algorithms are employed, then the landscape must not only be continuous but also smooth (derivatives are also continuous). Atomistic-based models were introduced as these are typically used to define the landscape—more details on electronic structure techniques will be provided in Chapter 2 on 1D structures. Even though the use of electronic structure techniques in the definition of the landscape, rather employed afterward in a second stage of improving approximate solutions, is becoming more popular, it is still beneficial to screen initial configurations. The basics of a range of popular global optimization algorithms that have been applied to predicting the atomic structures of nanoparticles, including nanoclusters, were described. Although global optimization software is available for non-profit-making organizations (e.g., academic research groups), the algorithms are relatively straightforward and so many research groups have developed in-house codes for structure prediction, which has led to more creativity as algorithms are adapted and combined. For example, Hamad et al. [96] have combined molecular dynamics-based simulated annealing with MCBH in their search for LMs of titania nanoclusters. MCBH was applied, after each drop in temperature, for a fixed number of MC steps and a fixed temperature in the Metropolis criterion (Equation 1.20). To increase the confidence of whether the lowest-energy LM configurations have been found, two independent genetic algorithms were also employed.

Nanoparticles are defined as particles that have a maximum cross section, in all directions, of the order of 10^{-9} m. The phase of a crystalline nanoparticle can change with particle size, and therefore, its characteristic structure and physical/electronic properties will also be particle size dependent and can be different to that observed for the bulk phase. Chemistry takes place at a surface, and the nanoscale, or size of nanoparticles, also implies that they have an extremely good surface-area-to-volume ratio. When there are more atoms associated with the surface than the interior, it is perhaps better to refer to structural types than bulk-like phases. For the very smallest nanoparticles, the so-called subnanoparticles, or nanoclusters, every additional atom can change the structural type, and therefore, the cluster community refers to this size regime using the expression *where every atom counts*. In this size regime the well-known quantum size effect—a decrease in the size of the particle is matched with an increase in the energy gap between the ground and excited electronic states (discussed in Chapter 2)—can fail as the change in the atomic structure has a more dominant effect than the particle's size [106]. For applications, I have concentrated on nanoclusters as these have the fewest variables and are where my interests lie. Key questions at this size regime are as follows: At what size will X of the lowest-energy configuration resemble the bulk phase? where X is the atomic structure or physical, electronic, and optical properties. Using bulk-like nanoclusters as secondary building units (SBUs), one can imagine growing the known bulk phase. Thus, another interesting question I am interested in answering is, what phase can be constructed from the relatively stable non-bulk-like nanocluster configurations? High symmetry nanoclusters of MgO, ZnO, GaN, and SiC, to name but a few, have been used as SBUs in the construction of nano- and microporous material References 100, 107–113.

Ideally we would like a cost function from which atomic structures can be obtained that match those adopted in nature. Currently, we do not have such a cost function, and the search for one is not helped by the fact that we still do not have an experimental technique that can provide atomic structures for nanoclusters. Assuming a particular cost function, or energy hypersurface, there is a community that is mainly interested in finding the lowest LM configurations. If, however, one is more interested in the lowest-energy structure, or GM, for a defined set of atoms, then establishing whether you have obtained a better GM than currently published is more problematic as there is no universal energy function. As shown for $(MgO)_n$ nanoclusters, a change in density function can lead to a change in ranking and also at least a slight refinement of the configurations. If a sensible energy function is employed in two independent studies, then even if the energy function is in the two studies, one may still expect the same structural types to be found in both. It is desirable to be able to readily match predicted structures that result from using different energy functions, particularly your newly predicted structures with those of others already published in the scientific literature. The WASP@N project is currently trying to fill this desire as well as to provide a searchable web database of predicted nanoclusters that have been published [89]. The search can be based on trying to find whether a newly generated structure is already within the database (e.g., based on connectivity arguments) or can simply find nanoclusters with a certain property (like composition, size, stoichiometry, dipole moment). Moreover, all nanoclusters uploaded into the database will also be refined using a chosen energy definition, and links between all published atomic configurations that relax to the same structure will be made available to users of this web database.

In this chapter, I have also concentrated on describing static lattice approaches, rather than molecular dynamics, to predicting nanoclusters. Phenotype moveclass operators described for nanoclusters can readily be adapted, or designed, for larger nanoparticles where the interior atoms can be assumed to adopt a bulk phase and therefore frozen in during the initial search using static lattice-based algorithms and, hence, reduce the computational cost of the second challenge. The LM configurations for nanoclusters of one compound can be data mined for another compound of the same stoichiometry [114], that is, used as initial configurations in a standard local optimization scheme. The type of investigations performed for bulk phases can also be applied to nanoclusters; for example, one can investigate the cation or anion ordering, or the effect of doping, or point defects on the structure and properties of the nanocluster. If the energy difference between the GM and other LMs is similar to thermal energy, then properties of the ensemble should be

Chapter 1

calculated—see Reference 25 where infrared spectra of two or more $(MgO)_n$ nanoclusters of the same size are combined in order to get a better match to that observed or Reference 49 where Boltzmann-weighted (Equation 1.19) superposition of spectra for $(Al_2O_3)_4$ is compared to experimental data (x-ray emission and absorption spectra as well as ultraviolet photoelectron spectrum and x-ray photoelectron spectrum) for bulk alumina.

The effect of the environment is one aspect I have not covered. However, the atomic coordinates of bare, isolated nanoclusters and, more generally, nanoparticles are a good starting point for investigating the effect of the environment on the structure and properties of nanoparticles. For a review of quantum mechanical studies of large metal, metal oxide, and metal chalcogenide nanoparticles and nanoclusters, see Reference 115.

Acknowledgments

First, I thank EPSRC for funding (EP/I03014X, EP/K038958, EP/L000202). Second, I am grateful to group members Alexey Sokol, Thomas Lazauskas and Matthew Farrow, and friend and colleague Christian Schön for useful discussions and feedback on my first draft. And, finally, I would like to thank all other authors of this book and, in particular, the editors Stefan T. Bromley and Martijn Zwijnenburg, for their patience and assistance.

References

1. Nanotechnology, from Wikipedia, https://en.wikipedia.org/wiki/Nanotechnology, last modified January 21, 2016.
2. Catlow, C. R. A., S. T. Bromley, S. Hamad, M. Mora-Fonz, A. A. Sokol, and S. M. Woodley. 2010. Modelling nano-clusters and nucleation. *Phys. Chem. Chem. Phys.* 12:786–811.
3. Hamad, S., S. M. Woodley, and C. R. A. Catlow. 2009. Experimental and computational studies of ZnS nanostructures. *Mol. Simul.* 35:1015–1032.
4. Zhang, H. Z., F. Huang, B. Gilbert, and J. F. Banfield. 2003. Molecular dynamics simulations, thermodynamic analysis, and experimental study of phase stability of zinc sulfide nanoparticles. *J. Phys. Chem. B* 107:13051–13060.
5. Zhang, H. Z., B. Chen, B. Gilbert, and J. F. Banfield. 2006. Kinetically controlled formation of a novel nanoparticulate ZnS with mixed cubic and hexagonal stacking. *J. Mater. Chem.* 16:249–254.
6. Burnin, A. and J. J. BelBruno. 2002. $Zn_nS_m^+$ cluster production by laser ablation. *Chem. Phys. Lett.* 362:341–348.
7. Twu, Y. J., C. W. S. Conover, Y. A. Yang, and L. A. Bloomfield. 1990. Alkali-halide cluster ions produced by laser vaporization of solids. *Phys. Rev. B* 42:5306–5316.
8. Kasuya, A., R. Sivamohan, Y. A. Barnakov, I. M. Dmitruk, T. Nirasawa, V. R. Romanyuk, V. Kumar et al. 2004. Ultra-stable nanoparticles of CdSe revealed from mass spectrometry. *Nat. Mater.* 3:99–102.
9. Schön, J. C. 1998. Structure prediction and modelling of solids: An energy landscape point of view. In *Proceedings of RIGI-Workshop 1998*, J. Schreuer (ed.). ETH Zurich, Zurich, Switzerland, pp. 75–93.
10. Schön, J. C. and M. Jansen. 2001. Determination, prediction, and understanding of structures, using the energy landscapes of chemical systems—Part I. *Zeitsch. Kristallog.* 216:307–325.
11. Schön, J. C. and M. Jansen. 2001. Determination, prediction, and understanding of structures, using the energy landscapes of chemical systems—Part III. *Zeitsch. Kristallog.* 216:361–383.
12. Schön, J. C. and M. Jansen. 2009. Prediction, determination and validation of phase diagrams via the global study of energy landscapes. *Int. J. Mater. Res.* 100:135–152.
13. Schön, J. C. 2015. Nanomaterials—What energy landscapes can tell us. *Proc. Appl Ceramics.* 9:157–168.
14. Wevers, M. A. C., J. C. Schön, and M. Jansen. 1999. Global aspects of the energy landscape of metastable crystal structures in ionic compounds. *J. Phys. Condens. Matter* 11:6487–6499.

15. Schön, J. C., M. A. C. Wevers, and M. Jansen. 2003. 'Entropically' stabilized region on the energy landscape of an ionic solid. *J. Phys. Condes. Matter* 15:5479–5486.
16. Doye, J. P. K. and D. J. Wales. 2002. Saddle points and dynamics of Lennard-Jones clusters, solids, and supercooled liquids. *J. Chem. Phys.* 116:3777–3788.
17. Neelamraju, S., J. C. Schön, K. Doll, and M. Jansen. 2012. Ab initio and empirical energy landscapes of $(MgF_2)_n$ clusters (n=3, 4). *Phys. Chem. Chem. Phys.* 14:1223–1234.
18. Jansen, M., I. V. Pentin, and J. C. Schön. 2012. A universal representation of the states of chemical matter including metastable configurations in phase diagrams. *Angew. Chem. Int. Ed.* 51:132–135.
19. Heard, C. J., R. L. Johnston, and J. C. Schön. 2015. Energy landscape exploration of sub-nanometre copper-silver clusters. *ChemPhysChem* 16:1461–1469.
20. Hoffmann, K. H. and P. Sibani. 1988. Diffusion in hierarchies. *Phys. Rev. A* 38:4261–4270.
21. Sibani, P., R. van der Pas, and J. C. Schön. 1999. The lid method for exhaustive exploration of metastable states of complex systems. *Comput. Phys. Commun.* 116:17–27.
22. Wales, D. J. 2003. *Energy Landscapes.* Cambridge University Press, Cambridge, U.K.
23. Sayle, T. X. T., C. R. A. Catlow, R. R. Maphanga, P. E. Ngoepe, and D. C. Sayle. 2005. Generating MnO_2 nanoparticles using simulated amorphization and recrystallization. *J. Am. Chem. Soc.* 127:12828–12837.
24. Sayle, T. X. T., F. Caddeo, N. O. Monama, K. M. Kgatwane, P. E. Ngoepeb, and D. C. Sayle. 2015. Origin of electrochemical activity in nano-Li_2MnO_3; stabilization via a 'point defect scaffold'. *Nanoscale* 7:1167–1180.
25. Haertelt, M., A. Fielicke, G. Meijer, K. Kwapien, M. Sierka, and J. Sauer. 2012. Structure determination of neutral MgO clusters-hexagonal nanotubes and cages. *Phys. Chem. Chem. Phys.* 14:2849–2856.
26. Santambrogio, G., M. Bruemmer, L. Woeste, J. Doebler, M. Sierka, J. Sauer, G. Meijer, and K. R. Asmis. 2008. Gas phase vibrational spectroscopy of mass-selected vanadium oxide anions. *Phys. Chem. Chem. Phys.* 10:3992–4005.
27. Shevlin, S. A. and S. M. Woodley. 2010. Electronic and optical properties of doped and undoped $(TiO_2)_n$ nanoparticles. *J. Phys. Chem. C* 114:17333–17343.
28. Doll, K., J. C. Schön, and M. Jansen. 2010. Ab initio energy landscape of LiF clusters. *J. Chem. Phys.* 133:024107.
29. Heiles, S. and R. L. Johnston. 2013. Global optimization of clusters using electronic structure methods. *Int. J. Quantum Chem.* 113:2091–2109.
30. Heiles, S., A. J. Logsdail, R. Schafer, and R. L. Johnston. 2012. Dopant-induced 2D-3D transition in small Au-containing clusters: DFT-global optimisation of 8-atom Au-Ag nanoalloys. *Nanoscale* 4:1109–1115.
31. Fernandez-Lima, F. A., O. P. VilelaNeto, A. S. Pimentel, C. R. Ponciano, M. A. C. Pacheco, M. A. C. Nascimento, and E. F. d. Silveira. 2009. A theoretical and experimental study of positive and neutral LiF clusters produced by fast ion impact on a polycrystalline LiF target. *J. Phys. Chem. A* 113:1813–1821.
32. Marom, N., M. Kim, and J. R. Chelikowsky. 2012. Structure selection based on high vertical electron affinity for TiO_2 clusters. *Phys. Rev. Lett.* 108:106801.
33. Farrow, M. R., Y. Chow, and S. M. Woodley. 2014. Structure prediction of nanoclusters; a direct or a pre-screened search on the DFT energy Landscape? *Phys. Chem. Chem. Phys.* 16:21119–21134.
34. Flikkema, E. and S. T. Bromley. 2003. A new interatomic potential for nanoscale silica. *Chem. Phys. Lett.* 378:622–629.
35. Woodley, S. M. 2009. Structure prediction of ternary oxide sub-nanoparticles. *Mater. Manuf. Processes* 24:255–264.
36. Wales, D. J. and J. P. K. Doye. 1997. Global optimization by basin-hopping and the lowest energy structures of Lennard-Jones clusters containing up to 110 atoms. *J. Phys. Chem. A* 101:5111–5116.
37. Woodley, S. M. 2008. The mechanism of the displacive phase transition in vanadium dioxide. *Chem. Phys. Lett.* 453:167–172.
38. Johnston, R. L. 2003. Evolving better nanoparticles: Genetic algorithms for optimising cluster geometries. *Dalton Trans.* (22):4193–4207.
39. Dick, B. G. and A. W. Overhauser. 1958. Theory of the dielectric constants of alkali halide crystals. *Phys. Rev.* 112:90.
40. Dove, M. T. 1993. *Introduction to Lattice Dynamics.* Cambridge University Press, Cambridge, U.K.

Chapter 1

41. Woodley, S. M., C. R. A. Catlow, J. D. Gale, and P. D. Battle. 2000. Development of a new force field for open shell ions: Application to modelling of LaMnO$_3$. *Chem Commun.* (19):1879–1880.

42. Woodley, S. M., P. D. Battle, C. R. A. Catlow, and J. D. Gale. 2001. Development of a new interatomic potential for the modeling of ligand field effects. *J. Phys. Chem. B* 105:6824–6830.

43. Burnin, A., E. Sanville, and J. J. BelBruno. 2005. Experimental and computational study of the Zn$_n$S$_n$ and Zn$_n$S$_n^+$ clusters. *J. Phys. Chem. A* 109:5026–5034.

44. Sanville, E., A. Burnin, and J. J. BelBruno. 2006. Experimental and computational study of small (n=1–16) stoichiometric zinc and cadmium chalcogenide clusters. *J. Phys. Chem. A* 110:2378–2386.

45. Bloomfield, L. A., C. W. S. Conover, Y. A. Yang, Y. J. Twu, and N. G. Phillips. 1991. Experimental and theoretical-studies of the structure of alkali-halide clusters. *Z. Phys. D-Atoms Mol. Clusters* 20:93–96.

46. Woodley, S. M., A. A. Sokol, and C. R. A. Catlow. 2004. Structure prediction of inorganic nanoparticles with predefined architecture using a genetic algorithm. *Z. Anorg. Allg. Chem.* 630:2343–2353.

47. Doye, J. P. K. and D. J. Wales. 1999. Structural transitions and global minima of sodium chloride clusters. *Phys. Rev. B* 59:2292–2300.

48. Woodley, S. M. 2013. Knowledge Led Master Code search for atomic and electronic structures of LaF$_3$ nanoclusters on hybrid rigid ion-shell model-DFT landscapes. *J. Phys. Chem. C* 117:24003–24014.

49. Woodley, S. M. 2011. Atomistic and electronic structure of (X$_2$O$_3$)$_n$ nanoclusters; n = 1–5, X = B, Al, Ga, In and Tl. *Proc. R. Soc. A: Math. Phys. Eng. Sci.* 467:2020–2042.

50. Rossi, G. and R. Ferrando. 2006. Global optimization by excitable walkers. *Chem. Phys. Lett.* 423:17–22.

51. Al-Sunaidi, A. A., A. A. Sokol, C. R. A. Catlow, and S. M. Woodley. 2008. Structures of zinc oxide nanoclusters: As found by evolutionary algorithm techniques. *J. Phys. Chem. C* 112:18860–18875.

52. Müller-Krumbhaar, H. 1988. Fuzzy-logic, m-spin glasses and 3-SAT. *Europhys. Lett.* 7:479–484.

53. Möbius, A., A. Neklioudov, A. DiazSanchez, K. H. Hoffmann, A. Fachat, and M. Schreiber. 1997. Optimization by thermal cycling. *Phys. Rev. Lett.* 79:4297–4301.

54. Kirkpatrick, S., C. D. Gelatt, and M. P. Vecchi. 1983. Optimization by simulated annealing. *Science* 220:671–680.

55. Cerny, V. 1985. Thermodynamical approach to the traveling salesman problem—An efficient simulation algorithm. *J. Optim. Theory Appl.* 45:41–51.

56. Hoffmann, K. H. and J. C. Schön. 2005. Kinetic features of preferential trapping on energy landscapes. *Found. Phys. Lett.* 18:171–182.

57. Fischer, A., K. H. Hoffmann, and J. C. Schön. 2011. Competitive trapping in complex state spaces. *J. Phys. A Math. Theor.* 44:075101.

58. Hoffmann, K. H. and J. C. Schön. 2013. Controlled dynamics on energy landscapes. *Eur. Phys. J. B* 86:220.

59. Spano, E., S. Hamad, and C. R. A. Catlow. 2003. Computational evidence of bubble ZnS clusters. *J. Phys. Chem. B* 107:10337–10340.

60. Wevers, M. A. C., J. C. Schön, and M. Jansen. 2001. Characteristic regions on the energy landscape of MgF$_2$. *J. Phys. A-Math. Gen.* 34:4041–4052.

61. Schön, J. C. 2015. G42+ *Manual*. MPI for Solid State Research, Stuttgart, Germany.

62. Woodley, S. M. and A. A. Sokol. 2012. From ergodicity to extended phase diagrams. *Angew. Chem. Int. Ed.* 51:3752–3754.

63. Woodley, S. M. 2012. Knowledge Led Master Code. http://www.ucl.ac.uk/klmc/Software/. Last accessed November 5, 2015.

64. Hartke, B. 2004. Application of evolutionary algorithms to global cluster geometry optimization. In *Applications of Evolutionary Computation in Chemistry*, R. L. Johnston (ed.). Springer, Berlin, Germany, pp. 33–53.

65. Press, W. H., S. A. Teukolsky, W. T. Vetterling, and B. P. Flannery. 1992. *Numerical Recipes in Fortran.* Cambridge University Press, Cambridge, U.K.

66. Pickard, C. J. and R. J. Needs. 2011. Ab initio random structure searching. *J. Phys. Condes. Matter* 23:053201.

67. Flikkema, E. and S. T. Bromley. 2004. Dedicated global optimization search for ground state silica nanoclusters: (SiO$_2$)$_N$ (N = 6–12). *J. Phys. Chem. B* 108:9638–9645.

68. Li, Z. Q. and H. A. Scheraga. 1987. Monte-Carlo-minimization approach to the multiple-minima problem in protein folding. *Proc. Natl. Acad. Sci. U.S.A.* 84:6611–6615.

69. Goedecker, S. 2004. Minima hopping: An efficient search method for the global minimum of the potential energy surface of complex molecular systems. *J. Chem. Phys.* 120:9911–9917.

70. Cheng, L. J., Y. Feng, J. Yang, and J. L. Yang. 2009. Funnel hopping: Searching the cluster potential energy surface over the funnels. *J. Chem. Phys.* 130:214112.

71. Oakley, M. T., R. L. Johnston, and D. J. Wales. 2013. Symmetrisation schemes for global optimisation of atomic clusters. *Phys. Chem. Chem. Phys.* 15:3965–3976.

72. Schön, J. C. 1996. Studying the energy hypersurface of multi-minima systems—The threshold and the lid algorithm. *Ber. Bunsen-Ges. Phys. Chem. Chem. Phys.* 100:1388–1391.

73. Schön, J. C., H. Putz, and M. Jansen. 1996. Studying the energy hypersurface of continuous systems—The threshold algorithm. *J. Phys. Condes. Matter* 8:143–156.

74. Woodley, S. M. and A. M. Walker. 2007. New software for finding transition states by probing accessible, or ergodic, regions. *Mol. Simul.* 33:1229–1231.

75. Henkelman, G. and H. Jonsson. 2000. Improved tangent estimate in the nudged elastic band method for finding minimum energy paths and saddle points. *J. Chem. Phys.* 113:9978–9985.

76. Miron, R. A. and K. A. Fichthorn. 2001. The Step and Slide method for finding saddle points on multidimensional potential surfaces. *J. Chem. Phys.* 115:8742–8747.

77. Banerjee, A., N. Adams, J. Simons, and R. Shepard. 1985. Search for stationary-points on surface. *J. Phys. Chem.* 89:52–57.

78. Wang, Y. C., J. A. Lv, L. Zhu, and Y. M. Ma. 2010. Crystal structure prediction via particle-swarm optimization. *Phys. Rev. B* 82:094116.

79. Fan, T. E., T. D. Liu, J. W. Zheng, G. F. Shao, and Y. H. Wen. 2015. Structural optimization of Pt-Pd-Au trimetallic nanoparticles by discrete particle swarm algorithms. *J. Mater. Sci.* 50:3308–3319.

80. Raczynski, P. and Z. Gburski. 2005. The search for minimum potential energy structures of small atomic clusters. Application of the ant colony algorithm. *Mater. Sci.* 23:599–606.

81. Cheng, J. and R. Fournier. 2004. Structural optimization of atomic clusters by tabu search in descriptor space. *Theor. Chem. Acc.* 112:7–15.

82. Fournier, R. 2010. Density-functional and global optimization study of copper-tin core-shell clusters. *Can. J. Chem. Rev. Can. Chim.* 88:1071–1078.

83. Dieterich, J. M. and B. Hartke. 2010. OGOLEM: Global cluster structure optimisation for arbitrary mixtures of flexible molecules. A multiscaling, object-oriented approach. *Mol. Phys.* 108:279–291.

84. Roberts, C. and R. L. Johnston. 2001. Investigation of the structures of MgO clusters using a genetic algorithm. *Phys. Chem. Chem. Phys.* 3:5024–5034.

85. Pareto, V. 1896. *Cours d'economie politique*. Librairie de l'Universite, Lausanne, Switzerland.

86. Woodley, S. M., P. D. Battle, J. D. Gale, and C. R. A. Catlow. 1999. The prediction of inorganic crystal structures using a genetic algorithm and energy minimisation. *Phys. Chem. Chem. Phys.* 1:2535–2542.

87. Woodley, S. M. 2004. Prediction of crystal structures using evolutionary algorithms and related techniques. In *Applications of Evolutionary Computation in Chemistry*. Springer, Berlin, Germany, pp. 95–132.

88. Deaven, D. M. and K. M. Ho. 1995. Molecular-geometry optimization with a genetic algorithm. *Phys. Rev. Lett.* 75:288–291.

89. Woodley, S. M. 2014. Database of published atomic structures of nanoclusters. http://www.ucl.ac.uk/klmc/Hive. Last accessed date November 5, 2015.

90. Chuchev, K. and J. J. Belbruno. 2005. Small, nonstoichiometric zinc sulfide clusters. *J. Phys. Chem. A* 109:1564–1569.

91. Ayuela, A., J. M. López, J. A. Alonso, and V. Luaña. 1993. Theoretical study of NaCl clusters. *Z. Phys. D—Atoms, Molecules and Clusters* 26:213–215.

92. de la Puente, E., A. Aguado, A. Ayuela, and J. M. Lopez. 1997. Structural and electronic properties of small neutral $(MgO)_n$ clusters. *Phys. Rev. B* 56:7607–7614.

93. Azpiroz, J. M., I. Infante, X. Lopez, J. M. Ugalde, and F. De Angelis. 2012. A first-principles study of II-VI (II = Zn; VI = O, S, Se, Te) semiconductor nanostructures. *J. Mater. Chem.* 22:21453–21465.

94. Wang, B., S. Nagase, J. Zhao, and G. Wang. 2007. Structural growth sequences and electronic properties of zinc oxide clusters $(ZnO)_n$ (n = 2–18). *J. Phys. Chem. C* 111:4956–4963.

95. Wootton, A. and P. Harrowell. 2004. Inorganic nanotubes stabilized by ion size asymmetry: Energy calculations for AgI clusters. *J. Phys. Chem. B* 108:8412–8418.

96. Hamad, S., C. R. A. Catlow, S. M. Woodley, S. Lago, and J. A. Mejias. 2005. Structure and stability of small TiO_2 nanoparticles. *J. Phys. Chem. B* 109:15741–15748.

97. Woodley, S. M., S. Hamad, J. A. Mejias, and C. R. A. Catlow. 2006. Properties of small TiO_2, ZrO_2 and HfO_2 nanoparticles. *J. Mater. Chem.* 16:1927–1933.

98. Woodley, S. M., S. Hamad, and C. R. A. Catlow. 2010. Exploration of multiple energy landscapes for zirconia nanoclusters. *Phys. Chem. Chem. Phys.* 12:8454–8465.

99. Walsh, A. and S. M. Woodley. 2010. Evolutionary structure prediction and electronic properties of indium oxide nanoclusters. *Phys. Chem. Chem. Phys.* 12:8446–8453.

100. Farrow, M., J. Buckeridge, C. Catlow, A. Logsdail, D. Scanlon, A. Sokol, and S. Woodley. 2014. From stable ZnO and GaN clusters to novel double bubbles and frameworks. *Inorganics* 2:248–263.

101. Doye, J. P. K., M. A. Miller, and D. J. Wales. 1999. The double-funnel energy landscape of the 38-atom Lennard-Jones cluster. *J. Chem. Phys.* 110:6896–6906.

102. Ziemann, P. J. and A. W. Castleman. 1991. Stabilities and structures of gas phase MgO clusters. *J. Chem. Phys.* 94:718–728.

103. Dong, R. B., X. S. Chen, X. F. Wang, and W. Lu. 2008. Structural transition of hexagonal tube to rocksalt for $(MgO)_{3n}$, 2 <= n <= 10. *J. Chem. Phys.* 129:044705.

104. Barnakov, Y. A., C. E. Bonner, A. Kasuya, Y. Noda, R. Sivamohan, R. Belosludov, Y. Kawazoe, I. Dmitruk, and V. Romanyuk. 2007. Recent advances in the studies of CdSe magic clusters. *Abstr. Pap. Am. Chem. Soc.* 233.

105. Woodley, S. B., A. A. Sokol, C. R. A. Catlow, A. A. Al-Sunaidi, and S. M. Woodley. 2013. Structural and optical properties of Mg and Cd doped ZnO nanoclusters. *J. Phys. Chem. C* 117:27127–27145.

106. Shevlin, S. A., Z. X. Guo, H. J. J. van Dam, P. Sherwood, C. R. A. Catlow, A. A. Sokol, and S. M. Woodley. 2008. Structure, optical properties and defects in nitride (III-V) nanoscale cage clusters. *Phys. Chem. Chem. Phys.* 10:1944–1959.

107. Liu, Z., X. Wang, J. Cai, G. Liu, P. Zhou, K. Wang, and H. Zhu. 2013. From the ZnO hollow cage clusters to ZnO nanoporous phases: A first-principles bottom-up prediction. *J. Phys. Chem. C* 117:17633–17643.

108. Woodley, S. M., M. B. Watkins, A. A. Sokol, S. A. Shevlin, and C. R. A. Catlow. 2009. Construction of nano- and microporous frameworks from octahedral bubble clusters. *Phys. Chem. Chem. Phys.* 11:3176–3185.

109. Watkins, M. B., S. A. Shevlin, A. A. Sokol, B. Slater, C. R. A. Catlow, and S. M. Woodley. 2009. Bubbles and microporous frameworks of silicon carbide. *Phys. Chem. Chem. Phys.* 11:3186–3200.

110. Bromley, S. T. 2007. A computational study into the viability of new molecular materials polymorphs based on fully-coordinated inorganic nanoclusters. *Crystengcomm* 9:463–466.

111. Sangthong, W., J. Limtrakul, F. Illas, and S. T. Bromley. 2008. Stable nanoporous alkali halide polymorphs: A first principles bottom-up study. *J. Mater. Chem.* 18:5871–5879.

112. Zwijnenburg, M. A. and S. T. Bromley. 2011. Structural richness of ionic binary materials: An exploration of the energy landscape of magnesium oxide. *Phys. Rev. B* 83:024104.

113. Carrasco, J., F. Illas, and S. T. Bromley. 2007. Ultralow-density nanocage-based metal-oxide polymorphs. *Phys. Rev. Lett.* 99:235502.

114. Sokol, A. A., C. R. A. Catlow, M. Miskufova, S. A. Shevlin, A. A. Al-Sunaidi, A. Walsh, and S. M. Woodley. 2010. On the problem of cluster structure diversity and the value of data mining. *Phys. Chem. Chem. Phys.* 12:8438–8445.

115. Fernando, A., K. L. D. M. Weerawardene, N. V. Karimova, and C. M. Aikens. 2015. Quantum mechanical studies of large metal, metal oxide, and metal chalcogenide nanoparticles and clusters. *Chem. Rev.* 115:6112–6216.

2. One-Dimensional Nanosystems

John Buckeridge and Alexey A. Sokol

2.1 Introduction 47

2.2 Long-Term Interactions 50

2.3 Energy and Electronic Structure Theory 52

2.4 Dynamical Properties 61

2.5 Structure and Phase Transitions in 1D 67

2.6 Charge and Heat Transport in Ideal Systems 70

2.7 Defect States 76

2.8 Conclusions 77

Acknowledgments 77

References 77

2.1 Introduction

The simplest example of matter, self-organized at nanoscale, rather than individual nanoparticles considered in Chapter 1, is provided by quasi 1D objects such as nanowires, rods, ribbons, or tubes. The key feature of these systems is the contrast between the confinement of the constituent electrons and all relevant quasiparticles, in two orthogonal directions, and the extension in the third, which leads to a coexistence and codependence of discrete and continuous properties as seen in electronic, vibrational (phonon), and magnetic (magnon) spectra that are both dispersive and oscillatory.

The unique features of 1D systems have spurred a wide interest in their fundamentals and led to a variety of applications. The combination of improvements in synthesis and characterization has resulted in the development of 1D-structure-based devices for optoelectronics, energy, device physics, nanomechanics, biomedicine, and nanochemistry. The full potential of nanowires including a wider context [1–30] is illustrated in Figure 2.1 (cf. recent reviews [31,32]).

The properties of particular 1D systems are determined by their atomic structure; thicker nanowires behave in many ways like bulk crystallites, whereas ultrathin wires and nanorods (segments of nanowire or elongated nanoparticles) have more pronounced discrete features. In direct analogy with hollow cage structures of nanoparticles described in Chapter 1, nanotubes proved to be readily formed and are well known, especially in

Chapter 2

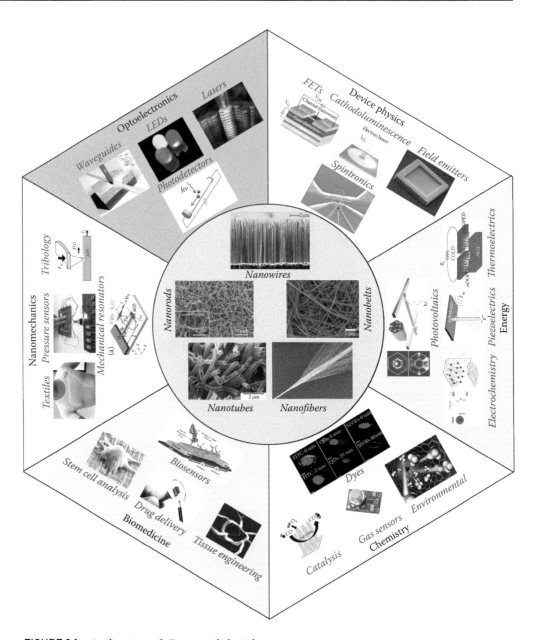

FIGURE 2.1   Applications of 1D materials [1–30].

their carbon form. The latter has been extensively reviewed, with a number of monographs devoted to them (for example, References 33–35) and does not warrant further discussion. Different methods of synthesis, preparation, and postsynthetic treatment allow complex structures involving the combination of two or more materials, such as nanorods capped or cladded with another material, to be formed, which have numerous advantages [25]. Key to the properties of such systems is how they interact with supports in devices; many nanorods and wires are formed directly on a substrate. An important feature to understand, therefore, in addition to the properties of 1D systems along their extended dimension, is the structure of interfaces. Furthermore, the ability to assemble branch points and tripods using different materials, which are useful for increasing the surface area of adsorber materials and designing nanoscale electrical circuits, involves a combination of interfaces between nanowires, as well as the interface with a substrate or contact. A different form of support is provided by materials with nanoporous architecture capable of hosting 1D nanowires, either singly or assembled in stacking sequences, thus giving rise to highly functional nanocomposite materials. Typically, materials formed by strongly bound inorganic compounds play the role of the support including zeolites [36] and semiconductors (e.g., silicon carbide [37] and zinc oxide [38]), but more recently, organic and hybrid metal organic frameworks have found their use [39,40].

Atomistic and electronic modeling is an essential component in understanding and predicting properties of 1D nanostructures. In this chapter, we will highlight the main contemporary approaches and outstanding examples of their application.

The central challenge to computational approaches in materials science is to bridge the gap between what can be accurately modeled and what are the real-life structures that can be synthesized in the laboratory.* The most successful current models have been developed for gas-phase molecules on one end of the scale and crystalline materials using 3D periodic boundary conditions (PBCs) on the other. The 1D systems of interest are of course neither; a major model and software development is, therefore, a prerequisite of further work, examples of which are discussed in this chapter.

The focus of our research into the structure and properties of 1D systems is localized states, or point defects, the charge of which is perhaps one of the most essential characteristics. An accurate treatment of charged defects, that is, those supporting trapped electrons or holes, is required to model processes of fundamental and applied interest, involving charge transfer and excitations. Another issue is the behavior of localized states in 1D systems of a large effective diameter, where a large number of atoms (e.g., $>10^3$) must be included in the model. Appropriate methods providing such a treatment for bulk are now routine, but are not applicable to lower-dimensional systems.

This chapter is arranged as follows: The electrostatics of systems extended in one dimensions is discussed first. The essential features of the electronic structure theory in 1D are introduced next. We then discuss the dynamical behavior, structure, phase stability, and transport in 1D. Next, kinetic phenomena are overviewed, leading to the studies of defects and wire or tube surface properties.

To illustrate some of the main concepts in this chapter, we will use a toy system—an infinite line of atoms with a basis of one or two atoms in a unit cell. The line can either be straight or buckled as shown in Figure 2.2, with a metallic and covalent character of bonding represented by Si and ionic dielectric behavior exhibited by ZnO and CdS. Its electronic and vibrational properties will be discussed in detail in Sections 2.3 and 2.4.

---

* In this respect, computational science is in a surprisingly good position as the experimentalists report synthesis and preparation of smaller and smaller high-quality nanostructures, for which modeling becomes feasible using modern high-performance computers.

Chapter 2

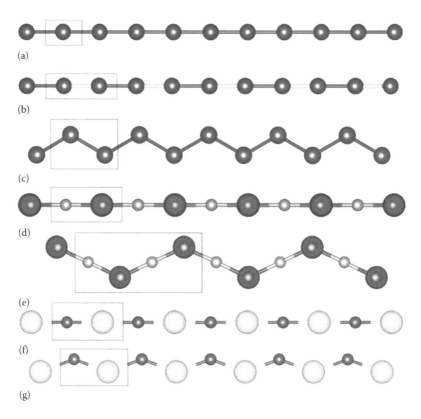

FIGURE 2.2 The infinite 1D systems: (a) linear equispaced Si configuration, (b) dimerized Si chain, (c) buckled saw-tooth Si configuration, (d, e) linear and buckled ZnO chains, and (f, g) linear and buckled CdS chains. The unit cell of each is highlighted with a dotted line. Si, blue; Zn, gray; O, red; Cd, violet; and S, yellow.

2.2 Long-Term Interactions

Energy evaluation, as introduced in Chapter 1, requires an accurate account of many interactions in the system including short- and long-range terms.* But in contrast to the finite nanoparticles and clusters, modeling 1D systems encounters one problem, which is common to all extended systems with long-range potentials—the conditional convergence of electrostatic series. *Conditional* does not mean that the convergence is just slow, but that the result of the calculation is mathematically ill defined, which we illustrate below using our toy system of a linear (nonbuckled) two-atomic chain with an interatomic distance between the nearest neighbors of d (see, e.g., Figure 2.2d). For clarity, in this following analysis we use the atomic system of units. An electrostatic potential on atom A with a negative charge q can be calculated as a series of contributions from more and more distant pairs of atoms symmetrically arranged around A, whose charge alternates between $+q$ and $-q$,

$$\varphi = \frac{2q}{d} \sum_{n=1}^{\infty} \frac{(-1)^{n+1}}{n}. \tag{2.1}$$

---

* Note, however, that some atomistic models, in particular for covalent and metallic systems, neglect completely the long-range interactions, for example, using bonding harmonic or embedded atom force fields, respectively.

The astute reader will have noticed that this infinite sum is a particular case of the Taylor expansion of $\ln(1 + x)$ around zero with $x = 1$:

$$\ln 2 = \sum_{n=1}^{\infty} \frac{(-1)^{n+1}}{n}. \tag{2.2}$$

Let us have a careful look at this series:

$$\sum_{n=1}^{\infty} \frac{(-1)^{n+1}}{n} = 1 - \frac{1}{2} + \frac{1}{3} - \frac{1}{4} + \frac{1}{5} - \frac{1}{6} \cdots. \tag{2.3}$$

We can rearrange its terms, for example, thus,

$$\left(1 - \frac{1}{2}\right) - \frac{1}{4} + \left(\frac{1}{3} - \frac{1}{6}\right) - \frac{1}{8} + \left(\frac{1}{5} - \frac{1}{10}\right) - \frac{1}{12} + \left(\frac{1}{7} - \frac{1}{14}\right) \cdots$$

$$= \frac{1}{2} - \frac{1}{4} + \frac{1}{6} - \frac{1}{8} + \frac{1}{10} - \frac{1}{12} + \frac{1}{14} - \cdots$$

$$= \frac{1}{2}\left(1 - \frac{1}{2} + \frac{1}{3} - \frac{1}{4} + \frac{1}{5} - \frac{1}{6} + \frac{1}{7} - \cdots\right) = \frac{1}{2}\ln 2, \tag{2.4}$$

which demonstrates that convergence of the series is conditional. The physical implication of this behavior is that neither the total energy of a 1D system is defined nor any related property such as the work required to remove or add a charged particle to the system in a particular location, that is, to create a point defect.

The origin of this unphysical result is usually related to an uncertainty introduced by assumption that the 1D system is infinite in length. If, instead of an infinite chain, we considered its finite neutral fragment (a rod) of length $2l$, then charges $\pm q$ that terminate the chain on both ends would define the overall rod dipole $\vec{P}$ and therefore its potential at any point sufficiently removed from the rod (i.e., where the atomic structure can be neglected). So we can associate with the system a dipolar density $\vec{p} = \vec{P}/2l = q\vec{e}$ (see Figure 2.3). By applying the cyclic boundary conditions to the system, we would recreate the original infinite chain, with a dipolar density, or polarization, $\vec{p}$, the value of which depends on the termination procedure. This arbitrary choice gives rise to an uncertainty in the potential reference in 1D, but in contrast to 3D bulk systems, the electric field is defined uniquely as the corresponding series converges absolutely.

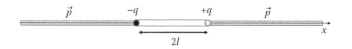

FIGURE 2.3 A model of an infinite 1D chain of dipoles with density p.

Chapter 2

In nature (or the laboratory), any such system is of course finite and terminated in a particular fashion, setting the values of both potential and field uniquely. If the *macroscopic* local field is nonzero, the system would be ferroelectric and would naturally acquire a domain structure, where the voltage the charge carrier is subjected to should be commensurate with usual physical constraints, for example, that of a characteristic band gap discussed in the next section. More often however, the field will be very close to zero (subject only to local fluctuations), and the potential will be uniquely determined by ordering terms in the electrostatic series such that each fractional sum corresponds to a fragment with a zero dipole. With the value of the potential thus defined, there is still however a problem due to the slow convergence of the series.

To tackle this problem of slow convergence, in an ingenious and very popular Ewald's approach [41], a Coulomb potential is split into two contributions: one short and the other long range. The short-range term is summed up directly over the lattice where the speed of convergence can be tuned with a careful choice of the potential form and parameters, customarily defined as the Coulomb potential of a Gaussian charge distribution. The long-range term can be easily evaluated in reciprocal space, where its Fourier transform is short ranged and therefore the corresponding sum converges quickly. Mathematical details of this technique and its application to systems of different dimensionalities can be found, for example, in References 42 and 43. Notably, an alternative approach implemented in a number of atomistic and electronic structure codes has been proposed by Saunders et al. [44], who exploited the Euler–MacLaurin summation formula rather than Fourier-based constructs.

Finally, we note that only few current codes tackle specifically 1D boundary conditions; however, there are many *ab initio* and atomistic (in particular, molecular dynamics, or MD) codes providing accurate treatment of any systems in 3D. Therefore, a popular choice for modeling lower-dimensional systems is a supercell approach, in which, for example, a wire is modeled within one unit cell surrounded in two directions by a layer of vacuum. Periodic repetition of this unit cell in all three dimensions results in an infinite stack of nanowires arranged on a 2D lattice. Although this system could be of interest as it reflects the real situation of nanowire synthesis in some cases, more often the resulting interactions between periodically repeated nanowires are completely fictitious. To improve the model, one should remove these interactions, which can be incorporated in the relevant code or done *a posteriori*, which is an area of continuing development; see, for example, Reference 45.

2.3 Energy and Electronic Structure Theory

Having established the electric potential, in the field of which move all charge carriers, we now discuss the energy states of such charge carriers. We begin by briefly recalling qualitative theories of the electronic structure of 1D periodic systems. This approach is commonly used in solid state texts as an introduction to band theories of crystalline solids, but as we concern ourselves with 1D rather than 3D, we feel that such a recall is beneficial for the reader.

In a 1D system characterized by a linear coordinate x, an electron in a stationary state is described by a wave function, $\psi(x)$, which is subject to the corresponding Schrödinger equation with an external general potential, $\hat{V}(x)$:

$$\hat{T}\psi(x) + \hat{V}(x)\psi(x) = E\psi(x), \tag{2.5}$$

where
$\hat{T} = -(\hbar^2/2m_e)\nabla^2$ is the kinetic energy operator
$\hbar$ is the reduced Planck's constant
m_e is the electron mass
E is the electron energy

For simplicity, we will assume here that the potential is local, working as an operator of multiplication by a function of argument x, with a hat dropped. We now introduce PBCs so that

$$V(x+L) = V(x), \tag{2.6}$$

with L being the period of potential (the unit cell parameter in 1D). Bloch's theorem states that solutions to Equation 2.5 should take the following form:

$$\psi_{nq}(x) = e^{iqx}u_{nq}(x), \tag{2.7}$$

where $u_{nq}(x)$ is a periodic function with the same periodicity as V. The two indices number possible solutions of Equation 2.5. One simple implication is

$$\psi_{nq}(x+L) = e^{iq(x+L)}u_{nq}(x). \tag{2.8}$$

Then for ψ to be a solution of Equation 2.5,

$$\psi_{nq}(x+L) = e^{iqL}\psi_{nq}(x). \tag{2.9}$$

The two indices numbering wave functions have their correspondence in the electronic energies, which form the band structure. The index q is not unique as

$$q, q \pm \frac{2\pi}{L}, q \pm 2\frac{2\pi}{L}\ldots \tag{2.10}$$

gives identical wave functions and energies. To describe the complete band structure, it is therefore sufficient to consider only the range of q $[-\pi/L, \pi/L]$, known as the reduced zone scheme.

In fact, spectra of elementary excitations of quasiparticles in extended systems are described with wavelike equations closely related in their form to Equation 2.5 and therefore have similarly structured solutions. An important example, considered in Section 2.4, is the case of phonons, which are quantized waves of atomic vibrations.

Further, for a given value of q, there is an infinite number of solutions, which we typically index with n from low energy to high. The lowest-energy solutions correspond to the bound electronic states of atoms or molecules, from which the unit cell (i.e., the smallest repeatable unit in the periodic system) is made up; they are occupied by the core and valence electrons and can be separated from or merge with the higher lying extended states. For an atom, the full electronic energy spectrum consists of two parts—lower-energy discrete and higher-energy continuous; for the periodic system, because of the variation of energy with q, both parts of the spectrum are continuous. However, the discrete nature of the bound states is retained in the spectra of the periodic system in the form of discrete bands (which may, however, overlap with each other) as will be seen below.

The origin of the discrete bands can best be described in terms of the simplest example of an electron confined within a 1D system, that is, the Kronig–Penney model, which consists of a periodic array of potential wells, each of width a, separated by barriers of width b and height V_0, as shown in Figure 2.4a. Given Bloch's theorem (Equation 2.8), the allowed solutions of Schrödinger's equation for an electron with energy E subject to such a potential are linear combinations of plane waves, the wavenumbers q of which are constrained by the boundary conditions:

$$\cos q(a+b) = \cos ka \cosh \kappa b + \frac{1}{2}(\kappa/k - k/\kappa)\sin ka \sinh \kappa b, \tag{2.11}$$

where

$$k = \sqrt{2m_e E}/\hbar$$
$$\kappa = \sqrt{2m_e(V_0 - E)}/\hbar$$

If we plot the right-hand side (RHS) of Equation 2.11 as a function of k (see Figure 2.4b), it is immediately obvious that, for certain values of k (or equivalently E), its magnitude is greater than 1, which is incompatible with the left-hand side, indicating that the equation at this value has no solution. Such values of k or E make continuous bands of forbidden states and are commonly referred to as energy gaps. When we consider the functional relationship between q and E (an inverse function of the conventional band structure), we observe alternating intervals of allowed and disallowed states, with one example (with $a = 6$ Å, $b = 4$ Å, and $V_0 = 6$ eV, solved using an in-house developed program KP-SOLVE*) given in Figure 2.4c. The range of q is confined within the reduced zone scheme, which is the 1D first Brillouin zone (BZ). The band structure can be used to derive other important features such as the electron (or hole) effective mass m^*, and the density of states (DOS), which shows band widths and gap values, on which many optical and transport properties depend.

A realistic potential in Equation 2.5, however, will be much more complicated, reflecting the chemical nature of the system, and, unfortunately, there are no simple closed solutions to Equation 2.5. To simplify the problem, a basis set is introduced, which can refer either to the behavior of the electron bound to a nucleus, as in an atom, or to the extended plane-wave-like nature of a nearly free electron. The former approach is employed by

---

* Note that the lowest band is underrepresented using a low-resolution scan we employed and is in fact a continuous, approximately straight horizontal line.

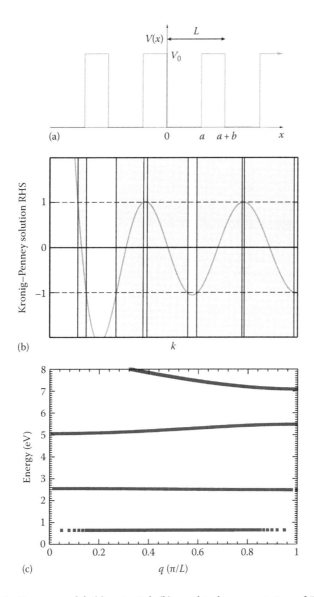

FIGURE 2.4 Kronig–Penney model: (a) potential, (b) graphical representation of Equation 2.11, with allowed solutions highlighted in gray, and (c) resulting dispersion E vs. q. Note that the lowest band is underrepresented using a low-resolution scan we employed and is in fact a continuous, approximately straight horizontal line.

tight binding, or model Hamiltonian methods, whereas the latter by model potential techniques. Such methods typically rely on empirical parameters fitted to reproduce experiment or, more recently, high-level *ab initio* calculations. A modern implementation of model potentials, in particular in the field of quantum chemistry of molecules on one hand and solid state physics on the other, takes the form of semilocal pseudopotentials, which account for the core electrons (see, for example, References 46–50). One can then separate the many-electron system into core and valence electrons, treating the valence electrons explicitly, which move in the field of not just nuclei but also core electrons.

Chapter 2

The band structure discussed so far relates to states that extend along the periodic spatial dimension of the 1D system. In the other two orthogonal directions, quantum confinement will determine the form of the electronic states. The degree to which confinement has an effect depends on the thickness of the nanowire, where the typical localization length of the corresponding state should be compared with the characteristic confining dimension.

The one-electron picture discussed so far is in many respects too simplistic when we consider real systems of many electrons, a number of more appropriate accurate theoretical approaches of various sophistication have been developed and implemented in commonly accessible software packages. The essential features of the electronic band structure, which we highlighted using a simple Kronig–Penney model, are retained for the majority of systems of interest. Key exceptions concern systems with critical two or more electron behaviors observed, for example, in the superconducting regime. As the electronic system in quantum mechanics is described by a many-electron wave function, the solution of a many-electron analog of Equation 2.5 becomes necessary but intractable. The applied studies typically employ one of the following levels of theory, which accounts for the quantum-mechanical many-electron interactions, that is, exchange and correlation:

1. Mean field and Hartree–Fock theories constitute the simplest approach to treating electron–electron interactions is to model each electron in the mean field due to all electrons in the system, or rather their charge density. The obvious fault with such an approach is an ensuing self-interaction of an electron that, perhaps, could be insignificant for systems in which there are very large numbers of electrons. Importantly, however, in close proximity to a given electron at a given position and time, we should expect depletion in the probability to find other electrons, due to both Coulomb repulsion and the Pauli exclusion principle. The former dynamical effect (and other related phenomena involving interactions with more than two electrons) is referred to as electron correlation and the latter, which concerns only same-spin particles, electron exchange. Mean field theory, although neglecting these interactions, only requires the solution of the one-electron Schrödinger equation (2.5) in an external local potential derived from the total charge density (known as the Hartree potential). By introducing an antisymmetric many-electron wave function, in an approach known as Hartree–Fock theory, one can derive a potential that contains a local (Hartree) term and a nonlocal (exchange) term, so that electron exchange is treated explicitly. The formulation of these terms leads to a fortuitous cancellation of the self-interaction energy. The drawback, however, is an increased complexity in the calculation due to the nonlocal term. Thus, to describe n electrons, we need n one-electron orbitals, a problem that becomes tractable using modern computers.

2. Semiempirical methods that originate from the Hartree–Fock theory using a minimal basis of localized atomic valence orbitals, where the matrix elements associated with three or more atomic centers are omitted and one- and two-center terms are approximated using simple analytical functions, which are then parameterized and fitted to experimental and/or *ab initio* data. The variability in the chemical environment is treated by these methods via a self-consistent field (SCF) procedure in full analogy with the parent *ab initio* approaches.

SCF versions of tight-binding approaches have also been developed and widely applied to the study of nanosized systems.

3. Density functional theory (DFT), where the total energy of the system is represented as a functional of the charge density. The pragmatic approach in the Kohn–Sham formulation of DFT is, in analogy with the mean field approach, to map the many-electron function to a single-particle orbital, representing an idealized electron moving in the mean field of all electrons (including itself) and nuclear (core) charges, now also including exchange and correlation effects. Although in principle exact, the potential (exchange and correlation) contribution from all the electrons is not generally known and therefore requires some form of an approximation. The simplest local density approximation assumes a full locality of the potential that only depends on the charge density at the point where it is probed, and is usually equated with the known potential of a homogeneous electron gas of the given density. Furthermore, sophisticated approximations include the first derivatives of the charge density (generalized gradient approximation, GGA), the second derivatives of the charge density and/or kinetic density (meta-GGA), and a fraction of exact Hartree–Fock exchange (hybrid DFT). In all of these methods occupied Kohn–Sham orbitals are used to construct the potential and energy [51]. Recently, even more sophisticated methods using unoccupied Kohn–Sham states have been advanced, termed double hybrids [52].

4. Many-electron theories, in contrast to DFT, include an explicit account of electron–electron interactions. In the first instance, the total energy can be constructed as a series of consecutive improvements over the Hartree–Fock approximation, taking into account correlation between pairs, trios, and higher-order combinations of particles in a perturbative or *exact* manner, leading to Møller–Plesset (MPn) and coupled cluster methods on one hand and configurational interaction and multiconfigurational SCF techniques on another. An analogous, closely related family of methods deals with the problem of one, two, and more particle spectral properties of the system by employing the formalism of Green's functions, which is also used for the treatment of localized defect states and transport. The most widely used approaches include the following: GW, as applied to calculations of ionization and attachment energies, thus allowing direct comparison with experimental photoelectron and secondary electron spectra; the random phase approximation (RPA), typically used in the same context as GW, but recently applied with significant success to calculating the total energy when perturbative single and double excitations are treated (both GW and RPA are also used for the calculation of the dielectric function); and, finally, the Bethe–Salpeter equation, which describes explicitly twoparticle interactions within a many-particle system and allows accurate modeling of excitonic spectra. All these approaches are very expensive with respect to computer resources but are becoming more accessible as the years pass by.

To highlight the main features of 1D band structures and DOS, we have calculated the band dispersion for a series of toy linear chain systems (see Figure 2.2) using hybrid DFT. The chains are assumed to repeat periodically in the x direction, with a unit cell length of a. For each system, we have either carried out full geometry optimization or, in the case of perfectly linear chains, optimized the structure with symmetry constraints.

All calculations have been performed using the VASP code [53–56], with the projector-augmented wave approach [57] to describe the interaction between the core (Si:[Ne], Cd:[Kr], Zn:[Ar], S:[Ne], O:[He]) and valence electrons, and the solid-corrected Perdew–Burke–Ernzerhof GGA exchange-correlation density functional with 25% exact Hartree–Fock exchange included (PBEsol0) [58–61]. Total energy convergence within 10^{-4} eV/atom has been achieved using a plane-wave cut-off energy of 1200 eV and a $48 \times 1 \times 1$, $24 \times 1 \times 1$, $12 \times 1 \times 1$ Monkhorst–Pack k-point mesh for the one-, two-, and four-atom unit cell cases, respectively. Calculations were deemed to be converged when the forces on all atoms were less than 0.01 eV Å^{-1}.

The calculated band structures and DOS are shown in Figures 2.5 through 2.7. In all cases we observe cosine-like dispersion, as expected from our discussion on the Kronig–Penney potential above. For the ionic bonding cases, we have indicated the orbital contributions to the DOS, which demonstrates the origin of the bands. We now discuss the particular cases in more detail.

For metallic and covalent bonding, we have chosen Si as a representative system. The perfectly linear case, Figure 2.2a, consists of a single Si atom in the unit cell, with a Si–Si bond length of 2.12 Å. This structure is unstable with respect to a Peierl's type of distortion (see Section 2.5), which can be traced to its unpaired electrons occupying orbitals perpendicular to the chain. The system is metallic, as can be seen from the band structure, Figure 2.4a, where the metallic band is shown in green. Two distortions can occur to break the symmetry and lead to electron pairing: dimerization (Figure 2.2b) and buckling (Figure 2.2c). In both cases, the unit cell doubles to two atoms per cell. We find that buckling, where the Si–Si bond length increases to 2.20 Å and the bonds form an angle of 117.11°, is the ground state, being 0.56 eV/atom more stable than the linear case (and 0.52 eV/atom more stable than the dimerized case). Looking at the band structures (Figure 2.5b and c), we observe band folding due to the increased unit cell sizes and clearly see a direct band gap opened up (0.46 eV for the dimerized chain, 0.49 eV for the buckled chain). The reduced symmetry of the buckled chain is reflected in the breaking of degeneracy of the bands.

For ionic bonding, we have considered two systems: CdS and ZnO. The main features we observe in the band structures (Figures 2.6 and 2.7) are a large band gap (approx. 2 to 5 eV) between anion p (making up the top of the valence band) and cation s orbitals (making up the bottom of the conduction band) and localized (low dispersion) orbitals deep in the valence bands (cation d and anion s orbitals). For CdS, the buckled chain structure (Figure 2.2g), with two atoms in its unit cell, is 0.18 eV/atom more stable than the linear chain (Figure 2.2f). Due to the buckling (with an angle of 140.56°), the CdS bond length shortens from 2.37 to 2.29 Å, which results in a greater anion–cation repulsion and a corresponding increase in the band gap (from 2.38 to 2.93 eV). On buckling, the breaking of degeneracy within the p and d orbitals is evident from the band structures.

For ZnO, a similar set of results to that of CdS was found, apart from one important difference: the buckling of the chain occurred so that the Zn ions remained linearly coordinated. This coordination means that the unit cell consists of four atoms (see Figure 2.2e), which leads to a folding of the bands due to the reduced reciprocal unit cell size. The buckling angle is 130.31°, and the Zn–O bond length reduces from 1.742 to 1.736 Å, with a resulting increase in band gap from 5.15 to 5.55 eV.

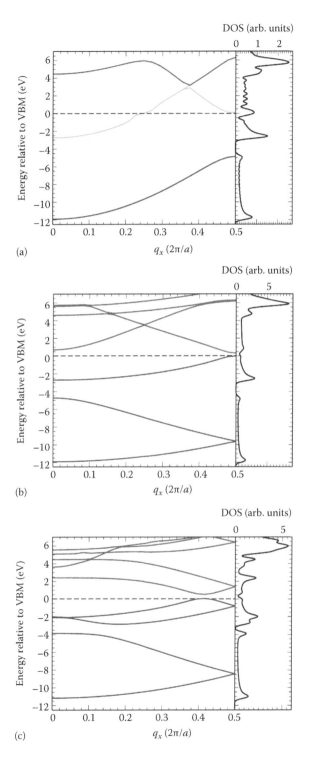

FIGURE 2.5 Band structure and density of states of the (a) linear, (b) dimerized, and (c) buckled Si chains. Here and Figures 2.6 and 2.7, blue or light gray lines are used for valence states (bands), red or charcoal gray for conduction states, and green for metallic half-filled bands.

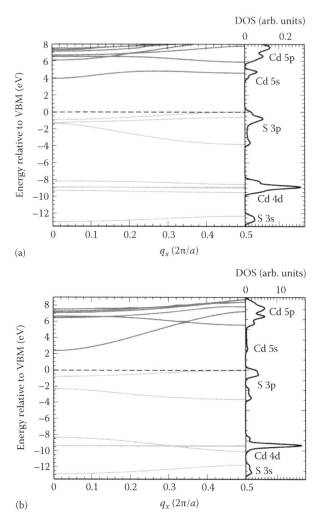

FIGURE 2.6 Band structure and density of states of the (a) buckled and (b) linear CdS chains.

A large number of studies have been performed to determine the band structure of various nanowire systems, a representative set of which we list next. Si nanowires are the most well studied, using computational approaches including $\mathbf{k} \cdot \mathbf{p}$ [62], tight binding [63], and DFT [64–66]. A comprehensive $\mathbf{k} \cdot \mathbf{p}$ study of core–shell nanowires for III–V systems has been performed by Pistol [67], including Γ-point energies and effective masses, while the band structure of InAs nanowires has been determined using a tight-binding approach by Lind [68]. Ballistic transport through the 1D subbands is included in many of these studies. In all cases, the calculated band structures display the basic features of those of our toy models, as is evident, for example, from a DFT study on Si nanowires by Nolan et al. [65]—see Figure 2.8. Experimental techniques used to study electronic bands in 1D nanosystems include resonant Raman spectroscopy (comparing the electronic structure of GaAs in wurtzite and zinc blende phases) [69], angle-resolved photoluminescence spectroscopy [70], photoluminescence (PL) [71], time-dependent PL and PL excitation [72], and conductance measurements [73].

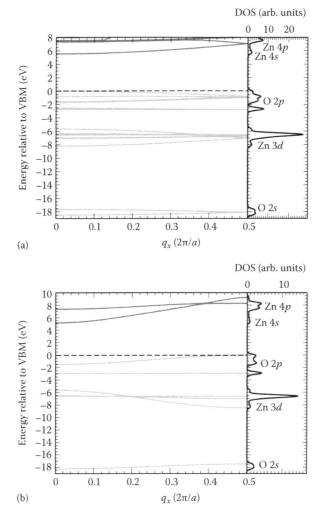

FIGURE 2.7   Band structure and density of states of the (a) buckled and (b) linear ZnO chains.

2.4   Dynamical Properties

The structure and properties of 1D systems in equilibrium are still of course affected by the thermal motion of atoms and, in particular, at low temperatures, by its quantum character, which should be expected based on the uncertainty principle. The fact that we are able to describe the electronic properties without recourse to the nuclear motion is a consequence of the dramatic difference in particle masses between an electron and even the lightest nucleus, that is, a proton. Such a separation of the electronic motion from the nuclear is described by the Born–Oppenheimer theory, which uses the ratio of these two masses as a small parameter $\sim O(10^{-3})$, about which the total energy of an electron-nuclear system is expanded. As the electronic energy can be determined for a given set of nuclear coordinates, it may be considered as a potential, in the field of which nuclei move, which forms the basis of the semiclassical method of interatomic potentials (see Chapter 1).

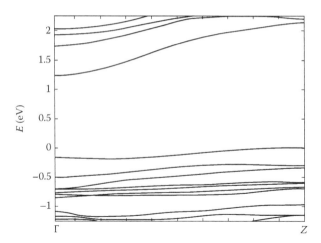

FIGURE 2.8 Calculated band structure of Si [100] oriented nanowires (with a square section, 10 Å across, OH terminated), using the PBE exchange and correlation density functional and a plane-wave basis set. (From Nolan, M. et al., *Nano Lett.*, 7, 34, 2007.)

The potential energy of any system is a smooth function of nuclear coordinates, about which it can be expanded into a Taylor series. Typically, the potential energy of nuclei (usually referred to as atoms in this context) is sufficiently well represented by harmonic wells with a possible inclusion of anharmonic corrections, which corresponds to retaining the quadratic and higher-order cubic terms (the first-order term drops out as a necessary condition of the system stability). On average, the atoms can be found at the bottom of these wells, and the first computational task in a study of a material is to determine positions of these minima, the object of geometry optimization discussed in full in Chapter 1.

When the coordinates of the minimum of interest are found, the thermal motion around the minimum can be described classically, in the harmonic approximation, as a system of coupled linear oscillators. The solution of the problem leads to a set of eigenmodes, within which the atomic system can vibrate. The corresponding quantum-mechanical description involves the quantization of such modes into particles of atomic vibration known as phonons. By pumping energy into a particular eigenmode, at a classical level, we increase an amplitude of the corresponding atomic oscillation, or, at a quantum-mechanical level, we generate more phonons of a particular eigenenergy (or frequency). The eigenmodes are in practice vectors of preexponential coefficients in a plane-wave representation of waves propagating through the system, which are subject to the same PBCs considered earlier for electrons and consequently an analogous Bloch theorem. For each atom in a unit cell there are three spatial degrees of freedom, so there are three corresponding unique phonon (vibrational) modes, implying a very large number for an extended system, which can however be conveniently enumerated using **q**-points in the reciprocal space, all within just one first BZ. If our unit cell contains M atoms, the number of possible modes due to just these atoms will be $3M$, which is the number of branches spanning the first BZ, with the energy (frequency) dependence on the **q**-point known as dispersion, analogous to the dispersion of electronic energy levels discussed in Section 2.3. The three lowest-frequency modes are termed the acoustic modes, as in the elastic limit they correspond to sound waves in the macroscopic system, and consist of in-phase motion of the atoms within the unit cell. Any other modes are termed optical and consist of motion where the center of mass of the atoms remains constant, implying a well-defined phase difference between their individual vibrations.

Similar to electrons, in 1D, the atomic degrees of freedom may be restricted, depending on the level of confinement. In the two orthogonal directions where periodicity is removed, the phonon modes become restricted or confined, which alters their dispersion. This lack of periodicity results in much larger unit cells, which in turn leads to an increase in the number of eigenmodes. Moreover, surface scattering leads to a lifetime or broadening of the modes that can be directly observed using Raman spectroscopic techniques. In order to demonstrate the basic concepts of phonon dispersion in 1D systems, we now discuss the dynamical properties of our toy 1D chain systems presented in Figure 2.2.

Phonon frequencies have been calculated with a frozen phonon approach, which requires knowledge of second derivatives of energy (force constants) with respect to atomic degrees of freedom in a suitably chosen supercell (24-atom unit cells in all cases discussed). The force constants were evaluated numerically, employing the method of finite atomic displacements with energy and forces calculated at the hybrid DFT level (introduced in our discussion of the electronic bands in Section 2.3). $\mathbf{q}$-point interpolation as implemented in the postprocessing program PHONOPY [74] was then used to determine the dynamical matrices and phonon dispersions.

We first discuss the ionic system CdS (Figure 2.9). In Section 2.2, we found that the buckled chain was lower in energy than the linear chain. For both the linear and buckled chains, as we have two atoms in the unit cell, there are six phonon modes, three acoustic and three optical. The transverse acoustic (TA) modes (shown in dispersion graphs in Figure 2.9) consist of perpendicular displacements of the chain and remain at low frequencies throughout the BZ. The longitudinal acoustic (LA) mode has linear dispersion close to the Γ point, the slope of which is the speed of sound (as in bulk). For the linear chain (Figure 2.9b), the transverse optical (TO) modes are imaginary, indicating that the system is unstable with reference to a distortion along these modes. As these modes consist of out-of-phase motion of the two atoms in the unit cell perpendicular to the chain, this distortion corresponds to the buckling of the chain, that is, a phase transition to the lower-energy buckled configuration. The TA modes in this case are degenerate, as expected for a linear chain. The longitudinal optical (LO) mode remains relatively dispersionless at approximately 500 cm^{-1}. For the ground state buckled system (Figure 2.9a), the dispersion becomes more complicated. The TA and TO modes are no longer degenerate, and the TO mode, corresponding to a *wagging* motion of the ions, softens to lower frequency (ranging from just above 0 to 56 cm^{-1} over the BZ). The LO mode also softens slightly. It is worth noting that the calculated Γ point frequencies of the optical modes (218 and 466 cm^{-1}) are in reasonable agreement with Raman measurements on CdS nanowires (301 and 598 cm^{-1}) [75], when one takes into account the red shift due to confinement, as our system is the limiting case of one atom thick wires. One of the TA modes, however, in this case becomes imaginary, which indicates that the system is unstable. We show how this instability is resolved in the following example of the ZnO chain.

The results for the ZnO chain display common features to those of CdS, apart from some important distinctions. We remind the reader that the ground state for this system is the buckled chain (Figure 2.2d) with a four-atom unit cell. For the linear chain

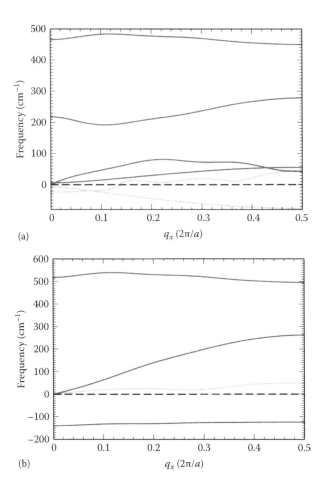

FIGURE 2.9 Phonon dispersion of the (a) buckled and (b) linear CdS chains. (TA bands are shown in light gray, while all other bands in dark gray.)

(Figure 2.2b), the modes observed are similar to those of the linear CdS chain, except for an increase in frequency of the LO and LA modes, reflecting the stronger Zn–O interaction, real TO modes (which range from ~170 to 120 cm^{-1} across the BZ), and lower-frequency TA modes. The TA modes remain close to zero up to $q_x = 0.6$ ($2\pi/a$), which may be an artifact of the calculation (these modes are low-frequency sound waves, and it is possible that their low frequency is within the numerical error of our simulation). The fact that all modes are real indicates that the linear chain is stable, in contrast to the CdS case. The lower-energy structure, that is, the four-atom unit cell buckled chain, is therefore arrived at by a first-order phase transition, in which an energetic barrier must be overcome. We will discuss phase transitions in more detail in Section 2.5. For the buckled chain (Figure 2.10a), we see folding of the phonon bands due to the increased unit cell size, resulting in a doubling in the number of bands, and a splitting of the transverse modes.

The case of the Si chain is somewhat more complicated. The linear chain is unstable with respect to a Peierl's distortion (see Section 2.5), meaning it can dimerize (Figure 2.2b), but the ground state consists of a buckled chain (Figure 2.2c). The instability of the

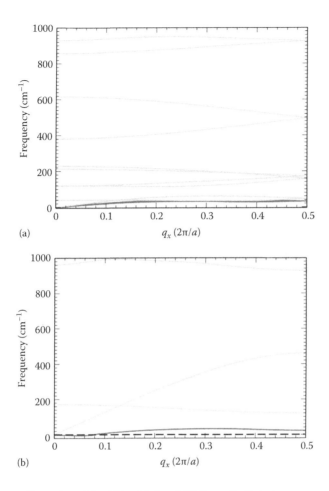

(a) $q_x\,(2\pi/a)$

(b) $q_x\,(2\pi/a)$

FIGURE 2.10   Phonon dispersion of the (a) buckled and (b) linear ZnO chains.

linear chain is reflected in its phonon dispersion (Figure 2.11a). As the unit cell consists of a single Si atom, there are only three modes, all of which are acoustic. The LA mode becomes imaginary away from the Γ point, corresponding to the dimerizing distortion. The TA mode at the edge of the BZ (shown in green) is also imaginary and corresponds to a buckling distortion. For the dimerized case (Figure 2.11b), the dispersion is similar to that of the CdS linear chain, as one would expect (as both are linear and unstable). In this case, there are optical modes, as the unit cell consists of two atoms. The (degenerate) TO modes are imaginary, reflecting the fact that the buckling distortion is energetically preferred by the system. For the buckled chain (Figure 2.11c), the TO modes are no longer imaginary, but one of them (the *wagging* mode) becomes mixed in with the TA modes. The oscillations visible in these modes' dispersion is a result of the insufficiently large size of supercell we could afford using modern supercomputers. The higher-frequency LO and TO modes are well separated at Γ, but mix as q_x varies across the BZ. Interestingly, at the zone boundary the modes become degenerate, reflecting the fact that the unit cell consists of one species, meaning that the mode distortions in different directions become equivalent when the phase between the unit cells is exactly opposite (which is the case at the zone boundary).

Chapter 2

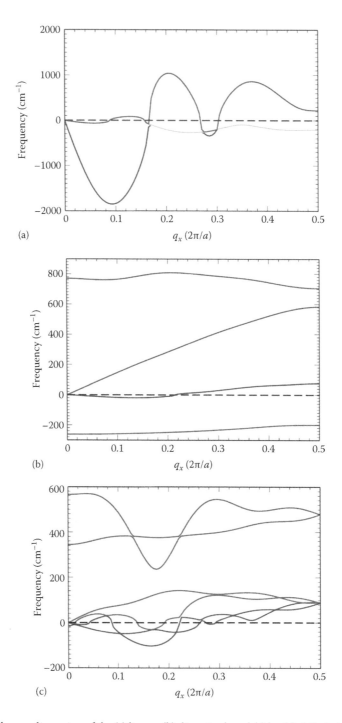

FIGURE 2.11   Phonon dispersion of the (a) linear, (b) dimerized, and (c) buckled Si chains.

The effect of quantum confinement on the phonon modes in 1D systems has been extensively studied in the literature, from which we provide some examples. Using Raman spectroscopy, the resulting shift and broadening of the optical modes due to confinement have been observed, for example, in nanowires of Si [76–78], GaAs [79,80], CdS [75], SiC [81], and GaN [82]. Other experimental techniques have also been used to study the confinement of phonon modes, including optical absorption in CdTe nanowires [83] and inelastic transitions between nanostructures as a probe in InAs nanowires [84]. Such modes were used to explain the low thermal conductivity and conversely high Seebeck coefficient of Si nanowires [85,86] by illustrating that confinement leads to surface scattering of optical modes, but efficient transmission of acoustic modes due to the wire diameter being less than the wave length of the phonon mode.

Theoretical studies on the dynamical properties of 1D systems have also been reported. The confinement-induced shift and broadening of optical modes were shown to be strongly affected by the surface structure by Richter et al. [87]. Thonhauser and Mahan [88] used a ball-and-spring model to study phonon modes in a hexagonally cross-sectioned Si nanowire, demonstrating that the mode boundary conditions consisted of zero stress on surface atoms for acoustic modes and zero displacement (or *clamped* boundary conditions) for optical modes. Their optical eigenmodes are shown in Figure 2.12. A tight-binding approach has been used to determine the electronic and phonon properties and hence the electron–phonon interaction and transport properties, of Si nanowires [89] and field effect transistors based on such nanowires [90]. The mobilities determined in this way were in good agreement with experiment. Atomistic calculations of phonon frequencies and molecular dynamical simulations of their transport have been performed on Si nanowires by Donadio and Galli [91], which confirmed the strong effect of the surface structure and wire diameter on the confined phonons.

An example of a calculated phonon dispersion (of a Si nanowire) [88] is given in Figure 2.13.

Finally, Mizuno et al. [92] have provided an extensive theoretical study of the possible confined modes in nanowires with square and rectangular cross sections.

2.5 Structure and Phase Transitions in 1D

The stability of a 1D system is no less surprising than the stability of graphene sheets, which have been the subject of intense research in recent years. One would naïvely expect thin nanowires to be quite brittle, but it has been shown, by experiment and molecular dynamics (MD) simulations, to often not be the case. The mechanical properties of nanowires are quite different to those of the corresponding bulk systems, mainly due to the proportional increase in surface area as size is reduced. Depending on the orientation and thickness of the nanowire, its yield strength can vary widely, but remain significantly higher than that of bulk. This effect is related to energetic barriers to slippage, which occur due to the stacking of the wire along the axis, as opposed to on the surface. Conversely, the Young's modulus is typically less sensitive to the wire dimensions, but some MD modeling indicates that it can vary for Au nanowires of a particular orientation. For a detailed review of the mechanical properties of nanowires, see Reference 93.

In the 1930s, Peierls pointed out that a 1D metallic chain would be unstable with respect to band gap formation. Considering a linear chain of atoms of separation a, with

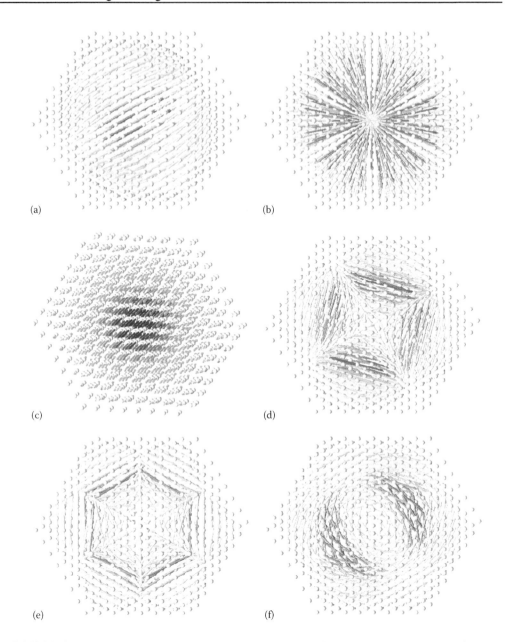

FIGURE 2.12   Optical eigenmodes in a Si nanowire as viewed along the [111] direction. (From Thonhauser, T. and Mahan, G.D., *Phys. Rev. B*, 69, 075213, 2004.)

one electron per atom, according to the Kronig–Penny model considered earlier, the system will have one half-filled band that would show metallic behavior. If the atomic spacing were to change so that each atom became simultaneously closer to one neighbor and further from the other, that is, if the symmetry were to spontaneously break as the system dimerizes, the increase in elastic energy due to the stretching of one bond may be offset by the energy gain in the formation of a covalent bond of σ character. In this case, the effective lattice spacing is doubled, and the band structure will now contain an energy gap, with a fully occupied valence band. The internal energy of the chain, which

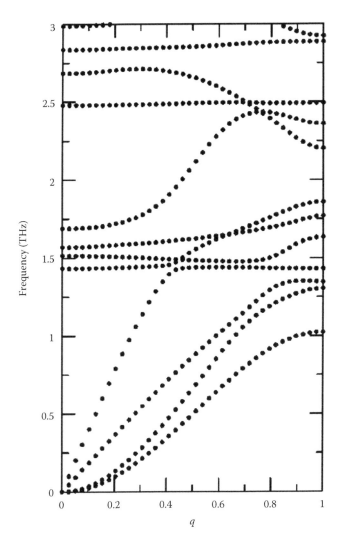

FIGURE 2.13 Phonon dispersion of a hexagonally cross-sectioned Si nanowire. (From Thonhauser, T. and Mahan, G.D., *Phys. Rev. B*, 69, 075213, 2004.)

includes the thermal energy due to atomic vibrations, will of course depend on the temperature. Below a critical temperature, such lattice distortions will be favorable, while at higher temperatures, metallic conductivity will be observed.

Different synthesis procedures lead to different 1D structures, such as wires, tubes, and ribbons that typically retain bulk-like atomic arrangements. Once the nanostructure is of sufficiently small diameter (<5–10 nm), however, the bulk-like characteristics are no longer realized. Determining the exact configuration at such sizes is challenging experimentally, leaving theory as the only viable approach at present.

The configuration of a nanowire, even with a given number of atoms, can present a great challenge for structural determination. Large progress is expected from applying global optimization methods, which, however, has not yet been realized. The absolute majority of the current modeling is done based on carving a model system from bulk

Chapter 2

or on construction of new architectures from small predefined nanostructures, usually obtained using global optimization techniques. In some cases, novel atomic configurations can result from numerical experiments termed uniaxial tensile loading followed by simulated annealing, where a bulk cut is strained in the extended direction and then allowed to relax. See Chapter 1 for an extended discussion of techniques involved in structure prediction and applied studies.

2.6 Charge and Heat Transport in Ideal Systems

The calculation of transport properties in solids is well established, with a variety of approaches available differing in complexity and accuracy. In a metal or semiconductor, electrons (holes) occupy only a fraction of the available conduction (valence) states, which is a necessary condition for energy and momentum transfer. Typically, when treating electron (or hole) carrier transport, the effective mass approximation is applied (see References 94 and 95, whose arguments we follow), within a one-particle picture, where interactions between carriers are considered negligible and the motion is described either semiclassically, via Newton's laws, or quantum mechanically, using the Schrödinger equation. Within this approximation, the effect of the periodic crystal potential is such that the carrier moves as if it were a free particle, but with a different mass $m = m^* m_e$, termed the effective mass. As m^* arises from the periodic potential, it can be derived from the crystal band structure.

In Section 2.3, we treated electrons as Bloch waves with well-defined quantum numbers $\vec{q}$ (wave numbers). Such a description implies that the electrons delocalize over the entire space. When considering transport, it is often more convenient to describe electrons as particles with well-defined positions and/or velocities/(quasi-)momenta (subject, however, to the Heisenberg uncertainty principle), which can be achieved by representing electrons as packets (groups) of Bloch waves, moving with a group velocity

$$\vec{v} = \nabla_{\vec{q}}\,\omega(\vec{q}) = \frac{1}{\hbar}\nabla_{\vec{q}} E(\vec{q}), \tag{2.12}$$

where $E(\vec{q})$ is the $\vec{q}$-space structure (dispersion) for the particular conduction band. Given an applied electric field, $\vec{\mathcal{E}}$, the work done on the particle of charge $-e$ is given by

$$\delta E = -e\vec{\mathcal{E}} \cdot \vec{v}\delta t \tag{2.13}$$

From Equation 2.12 we can derive

$$\delta E = \hbar \vec{v} \cdot \delta \vec{q}, \tag{2.14}$$

and combining these equations, we find

$$\hbar \dot{\vec{q}} = -e\vec{\mathcal{E}}. \tag{2.15}$$

Taking the time derivative of Equation 2.12, for each component i ($= x, y, z$)

$$\dot{v}_i = \frac{1}{\hbar} \sum_j \frac{\partial^2 E}{\partial q_i \partial q_j} \dot{q}_j, \tag{2.16}$$

from which, taking Equation 2.15, we derive

$$\dot{\vec{v}}_i = \frac{1}{\hbar^2} \sum_j \frac{\partial^2 E}{\partial q_i \partial q_j} (-e\vec{\mathcal{E}})_j. \tag{2.17}$$

Equation 2.16 is analogous to the classical Newton's equation, with mass given by the tensor:

$$\left(\frac{1}{m^\star}\right)_{ij} = \frac{1}{\hbar^2} \frac{\partial^2 E}{\partial q_i \partial q_j}, \tag{2.18}$$

which is symmetric and can therefore be transformed to principle axes.

At equilibrium, there is no net displacement $\vec{r}$ of carriers. One then considers the case when there is an applied field $\vec{\mathcal{E}}$, which accelerates the particle subject to occasional scattering processes that in turn slow it down (if such scattering processes were not present, the acceleration due to $\vec{\mathcal{E}}$ would result in an infinite velocity and hence infinite current). Thus, naturally emerge two time scales: over one, the particles accelerate between collisions, and over the other, subject to many collisions the particles move with an average constant velocity—see below. The slowing down or drag will be proportional to the particle velocity and inversely proportional to the average time τ between collisions. Such a drag force is also proportional to the particle mass, which affects the acceleration between the collisions and therefore the average velocity. The appropriate Newton's equation (over the longer time scale) is

$$m^\star \frac{d^2\vec{r}}{dt^2} + \frac{m^\star}{\tau} \frac{d\vec{r}}{dt} = -e\vec{\mathcal{E}}, \tag{2.19}$$

where τ is termed as the scattering relaxation time. As the particle is accelerated, the velocity and hence the drag increase until a steady state is achieved, in which the drag term counters precisely the applied field so that on average the particles move with a drift velocity given by

$$\vec{v}_d = -e\vec{\mathcal{E}} \frac{\tau}{m^\star}, \tag{2.20}$$

from which the current density $\vec{J} = -ne\vec{v}_d$ (n is the charge carrier density) and, hence, the conductivity σ can be determined (from $\vec{J} = \sigma\vec{\mathcal{E}}$ where a scalar conductivity is defined as one of the principal values of tensor σ). A more useful property, however, to calculate is the mobility μ, defined as the constant (tensor) of proportionality between $\vec{v}_d$ and $\vec{\mathcal{E}}$

Chapter 2

(i.e., $\vec{v}_d = \mu \vec{\mathcal{E}}$), as it is independent of the carrier density, which can vary widely in semi-conductors with temperature or doping level. We then have

$$\mu = e\frac{\tau}{m^{\star}}. \tag{2.21}$$

As shown earlier, $m^{\star}$ can be derived from the crystal band structure (it could also be taken from experiment or used as a parameter to be fitted to other experimental results). The complications to determining μ come from the property τ.

The simplest approach is to treat τ as a constant, derived phenomenologically, for example, from measurements of resistivities, which, however, limits the predictability of the model. As we consider transport in the periodic system, scattering occurs from breaks in periodicity, that is, lattice vibrations, defects, and surfaces/interfaces. Each of these scattering processes has a characteristic rate τ_i, the reciprocal of which is the scattering rate R_i, which, according to Matthiessen's rule, can be summed together to determine a total scattering rate R. Different approaches can be taken to calculate these scattering rates, with the most accurate being the determination of appropriate wave functions and applying Fermi's golden rule, including the relevant scattering potential (e.g., electron–phonon, electron–electron, screened Coulomb).

The mobility was introduced in order to characterize the transport properties of an average single carrier, which in turn are determined by the statistical properties of scatterers. For example, the concentration of charged impurities in a system can be considered as constant over a wide range of temperatures above a certain threshold, whereas the number of acoustic vibrations is strongly dependent on temperature (following Bose–Einstein statistics). The effect of temperature on the mobility then consists of two contributions: the number of scattering centers themselves and the energy-dependent scattering of the carriers (which will be reflected in the form of the wave functions and scattering potential used in Fermi's golden rule). Such temperature effects are of a more fundamental nature than those that occur purely based on the variation of carrier density.

Heat transfer involves the redistribution of thermal energy from one region of the system to a neighboring region, in which the population (energy distribution) of local electronic states and/or phonon modes is modified. In contrast to charge transport, therefore, two mechanisms contribute to heat transport: electronic heat transport and heat transport by phonons. We note that phononic thermal transport in 1D inorganic nanomaterials is dealt with in more detail in Chapter 8. The electronic contribution dominates in metals and degenerate semiconductors. For a temperature gradient, ∇T, we have

$$\vec{J}_{th} = -\kappa \nabla T, \tag{2.22}$$

where

$\vec{J}_{th}$ is the total thermal current (i.e., the net thermal energy transported across unit area in unit time)

κ is the thermal conductivity

The negative sign indicates transport from high to low T

Neglecting mutual scattering between the two types of particles involved, the two different contributions can be determined separately and summed together, that is, $\kappa = \kappa_e + \kappa_p$, where κ_e and κ_p are the thermal conductivities associated with electrons and phonons, respectively.

As the thermal energy in the system is determined by the internal energy with contributions from the potential, vibrational, and charge carrier terms, a key property to consider is the specific heat capacity $c_v = \partial U/\partial T$, where U is the internal energy. Using the chain rule, we have $\nabla U = c_v \nabla T$, which, from Equation 2.22 and the fact that thermal currents relate to changes in U as a function of position, indicates that c_v will appear in an expression for κ. For electrons, we have

$$\kappa_e = \frac{1}{3} v^2 \tau c_v, \tag{2.23}$$

where v is the average (drift) velocity.

For phonons, the situation is somewhat more complicated. Phonons, which are the quanta of harmonic lattice vibrational modes, are derived under the harmonic approximation, while thermal transport effects are anharmonic in nature. Similar to how we treated electrons earlier, we can build a picture of phonon Bloch wave packets propagating through the system, redistributing the thermal energy by scattering mechanisms, in which the initial packet can annihilate (reducing the phonon population at the source) while a new phonon packet(s) is created (increasing the local phonon population at the sink). We therefore, as a first approximation, consider phonons as particles that can interact with each other, with a drift velocity v_{ph} and average scattering times τ_{ph}, which are primarily determined by interactions with other phonons, electrons, or defects. The heat capacity in this case can be obtained within the usual quasiharmonic approximation. This description yields an analogous expression for κ_p to that given in Equation 2.23 for κ_e. The relevant τ can be calculated in a similar fashion to that discussed previously for the charge carriers. Such calculations are key to determining the highly significant thermoelectric properties of materials [96].

Transport in 1D systems is different from what we know for the bulk systems as the following applies:

1. The underlying one-particle spectra, both electron and phonon, have essential 1D features from effects such as quantum confinement and symmetry breaking (see Sections 2.3 through 2.5) with a direct impact on the carrier concentrations on one hand and on the relaxation times on the other.
2. The surface-to-bulk ratio becomes significant and, for example, surface states could be either dominant scatterers or conduction channels, which would determine the values of the relaxation time but would also require consideration of the current inhomogeneity in the dimensions normal to the wire axis.
3. The current may display intrinsic quantum properties if the wire dimensions are such that the length of the wire is less than the mean free path of the charge carriers, so that scattering events become unlikely and a ballistic transport occurs.

Chapter 2

Next, we will consider a few example studies of realistic 1D systems, for which both theory and modeling have been successfully applied. All three factors peculiar to 1D may combine or remain hidden but invariably at least one would play an important role.

The electron mobility in a Si nanowire has been modeled as affected by fluctuations in the electron waves due to the surface roughness and by electron–phonon scattering via differing intra- and intervalley mechanisms by Jin et al. [97]. The electron wave functions were calculated by self-consistently solving the Schrödinger and Poisson equations, with 1D effects leading to electron energy subband formation. The surface effects introduced in the model included the Coulomb potential due to charge fluctuations and surface polarization and the resultant nonparabolicity of the energy subbands. Fermi's golden rule is used to calculate the scattering rates from the different scattering mechanisms (surface roughness, Coulomb, and phonon), and a Kubo–Greenwood formulism is used to determine the mobility. In Figure 2.14, the resulting mobility as a function of wire diameter d and effective electric field E_{eff} perpendicular to the wire axis is shown, highlighting the different contributions from the different scattering mechanisms. As expected, the surface effects dominate for smaller d.

In Reference 98, Zou and Balandin calculate the heat conduction due to phonon transport in a Si nanowire. The conductivity is assumed to be dominated by acoustic phonon transport. The effect of confinement on the phonon frequencies and scattering due to phonon–phonon interactions (including three-phonon *Umklapp* or flip-over processes and phonon mass differences), boundary (i.e., surface) effects, and electron–phonon interactions are included in the model. They found that confinement significantly softened the acoustic mode dispersion, while boundary scattering, which dominated over a wide range of phonon frequencies, led to approximately an order of magnitude reduction in the thermal conductivity, relative to bulk. Even without boundary scattering, they found significant reductions in the thermal conductivity, indicating the important role of quantum confinement in the transport properties.

When determining the scattering of electrons by phonons (or *vice versa*), the interaction Hamiltonian is characterized by parameters known as deformation potentials (relating the degree to which the electronic energies change due to the interaction), which depend on the strain introduced by the ionic motion within the phonon mode. As a result, different phonons (LA, TA, LO, TO) have different deformation potentials associated with them. Calculations of electronic structure as a function of strain introduced by phonons can be used to determine the deformation potentials, which are material specific. Murphy-Armando et al. [99] have calculated how the electron–phonon interaction in Si changes for the case of 1D nanowires, in comparison to bulk. Using first-principles methods to determine the electronic structure, they derived the deformation potentials for different wire thicknesses and orientations, considering termination in the confinement directions by H atoms and by hydroxyls. It was found that, although the surface structure and chemistry strongly modified the band structure, they had little effect on the deformation potentials. The reduced dimensionality of the wire, however, was found to alter the deformation potentials fundamentally, leading to them being strongly dependent on the strain direction. In addition, the orientation of the wire affected the deformation potentials, with [110] wires being more anisotropic than [100] and bulk, resulting in suppression of breathing mode scattering and increased electron mobility.

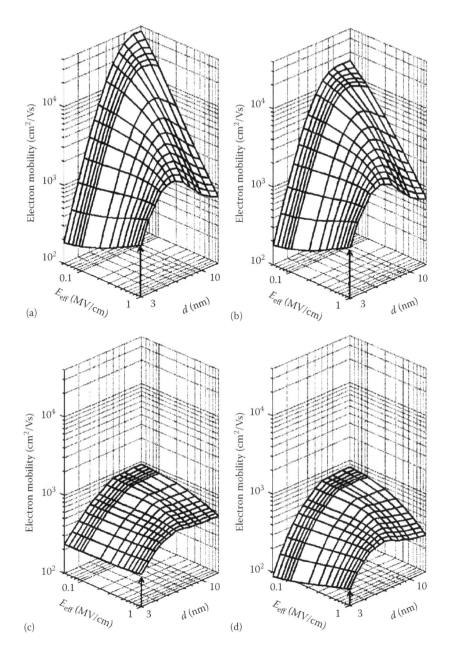

FIGURE 2.14 Calculated mobility as a function of wire diameter and effective field perpendicular to wire axis, highlighting the different contributions from surface roughness (SR), Coulomb (C), and phonon (ph) scattering. (a) μ_{SR}, (b) μ_{SR+C}, (c) μ_{ph}, and (d) μ_{tot}. (From Jin, S. et al., *J. Appl. Phys.*, 102, 083715, 2007.)

In reference to point (3) earlier, an extreme example is provided by the so-called 1D molecular wires, in which, for very short lengths (<5 nm), electron transport occurs mainly by quantum tunneling effects, while at longer lengths by a thermally induced *hopping*. An example is shown in Figure 2.15 [100]. We do not discuss in detail the transport in such systems, as it is debatable as to whether they truly are 1D

Chapter 2

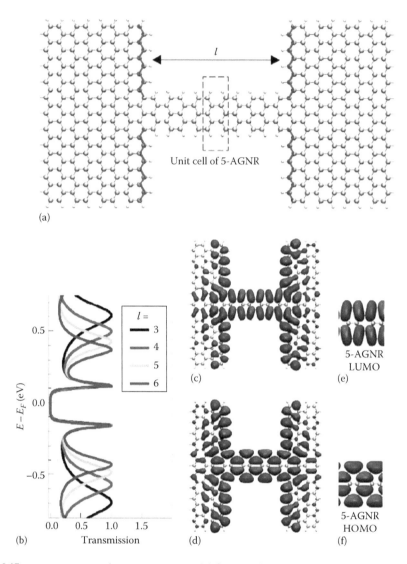

FIGURE 2.15 A computational transmission model for a molecular wire. (a) Structure, (b) transmission as a function of energy above Fermi Level, (c) and (e) lowest unoccupied molecular orbital (LUMO); (d) and (f) highest occupied molecular orbital (HOMO). (From Khoo, K.H. et al., *Phys. Chem. Chem. Phys.*, 17, 77, 2015.)

(as opposed to 0D nanostructures). For a recent review on calculations of electron transport in such systems, see Reference 100.

2.7 Defect States

Similar to extended 3D systems, the structure in 1D can order at different length scales. Irrespective of a long-range order, or *crystallinity*, any extended solid system with homogeneous distribution of atoms (characterized by constant linear density) is expected to exhibit some form of a short-range order, with atoms distributed around any given center in a pattern, which would be repeated throughout the system. Breaking such a pattern gives rise to point defects, intrinsic in the form of vacancies, interstitial atoms, and antisites (arsenic occupying, e.g., a gallium site in GaAs); extrinsic, or impurities occupying regular or interstitial sites (e.g., silicon at a gallium site in GaAs); and their complexes. Following the same entropic, or free energy arguments exploited in the theory of phase stability, finite concentrations of intrinsic defects are unavoidable in any given macroscopic sample, whereas impurities would be introduced during synthesis as the source material is practically never absolutely pure. Importantly, however, both intrinsic and extrinsic defects could be introduced into the system of interest intentionally to modify physical and chemical properties of the material, for example, its optical spectra, resistivity, or (photo-)catalytic activity. By disrupting the atomic structure, point defects introduce localized electron and vibrational states and act as scatterers for free charge carriers and phonons.

Modeling of defect states in nanowires is still in its infancy and we refer the interested reader to recent publications for pertinent examples and a more detailed review of the topic [101–104].

2.8 Conclusions

In conclusion we would like to point the interested reader to the follow-up chapters in this book, where a number of topics, which we only touched, are discussed in depth. There remain in this field many open questions and problems to attack by new researchers, of which an especially curious one is: why do 1D systems exist, can be made, and persist for macroscopic times, at all? Accurate predictive modeling of nonequilibrium processes in such complex systems is only beginning to be developed and promises novel exciting discoveries.

Acknowledgments

For many useful insights, discussions and support we are indebted to our colleagues and collaborators S.T. Bromley, C.R.A. Catlow, A. Chutia, M.R. Farrow, R. Galvelis, N. Jiang, T. Lazauskas, Z. Raza, D.O. Scanlon, S.A. Shevlin, A. Walsh, and S.M. Woodley.

References

1. Bertness, K. A., N. A. Sanford, J. M. Barker, J. B. Schlager, A. Roshko, A. V. Davydov, and I. Levin. 2006. Catalyst-free growth of GaN nanowires. *Journal of Electronic Materials* 35:576–580.
2. Kim, D. K., P. Muralidharan, H.-W. Lee, R. Ruffo, Y. Yang, C. K. Chan, H. Peng, R. A. Huggins, and Y. Cui. 2008. Spinel $LiMn_2O_4$ nanorods as lithium ion battery cathodes. *Nano Letters* 8:3948–3952.
3. Chen, X. J., B. Gayral, D. Sam-Giao, C. Bougerol, C. Durand, and J. Eymery. 2011. Catalyst-free growth of high-optical quality GaN nanowires by metal-organic vapor phase epitaxy. *Applied Physics Letters* 99:251910.
4. Prado-Gonjal, J., R. Schmidt, and E. Morán. 2015. Microwave-assisted routes for the synthesis of complex functional oxides. *Inorganics* 3:101.
5. Nakane, K. and N. Ogata. 2010. Photocatalyst nanofibers obtained by calcination of organic-inorganic hybrids. DOI: 10.5772/8155. Available from: http://www.intechopen.com/books/nanofibers/photocatalyst-nanofibers-obtained-by-calcination-of-organic-inorganic-hybrids.
6. Treutlein, P., D. Hunger, S. Camerer, T. W. Hänsch, and J. Reichel. 2007. Bose-Einstein condensate coupled to a nanomechanical resonator on an atom chip. *Physical Review Letters* 99:140403.

Chapter 2

7. Curreli, M., C. Li, Y. Sun, B. Lei, M. A. Gundersen, M. E. Thompson, and C. Zhou. 2005. Selective functionalization of In$_2$O$_3$ nanowire mat devices for biosensing applications. *Journal of the American Chemical Society* 127:6922–6923.

8. Céline, M., C. Caroline, M. Eléonore, M. Jean-François, C. Alexandre, and S. Jean-Pierre. 2013. Improvements in purification of silver nanowires by decantation and fabrication of flexible transparent electrodes. Application to capacitive touch sensors. *Nanotechnology* 24:215501.

9. Wang, N., C. Gao, F. Xue, Y. Han, T. Li, X. Cao, X. Zhang, Y. Zhang, and Z. L. Wang. 2015. Piezotronic-effect enhanced drug metabolism and sensing on a single ZnO nanowire surface with the presence of human cytochrome P450. *ACS Nano* 9:3159–3168.

10. Myung-Gyu, K., J. L. Henri, and S. Fred. 2013. Stable field emission from nanoporous silicon carbide. *Nanotechnology* 24:065201.

11. Hill, M. T., M. Marell, E. S. Leong, B. Smalbrugge, Y. Zhu, M. Sun, P. J. van Veldhoven, E. J. Geluk, F. Karouta, and Y.-S. Oei. 2009. Lasing in metal-insulator-metal sub-wavelength plasmonic waveguides. *Optics Express* 17:11107–11112.

12. Feng, Y. and X. Zheng. 2010. Plasma-enhanced catalytic CuO nanowires for CO oxidation. *Nano Letters* 10:4762–4766.

13. He, Y., Y. Zhong, F. Peng, X. Wei, Y. Su, S. Su, W. Gu, L. Liao, and S.-T. Lee. 2011. Highly luminescent water-dispersible silicon nanowires for long-term immunofluorescent cellular imaging. *Angewandte Chemie International Edition* 50:3080–3083.

14. Wang, Q., J. Sun, and H. Liu. 2011. Semiconducting oxide nanowires: Growth, doping and device applications. In *Nanowires—Implementations and Applications*, A. Hashim (ed.), InTech, Rijeka, Croatia/Shanghai, China, pp. 59–98.

15. Shalin, A. S., P. Ginzburg, P. A. Belov, Y. S. Kivshar, and A. V. Zayats. 2014. Nano-opto-mechanical effects in plasmonic waveguides. *Laser & Photonics Reviews* 8:131–136.

16. Wang, X., W. Tian, M. Liao, Y. Bando, and D. Golberg. 2014. Recent advances in solution-processed inorganic nanofilm photodetectors. *Chemical Society Reviews* 43:1400–1422.

17. Guo, C. X., Y. Dong, H. B. Yang, and C. M. Li. 2013. Graphene quantum dots as a green sensitizer to functionalize ZnO nanowire arrays on F-doped SnO$_2$ glass for enhanced photoelectrochemical water splitting. *Advanced Energy Materials* 3:997–1003.

18. Kim, S.-K., K.-D. Song, T. J. Kempa, R. W. Day, C. M. Lieber, and H.-G. Park. 2014. Design of nanowire optical cavities as efficient photon absorbers. *ACS Nano* 8:3707–3714.

19. Sun, C., J. Shi, and X. Wang. 2010. Fundamental study of mechanical energy harvesting using piezoelectric nanostructures. *Journal of Applied Physics* 108:034309.

20. Cheran, L.-E., A. Cheran, and M. Thompson. 2014. Chapter 1: Biomimicry and materials in medicine. In *Advanced Synthetic Materials in Detection Science*, S. M. Reddy (ed.), The Royal Society of Chemistry, London, pp. 1–25.

21. Pennelli, G. 2014. Review of nanostructured devices for thermoelectric applications. *Beilstein Journal of Nanotechnology* 5:1268–1284.

22. Hofmann, C. E. and H. A. Atwater. 2008. A plasmonic 'bull's-eye' nanoresonator. *In SPIE* 1–3. DOI: 10.1117/2.1200802.1088.

23. Yan, L., J. Zhang, C.-S. Lee, and X. Chen. 2014. Micro- and nanotechnologies for intracellular delivery. *Small* 10:4487–4504.

24. Feng, P., F. Shao, Y. Shi, and Q. Wan. 2014. Gas sensors based on semiconducting nanowire field-effect transistors. *Sensors* 14:17406.

25. Brian, P., O. A. Carlos, C. Chang-Hee, and A. Ritesh. 2014. Tailoring light–matter coupling in semiconductor and hybrid-plasmonic nanowires. *Reports on Progress in Physics* 77:086401.

26. Tang, J. and K. L. Wang. 2015. Electrical spin injection and transport in semiconductor nanowires: Challenges, progress and perspectives. *Nanoscale* 7:4325–4337.

27. Lim, Z. H., Z. X. Chia, M. Kevin, A. S. W. Wong, and G. W. Ho. 2010. A facile approach towards ZnO nanorods conductive textile for room temperature multifunctional sensors. *Sensors and Actuators B: Chemical* 151:121–126.

28. Zou, D., Z. Lv, X. Cai, and S. Hou. 2012. Macro/microfiber-shaped electronic devices. *Nano Energy* 1:273–281.

29. Tian, B., J. Liu, T. Dvir, L. Jin, J. H. Tsui, Q. Qing, Z. Suo, R. Langer, D. S. Kohane, and C. M. Lieber. 2012. Macroporous nanowire nanoelectronic scaffolds for synthetic tissues. *Nature Materials* 11:986–994.

30. Kheireddin, B. A., V. Narayanunni, and M. Akbulut. 2012. Influence of shearing surface topography on frictional properties of ZnS nanowire-based lubrication system across ductile surfaces. *Journal of Tribology* 134:022001.

31. Cademartiri, L. and G. A. Ozin. 2009. Ultrathin nanowires—A materials chemistry perspective. *Advanced Materials* 21:1013–1020.

32. Li, H., X. Wang, J. Xu, Q. Zhang, Y. Bando, D. Golberg, Y. Ma, and T. Zhai. 2013. One-dimensional CdS nanostructures: A promising candidate for optoelectronics. *Advanced Materials* 25:3017–3037.

33. Harris, P. J. F. 2009. *Carbon Nanotube Science.* Cambridge University Press, Cambridge, U.K.

34. Javey, A. and J. Kong (eds.) 2009. *Carbon Nanotube Electronics.* Springer, New York.

35. Tománek, D. 2014. Guide through the nanocarbon jungle. *Buckyballs, Nanotubes, Graphene and Beyond.* Morgan & Claypool Publishers, San Rafael, CA, p. 93.

36. Li, C. P., X. H. Sun, N. B. Wong, C. S. Lee, S. T. Lee, and B. K. Teo. 2002. Ultrafine and uniform silicon nanowires grown with zeolites. *Chemical Physics Letters* 365:22–26.

37. Gurwitz, R., G. Tuboul, B. Shikler, and I. Shalish. 2012. High-temperature gold metallization for ZnO nanowire device on a SiC substrate. *Journal of Applied Physics* 111:124307.

38. Perillat-Merceroz, G., R. Thierry, P. H. Jouneau, P. Ferret, and G. Feuillet. 2012. Compared growth mechanisms of Zn-polar ZnO nanowires on O-polar ZnO and on sapphire. *Nanotechnology* 23:125702.

39. Khaletskaya, K., J. Reboul, M. Meilikhov, M. Nakahama, S. Diring, M. Tsujimoto, S. Isoda et al. 2013. Integration of porous coordination polymers and gold nanorods into core–shell mesoscopic composites toward light-induced molecular release. *Journal of the American Chemical Society* 135:10998–11005.

40. Volosskiy, B., K. Niwa, Y. Chen, Z. Zhao, N. O. Weiss, X. Zhong, M. Ding, C. Lee, Y. Huang, and X. Duan. 2015. Metal-organic framework templated synthesis of ultrathin, well-aligned metallic nanowires. *ACS Nano* 9:3044–3049.

41. Ewald, P. P. 1921. The calculation of optical and electrostatic grid potential. *Annals of Physics, Berlin* 64:253–287.

42. Mertins, F. 1999. Potentials in low-dimensional and semi-infinite crystals. *Annals of Physics, Berlin* 8:261–300.

43. Minary, P., J. A. Morrone, D. A. Yarne, M. E. Tuckerman, and G. J. Martyna. 2004. Long range interactions on wires: A reciprocal space based formalism. *Journal of Chemical Physics* 121:11949–11956.

44. Saunders, V. R., C. Freyriafava, R. Dovesi, and C. Roetti. 1994. On the electrostatic potential in linear periodic polymers. *Computer Physics Communications* 84:156–172.

45. Castro, A., E. Rasanen, and C. A. Rozzi. 2009. Exact Coulomb cutoff technique for supercell calculations in two dimensions. *Physical Review B* 80:033102-1-4.

46. Abarenko, I. V. and V. Heine. 1965. Model potential for positive ions. *Philosophical Magazine* 12:529.

47. Kleinman, L. and D. M. Bylander. 1982. Efficacious form for model pseudopotentials. *Physical Review Letters* 48:1425–1428.

48. Austin, B. J., L. J. Sham, and V. Heine. 1962. General theory of pseudopotentials. *Physical Review* 127:276.

49. Hamann, D. R., M. Schluter, and C. Chiang. 1979. Norm-conserving pseudopotentials. *Physical Review Letters* 43:1494–1497.

50. Wadt, W. R. and P. J. Hay. 1985. Ab initio effective core potentials for molecular calculations. Potentials for main group elements Na to Bi. *The Journal of Chemical Physics* 82:284–298.

51. Martin, R. M. 2004. *Electronic Structure. Basic Theory and Practical Methods.* Cambridge University Press, Cambridge, U.K.

52. Zhao, Y., B. J. Lynch, and D. G. Truhlar. 2004. Doubly hybrid meta DFT: New multi-coefficient correlation and density functional methods for thermochemistry and thermochemical kinetics. *Journal of Physical Chemistry A* 108:4786–4791.

53. Kresse, G. and J. Furthmüller. 1996. Efficiency of ab-initio total energy calculations for metals and semiconductors using a plane-wave basis set. *Computational Materials Science* 6:15–50.

54. Kresse, G. and J. Furthmüller. 1996. Efficient iterative schemes for *ab initio* total-energy calculations using a plane-wave basis set. *Physical Review B* 54:11169–11186.

55. Kresse, G. and J. Hafner. 1993. *Ab initio* molecular dynamics for liquid metals. *Physical Review B* 47:558–561.

56. Kresse, G. and J. Hafner. 1994. *Ab initio* molecular-dynamics simulation of the liquid-metal–amorphous-semiconductor transition in germanium. *Physical Review B* 49:14251–14269.

Chapter 2

57. Blöchl, P. E. 1994. Projector augmented-wave method. *Physical Review B* 50:17953–17979.
58. Ernzerhof, M. and G. E. Scuseria. 1999. Assessment of the Perdew–Burke–Ernzerhof exchange-correlation functional. *The Journal of Chemical Physics* 110:5029–5036.
59. Perdew, J. P., K. Burke, and M. Ernzerhof. 1996. Generalized gradient approximation made simple. *Physical Review Letters* 77:3865.
60. Perdew, J. P., A. Ruzsinszky, G. I. Csonka, O. A. Vydrov, G. E. Scuseria, L. A. Constantin, X. Zhou, and K. Burke. 2008. Restoring the density-gradient expansion for exchange in solids and surfaces. *Physical Review Letters* 100:136406.
61. Adamo, C. and V. Barone. 1999. Toward reliable density functional methods without adjustable parameters: The PBE0 model. *The Journal of Chemical Physics* 110:6158–6170.
62. Gnani, E., S. Reggiani, A. Gnudi, P. Parruccini, R. Colle, M. Rudan, and G. Baccarani. 2007. Band-structure effects in ultrascaled silicon nanowires. *IEEE Transactions on Electron Devices* 54:2243–2254.
63. Nehari, K., N. Cavassilas, J. L. Autran, M. Bescond, D. Munteanu, and M. Lannoo. 2006. Influence of band structure on electron ballistic transport in silicon nanowire MOSFET's: An atomistic study. *Solid-State Electronics* 50:716–721.
64. Scheel, H., S. Reich, and C. Thomsen. 2005. Electronic band structure of high-index silicon nanowires. *Physica Status Solidi (B)* 242:2474–2479.
65. Nolan, M., S. O'Callaghan, G. Fagas, J. C. Greer, and T. Frauenheim. 2007. Silicon nanowire band gap modification. *Nano Letters* 7:34–38.
66. Hong, K.-H., J. Kim, S.-H. Lee, and J. K. Shin. 2008. Strain-driven electronic band structure modulation of Si nanowires. *Nano Letters* 8:1335–1340.
67. Pistol, M. E. and C. E. Pryor. 2008. Band structure of core-shell semiconductor nanowires. *Physical Review B* 78:115319.
68. Lind, E., M. P. Persson, Y.-M. Niquet, and L. E. Wernersson. 2009. Band structure effects on the scaling properties of [111] InAs Nanowire MOSFETs. *IEEE Transactions on Electron Devices* 56:201–205.
69. Ketterer, B., M. Heiss, E. Uccelli, J. Arbiol, and A. Fontcuberta i Morral. 2011. Untangling the electronic band structure of Wurtzite GaAs nanowires by resonant Raman spectroscopy. *ACS Nano* 5:7585–7592.
70. Yeom, H. W., Y. K. Kim, E. Y. Lee, K. D. Ryang, and P. G. Kang. 2005. Robust one-dimensional metallic band structure of silicide nanowires. *Physical Review Letters* 95:205504.
71. Brus, L. 1994. Luminescence of silicon materials: Chains, sheets, nanocrystals, nanowires, microcrystals, and porous silicon. *The Journal of Physical Chemistry* 98:3575–3581.
72. Perera, S., K. Pemasiri, M. A. Fickenscher, H. E. Jackson, L. M. Smith, J. Yarrison-Rice, S. Paiman, Q. Gao, H. H. Tan, and C. Jagadish. 2010. Probing valence band structure in wurtzite InP nanowires using excitation spectroscopy. *Applied Physics Letters* 97:023106.
73. Lu, W., J. Xiang, B. P. Timko, Y. Wu, and C. M. Lieber. 2005. One-dimensional hole gas in germanium/silicon nanowire heterostructures. *Proceedings of the National Academy of Sciences of the United States of America* 102:10046–10051.
74. Togo, A., F. Oba, and I. Tanaka. 2008. First-principles calculations of the ferroelastic transition between rutile-type and $CaCl_2$-type SiO_2 at high pressures. *Physical Review B* 78:134106.
75. Lee, K.-Y., J.-R. Lim, H. Rho, Y.-J. Choi, K. J. Choi, and J.-G. Park. 2007. Evolution of optical phonons in CdS nanowires, nanobelts, and nanosheets. *Applied Physics Letters* 91:201901.
76. Wang, R.-P., G.-W. Zhou, Y.-L. Liu, S.-H. Pan, H.-Z. Zhang, D.-P. Yu, and Z. Zhang. 2000. Raman spectral study of silicon nanowires: High-order scattering and phonon confinement effects. *Physical Review B* 61:16827–16832.
77. Adu, K. W., H. R. Gutiérrez, U. J. Kim, G. U. Sumanasekera, and P. C. Eklund. 2005. Confined phonons in Si nanowires. *Nano Letters* 5:409–414.
78. Adu, K. W., Q. Xiong, H. R. Gutierrez, G. Chen, and P. C. Eklund. 2006. Raman scattering as a probe of phonon confinement and surface optical modes in semiconducting nanowires. *Applied Physics A* 85:287–297.
79. Shi, W. S., Y. F. Zheng, N. Wang, C. S. Lee, and S. T. Lee. 2001. Oxide-assisted growth and optical characterization of gallium-arsenide nanowires. *Applied Physics Letters* 78:3304–3306.
80. Spirkoska, D., G. Abstreiter, and A. F. I. Morral. 2008. Size and environment dependence of surface phonon modes of gallium arsenide nanowires as measured by Raman spectroscopy. *Nanotechnology* 19:435704.

81. Shi, W., Y. Zheng, H. Peng, N. Wang, C. S. Lee, and S.-T. Lee. 2000. Laser ablation synthesis and optical characterization of silicon carbide nanowires. *Journal of the American Ceramic Society* 83:3228–3230.

82. Chen, C.-C., C.-C. Yeh, C.-H. Chen, M.-Y. Yu, H.-L. Liu, J.-J. Wu, K.-H. Chen, L.-C. Chen, J.-Y. Peng, and Y.-F. Chen. 2001. Catalytic growth and characterization of gallium nitride nanowires. *Journal of the American Chemical Society* 123:2791–2798.

83. Lo, S. S., T. A. Major, N. Petchsang, L. Huang, M. K. Kuno, and G. V. Hartland. 2012. Charge carrier trapping and acoustic phonon modes in single CdTe nanowires. *ACS Nano* 6:5274–5282.

84. Weber, C., A. Fuhrer, C. Fasth, G. Lindwall, L. Samuelson, and A. Wacker. 2010. Probing confined phonon modes by transport through a nanowire double quantum dot. *Physical Review Letters* 104:036801.

85. Boukai, A. I., Y. Bunimovich, J. Tahir-Kheli, J.-K. Yu, W. A. Goddard III, and J. R. Heath. 2008. Silicon nanowires as efficient thermoelectric materials. *Nature* 451:168–171.

86. Chen, R., A. I. Hochbaum, P. Murphy, J. Moore, P. Yang, and A. Majumdar. 2008. Thermal conductance of thin silicon nanowires. *Physical Review Letters* 101:105501.

87. Richter, H., Z. P. Wang, and L. Ley. 1981. The one phonon Raman spectrum in microcrystalline silicon. *Solid State Communications* 39:625–629.

88. Thonhauser, T. and G. D. Mahan. 2004. Phonon modes in Si [111] nanowires. *Physical Review B* 69:075213.

89. Zhang, W., C. Delerue, Y.-M. Niquet, G. Allan, and E. Wang. 2010. Atomistic modeling of electron-phonon coupling and transport properties in n-type [110] silicon nanowires. *Physical Review B* 82:115319.

90. Luisier, M. and G. Klimeck. 2009. Atomistic full-band simulations of silicon nanowire transistors: Effects of electron-phonon scattering. *Physical Review B* 80:155430.

91. Donadio, D. and G. Galli. 2009. Atomistic simulations of heat transport in silicon nanowires. *Physical Review Letters* 102:195901.

92. Seiji, M. and N. Norihiko. 2009. Acoustic phonon modes and dispersion relations of nanowire superlattices. *Journal of Physics: Condensed Matter* 21:195303.

93. Li, H. and F. Sun. 2012. Recent advances in mechanical properties of nanowires. In *Nanowires—Recent Advances*, X. Peng (ed.), InTech, Rijeka, Croatia, pp. 371–394.

94. Ibach, H. and H. Lüth. 2009. Dynamics of atoms in crystals. In *Solid-State Physics*. Springer, Berlin, Heidelberg, pp. 83–112.

95. Yu, P. Y. and M. Cardona. 2005. *Fundamentals of Semiconductors*. Springer, Berlin, Germany.

96. Skelton, J. M., S. C. Parker, A. Togo, I. Tanaka, and A. Walsh. 2014. Thermal physics of the lead chalcogenides PbS, PbSe, and PbTe from first principles. *Physical Review B* 89:205203-1-10.

97. Jin, S., M. V. Fischetti, and T.-W. Tang. 2007. Modeling of electron mobility in gated silicon nanowires at room temperature: Surface roughness scattering, dielectric screening, and band nonparabolicity. *Journal of Applied Physics* 102:083715.

98. Zou, J. and A. Balandin. 2001. Phonon heat conduction in a semiconductor nanowire. *Journal of Applied Physics* 89:2932–2938.

99. Murphy-Armando, F., G. Fagas, and J. C. Greer. 2010. Deformation potentials and electron–phonon coupling in silicon nanowires. *Nano Letters* 10:869–873.

100. Khoo, K. H., Y. Chen, S. Li, and S. Y. Quek. 2015. Length dependence of electron transport through molecular wires—A first principles perspective. *Physical Chemistry Chemical Physics* 17:77–96.

101. Rurali, R. and X. Cartoixà. 2009. Theory of defects in one-dimensional systems: Application to al-catalyzed Si nanowires. *Nano Letters* 9:975–979.

102. Buckeridge, J., S. T. Bromley, A. Walsh, S. M. Woodley, C. R. A. Catlow, and A. A. Sokol. 2013. One-dimensional embedded cluster approach to modeling CdS nanowires. *Journal of Chemical Physics* 139:124101-1-11.

103. Wang, Z., J. Li, F. Gao, and W. J. Weber. 2010. Defects in gallium nitride nanowires: First principles calculations. *Journal of Applied Physics* 108:044305.

104. Wrasse, E. O., P. Venezuela, and R. J. Baierle. 2014. Ab initio study of point defects in PbSe and PbTe: Bulk and nanowire. *Journal of Applied Physics* 116:183703.

Chapter 2

3. Two-Dimensional Nanosystems

Benjamin J. Morgan

3.1   Introduction 83

3.2   Two-Dimensional Nanosystems 85
    3.2.1   Free-Standing Thin Films 85
        3.2.1.1   Top-Down Model of Free-Standing Thin Films 87
        3.2.1.2   Bottom-Up Construction of Free-Standing Thin Films 88
    3.2.2   Supported Thin Films 90
    3.2.3   Heterostructures 90

3.3   Computational Modeling of 2D Nanostructures 91
    3.3.1   Coulombic Interactions in 3D and 2D Systems 93
    3.3.2   Free-Standing Thin Films 96
        3.3.2.1   Choice of Bulk Phase 96
        3.3.2.2   Choice of Surface Orientation and Termination: Tasker
                Classification of Surfaces 97
        3.3.2.3   Nonpolar Thin Films: Parity and In-Plane Symmetry 98
        3.3.2.4   Polar Thin Films 99
        3.3.2.5   Choice of Surface Reconstruction 101
    3.3.3   Supported Thin Films 103
        3.3.3.1   Substrate Polarization and Charge Transfer 103
        3.3.3.2   Epitaxial Relationships 103
    3.3.4   Modeling Heterostructures 108

3.4   Assessment of Geometric Stability 108

References 109

3.1   Introduction

Bulk material properties, such as melting temperature, electrical conductivity, and molar free energy, are *intensive*: they do not depend on the extent of material considered and are fixed by relevant thermodynamic variables. In macroscopic bulk systems, the proportion of atoms close to surfaces is vanishingly small, and the effect of surfaces or interfaces on bulk material properties can generally be neglected. In contrast, nanoscale systems such

Chapter 3

as nanocrystals, thin films, and heterostructures are defined, in part, by small distances between surfaces or interfaces, and a significant proportion of their constituent atoms reside in environments that are not bulk-like. The contribution from these regions to material properties is no longer negligible, and nanosystem behaviors can consequently differ significantly from those of corresponding bulk phases.

In many cases, nanosystem material properties diverge from bulk values as characteristic length scales are reduced. One approach to understanding this phenomenon is to consider the contributions made by ideal *bulk* and *surface* properties and then combine these in proportions that depend on the system dimensions. It is the presence of surfaces and interfaces that distinguish nanosystems from their bulk counterparts, and a number of nanoscale phenomena can be considered as a consequence of size-dependent surface contributions to material properties. Examples include changes in the relative stabilities of competing solid phases upon nanosizing and increases in melting temperatures in nanosystems relative to the bulk materials, which have each been explained by considering surface contributions to free energies as a function of size [1,2].

Some nanoscale behaviors, however, cannot be explained as direct consequences of increasing proportions of surface regions; instead, a more detailed description of nanostructuring is necessary. At small length scales a formal identification of surface and bulk regions may break down, along with models that rely on this distinction. Alternatively, the presence of surfaces may remove an underlying bulk crystal symmetry, allowing nanosystems to undergo electronic or geometric symmetry breaking. This can produce quantum confinement effects, which modify semiconductor band gaps and lead to size-dependent optical and electronic properties [3], phenomena such as quantum wells or 2D electron and hole gases [4–7], or spontaneous phase transitions in thin films [8]. Whether the properties of a specific nanosystem arises as a continuous deviation from bulk behavior due to a simple increased surface contribution or due to the entire system being affected in some nontrivial manner, understanding nanomaterial structures is key to explaining, and perhaps controlling, this behavior.

This chapter discusses the use of computational modeling to study structures of inorganic 2D nanosystems.* Three-dimensional bulk phases, when modeled, are usually considered to be infinite in all three dimensions. Two-dimensional nanosystems analogously can be considered to be infinite in only two dimensions, since surfaces or interfaces restrict the system length in the perpendicular third dimension.

A model 2D nanosystem may be conceived by cleaving a bulk crystal to form a planar free-standing thin film delineated by a pair of surfaces. If the distance between surfaces, or film *thickness*, is small, then the material properties can differ from those of the parent bulk crystal, and the system can be considered *nanoscopic*. Let us imagine this film displays some behavior that distinguishes it from the analogous bulk material and qualifies it for further study. Perhaps we wish to optimize some particular material property and are interested as to whether changing the structure—for example, varying the film thickness—could help us understand the physical mechanisms behind the size-dependent behavior, or even suggest a process of synthetic control. One experimental strategy is to synthesize and characterize films across a range of thicknesses, to empirically assess

---

* Because we are considering *inorganic* nanosystems, the most widely studied 2D nanomaterial, graphene, will not be discussed here. A number of reviews are available for the interested reader, including References 9 and 10.

the relationship between the property of interest and the chosen structural parameter. This approach is direct, but not without potential shortcomings. It might not be possible to produce monodisperse samples for characterization, or samples might form with a variety of surface terminations—this introduces a second structural parameter that now must be considered in our analysis—or the structural features of these films might change discontinuously with thickness. Unless we understand of the structural variation in our samples, a direct link between the nanoscale phenomena of interest and a particular structural parameter—the film thickness—can remain elusive.

A system modeled computationally is structurally well defined, and the consequences of any changes in structure can, in theory, be calculated to arbitrary precision.* The structures of bulk crystals are fully defined by their unit cell geometries and the internal coordinates of their atomic basis. Two-dimensional nanomaterials have surfaces or interfaces that break translational symmetry in one dimension. To completely define a structure now requires not only the geometric arrangement of atoms and their periodic arrangement in two dimensions but also parameters such as the thickness in the third dimension, the orientation of the crystal lattice with respect to the surfaces or interfaces, and the details of surface termination, such as those arising from surface reconstruction or at an interface, which is the epitaxial relationship with a secondary phase.

A secondary benefit of computational modeling is the ability to consider structures that are experimentally unknown, unstable, or otherwise hard to isolate. Studies performed *in silico* afford precise control over structural variation; for example, structural parameters that are strongly correlated in experimental samples may be varied independently in theoretical models. The power of computational studies of *hypothetical* structures has been demonstrated by cases where previously unknown phases of materials were predicted by simulation and only subsequently observed in experiment.[†]

3.2   Two-Dimensional Nanosystems

A 2D nanosystem has parallel planar surfaces or interfaces with a small *nanoscale* separation. This spatial restriction in one dimension differentiates 2D nanosystems from 3D bulk phases and is the reason for their interesting behavior. Depending on whether these planar interfaces are vacuum-exposed surfaces or solid–solid interfaces, a 2D nanosystem can be placed into one of three classes: a free-standing thin film, a supported thin film, or a heterostructure (see Figure 3.1). Each of these classes will be described in turn, with a focus on introducing relevant structural features and concepts. Further details on modeling each class of 2D nanosystem will follow in Section 3.3.

3.2.1   Free-Standing Thin Films

Free-standing thin films are 2D sheets with a pair of vacuum-exposed parallel surfaces and provide idealized models of thin film structures. Experimental thin films are usually suspended in liquid solvent or grown on a solid support, and simulations of free-standing thin films typically neglect interactions such as solvent adsorption, epitaxial

---

* The accuracy of calculated properties will, however, depend on the chosen theoretical approach.
† For example, Čančarević et al. predicted the existence of a metastable wurtzite phase of LiCl (Reference 11), which prompted the synthesis of this, then unknown, phase as an epitaxially strained thin film (Reference 12).

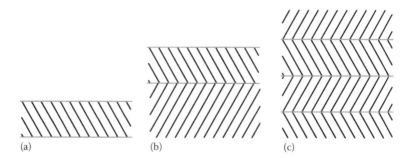

FIGURE 3.1 Two-dimensional nanosystems have parallel interfaces that restrict the characteristic length in one dimension. Classification into three classes is possible according to whether none, one, or both of these interfaces are free (vacuum-exposed) surfaces or solid–solid interfaces: (a) Free-standing thin films, (b) supported thin films, and (c) heterostructures.

strain, or substrate polarization, present in real systems. Where these interactions are significant models of free-standing thin films may not give good descriptions of experimental behavior in systems, in particular where interaction strengths change *qualitatively* with nanoscale morphology. For experimental systems with weak interfacial interactions, however, neglecting the presence of a second phase can make free-standing thin films appealing models for studying relationships between material properties and specific structural features, such as film thickness or surface termination.

Although the structure of a 2D nanosystem is completely represented by a full set of atomic coordinates, a schematic description of nanoscale structure is often conceptually useful: this might focus on film thickness, surface termination or reconstruction, or the arrangement of component layers within a film. For free-standing thin films, different structural schemes can be developed by considering a generic 2D nanofilm constructed from simpler reference systems (Figure 3.2). One reference is a bulk crystal. Cleaving this crystal to form a parallel pair of free surfaces generates a 2D thin film in a

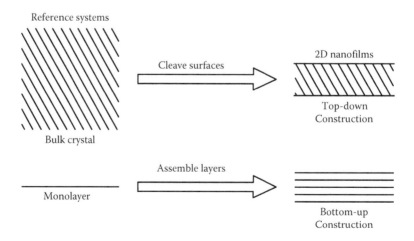

FIGURE 3.2 Two conceptual approaches to generating free-standing thin films from reference systems: by cleaving a bulk crystal (a top-down approach) or by stacking atomically thin monolayers (a bottom-up approach).

top-down process. An alternate bottom-up construction starts from a reference system of atomically thin lamellar sheets, which are stacked to form a thin film. When considering specific free-standing thin films, either one of these two models can be useful, and each emphasizes a different set of metrics that schematically describe structural differences between individual nanostructures.

3.2.1.1 Top-Down Model of Free-Standing Thin Films

A top-down construction of a free-standing thin film uses a bulk crystal as the starting reference. Cleaving this crystal to form a pair of parallel surfaces produces a 2D film. If the surfaces are well separated, then the center of the slab is expected to remain bulk-like, having a local structure highly similar to the corresponding parent bulk crystal. The atomic structure close to the surfaces, however, may differ significantly because of atoms moving away from their bulk positions or a spontaneous reconstruction of the surface.

For relatively thick films, the local geometry at the center of the film can be expected to be converged with respect to increases in film thickness. In this regime, material properties will depend only on the relative proportion of surface and bulk regions. The distinction between bulk and surface regions suggests the complete thin film structure can be defined by describing the structures of each region and the relative thickness of the bulk and surface regions (Figure 3.3a). If the film thickness is decreased, however, and the separation between the two surfaces falls below some threshold distance, then the surface regions will overlap: the distortions from bulk geometry that characterize the surface regions now extend through the entire film—Figure 3.3a through c—and the distinction between surface and bulk regions breaks down. If the deviations from bulk geometry at the center of the film are not severe, it can still be useful to describe the structure of the nanofilm in terms of local deviations from bulk structural parameters: for example, bulk interatomic spacing and lattice parameters. With further decreases of film thickness, however, structural perturbations at the center of the slab will become more significant, and rather than continuing to describe the structure by reference to the corresponding bulk phase, it may be simpler to consider the film as a distinct structural phase.

When the top-down model is appropriate, the distinction between bulk and surface regions suggests a simple schematic description of the structure with independent bulk and surface components. First, we consider the bulk region at the interior of the film. Different thin film structures can be constructed by considering different reference

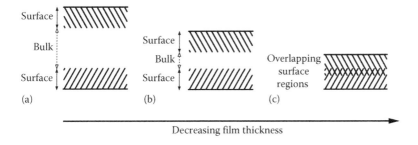

(a)   (b)   (c)

Decreasing film thickness

FIGURE 3.3 (a) For relatively thick films, it can be useful to consider separately the structures of surface and bulk regions. (b) Providing the film is thick enough that the center can still be considered bulk-like, then the system is described by combining the surface and bulk regions in proportions that depend on the film thickness. (c) For thinner films the surface regions overlap, and this conceptual model breaks down.

Chapter 3

bulk phases. The free energy of a thin film depends on both bulk and surface contributions, and structures that are metastable or even unstable in bulk crystals may be stable within a nanofilm if their surface energies are favorable. The range of potential thin film structures may therefore show much more variety than in the analogous bulk material. Second, we consider the surface structure. For a given bulk phase and thickness of film, significant structural modification of the surface regions is possible. Cleaving the parent bulk crystal along different crystallographic planes generates a range of surfaces and also changes the relative orientation of the bulk region with respect to the surface plane. Surface-generating slices can intersect the unit cell at different positions, and depending on the complexity of the bulk unit cell a number of differently terminated surfaces might be generated with the same surface orientation [13]. Finally, if the surface reconstructs, then a number of alternative termination structures might need to be considered.

3.2.1.2 Bottom-Up Construction of Free-Standing Thin Films

The top-down construction of 2D thin films represents a reduction of length scales in one dimension from macro- to nanoscopic. This conceptual model implies layers of the film are distinguishable as bulk and surface regions; this is a reasonable assumption for relatively thick films, where the center of the slab is bulk-like. For very thin films, however, we can expect the two surface regions to interact, and it may no longer be appropriate to consider distinct surface and bulk regions (Figure 3.3).

When films are very thin—perhaps only a few interatomic distances across—a more natural reference system is monolayer 2D sheets. These sheets are infinite in two dimensions but only one atom thick in the third, exemplified by the graphene allotrope of carbon, which has the familiar hexagonal lattice motif of graphite while being only a single atom thick.

Inorganic monolayer films have been known since 1966, when single layers of MoS_2 were isolated by Frindt [14,15]. Bulk MoS_2 has a lamellar structure, consisting of 2D layers held together by van der Waals forces, and these can be easily separated by chemical exfoliation. Interest in atomically thin films has risen following the discovery of graphene and the explosion of associated research [16], and a number of other inorganic lamellar materials, including WS_2, $NbSe_2$, and hexagonal BN, have since been used as sources for monolayer nanofilms.

Stacking component monolayers allows the construction of multilayer films. Instead of describing the general structure by referring to *bulk* and *surface* regions—distinctions that have no meaning in an atomically thin film and that are of limited value in films only a few layers thick—the bottom-up model suggests we focus on a different set of parameters for characterizing film structures. The fundamental structural component is now the 2D structure of the component monolayer films. For example, monolayer hexagonal BN is an inorganic structural analog of graphene, and h-BN thin films can be considered as layers of these hexagonal sheets. Monolayers of MoS_2, WS_2, $NbSe_2$, and related materials share this hexagonal net structure, but with anion dimers oriented perpendicular to the plane of the sheet, and thin films of these materials can also be understood using a motif of stacked *graphitic* layers. A number of monolayer nanosheets with more complex structures have been identified by chemical exfoliation of different inorganic solids [17].

By changing the relative stacking arrangement of stacked monolayers, a single mono-layer geometry can be used as the reference for a variety of structurally distinct multi-layer thin films. An example is given by comparing the bulk wurtzite and zinc blende structures found in materials such as ZnO, CdS, and SiC. Wurtzite and zinc blende can both be considered to be constructed from close-packed planes of cations and anions. These planes consist of hexagonal sheets of atoms that are structurally related to the planar layers in h-BN, but with each atomic species moved out of the shared plane: one atomic species is displaced along +z and the other along −z, which causes the sheets to pucker (Figure 3.4a). This gives a structure where every six-membered hexagon adopts a geometry similar to the *chair* conformation of cyclohexane [18,19].

By stacking these puckered hexagonal sheets, both the wurtzite and zinc blende structures can be formed. The relative orientation of each sheet with respect to its neighbors determines whether a particular stacking sequence produces wurtzite or zinc blende. In zinc blende all the sheets have the same orientation, whereas in wurtzite alternating sheets are reflected in a plane perpendicular to the close-packed layers, and these different relative orientations of adjacent hexagonal sheets produce the different characteristic stacking sequences of each phase: zinc blende (ABC) and wurtzite (AB) (Figure 3.4b).

In both wurtzite and zinc blende, the *relative* orientation of any two neighboring layers is the same throughout each crystal: neighboring sheets always have a symmetric orientation in zinc blende and always have an antisymmetric orientation in wurtzite. Combinations of these two patterns generate a spectrum of alternate polytypes with longer repeat units along the close-packed axis. Bulk SiC [20] and nanostructured AgI [21,22] both provide examples of this *stacking fault* polytypism.

These examples illustrate how different structures can be constructed by repeating a single monolayer structure and only varying the relative orientation of adjacent sheets.

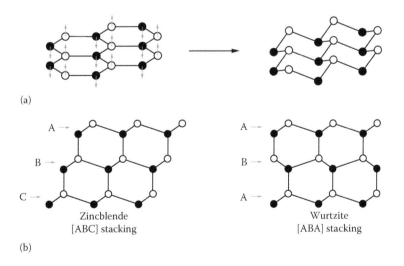

(a)

(b)

Zincblende
[ABC] stacking

Wurtzite
[ABA] stacking

FIGURE 3.4 (a) Displacing alternate atoms of a planar *graphitic* hexagonal-net monolayer along ±z produces a puckered sheet. (b) Stacking these puckered monolayers can produce the zinc blende or wurtzite structures, depending on the relative orientation of adjacent monolayers. The B layer in the wurtzite structure is a reflected copy of the B layer in the zinc blende structure.

Chapter 3

Even greater structural variety can be generated by simply changing the structure of each component layer or by combining layers with two or more different structures.

3.2.2 Supported Thin Films

Free-standing thin films are often used as model systems for computational modeling of 2D nanostructures and may provide good approximations of experimental systems where only weak interactions with a second phase are expected.* A second class of 2D nanosystem is *supported* thin films. These consist of a thin film positioned on top of a second solid phase, forming a solid–solid interface that is now considered explicitly (Figure 3.1b). The presence of a second solid phase means interfacial energies, and their dependence on parameters such as film orientation and surface termination may differ significantly from vacuum surface energies, and supported thin films can behave very differently to corresponding unsupported free-standing thin films. In some cases, differences in surface energies between supported and unsupported thin films can lead to different structures being preferentially adopted in each case [23,24].

The presence of a supporting substrate can affect the surface or interface energy of a thin film through a number of mechanisms. Even if the substrate is considered as a structureless dielectric continuum, atoms close to the thin film–substrate interface will experience a different electrostatic (Madelung) potential from those at a vacuum surface, leading to different structural relaxations. The substrate may also polarize in response to the presence of the supported thin film, or electrons may be transferred across the interface [25]: electronic dipoles are formed that can stabilize dipolar thin film structures that would be unstable in an analogous free-standing film. The atomic structure at the interface can be important, because the relative alignment of the substrate and thin film crystal lattices can have a significant effect on the interfacial energy: often preferred *epitaxial* relationships exist where the two crystalline lattices are aligned. The orientation and surface termination of the supporting phase can therefore strongly influence the stability of differently oriented thin film lattices or even favor particular structures.[†] Exact epitaxial coincidence between the substrate and supported thin film lattices is usually not possible, and a coincident lattice relationship is achieved by straining the supported thin film. While small epitaxial strains may be accommodated without a significant distortion to the thin film structure, a large epitaxial mismatch can lead to the formation of dislocations or grain boundaries [26,27], or even drive a transition to a different crystalline phase in the supported thin film [28,29].

3.2.3 Heterostructures

The third class of 2D nanomaterials considered here are heterostructures (Figure 3.1c). These are, perhaps incongruously, *bulk* materials consisting of two or more solid phases arranged in layers. Each component layer is bulk-like (considered infinite) in two dimensions but is bounded in the third by a pair of solid–solid interfaces. From the perspective

---

* Many experimentally known unsupported thin films are formed by chemical exfoliation of lamellar bulk phases, and strong surface–solvent interactions play a critical role in this process.
† For example, thin films of ZnO of specific thickness are predicted to adopt different structures depending on whether they are free standing or epitaxially supported and strained: cf. References 8 and 24.

of individual layers, these interfaces break translational symmetry, and heterostructures with sufficiently small interfacial separations can show nanoscale behavior, mirroring the transition in thin films from bulk to nanoscale behavior with decreasing thickness. Because the phase boundaries are solid–solid interfaces, heterostructures are closely related to supported thin films and may be considered as arrays of epitaxially layered thin films, with the added complication of epitaxial relationships at *both* interfaces. Heterostructures often cannot be described simply as a composite of their component bulk materials, because the presence of internal interfaces can cause behaviors not observed in the parent bulk phases. This can be a pure interface effect, where the deviations from bulk behavior are due to the nonnegligible contribution of near-interface regions or new behavior due to the broken translational symmetry introduced by the heterointerfaces [4,30–32].

Some bulk materials can be considered to be constructed from 2D layers that are chemically identical but structurally different. For example, SiC and AgI both can adopt zinc blende and wurtzite structures, as well as a number of polytypes constructed by alternating between these phases along the $[0001]_{B4}/[111]_{B3}$ direction of stacked close-packed layers [21,22,31–33]. These polytypes can be considered a special class of heterostructures, with stacking faults that are equivalent to internal heterointerfaces, and may exhibit behaviors more commonly associated with conventional heterostructures due to symmetry breaking [32,34].

3.3   Computational Modeling of 2D Nanostructures

Structures of 2D nanosystems can be discussed in general or specific terms: general characteristics include parameters such as domain thickness, lattice orientation, and surface terminations and broadly can be categorized as defining *morphology*. Specific details depend on the precise geometric configurations of atoms, for example, the structures of preferred surface reconstructions or epitaxial relationships across solid–solid interfaces. General morphological descriptions are useful for relating qualitative variations in behavior to structural trends, such as how film thickness affects a preference for particular surfaces being exposed. Specific geometries allow a quantitative understanding of behavior and often are necessary for further modeling of specific material properties.

Modeling structures is, in general, a problem of predicting stable atomic geometries. Distinct structures correspond to different local energy minima. Fully first-principles structure prediction seeks to identify globally stable, or metastable, phases knowing only the system stoichiometry and relevant thermodynamic conditions. This is a difficult problem that remains unsolved in general terms and forms a field of research in its own right [35].

Modeling the structures of 2D nanomaterials may appear to be even more challenging than modeling structures of bulk crystals. Surfaces and interfaces break translational symmetry and can have complex local structures, and calculations often must consider large unit cells. Two-dimensional nanostructures can have a large number of competing metastable phases that may need to be considered for a global description of structural variation. However, because reorganizing surfaces or interfaces is often associated with substantial kinetic barriers [36], global searches over configurational space can be computationally inefficient.

Chapter 3

Fortunately, instead of considering all potential structures for any particular system, we are often interested in addressing a specific question about the structure: what is the preferred phase for a free-standing film with a particular thickness? Is this preference affected by possible surface reconstructions? What happens for an analogous supported film, where epitaxial relationships with the substrate might be significant? Each of these questions can be explored by modeling a restricted set of structures, with the scope determined by the details of the question under consideration. Following the example of studying the relationship between film thickness and phase stability, one strategy would be to model idealized free-standing thin films with a range of bulk phase structures and thicknesses, so that their stabilities could be directly compared. Similar calculations allowing for additional structural variation—epitaxial interactions and surface reconstructions— might then be considered as the scope of the enquiry increases.

Modeling structures of 2D nanomaterials often consists of two stages. First, one or more candidate structures must be proposed, taking into account general structural factors under consideration, such as nanostructure thickness, crystal lattice orientation, surface terminations, and epitaxial relationships. Second, these structures are optimized to locate local energy minima corresponding to stable structures. If the structure can be simply related to a well-defined bulk phase or good experimental structural data are available, then proposing an approximate starting structure for subsequent optimization can be straightforward. More complex structures, however, might not be readily generated heuristically and may not be well characterized from experimental data. In such cases, techniques that can explore the global energy landscape may be useful, such as simulated annealing or applying genetic algorithms.

Formally, the stabilities of experimental systems depend on their relative *free energies*. Calculating free energies, however, is often computationally expensive and technically involved. A common practice is to approximate free energies with *potential energies* for structures at zero temperature. This neglects PV enthalpic terms as well as entropic contributions from lattice vibrations and any configurational disorder, although the configurational entropy, if known, may be included as a correction to the calculated energies. Often we are interested in *relative* energies, and differences in vibrational entropies and PV terms between candidate solid structures can typically be assumed to be negligible.* Considering zero-temperature potential energies instead of free energies allows stable structures to be associated with potential energy minima as a function of atomic coordinates. These minima can be found by optimizing an approximate starting structure through a process of energy minimization. Within this scheme the system energy is (locally) minimized iteratively by adjusting ion positions until the forces on all ions are zero.†

Performing a geometry optimization of some starting structure therefore depends on the ability to calculate the potential energy as a function of atomic coordinates. A number of different methods can be used, from simple atomistic descriptions that

---

* It may be necessary to account for vibrational motion in some cases to obtain good agreement between experimental and computational structural data. For example, Harrison et al. have discussed how energy-minimized zero-temperature structures for the TiO_2 (110) surface can agree poorly with experimental data because the effect of soft anharmonic vibrational modes, which shift average surface ion positions, is neglected: see Reference 37.

† Within the numerical accuracy of the calculation, configurations with atomic forces all equal to zero may be points of inflection rather than true minima. Testing for this possibility is discussed in Section 3.4.

describe the potential energy of the atoms according to a classical interatomic force field to electronic structure methods that approximately solve the Schrödinger equation. The appropriate choice for a given simulation study will depend on the desired accuracy of the result—what is the purpose of the calculation—and the computational effort: itself dependent on additional factors such as the size of each system and the cost of optimizing a single structure versus sampling a large number of competing structures. General strategies for structural optimization are discussed at length in Reference 35. The focus here is on modeling the structures of 2D nanostructures, and this discussion will therefore emphasize cases where the structures of specific 2D nanosystems might introduce complications that we would prefer to be aware of.

A direct minimization of the potential energy will usually only find the local energy minimum. In general, we have no way of knowing whether this is also the *global* energy minimum and should therefore be careful about interpreting a structure obtained from a single energy minimization as being a representative preferred structure. Additional geometry minimizations from different starting configurations will provide more information: these will either converge on the same minimum, in which case we can increase our confidence that we have found the minimum energy geometry for a reasonable volume of configurational phase space, or will identify alternate competing minima, which can then be directly compared with the previous results. Alternatively, methods that search configuration space more efficiently and also reduce the bias from the choice of starting configurations, such as simulated annealing or genetic algorithms, might be more effective.

3.3.1 Coulombic Interactions in 3D and 2D Systems

In computer simulations of 3D bulk materials, we face the problem of how best to represent a macroscopic system using a model that contains tens to (perhaps) millions of atoms. To avoid surface artifacts from the relatively small simulation size, it is common to employ *periodic boundary conditions*: the simulation cell is considered to be periodically repeated in all directions to form an infinite lattice of *image* cells, and our macroscopic bulk material is represented as a (periodic) infinite system (Figure 3.5a). Providing the simulation cell is moderately sized, short-ranged interactions between atoms decay on a length scale smaller than the dimensions of the cell, and can be calculated as a real-space direct sum that considers only the neighboring periodic images.* Electrostatic interactions are long ranged, however, and cannot be simply truncated [39]. For an array of point charges, if the Coulomb energy is calculated as a direct sum over periodic image cells, then the result is conditionally convergent and depends on the order of the sum over image cells. Even for schemes that give the correct bulk limit, this summation converges very slowly. An efficient solution is provided by the Ewald method [40]: the slow-to-converge direct sum of Coulomb terms over periodic cells is replaced by two equivalent sums that each converge quickly, that is, one in real space and one in reciprocal space.†

---

* Atomistic models that use an explicit interatomic potential will usually apply the minimum-image convention: short-ranged interactions are truncated at a distance where each atom only interacts with the closest periodic image of another atom. See, for example, Reference 38.

† Implementation details can be found in standard textbooks on molecular simulation; for example, References 38 and 41.

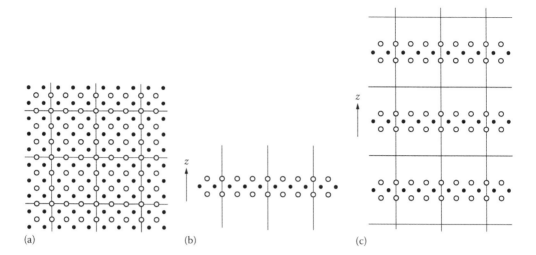

FIGURE 3.5 (a) Three-dimensional bulk calculations commonly use periodic boundary conditions, where the simulation cell is replicated infinitely in three dimensions. (b) When modeling a 2D system, we can choose to replicate the simulation cell in only two dimensions to give a true 2D system. (c) Alternatively, we can construct a pseudo-2D calculation within 3D boundary conditions, where a vacuum gap between periodic images along z negates their interactions.

Two-dimensional nanosystems are usually considered bulk-like (infinite) in two dimensions but finite in the third. By analogy to the 3D periodic boundary conditions used when modeling bulk materials, a simple approach to modeling a 2D system uses 2D periodic boundary conditions. The simulation cell is considered to be replicated in only two directions, x and y, and the long-ranged Coulomb energy is calculated from a 2D Ewald sum (Figure 3.5b) [42,43]. This approach gives an exact description of a true 2D system, such as a free-standing thin film. However, the 2D Ewald sum is much more computationally demanding than the 3D counterpart and may be prohibitively expensive.*

An alternative approach that allows the more-efficient 3D Ewald sum to be used is to construct a pseudo-2D system within 3D periodic boundary conditions, by including a vacuum layer that separates periodic images along the z direction to give a slab geometry (Figure 3.5c). For a pseudo-2D calculation such as this, we must bear in mind that rather than our intended model of an isolated slab, we are now modeling an infinite array of slabs, and spurious slab–slab interactions between periodic images along the z direction have the potential to introduce calculation artifacts. As the vacuum spacing between periodic images increases in proportion to the thickness of the slab, the results will converge to those of an isolated 2D system. The question then is how large a vacuum separation should we expect to need for reasonable convergence of the Coulomb energy with respect to the vacuum region width?

Unless the modeled system can be arranged into charge neutral layers (cf. Section 3.3.2.2), the slab will possess a nonzero dipole normal to the surface. Periodic images of

---

* Employing a 2D Ewald sum may not simply be a question of computational expense, because a particular modeling study might require the use of a simulation code for which the two-dimensional Ewald sum is not implemented.

the slab will behave as a stack of parallel plate capacitors, with a dipole–dipole interaction between periodic images that decays slowly with the separation between slabs. For dipolar slabs, therefore, even relatively large slab–slab separations may not give the correct limiting behavior.

Although periodic boundary conditions allow us to describe an infinite array of image cells, there is still, perhaps counterintuitively, the question of what happens at the *surface* of any set of cells. For a calculation cell with a polar charge distribution, the outer boundary of any contiguous set of image cells will be charged, and this is associated with a depolarizing field that opposes the formation of polar charge distributions. For an infinite 3D periodic system, this depolarizing field is an artifact of our model. This can be seen by considering transposing the atomic coordinates with respect to the simulation cell. The total system energy should be invariant with respect to a shift of the coordinate system. Such a transposition, however, can produce configurations with different *cell* dipoles, giving different formal surface charges and different associated polarization energies. Under 3D boundary conditions, used for normal bulk calculations, this can be resolved by considering the periodic system as embedded in a medium with an infinite dielectric constant.* The $k = 0$ term in the 3D sum over images can now be ignored, with this equivalent to cancelling the surface-charge dipole.

In a pseudo-2D calculation, the position of the cell boundaries in the z direction *is* meaningful. Now it is correct to include the depolarizing field associated with the surface charge along z; neglecting this depolarizing field artificially stabilizes structures with dipoles along z. In the limit of an infinitely thin slab in the z direction, the correct limiting behavior can be recovered by a correction term that restores the interaction of the depolarizing field, by adding the following terms to total energy and the forces on each ion, i†:

$$U_c = +\frac{2\pi}{V} M_z^2,$$
(3.1)

$$F_i = -\frac{4\pi q_i}{V} M_z,$$
(3.2)

where M_z is the net dipole moment of the simulation cell in the z direction [44]. The zero-surface-charge boundary conditions used in a standard 3D Ewald summation are equivalent to assuming $M_z = 0$.

Real 2D nanosystems are not infinitely thin, and higher-order terms are necessary for a full correction to the spurious electrostatic slab–slab interaction. The lowest order correction—given earlier—is usually sufficient if the vacuum gap between slabs is approximately three to five times the thickness of the slab [45,46]. For a specific set of calculations, it is critical to check that the Coulomb energy is converged with respect to the vacuum thickness.

---

* The condition that the embedding medium has infinite dielectric constant is called *conducting, metallic,* and *tinfoil* boundary conditions.
† Including the effect of the depolarizing field due to surface charge dipoles is known as applying *insulating* or *vacuum* boundary conditions.

Up to this point we have been considering systems of point charges. In electronic structure calculations, both point charges (nuclei) and a continuous charge distribution need to be considered, for which equivalent corrections for erroneous cell–cell interactions are possible. A popular approach is the Makov–Payne correction [47], which introduces a correction term in the electrostatic potential of

$$\Delta\phi(r) = \frac{4\pi}{3V}\mathbf{r}\int_{\text{cell}} d^3r'\mathbf{r}'\rho(r'),$$

(3.3)

for spherical boundary conditions. In electronic structure calculations, as for atomistic (point-charge) descriptions it is important to check system energies are converged with respect to slab separations whatever correction method is chosen.

Finally, for this discussion of long-ranged electrostatics, although heterostructures can be considered 2D systems, particularly when interlayer separations are small, it should be remembered that these are *bulk* systems and the discussion earlier does not apply: standard 3D boundary conditions and Ewald summations are appropriate.

3.3.2 Free–Standing Thin Films

The top-down model of free-standing thin films connects the structures of 2D films to those of reference bulk systems. By considering thin film construction as cutting planar surfaces in a bulk phase, this conceptual model suggests that the general structures of free-standing thin films can be usefully discussed in terms of their relationship to the parent bulk phase: specifically the relevant bulk structure, the orientation and position of the surfaces relative to the bulk lattice, and any differences between an ideal cleaved surface and the surfaces expressed by thin films.

3.3.2.1 Choice of Bulk Phase

The top-down model is most appropriate where thin films are sufficiently thick that they remain bulk-like at their centers. Under these conditions, it is not unreasonable to expect bulk and thin film systems that share their chemical composition to share the same structures and therefore to consider those bulk structures as the most plausible candidates for corresponding thin films. As characteristic length scales are reduced, however, surface energies become increasingly significant, and structures that are disfavored in macroscopic bulk samples may be competitive or even favored in thin films [48]. For example, two common polymorphs of TiO_2 are rutile and anatase, with rutile the thermodynamically favored phase in the bulk [49]. In nanocrystals, however, anatase is predominantly observed, and this has been attributed to lower surface energies for anatase than rutile under typical synthesis conditions [50,51]. Similar behavior occurs for ZnS, where the bulk phase adopts the zinc blende structure under ambient conditions, but nanocrystals usually form with the wurtzite structure, known as a high-temperature phase in bulk samples [52]. In some cases, the favored structures for thin films are not known for the corresponding bulk material. For example, free-standing thin films of ZnO below a critical thickness have been predicted to adopt BCT or h-BN structures [8], neither of which are known experimentally as stable bulk ZnO phases, but are observed in bulk β-BeO and the hexagonal form of BN. In general, thin films may adopt a wide range of structural motifs. It should not be assumed that favored structures will be closely related to the corresponding bulk phases of the same material—particularly in the case of very thin films, where surface contributions will be most significant—and *data-mining* structures from a range of materials can be a useful strategy for generating approximate thin film structures that can be compared directly following geometry optimization.

3.3.2.2 Choice of Surface Orientation and Termination: Tasker Classification of Surfaces

Idealized free-standing thin film structures can be generated by cleaving a bulk crystal along parallel planes. For any crystal this requires selecting one of many potential surface planes, and each choice produces a thin film with a distinct structure. Faced with this diverse array of possible structures, it can be useful to classify thin films according to their symmetries, and to use this as a framework for thinking about how the choice of surface termination affects the structure and stability of the entire thin film.

The structure of a thin film with ideal planar surfaces can be classified by considering the structure as xy sheets of atoms stacked perpendicular to the surface plane. In a stoichiometric material, these sheets are charge neutral or they can be collected into repeat units along z that are charge neutral. By considering the symmetry of each charge neutral unit, the surfaces can be classified as one of three types, as described by Tasker [53] (Figure 3.6).

If individual xy sheets have no net charge—this is the case if they contain a stoichiometric ratio of atoms—then the film surfaces will be uncharged. These surfaces are nonpolar and are classed as type 1. Examples are the $\langle 100 \rangle$ and $\langle 110 \rangle$ surfaces of the rock salt structure of materials such as MgO, NaCl, and NiO. A thin film with type 1 surfaces has no net dipole perpendicular to its surfaces. If individual xy sheets are charged, however, then the thin film character depends on the relative charges and positions of these sheets when collected into repeat units that are charge neutral. These repeat units are either symmetric or asymmetric with respect to reflection in the xy plane. A symmetric repeat unit has a zero dipole, and any collection of subunits will be nonpolar. Surfaces at the

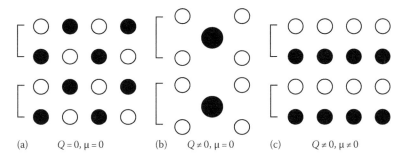

(a)	(b)	(c)
$Q = 0, \mu = 0$	$Q \neq 0, \mu = 0$	$Q \neq 0, \mu \neq 0$

FIGURE 3.6 Classification of surfaces according to Tasker. (a) Each sheet of atoms is charge neutral, and repeat units are therefore apolar. Surfaces are type 1. (b) Sheets of atoms are charged, but can be collected in apolar repeat units. Surfaces are type 2, and there is no net dipole perpendicular to the surface. (c) Sheets of atoms are charged, and the repeat unit is polar. A net dipole perpendicular to the surface exists. Surfaces are type 3 and are electrostatically unstable. (Adapted from Noguera, C., *J. Phys. Condens. Matter*, 12, R367, 2000.)

Chapter 3

edges of these nonpolar units are classified as type 2. Examples include the anion terminated $\langle 100 \rangle$ surface of fluorite structured materials, such as CaF_2 and CeO_2. If individual sheets are charged and cannot be collected into a *symmetric* repeat unit, then each subunit will possess a dipole moment perpendicular to the surface. Now there is a nonzero dipole moment perpendicular to the surfaces across the entire film. Surfaces are classed as type 3 and are both charged and polar. Examples of type 3 surfaces are the $\langle 111 \rangle$ surfaces of rock salt, the $\langle 111 \rangle$ surfaces of zinc blende structured materials such as ZnS, and the $\langle 0001 \rangle$ surfaces of wurtzite structured materials such as ZnO.

3.3.2.3 Nonpolar Thin Films: Parity and In-Plane Symmetry

The transition from bulk to 2D nanosystem corresponds to the removal of translational symmetry in one direction. One possible consequence of this is the potential emergence of dipoles perpendicular to the surface of the slab, corresponding to Tasker type 3 surfaces. A nondefective thin film is crystalline in two dimensions, and the symmetry of the *in-plane* repeat unit may also be significant. If a film is *nonpolar*, its two surfaces are equivalent and related by some symmetry operation: either a reflection in a mirror plane parallel to the surface plane, or a 180° rotation around a C_2 axis parallel to the surface plane, or both. For all but the very simplest bulk crystal unit cells, only one of these two symmetry operations interchanges the two surfaces, with consequences for the symmetry of the 2D unit cell that defines the film structure.

Figure 3.7 shows an example of two free-standing films cleaved from a perfect rock salt (B1) crystal. Films *a* and *b* both have type 1 nonpolar (100) surfaces and possess either an odd or even number of total layers, giving odd or even parity. For film *a* the surfaces are related by a reflection; for film *b* the surfaces are related by a rotation. The in-plane

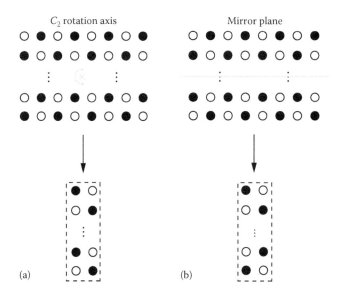

(a)   (b)

FIGURE 3.7 For some crystal lattices, the symmetry operations relating the surfaces of a nonpolar thin film depend on whether an even or odd number of layers are considered. (a) Here, for a (001)-oriented rock salt film, even numbers of layers correspond to surfaces being related by a C_2 rotation. (b) Odd numbers of layers correspond to surfaces being related by a mirror plane. Each case results in a different in-plane 2D space group for the thin film.

2D unit cells in each slab have different symmetries, and this can be expected to correspond to differences in geometric relaxations, phonons, and other structural characteristics of these films. The set of relevant symmetry operations for each film depends on whether the slab has an odd or even number of layers, and this can lead to an even–odd oscillation in properties such as surface energies and optimized surface geometries. The structures of both odd- and even-number-of-layers slabs will converge on the same *bulk* values for sufficiently thick films, but where there are large surface relaxations or low-frequency phonons for one or both classes of film, this convergence may require a very large number of slabs. An example of this even–odd layer behavior is found in stoichiometric (110)-oriented films of rutile TiO_2, where surface energies and electronic structure properties oscillate significantly with the number of layers, even for moderately thick films [55–58].

3.3.2.4 Polar Thin Films

The class of surface presented by a thin film has implications for the convergence of the surface energy with increasing film thickness and for the expected distance from the surface at which internal structures converge to the bulk geometry. For nonpolar thin films with either type 1 or type 2 surfaces, the surface energy converges quickly to bulk values with increasing film thickness. In films more than a few sheets thick, the electrostatic (Madelung) energy approaches bulk values within a few atomic layers of the surface, and the structures of internal sheets quickly converge to the structure of the corresponding bulk. For polar films with type 3 surfaces, however, the existence of a nonzero dipole perpendicular to the surface plane means the surface energy *diverges* with increasing film thickness. The electric field associated with this dipole destabilizes the film structure, and as a consequence the dipole will usually be quenched by some process involving electronic or geometric reorganization [54,59].

The complex issue of surface dipole quenching in polar materials is illustrated by bulk and thin film ZnO. The ZnO lattice is polar along the *c* direction, and cleaving the crystal perpendicular to *c* produces two polar ⟨0001⟩ surfaces: one oxygen terminated and one zinc terminated. These are polar type 3 surfaces and would be expected to have divergent surface energies. ZnO samples, however, predominantly exhibit ⟨0001⟩ surfaces, which appear at odds with the expectation from an ionic model that these polar surfaces should be high in energy and disfavored.

One mechanism by which the surface dipoles of polar thin films may be quenched is by surface layers accommodating vacancies or adatoms—if the system is considered to exist in a residual experimental atmosphere then foreign species may also adsorb at the surface. Changing the stoichiometry of the surface layers produces a surface effective charge that opposes the net dipole moment due to the internal thin film structure. These surface defects may adopt ordered configurations, producing exotic surface motifs such as terraces, islands, or holes, which can be observed in surface microscopy images [60,61].

Following the example of ZnO ⟨0001⟩ surfaces, a combination of experimental and theoretical studies has demonstrated that the Zn-terminated ⟨0001⟩ surface is stabilized by the removal of zinc and oxygen atoms from the first and second layers, resulting in the formation of triangular-shaped reconstructions [60–62]. The O-terminated surface is similarly predicted to show a large range of ordered nonstoichiometric surfaces, with the specific structure sensitive to atmospheric conditions [63–65].

Chapter 3

The electrostatic field of a polar thin film can also be quenched by an electronic redistribution that compensates the surface charges. Electrons can flow from one surface to the other, in doing so forming electron and hole surface states [7,59,66]. Electronic charge redistribution can only be observed in computational modeling if the modeling technique can describe appropriate electronic degrees of freedom; for example, electronic structure methods or atomistic models that allow charge equilibration. The use of the Born–Oppenheimer approximation in electronic structure calculations means that at each step in a geometry optimization the nuclear coordinates are fixed, yet the electrons are permitted to redistribute. This favors quenching the dipole in a polar film by electronic charge redistribution—if possible—even if an alternative mechanism involving structural reorganization would result in a lower energy structure [8]. When predicting ground-state geometries for nominally polar thin films, therefore, care must be taken.

In a polar thin film, the nonzero dipole perpendicular to the surfaces produces an electric field across the film. One effect of this electric field is a shift of band energies at different positions in the film, with the largest relative shifts for the surface layers. In a nonmetallic thin film, a transfer of charge from one surface to the other depopulates states at the top of the valence band at the first surface and populates states at the bottom of the conduction band at the second surface. Usually, this has a large associated energy cost, because electrons must be raised in energy across the band gap. In a polar thin film, however, because the band edges are shifted by the dipole-associated electric field, it is possible for the conduction band minimum at one surface to move *below* the valence band maximum of the other surface. Now charge transfer between the surfaces is favorable, and hole and electron surface states form spontaneously. For an idealized polar film, the film dipole scales linearly with thickness, and so the thicker the film, the larger the difference in relative band positions between the two surfaces. Dipole quenching through charge transfer between surfaces therefore only occurs for relatively thick films. For thinner films, the dipole-associated electric field may be too small to close the band gap, and dipole quenching will occur by some other mechanism, or the film may demonstrate *uncompensated polarity*, where it remains stable with a nonzero dipole [67].

A further alternative mechanism by which a thin film dipole may be quenched is if the entire film reconstructs to a nonpolar structure. This behavior is predicted for (111)-oriented MgO thin films. The (111) MgO surface is polar (type 3) and in bulk samples is stabilized through reconstruction [68]. In very thin films, however, a nonpolar graphitic h-BN structure has been predicted to be more stable than any structure derived from a rock salt MgO (111) surface [69].* Analogous behavior is predicted for thin films for a number of materials that adopt the wurtzite structure in the bulk [71,72]. For example, polar ⟨0001⟩ ZnO thin films have been studied by Harding et al. using density functional theory calculations. These predicted that films above a critical thickness undergo surface–surface-charge transfer and the formation of metallic surfaces,† but films below this critical thickness, however, undergo a *structural* relaxation to a graphitic planar structure—the same h-BN structure predicted for (111)-oriented MgO thin films.

---

* The hexagonal h-BN structure has been predicted to be metastable in bulk MgO, where it is labeled h-MgO, and to occur as an intermediate in wurtzite → rock salt pressure-driven phase transitions. See Reference 70.

† This metallization of polar surfaces in ZnO had been previously predicted by Wander et al. [66].

In thin films of materials that adopt the wurtzite structure in the bulk, the planar h-BN structure can be formed simply by flattening the six-membered rings in each (0001) layer. When this occurs in a thin film that starts with a (0001)-oriented wurtzite structure, it can be considered a mechanical relaxation in response to the surface dipole. The transition from wurtzite (polar) to h-MgO (nonpolar) is not the complete story however. Although this relaxation does remove the surface dipole and lower the overall energy of these films, the tetrahedral coordination, favored by these same materials in the bulk, is disrupted: the h-BN structure has a [3+2] trigonal bipyramidal coordination. An alternative metastable structure predicted theoretically for a number of nominally tetrahedrally coordinated materials is BCT [73–77].* BCT is nonpolar and has a four-coordinate structure. BCT-structured thin films are nonpolar, negating the dipole associated with the [0001]-oriented wurtzite structure, and have coordination geometries closer to tetrahedra of the bulk wurtzite ground state. This combination of a nonpolar structure and favorable coordination geometry means that BCT-structured free-standing thin films are predicted to be lower in energy than both h-BN and wurtzite for film thicknesses from 8 to 54 layers [8].†

The preferred stability of BCT compared to wurtzite and graphitic h-BN in ZnO thin films over a critical thickness range illustrates two points.‡ First, this is an example where a thin film structure generated in a top-down process—by cleaving planar surfaces from a thermodynamic bulk structure—is unstable, in this case because of the polar nature of the lattice and the thin film dipole that results. The dipole is quenched by reconstructing the entire film, and the resulting structure can be thought of as generated in an alternate top-down construction that starts from a structure only *metastable* in the bulk. Second, direct geometry optimizations started from ideally cleaved (0001) wurtzite thin films did not find the ground-state BCT structure, but instead were trapped in the h-BN structure [69,70]. This shows the potential drawbacks of using single starting geometries when modeling competing structures of thin films, especially in cases where significant structural changes are likely.

3.3.2.5 Choice of Surface Reconstruction

One mechanism for quenching dipoles in polar thin films, as discussed earlier, is the presence of vacancies or adatoms at thin film surfaces. If these surface defects are periodically arranged—and we exclude possible adsorption of extrinsic atomic species—then this mechanism can be considered a simple reconstruction. For example, specific patterns of surface vacancies produce *microfaceting* of the surface.

The simplest form of microfaceting consists of a simple reconstruction where some proportion of atoms at one surface are formally moved to the other surface, and this redistribution of surface ions has been used in simulations of polar films or nanostructures as a strategy for generating stable nonpolar models [79,80]. Microfaceting and other surface reconstructions are not restricted to polar surfaces. The (110) surface of MgO is nonpolar (Figure 3.8a). Watson et al. have shown, however, that the surface energy can

---

* So-called because this structure has the same topology as the silicate Mg-BCTT.
† Similarly, Demiroglu and Bromley have identified a wide range of additional low-energy nonpolar ZnO polymorphs based on hexagonal nets that are metastable in free-standing thin films and have relative energies that are sensitive to the thickness of the film and epitaxial strain [78].
‡ h-BN is predicted to be the *ground* state for very thin films of up to six layers.

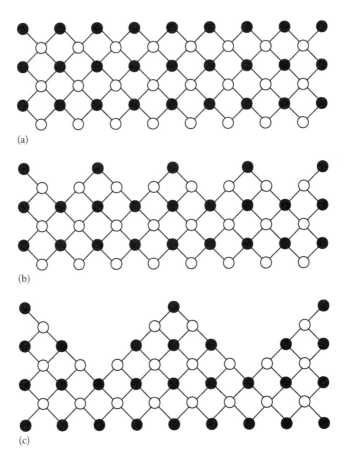

FIGURE 3.8 Alternative terminations for the rock salt (B1) (110) surface. (a) Unfaceted surface, (b) $(a\sqrt{2})\sqrt{2}$ faceted, and (c) $a\sqrt{2}$ faceted. (Adapted from Watson, G.W. et al., *Faraday Trans.*, 92, 433, 1996.)

be lowered considerably by removing alternate rows of MgO to give a $(a\sqrt{2})/2$ faceted surface (Figure 3.8b) [68]. Removing more rows gives an even more stable $a\sqrt{2}$ micro-faceted surface (Figure 3.8c). This can be considered a collection of stepped (100) planes, and the low surface energy can be explained by observing that the planar (100) surface is very low in energy.

The (111) surface of MgO is dipolar (Tasker type 3) and has a divergent surface energy. In a modeled (111)-oriented thin film, shifting one-half of the oxygen ions from the top of the slab to the bottom produces a nonpolar thin film, but the resultant surface energy is very high, and these microfaceted surfaces are unstable with respect to reconstruction into (110) planes. Alternate microfaceted surfaces with lower energies do exist, with the surface reconstructed into three-sided pyramids, with each pyramid side a (100) plane. A wide range of surface reconstructions for rock salt–structured (111) surfaces have been considered using computational modeling [69,81].

Simple reconstructions, such as microfaceting, can usually be proposed following a simple examination of the surface. More complex reconstructions that form even more stable structures may exist, however, and identifying these typically requires the use of more sophisticated structural modeling techniques that can perform a global search of geometries. For example, Zhu et al. have used an evolutionary algorithm to model reconstructions of the polar (111) MgO surface—among others—and found a variety of surface reconstructions that are not simply related to the ideal (111) surface and may not have been found by simple inspection [82]. A similar approach using genetic algorithms has been demonstrated by Deacon-Smith et al. to identify low-energy reconstructions of the polar surface of the perovskite $KTaO_3$ [83].

3.3.3 Supported Thin Films

Supported thin films share many features with unsupported thin films, as well as many of the issues relevant to their modeling. An important difference, however, is the presence of the supporting phase. This can affect the electrostatic behavior of thin films, either through polarization of the substrate or charge transfer between the substrate and the supported thin film [23,25,69]. The structure of the substrate at the interface is also important to consider. The possibility of favored epitaxial relationships between the two lattices can have a significant effect on the stability of particular orientations of the supported thin film and can also affect the thin film structure through epitaxial strain.

3.3.3.1 Substrate Polarization and Charge Transfer

When a thin film is placed on a supporting substrate, electrons may transfer between the substrate and the thin film. For example, Goniakowski and Noguera have discussed the predicted transfer of electrons between MgO (111) thin films and metal substrates [23]. The transfer of these charges polarizes the metallic substrate, which compensates for the polarity at the interface and stabilizes the polar structure of the supported thin films.* A similar process can occur even if the supported thin film *is not* polar: charge transfer between a nonpolar thin film and the substrate can cause rumpling of the thin film to produce a polar structure and an enhancement of the film adhesion energy [25].

3.3.3.2 Epitaxial Relationships

The effects of electronic redistribution between a substrate and supported thin film on the film structure are most significant when electrons in the substrate are relatively mobile, that is, for metallic supports. A second interaction mechanism that exists between any crystalline substrate and supported thin film is the potential for preferred epitaxial relationships across the interface. In general, relative orientations of the substrate and thin film lattices are preferred if there is epitaxial matching—the thin film lattice is oriented and/or strained such that a coincident site lattice exists between the thin film and the substrate (Figure 3.9). Favorable epitaxial relationships

---

* Analogous charge transfer and polarization occurs when metal clusters or films deposited on a polar oxide surface; see, for example, Reference 84.

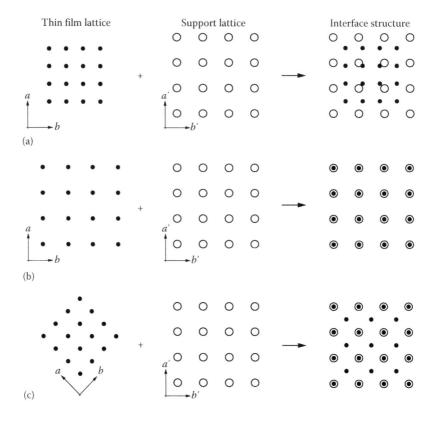

FIGURE 3.9 (a) We can expect thin films to have different lattice parameters to supporting substrates, producing an incoherent epitaxial relationship. In this example, a coherent interface can be recovered in two ways: (b) straining the film structure (expanding the lattice) allows a 1×1 coherent interface; (c) alternately, rotating the film by 45° gives a strain-free coherent $\sqrt{2} \times \sqrt{2}$ interface.

are expected in a number of supported thin film systems, and identifying preferred thin film orientations can be a necessary step in modeling realistic thin film–substrate interface structures.

One approach to finding favorable epitaxially matched relative lattice orientations is to use near-coincident-site-lattice theory [85]. Consider a heteroepitaxial interface. If the lattice parameters of the two regions are incommensurate then it is not possible to find an exact coincidence by only rotating one lattice relative to the other. Instead, some lattice misfit will remain and we can, at best, achieve near coincidence between the two crystal lattices. The near-coincident-site-lattice theory of Sayle et al. can be used to find lattice relationships that correspond to near coincidence, and then to quantify the degree of lattice mismatch between different potential epitaxial relationships. This procedure will be explained for the case where both systems have a cubic lattice, and an equivalent approach has been described for hexagonal lattices [85] (Figure 3.10).

The interfacial surfaces of two cubic systems can be considered as 2D square lattices with lattice parameters a_1 and a_2. A 2D coincidence site lattice can be constructed if it is possible to rotate one lattice with respect to the second about an axis normal to the

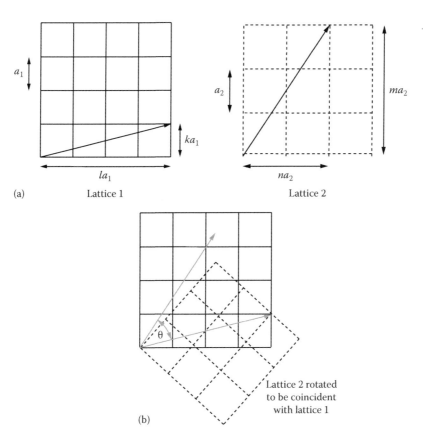

(a)   Lattice 1

(b)

FIGURE 3.10   (a) Schematic representation of two cubic lattices with different lattice parameters. (b) By rotation one lattice relative to the other, it may be possible to bring the two into coincidence. (Adapted from Sayle, T.X.T. et al., *Philos. Mag. A*, 68, 565, 1993.)

interfacial plane until three lattice sites in the two materials have common positions. This occurs when

$$\left(\frac{a_1}{a_2}\right)^2 = \left(\frac{m^2 + n^2}{k^2 + l^2}\right), \tag{3.4}$$

where k, l, m, and n are integers. The rotation angle θ that brings the two lattices into exact coincidence is given by

$$\theta = \tan^{-1}\left(\frac{n}{m}\right) \pm \tan^{-1}\left(\frac{l}{k}\right). \tag{3.5}$$

The density of planar coincidence sites is denoted Γ and defined as $1/\Sigma_i^p$, where Σ_i^p is the planar reciprocal coincidence density at the interface, given by[*]

$$\Sigma_1^p = (m^2 + n^2); \tag{3.6}$$

[*] The superscript p denotes *planar* or 2D coincidence sites, rather than the more standard 3D coincidence sites. See Reference 86.

$$\Sigma_2^p = (l^2 + k^2).\tag{3.7}$$

If Equation 3.4 can be satisfied exactly, then exact lattice coincidence is possible. However, for heteroepitaxial systems Equation 3.4 can never be satisfied. Because the two lattices are incommensurate, the left-hand side of this equation will be irrational, whereas the right-hand side must be rational. Satisfying Equation 3.4 exactly is only possible if one or both lattice parameters are changed, by expanding or contracting with respect to the original lattice parameter and introducing some degree of lattice strain. The misfit between the unstrained lattices is defined as the minimum deviation from the exact coincidence condition:

$$F = 2 \frac{\left| a_1 \left(\Sigma_1^p\right)^{\frac{1}{2}} - a_2 \left(\Sigma_2^p\right)^{\frac{1}{2}} \right|}{\left| a_1 \left(\Sigma_1^p\right)^{\frac{1}{2}} + a_2 \left(\Sigma_2^p\right)^{\frac{1}{2}} \right|}.\tag{3.8}$$

It is always possible to find a low value of F, corresponding to a mismatched interface with a small strain energy, by increasing the size of Σ^p. However, this can require very large interfacial unit cells, which might not be possible to model, and can also produce very low value of the coincident site density, Γ, which has been suggested to indicate low stability interfaces [87]. In practice, F can be used to screen possible epitaxial relationships by requiring the lattice mismatch to be lower than some upper limit (Figure 3.11).

In the mechanism described by Sayle et al., one lattice rotates relative to the other about an axis normal to the interface. An alternative mechanism has been described by Erwin et al. where an epitaxial interface forms between two materials with strongly

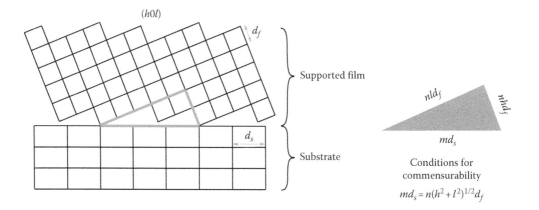

FIGURE 3.11 Schematic view of a commensurate interface between a film with lattice constant d_f and a substrate with lattice constant d_s. The interface plane is ($h0l$) with respect to the film. The condition for commensurability is that a coincidence site lattice, defined by a pair of integers (m, n), exists such that $md_s = n(h^2 + l^2)^{1/2} d_f$. (Adapted from Erwin, S.C. et al., *Phys. Rev. Lett.*, 107, 026102, 2011.)

mismatched lattice constants by one material tilting out of the interface plane— equivalent to a rotation about an axis *parallel* with the interface [88].

The unit cell of a film with orientation $(h0l)$ has length

$$L_f = (h^2 + l^2)^{1/2} d_f. \tag{3.9}$$

For a commensurate interface between film and substrate, there must exist a coincident site lattice, defined by a pair of integers (m, n), such that $md_s = nL_f$. Exactly satisfying this relationship may not be possible for arbitrary lattices. However, again by allowing the film to be strained, a relaxed condition can be written of

$$md_s = nL_f(1 + \epsilon_{xx}), \tag{3.10}$$

where ϵ_{xx} is the compressive or tensile strain along x. Erwin et al. have used this model to identify epitaxial growth modes under the constraints that (1) the misfit strain ϵ_{xx} should be as small as possible and (2) the period m of the coincident site lattice should be as small as possible. The second restriction is analogous to the suggestion that low values of Γ in the model of Sayle et al. correspond to low stability interfaces.

The techniques discussed above attempt to seek favorable epitaxial relationships by identifying low-strain and low-mismatch orientations and assuming these low in energy. An alternative approach that can be particularly useful when considering *incoherent* interfaces, characterized by disorder such as dislocations and grain boundaries, is to generate thin film geometries by dynamical simulation. An example of this technique is the *amorphization and recrystallization* described by Sayle et al. [89,90]. The process consists of constraining the supported film under high tension or compression and performing dynamical simulations at high temperature. This causes the thin film to undergo a transition to an amorphous phase. By performing subsequent simulations at lower temperatures, the thin film then recrystallizes under the influence of the support and is free to adopt any preferred interface geometry. A direct temperature quench of the amorphous thin film will usually produce a highly disordered film structure containing multiple defects and dislocations, because a certain degree of amorphous disorder can be *frozen in*. More ordered thin film structures can be obtained by using techniques that allow the system to explore the potential energy surface during the energy minimization, for example, simulated annealing or some alternative global optimization technique [91–93].

Lattice mismatch between a substrate and a supported thin film strains the film, and we may be interested in modeling the effects of this strain. If films can be considered as relatively thick, then a simple approach is to approximate the system as bulk-like and model *bulk* structures with a range of strains applied in only two directions—that is, within the xy plane—while the lattice is free to relax along the perpendicular z direction. Examples of this technique include studies of the cation site preference in epitaxially strained spinels [94] and of the potential for epitaxial strain to stabilize competing structural polymorphs in thin films [28]. An alternative approach is to model the strained interface explicitly [24]. This is more realistic, especially for very thin films, but the range of strains that can be considered are limited by the available substrates and allowed epitaxial orientations.

Chapter 3

3.3.4 Modeling Heterostructures

Many of the issues relating to the modeling of supported thin films also apply to nanoscale heterostructures. In particular, care must be taken in constructing appropriate interfaces, taking into account epitaxial relationships or generating structures using global energy minimization techniques. Calculations of heterostructures, however, are fundamentally different to those discussed earlier for free-standing and supported thin films, because we are now modeling *bulk* materials. Because of this, it may appear that the issues discussed earlier relating to polarity are no longer relevant: for example, we are now able to use standard 3D periodic boundaries and the 3D Ewald sum. Furthermore, if a material is periodic in three dimensions, then the simulation cell cannot support a nonzero electric field. In heterostructures, however, the translational symmetry is broken along the z direction relative to a homogeneous bulk crystal, and it is now possible for the lattice to support *local* electric fields and locally polarize. In semiconductor heterostructures, this produces phenomena such as 2D electron and hole gases [4–7], and analogous polarization exists in ionic systems [32].

3.4 Assessment of Geometric Stability

This chapter has focused on the concepts and techniques that pertain to structural modeling of 2D nanostructures. The fundamental problem faced in all cases is the calculation of stable atomic geometries. Because a stable structure corresponds to a minimum of the potential energy surface—within the approximation that PV and nonzero temperature contributions can be neglected—these geometries can be found by performing energy minimizations with respect to nuclear coordinates.

In practice, therefore, we will usually identify a successful energy minimization by finding a geometry for which all the atomic forces are zero. Atomic forces may also be zero, however, at maxima and points of inflection on our potential energy surface. Although we can likely neglect the former case, it is prudent to check that optimized geometries do not correspond to points of inflection and are truly energy minima.

One test is to calculate phonon frequencies for the optimized structure. If any phonons have imaginary frequencies, then the geometry does not correspond to a minimum and can be expected to relax away from this structure at any nonzero temperature. For example, when modeling thin films of materials such as ZnO that adopt the wurtzite structure in the bulk, calculations that use ⟨0001⟩-oriented wurtzite thin films as starting points for geometry optimizations can become trapped in a metastable planar h-MgO structure [72,95]. The h-MgO structure is relatively high in symmetry, however, and unless stabilized by in-plane strain [24], in fact corresponds to a point of inflection rather than a (local) minimum on the potential energy surface, shown by the existence of an imaginary frequency vibrational mode under ambient conditions [8].

An alternative approach is to perform some length of molecular dynamics simulation starting from the candidate minimum energy structure. At low temperatures, molecular dynamics allows us to explore the potential energy surface proximate to the optimized geometry. If we have in fact started with a point of inflection, then we can expect the system to quickly fall into a lower energy state. This technique is less well defined than a vibrational analysis, because it is possible for a system at nonzero temperature to overcome small kinetic barriers, and true (local) minima may therefore be predicted to be *unstable*. Equally, if the goal of our study is to find *global* potential energy minima, then this behavior is merely fortuitous and may be considered a form of simulated annealing.

References

1. S. Tolbert and A. P. Alivisatos, *Science* **265**, 373 (1994).
2. H. Zhang, B. Gilbert, F. Huang, and J. F. Banfield, *Nature* **424**, 1025 (2003).
3. J. Maier, *Adv. Mater.* **21**, 2571 (2009).
4. E. E. Mendez and G. Bastard, *Phys. Today* **46**, 34 (1993).
5. B. Ridley, O. Ambacher, and L. Eastman, *Semicond. Sci. Technol.* **15**, 270 (2000).
6. A. Ohtomo and H. Y. Hwang, *Nature* **427**, 423 (2004).
7. J. Goniakowski, F. Finocchi, and C. Noguera, *Rep. Prog. Phys.* **71**, 016501 (2007).
8. B. J. Morgan, *Phys. Rev. B* **80**, 174105 (2009).
9. D. W. Boukhvalov, *RSC Adv.* **3**, 7150 (2013).
10. A. K. Geim, *Science* **324**, 1530 (2009).
11. Ž. P. Čančarević, J. C. Schön, and M. Jansen, *Chem. Asian J.* **3**, 561 (2008).
12. A. Bach, D. Fischer, and M. Jansen, *Z. Anorg. Allg. Chem.* **635**, 2406 (2009).
13. B. J. Morgan, J. Carrasco, and G. Teobaldi, in submission.
14. R. F. Frindt, *J. Appl. Phys.* **37**, 1928 (1966).
15. P. Joensen, R. F. Frindt, and S. R. Morrison, *Mater. Res. Bull.* **21**, 457 (1986).
16. K. S. Novoselov, D. Jiang, F. Schedin, T. J. Booth, V. V. Khotkevich, S. V. Morozov, and A. K. Geim, *Proc. Natl. Acad. Sci.* **102**, 10451 (2005).
17. M. Osada and T. Sasaki, *Adv. Mater.* **24**, 210 (2012).
18. M. Wilson and P. A. Madden, *J. Phys. Condens. Matter.* **14**, 4629 (2002).
19. B. J. Morgan and P. A. Madden, *Phys. Chem. Chem. Phys.* **8**, 3304 (2006).
20. H. Jagodzinski, *Acta Crystallogr.* **7**, 300 (1954).
21. J. Lee, S. Adams, and J. Maier, *Solid State Ionics* **136**, 1261 (2000).
22. Y.-G. Guo, J.-S. Lee, Y.-S. Hu, and J. Maier, *J. Electrochem. Soc.* **154**, K51 (2007).
23. J. Goniakowski, L. Giordano, and C. Noguera, *Phys. Rev. B* **81**, 205404 (2010).
24. D. Wu, M. G. Lagally, and F. Liu, *Phys. Rev. Lett.* **107**, 236101 (2011).
25. J. Goniakowski and C. Noguera, *Phys. Rev. B* **79**, 155433 (2009).
26. N. Y. Jin-Phillipp, N. Sata, J. Maier, C. Scheu, K. Hahn, M. Kelsch, and M. Rühle, *J. Chem. Phys.* **120**, 2375 (2004).
27. D. C. Sayle and G. W. Watson, *Phys. Chem. Chem. Phys.* **2**, 5491 (2000).
28. B. J. Morgan, *Phys. Rev. B* **82**, 153408 (2010).
29. T. J. Pennycook, M. J. Beck, K. Varga, M. Varela, S. J. Pennycook, and S. T. Pantelides, *Phys. Rev. Lett.* **104**, 115901 (2010).
30. O. Ambacher, J. Smart, J. Shealy, N. Weimann, K. Chu, M. Murphy, W. Schaff et al., *J. Appl. Phys.* **85**, 3222 (1999).
31. P. Käckell and F. Bechstedt, *Mater. Sci. Eng. B* **37**, 224 (1996).
32. B. J. Morgan and P. A. Madden, *Phys. Rev. Lett.* **107**, 206102 (2011).
33. A. Fissel, *Phys. Rep.* **379**, 149 (2003).
34. B. J. Morgan and P. A. Madden, *J. Phys. Condens. Matter* **24**, 275303 (2012).
35. A. R. Oganov, *Modern Methods of Crystal Structure Prediction*. Wiley-VCH, New York (2011).
36. A. B. Herhold, C.-C. Chen, C. S. Johnson, S. H. Tolbert, and A. P. Alivisatos, *Phase Transit* **68**, 1 (1999).
37. N. M. Harrison, X.-G. Wang, J. Muscat, and M. Scheffler, *Faraday Discus.* **114**, 305 (1999).
38. D. Frenkel and B. Smit, *Understanding Molecular Simulation—From Algorithms to Applications.* Academic Press, San Diego, CA (1996).
39. P. J. Steinbach and B. R. Brooks, *J. Comp. Chem.* **15**, 667 (1994).
40. P. P. Ewald, *Ann. Phys.* **64**, 253 (1921).
41. M. P. Allen and D. J. Tildesley, *Computer Simulations of Liquids.* Oxford University Press, Oxford, U.K. (1987).

Chapter 3

42. D. M. Heyes, M. Barber, and J. H. R. Clarke, *J. Chem. Soc. Faraday Trans. 2* **73**, 1485 (1977).
43. A. Grzybowski, E. Gwóźdź, and A. Bródka, *Phys. Rev. B* **61**, 6706 (2000).
44. E. R. Smith, *Proc. R. Soc. Lond. A* **375**, 475 (1981).
45. I.-C. Yeh and M. L. Berkowitz, *J. Chem. Phys.* **111**, 3155 (1999).
46. P. S. Crozier, R. L. Rowley, E. Spohr, and D. Henderson, *J. Chem. Phys.* **112**, 9253 (2000).
47. G. Makov and M. C. Payne, *Phys. Rev. B* **51**, 4014 (1995).
48. A. Navrotsky, *Proc. Natl. Acad. Sci.* **101**, 12096 (2004).
49. J. Muscat, V. Swamy, and N. M. Harrison, *Phys. Rev. B* **65**, 224112 (2002).
50. A. S. Barnard and L. A. Curtiss, *Nano Lett.* **5**, 1261 (2005).
51. A. S. Barnard and H. Xu, *ACS Nano* **2**, 2237 (2008).
52. S. B. Qadri, E. F. Skelton, D. Hsu, A. D. Dinsmore, J. Yang, H. F. Gray, and B. R. Ratna, *Phys. Rev. B* **60**, 9191 (1999).
53. P. W. Tasker, *J. Phys. C Solid State Phys.* **12**, 4977 (1979).
54. C. Noguera, *J. Phys. Condens. Matter* **12**, R367 (2000).
55. S. P. Bates, G. Kresse, and M. J. Gillan, *Surf. Sci.* **385**, 386 (1997).
56. T. Bredow, L. Giordano, F. Cinquini, and G. Pacchioni, *Phys. Rev. B* **70**, 035419 (2004).
57. P. Murugan, V. Kumar, and Y. Kawazoe, *Phys. Rev. B* **73**, 075401 (2006).
58. T. He, J. L. Li, and G. W. Yang, *ACS Appl. Mater. Int.* **4**, 2192 (2012).
59. C. Noguera and J. Goniakowski, *Chem. Rev.* **113**, 4073 (2013).
60. G. Kresse, O. Dulub, and U. Diebold, *Phys. Rev. B* **68**, 245409 (2003).
61. O. Dulub, U. Diebold, and G. Kresse, *Phys. Rev. Lett.* **90**, 16102 (2003).
62. H. Meskine and P. A. Mulheran, *Phys. Rev. B* **84**, (2011).
63. J. V. Lauritsen, S. Porsgaard, M. K. Rasmussen, M. C. R. Jensen, R. Bechstein, K. Meinander, B. S. Clausen et al., *ACS Nano* **5**, 5987 (2011).
64. R. Wahl, J. V. Lauritsen, F. Besenbacher, and G. Kresse, *Phys. Rev. B* **87**, 085313 (2013).
65. S. E. Huber, M. Hellström, M. Probst, K. Hermansson, and P. Broqvist, *Surf. Sci.*, **628**, 50–61 (2014).
66. A. Wander, F. Schedin, P. Steadman, A. Norris, R. McGrath, T. S. Turner, G. Thornton, and N. M. Harrison, *Phys. Rev. Lett.* **86**, 3811 (2001).
67. J. Goniakowski, C. Noguera, and L. Giordano, *Phys. Rev. Lett.* **98**, 205701 (2007).
68. G. W. Watson, E. T. Kelsey, N. H. de Leeuw, D. J. Harris, and S. C. Parker, *Faraday Trans.* **92**, 433 (1996).
69. J. Goniakowski, C. Noguera, and L. Giordano, *Phys. Rev. Lett.* **93**, 215702 (2004).
70. S. Limpijumnong and W. R. L. Lambrecht, *Phys. Rev. B* **63**, 104103 (2001).
71. C. L. Freeman, F. Claeyssens, N. L. Allan, and J. D. Harding, *Phys. Rev. Lett.* **96**, 066102 (2006).
72. F. Claeyssens, C. L. Freeman, N. L. Allan, Y. Sun, M. N. Ashfold, and J. H. Harding, *J. Mater. Chem.* **15**, 139 (2005).
73. S. Hamad and C. R. A. Catlow, *J. Cryst. Growth* **294**, 2 (2006).
74. B. J. Morgan and P. A. Madden, *J. Phys. Chem. C* **9**, 2355 (2007).
75. J. Wang, A. J. Kulkarni, K. Sarasamak, S. Limpijumnong, F. J. Ke, and M. Zhou, *Phys. Rev. B* **76**, 172103 (2007).
76. B. J. Morgan, *Phys. Rev. B* **78**, 024110 (2008).
77. S. Hamad, S. M. Woodley, and C. R. A. Catlow, *Mol. Simul.* **35**, 1015 (2009).
78. I. Demiroglu and S. T. Bromley, *Phys. Rev. Lett.* **110**, 245501 (2013).
79. D. O. Scanlon, A. Walsh, B. J. Morgan, M. Nolan, J. Fearon, and G. W. Watson, *J. Phys. Chem. C* **111**, 7971 (2007).
80. S. Chaudhuri, P. Chupas, B. J. Morgan, P. A. Madden, and C. P. Grey, *Phys. Chem. Chem. Phys.* **8**, 5045 (2006).
81. A. Wander, I. Bush, and N. Harrison, *Phys. Rev. B* **68**, 233405 (2003).
82. Q. Zhu, L. Li, A. R. Oganov, and P. B. Allen, *Phys. Rev. B* **87**, 195317 (2013).
83. D. E. E. Deacon-Smith, D. O. Scanlon, C. R. A. Catlow, A. A. Sokol, and S. M. Woodley, *Adv. Mater* **26**, 7252–7256 (2014).
84. J. Goniakowski and C. Noguera, *Phys. Rev. B* **66**, 085417 (2002).
85. T. X. T. Sayle, C. R. A. Catlow, D. C. Sayle, S. C. Parker, and J. Harding, *Philos. Mag. A* **68**, 565 (1993).
86. Y. Gao, P. Shewmon, and S. A. Dregia, *Scr. Metall.* **22**, 1521 (1988).
87. D. G. Brandon, B. Ralph, S. Ranganathan, and M. S. Wald, *Acta Metall.* **12**, 813 (1964).
88. S. C. Erwin, C. Gao, C. Roder, J. Laehnemann, and O. Brandt, *Phys. Rev. Lett.* **107**, 026102 (2011).

89. D. C. Sayle, C. R. A. Catlow, N. Dulamita, M. J. F. Healy, S. A. Maicaneanu, B. Slater, and G. W. Watson, *Mol. Simul.* **28**, 683 (2002).
90. D. C. Sayle and R. L. Johnston, *Curr. Opin. Solid State Mater. Sci.* **7**, 3 (2003).
91. D. C. Sayle, *J. Mater. Chem.* **9**, 2961 (1999).
92. T. X. T. Sayle, S. C. Parker, and D. C. Sayle, *J. Mater. Chem.* **16**, 1067 (2006).
93. K. Doll, J. C. Schön, and M. Jansen, *Phys. Chem. Chem. Phys.* **9**, 6128 (2007).
94. D. Frisch and C. Ederer, *Appl. Phys. Lett.* **99**, 081916 (2011).
95. C. L. Freeman, F. Claeyssens, N. L. Allan, and J. H. Harding, *J. Cryst. Growth* **294**, 111 (2006).

Chapter 3

4. Nanocluster-Assembled Materials

Elisa Jimenez-Izal, Jesus M. Ugalde, and Jon M. Matxain

4.1 Introduction 113
    4.1.1 Nanoclusters as Building Blocks. 114
    4.1.2 Conditions for a Successful Assembly 115
    4.1.3 Properties of the Resulting Solids. 117
    4.1.4 Metastability 119

4.2 General Methods. 120
    4.2.1 Modeling 3D Cluster-Assembled Materials 122
    4.2.2 Local Minima Geometry Optimizations. 123
    4.2.3 Global Minimum Geometry Optimizations and Metastability 124

4.3 Nanocluster-Assembled Compounds 126
    4.3.1 Nanoclusters of 1:1 Stoichiometry. 127
    4.3.2 Other Binary and Ternary Compounds. 136
        4.3.2.1 XO_2 and Other Oxides 136
        4.3.2.2 Other Interesting Binary and Ternary
              Cluster-Assembled Solids 140

4.4 General Conclusions. 143

Acknowledgments. 144

References 144

4.1 Introduction

Nanoscience and nanotechnology have grown exponentially after the discovery of buckminsterfullerene (C_{60}) [1] and the observation of pronounced intensities for certain cluster sizes in the mass spectra of metal clusters [2]. These intensity peaks are associated to clusters with enhanced stability, named *magic clusters*. Advances in the degree of understanding of the physicochemical properties of these materials have had an enormous impact in the development of accurate deliberate structure control techniques over the past two decades. Thus, clusters with precisely defined atomic compositions can now be produced in large amounts both in gas phase and in solution. Where experimental techniques are not available, computational theory becomes an essential tool as a predictive basis for guiding experiment. Indeed, the improvement of computational methods and capabilities allows the prediction and the better understanding of new materials, avoiding, in some cases, the high cost of experimental random trial-and-error search.

Chapter 4

Nonetheless, it should be pointed out that nanoscience and nanotechnology are still in their infancy. After the discovery of nanoclusters and their study, the next step to consider is their assembly to form a solid, but strategic cooperative research is needed before their full potential can be realized. The importance of nanoclusters as building blocks can hardly be underestimated, as it will enable us to create new solid phases by using rational designs. Nanoclusters are characterized for having distinct properties from bulk, mainly due to two factors: high surface-to-volume ratio and quantum confinement effects. This second effect appears when the wavelength of the electrons is in the order of the size of the material in which lie, and the quantum effects rule the behavior and properties of the system. As a result electrons are confined in a small region of space, modifying the optoelectronic properties of the nanocluster. Since the properties vary with the size and composition, in principle, they could be tailored at will. Then, it is expected that solids made of clusters will own novel and tunable properties too. This fact broadens the horizon of possibilities of obtaining a wide range of materials and to create new polymorphs with the desired features.

In this chapter, we will focus on the new 3D solids that may be built by assembling such nanostructures. The 3D assembling of nanoclusters opens the possibility of building new solids that differ significantly in structure and properties from their bulk species. In this way, new and amazing strategies could be envisaged to construct novel materials. One could imagine assemblies of any kind of material and although resulting materials are mostly crystalline, assemblies of amorphous clusters have also been proposed for alkali metal halides [3]. In this chapter, we will briefly review diverse methods and approaches on this topic along with a number of cluster-assembled materials. We will mainly focus on the design of fullerene-like cluster-assembled materials of inorganic compounds with diameters on the order of 1–2 nm (up to tens of atoms), but other examples will also be highlighted.

4.1.1 Nanoclusters as Building Blocks

One of the most important targets of modern science is the design of new materials with desired properties. It was observed that, generally, a cluster's structure is not the same as that of the corresponding crystal. Moreover, the structure of clusters dramatically depends on the number of atoms. The addition of just one atom can dramatically change the cluster's geometry and electronic structure. It has been found that certain nanoclusters show enhanced stability with respect to similar-sized clusters, instead of having a linear size distribution of clusters with the degree of aggregation. These clusters are magic clusters and they behave as a whole unit, in the same way an atom does. Then, it is not senseless to think of the possibility of assembling them into a periodic solid. Indeed, some cluster-assembled solids have been already synthesized. As mentioned before, the experimental detection of C_{60} was a groundbreaking discovery. Likewise, one of the first solids obtained via cluster coalescence was the fullerite built up of C_{60} fullerenes [4]. In this new molecular solid, the C_{60} nanoclusters are weakly bonded by van der Waals interactions in a face-centered cubic motif. But the most important aspect of this new polymorph is that the structure of each fullerene is kept and its molecular properties preserved. So the clusters, after assembling them into the solid, do retain their identity as it happens with the atoms in the traditional solids. The synthesis of fullerite crystals confirmed that the assembly of clusters into solids is a realistic possibility and that the properties of the resulting phases will differ from those of the traditional ones.

We must emphasize that although this standpoint is innovative, it is very similar to what happens in the well-known molecular solids, where molecules act as building units. The clearest example is ice, whose building blocks are water molecules. The difference, however, is that water molecules occur in nature and are stable, while clusters produced in the laboratories are often metastable toward further growth.

In addition, Jena et al. [5–10] and Castleman et al. [11–12] showed that some clusters mimic the chemical behavior of elements in the periodic table and hence can be regarded as *superatoms*, providing an unprecedented ability to design novel materials. The superatom suggestion is that free electrons in the cluster occupy a new set of orbitals that are defined by the entire group of atoms rather than by each individual atom separately [13]. Thus, this model envisions a cluster of atoms as a single large atom. The chemistry of elements in the periodic table is already established from quantum mechanics, in spite of some recent debate on the issue [14]. Thus, it is well known that noble gas atoms possess closed electronic shells and consequently are inert. On the other hand, alkali and halogen atoms are very reactive as they have one excess electron and one less electron, respectively, from achieving the electronic closure. Jena's idea is based on the fact that some clusters mimic the chemistry of atoms [15]. They can be seen as human-made superatoms comprising a *new* 3D periodic table. Therefore, they will be able to yield new solids from assembly.

To better understand this conception, the case of Al_{13}^- nanocluster will be used as illustrative example. Al_{13}, with 39 valence electrons, is one electron short to achieve the electronic shell closure according to the jellium model [16] and its calculated electron affinity is comparable to that of chlorine [17]. Conversely, Al_{13}^- anionic cluster exhibits a large HOMO-LUMO gap and high stability, which is strengthened by its geometry, namely, a perfect icosahedron with one aluminum atom at the center [18]. Along with the outstanding stability, Al_{13}^- also shows enhanced chemical inertness. In fact Leuchtner et al. [11] observed that Al_{13}^- was much less reactive toward oxygen than Al_{13}. Additionally, it was found that an ionic saltlike cluster is obtained when potassium bounds to Al_{13} structure [19]. These findings support the idea of clusters mimicking atoms. Certainly, it seems that Al_{13} mimics the chemistry of halogen atoms, while Al_{13}^- emulates that of a nonreactive or noble gas element. In the same vein, Al_{14} was found to behave like an alkaline earth metal [20]. Based on this understanding diverse solids made of aluminum nanoclusters have been suggested and will be discussed in Section 4.3.

It is noteworthy that, as Jena pointed out [10], although the previously mentioned C_{60} *does not fit the superatom definition strictly, it embodies the spirit of the superatom concept as building blocks of novel cluster assembled materials.*

4.1.2 Conditions for a Successful Assembly

One of the main challenges to face in the assembly of clusters is that clusters are metastable toward further growth. In order to avoid this problem, several methods are used nowadays to stabilize nanostructures as they are assembled. One of these methods is to embed clusters in matrices. For instance, when the embedding medium is a zeolite (microporous aluminosilicate minerals), after clusters are trapped into them, they are completely isolated from other clusters. However, the embedded clusters themselves form a new type

Chapter 4

of bulk, with novel properties, which may be useful specially for catalytic activity [21]. For assembling clusters strictly speaking, the use of ligands and the assembly via organic frameworks [22–24], as well as the deposition on noninteracting substrates [25–27], are being extensively investigated. However, the properties of individual clusters are inherently tied to their stabilizing groups, deposition surface, and so on, and then they are likely to be modified by the environment in which they are embedded. So, as we pointed out in the introduction, we will focus on the assembly of bare small atomically precise inorganic nanoclusters. For other topics, the reader can see the review by Jena et al. [28].

The ultimate goal of building cluster-assembled materials is, once specified a desired set of properties, to predict one or more sets of superatom building blocks that could be synthesized and self-assembled into a material with such properties. Clusters with superalkali or superhalogen character, for instance, are expected to be worthwhile since they will offer properties not available from standard atomic building blocks [10]. Nevertheless, building up such materials displays new challenges, related to making and characterizing a new set of superatomic building blocks along with understanding the rules that govern their assembly. Of crucial importance at this point is the identification of highly stable building blocks, as many clusters at these small sizes have lifetimes too short for a controlled assembly into materials. One of the key aspects in this field is that the nanoclusters must retain their character in the assembled solid, as it happens in the fullerite crystals made of C_{60} fullerenes. To make this happen, nanoclusters must be energetically stable, since most stable structures will be chemically inert with respect to retaining their integrity upon interacting with other species. Thus, good candidates for being used as building units should have a closed-electronic shell and a large HOMO-LUMO gap, which is indicative for the thermodynamic stability. In this vein, nanoclusters exhibiting *magic number* behavior may be appropriate as monomers due to their high stability with respect to similar size and composition clusters. In addition, Jena et al. [29] highlighted that the binding energy of the nanoclusters should be larger than the cohesive energy of the resulting solid phase, because otherwise coalescence may occur. Likewise, the geometry of the building block is an important factor in this field, as the most symmetric compounds will make the assembly into a periodic structure easier. Fullerenes are fairly convenient structural materials for the formation of nanostructures because they are perfectly monodisperse and spherical. Other suitable materials could be the $(XY)_{12}$ spheroidal nanoclusters composed of different elements. This structure is highly symmetric, which will favor the 3D assembly and will be addressed in Section 4.3.

Additionally, the study of the assembling of hydrogen-doped icosahedral aluminum clusters, $Al_{13}H$, showed that the relative orientation of the clusters with respect to each other is a very important factor that must be taken into account [30]. Gong et al. [31] reached the same conclusion when they analyzed hypothetical solids based on $C@Al_{12}$ and $Si@Al_{12}$ structures. In fact, it was found that the most stable configuration for a periodic structure was a NaCl-like structure, with icosahedra in alternated orientation of 90° with all its nearest neighbors (see Figure 4.1).

Finally, it is desirable that clusters interact weakly with each other to avoid the coalescence. This fact is especially important in the case of metal clusters because the interactions are more delocalized. Then, although clusters with large band gaps are less reactive than others, in metallic clusters the overlap between orbitals of neighbor clusters may lead to substantial interaction [32]. As an example of this, Hakkinen et al. performed ab

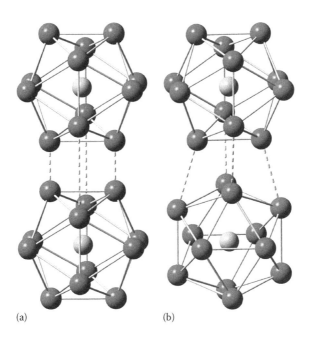

(a)   (b)

FIGURE 4.1 C@Al$_{12}$ dimer with the (a) unstable and (b) most stable configuration. Al and C atoms in dark and light, respectively.

initio molecular dynamics calculations on *magic* Na$_8$ clusters in different environments [32]. The cluster remained *magic* only when adsorbed to the insulating NaCl (001) surface, while it collapsed in Na(110) surface and interacting with other Na$_8$ cluster. They concluded that clusters with purely metallic intracluster bonding are not good candidates to form cluster-assembled structures, and that the concept of magic cluster is only meaningful after considering the effect of the environment the cluster is exposed to.

In some other cases, the building block itself is an anionic cluster, which must be surrounded by counterions in order to achieve the electrical neutrality. In these cases counterions may play an important coating role, stabilizing the cluster-assembled solids. Alonso et al. [33] studied the requirements that the clusters must fulfill to ensure a successful assembly, apart from the need of the high intrinsic stability of the monomers. With this purpose they analyzed existing PbA (A = alkali metal) crystals formed by tetrahedral Pb$_4^{4-}$ anionic nanoclusters separated by each other by alkali cations. Surprisingly, the cluster with the largest HOMO-LUMO gap in this family, namely, Pb$_4$Li$_4$, is the only one that does not form a cluster solid. The reason is that lithium, being the smallest alkali metal, does not prevent a strong cluster–cluster interaction, so the tetrahedron opens up forming a butterfly-like structure. The rest of the compounds (Pb$_4$Na$_4$, Pb$_4$K$_4$, Pb$_4$Rb$_4$, and Pb$_4$Cs$_4$), conversely, do form a cluster-assembled solid due to the coating effect of the large-sized alkali cations.

4.1.3 Properties of the Resulting Solids

The 3D assembling of nanoclusters broadens the possibility of finding novel solids whose properties differ significantly from the traditional ones. The importance of finding diverse polymorphs arises mainly because of their different properties (packing, thermodynamics,

Chapter 4

spectroscopy, kinetics-like dissolution rate, stability, and mechanical properties), which give rise to various applications. McCrone stated that "every compound has different polymorphic forms and, in general, the number of forms known for a given compound is proportional to the time and money spent in research on that compound" [34]. In agreement with this assertion, Zwijnenburg et al. [35] showed recently that for MX compounds (AgI, ZnO, ZnS, CdS, GaN, GaP, and SiC), there exists a dense spectrum of as yet undiscovered polymorphs, the majority of which lie only moderately higher in energy than the experimentally observed phases. Furthermore, they suggested that this is probably a general phenomenon for all the inorganic solids. Later [36], they explored the chemical and structural analogy between nanoporous zeolites and cosubstituted binary MX materials (on ZnO-derived compounds mainly) and found numerous low-energy structures.

Typical example of polymorphism are the carbon allotropes, graphite and diamond, which, despite having the same chemical composition, possess very different properties and therefore have been found suitable for very diverse applications. Additionally, recent developments in nanoscience and nanotechnology have made possible the synthesis of the fullerite solid made of C_{60} building blocks. Besides its outstanding mechanical strength, comparable to diamond, it possesses intrinsic semiconductor properties unlike diamond (a wide band gap semiconductor) and graphite (a semimetal), showing great prospects for practical applications. This example illustrates the wide range of possibilities that nanoscience opened in the design of new materials, as was first envisioned by Feynmann in 1959: "There is plenty of room at the bottom." In fact, fullerene-based materials have provided both proof of principle and inspiration for the development of further cluster-assembled materials [37].

There may be some advantages in materials formed via the assembly of clusters rather than employing the ones simply produced using the elements themselves [12]. For instance, it may be possible to synthesize materials with more than a single functionality, like detection followed by destruction. Besides, cluster-assembled materials will differ from the corresponding atom-assembled solids in several basic aspects [29]. First, in the cluster-assembled materials, there may be two or more different chemical bond types, one between the atoms forming the clusters and another one between atoms of different clusters. This is exactly what happens in the fullerite crystals: carbon atoms belonging to the same nanocluster are linked together by strong covalent bonds but the clusters are held together with van der Waals forces. Accordingly, in the cluster-assembled materials there are two length scales, namely, intracluster and intercluster scales, while in the traditional solids, there is just one length scale setup by their lattice constant. On the other hand, in the cluster-assembled materials the energy bands arise from the overlap between molecular orbitals. Conversely, the energy bands in the atom-assembled compounds are due to the overlap between atomic orbitals. Together with this fact, in the materials assembled from clusters, there are two kinds of vibrational modes (intra- and intercluster modes), whereas in the traditional solids, vibrations of the atoms lead to acoustic and phonon modes. All these basic differences will provide cluster-assembled materials novel optical, chemical, magnetic, electrical, thermal, and chemical properties.

The characteristic high surface-to-volume ratio of nanoclusters will give rise to nanoporous solids. This is important since the presence of pores in a material can render itself all sorts of useful properties that nonporous materials would not have. The surface area of a solid increases when it becomes nanoporous and, as a consequence, it improves catalytic, absorbent, and adsorbent properties, making them suitable for applications such as photovoltaic solar cells [38], biosensors [39,40], molecular sieves [41–43], and heterogeneous catalysis [44,45]. Moreover, nanoporous solids are attractive materials for energetically efficient and environmentally friendly catalytic and adsorption separation processes [46]. They are highly useful in industries for removal of SO_2 or NO_x emission as well as for energy storage and environmental separation technologies [47]. The size, shape, volume, and uniformity in the distribution of the pores directly relates to the ability of the bulk to perform the desired function in a particular application [48]. Prototypical porous materials are zeolites, which are mainly, although not only, alumina-silica crystalline materials widely used in industry. A large variety of zeolite architecture exists [49] with channels and cages within the micropore size, and a wide range of interconnected pore widths and shapes is possible.

4.1.4   Metastability

Another important feature of cluster-assembled materials is that they are, in general, metastable. At this point, the theoretical calculations constitute an important predictive tool for the identification of energetically feasible structures. Then, as Jena pointed out [10], "what is really needed for a paradigm shift in synthesizing cluster assembled materials is finding clever techniques where 'bottom up' design meets with 'top down' synthesis."

It appears valuable to draw attention to the concept of metastability and underline its importance. When analyzing the potential energy surface of a system, different minima will be found. The energetically most stable minimum is called the global minimum, that is, the most stable geometrical arrangement for that particular system. All other minima are called local minima and their stability depends on the energy barriers that separate these minima from others. At high temperatures, the system might have enough energy to overcome these barriers, and eventually it would end up into the global minimum arrangement. However, at lower temperatures some barriers would be large enough as to prevent this transition. Then, these local minima arrangements become stable for longer periods, that is, they are metastable (recall that lifetimes are related to energy barriers). From this point of view, *stable* stands for the structural arrangement that remains unaltered under any temperature, while metastable stands for structures that are stable for a period of time under certain conditions. In other words, *stable* is connected to thermodynamical stability, while *metastable* to kinetic stability. Therefore, kinetic control of processes such as chemical reactions and solid formation can lead to metastable structures while thermodynamical control will lead to the stable structure. Both stable and metastable structures may be found in nature. Considering the solid phases of carbon at standard conditions, graphite is the thermodynamical end product. Thus, graphite is the global minima because it is the most stable geometrical arrangement under standard conditions. Oppositely, diamond and fullerites are local minima of the global energy surface but their lifetimes are long enough for their technological applications.

The role of the kinetic reaction control in organic and bioorganic processes is obvious. Otherwise, at room temperature without kinetic control, the molecules of our body would end up into the thermodynamically most stable products. Kinetic control, therefore, plays a very important role in the presence of life in the Earth. In the case of solid-state synthesis, however, the high temperatures and the long reaction times make the

Chapter 4

thermodynamical control preferred over the kinetic control [50]. Moreover, the energy barriers between different phases are much smaller in inorganic compounds and, in general, solid structures have been thought to have one or few polymorphs, the experimental search for different new metastable polymorphs being scarce [51]. Nevertheless, depending on the temperature and the energy barriers that should be gained, metastable arrangements could be sufficiently long lived for their experimental detection, or even further, for their technological applications. In this respect, Feldman claims [50] that chemist should carefully design the synthesis strategy in order to reach metastable solids.

However, it is notable that new experimental techniques are being developed in the last years in order to produce metastable solids, for instance, the triaxial tensile strength (negative pressure). LiBr, LiCl, and LiI [52–54] solids were synthesized in their metastable wurtzite structure. This arrangement can be formed by depositing the elements (Li and halogens) atom by atom onto a substrate at low temperature. In this way, conditions of negative pressure via lattice mismatch effects are created. Other methods that may be used to obtain metastable polymorphs are based on a drumbbell shaped structure directing agents [55] which can be assembled or dissassembled via the amphoteric template strategy [56].

Summarizing, cluster-assembled materials are of great importance due to the promising novel properties that they provide, chiefly to face the enormous future challenges like ecological benignity or energetic efficiency. Fullerene-based materials have provided both the proof of principle and the inspiration for the development of further cluster-assembled materials. The new solids will have basic differences with respect to atom-assembled materials and besides, they could be tailored by changing the nature of the nanocluster (size and composition). Since the resulting solids will be metastable, although with long enough lifetimes, computation and experiment should go hand in hand to achieve this amazing goal.

The rest of this chapter is organized as follows. In Section 4.2, a general overview of the available methods for studying cluster-assembled materials may be read. Then, in Section 4.3, we discuss the structures and metastability (if possible) of a large number of binary and ternary inorganic compounds. Finally, in Section 4.4, the general conclusions are summarized.

4.2 General Methods

In this section, a general overview of the different methodologies used to characterize cluster-assembled materials is presented. Concretely, the general procedures for predicting new metastable structures will be reviewed, along with methods to study the metastability of such structures.

First of all, we will focus on two different ways of approaching the assembly of clusters in 3D. On the one hand, the bottom-up approach which consists in increasing systematically the number of clusters of the assembly will be discussed. This might be seen as building the assembly brick by brick (see Figure 4.2). For larger enough assemblies, its properties should be sufficiently converged to those of its corresponding solid phase. On the other hand, solid-state physics techniques that rely on using periodic boundary conditions on a properly selected unit cell will also be reviewed. Finally, we will focus on methods to assess the metastability of the various polymorphs of interest.

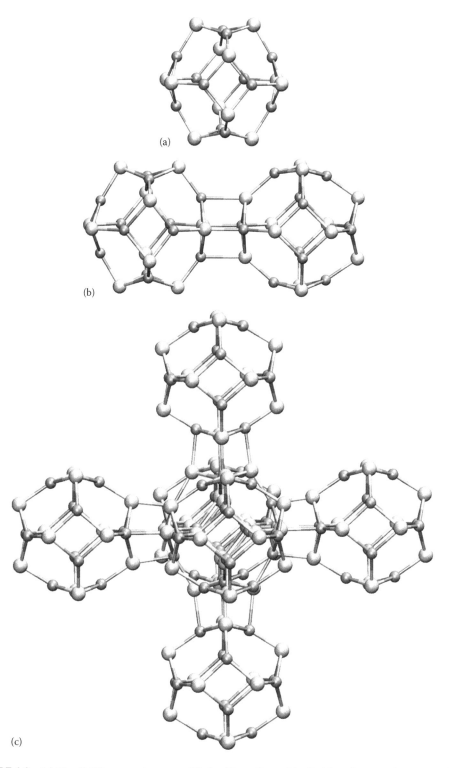

(a)

(b)

(c)

FIGURE 4.2 (a) The $(MX)_{12}$ cage structure, (b) the dimer formed by linking the cages by squares (also the unit cell for PBC calculations), and (c) a cluster of seven assembled cages (example for CA). M and X atoms light and dark, respectively.

4.2.1   Modeling 3D Cluster-Assembled Materials

There are two main approaches to model cluster-assembled solid structures: (1) by using periodic boundary conditions (PBCs or periodic, hereafter) and (2) by using the so-called cluster approach (CA, hereafter). In the first approach, the infinite solid structure is modeled by the use of a repeating unit cell, which is replicated all along the three dimensions by the use of appropriate translational vectors that define the repetition pattern along the three dimensions. The resulting structure will be a cluster-assembled solid with *infinite* clusters. In the CA approach, the assembly of clusters is studied by systematically adding more clusters to the finite initial seed structure. In other words, a cluster of clusters is built. When a sufficiently large number of clusters are included, both the CA and the PBC approach would lead to the similarly converged results. It is worth mentioning that this limit is not achieved by the CA, since it is not possible to add an infinite number of clusters. Hence, both methods are complementary. For instance, PBC can be used to model the final solid, while CA may be used to model the formation of the solid.

Let us visualize this with an example: A solid structure built by the assembly of $(MX)_{12}$ cage structures. First of all, let us focus on the structure of this cluster (see Figure 4.2). This is a cagelike hollow cluster, very close to a truncated icosahedron. It is formed by eight hexagons and six squares, two of each located perpendicular to each of the three coordinate axes, that is, the three Cartesian axes. A simple way of envisioning a possible assembly of such structures is by linking clusters along all three Cartesian coordinates, leading to a cubic structure.

In the CA the assembly is analyzed by adding one more $(MX)_{12}$ cluster at a time on all three x, y, z coordinate directions. In Figure 4.2, the assembly of two and seven $(MX)_{12}$ cages (six bonded to a central cage) may be observed. In this way, starting structures for further geometry optimization calculations are generated. According to Figure 4.2 and as shown in different studies for binary compounds, for instance, in [57], the link between cages is realized by the formation of new M–X bonds (not the less stable M–M and X–X bonds). In order to choose a proper unit cell for PBC calculations, therefore, two cages must be used. Otherwise, by the simple translation of one cage structure, the resulting solid would be formed by the energetically less stable M–M and X–X linked cages. Therefore, the unit cell must be a $(MX)_{12}$ dimer. Once the unit cell is chosen, the translational vectors are defined so that the intercluster M–X distances within the cell agree with the intercell M–X distances. In our case, the translational vectors necessary to build a cubic cell are $(0, a, a)$, $(a, 0, a)$, and $(a, a, 0)$. In this way, a starting geometry is built to perform the PBC calculations. Once these initial geometries are generated, the structure of the resulting solid must be optimized, both in the PBC and CA. Notice that in the CA, the optimized geometries are those corresponding to the whole cluster, while for the PBC calculations, in addition to the intracell geometries, the cell parameters (a, in our case) must be optimized as well.

There are two main approaches to perform such geometry optimizations: (1) local minima optimizations and (2) global minimum optimizations, labeled as LMOs and GMOs hereafter. In the first case, the minimum of the potential energy surface (PES or equivalent *energy landscape* hereafter), defined within the Born–Oppenheimer approximation [58], that is, located closest to the starting geometry, is found. Therefore, if one is interested in studying the whole PES in order to characterize the different polymorphs, one should generate different starting geometries. Since it is very unlikely to know a priori all important polymorphs (most still remain unknown), different techniques have been developed to generate a wide number of different structures. In principle, these techniques were developed to search for the lowest-lying minimum in the PES, named global minimum. Hence, these techniques are called GMOs and are useful to locate not only the global minimum but also additional local minima. Both approximations are briefly discussed in the next subsection.

4.2.2   Local Minima Geometry Optimizations

In order to perform local geometry optimizations, gradients of the PES are followed until stationary points (null gradients) are achieved. Usual methods are conjugate-gradients, Newton–Raphson, and so on, which are routinely implemented in standard software (see Reference 59 for more details). To calculate the PES, routinely, ab initio (first-principles) calculations are used, due to their accuracy. However, in order to save computation time, other less accurate methods, such as those employing potentials, can be used to model the M–X bonding, and in this way, classical or semiclassical calculations can be performed. In GMO, the use of ab initio methods may be prohibitive. Thus, the computationally more affordable (semi)classical methods can serve to generate starting geometries, which are reoptimized later by using more accurate first-principles methods. Therefore, ab initio methods will be considered in this section, while the others will be mentioned in Section 4.2.3.

First-principle, or ab initio, calculations are calculations based on quantum theory. Many ab initio methods have been developed since quantum theory was proposed in the 1920s, based on the wavefunction, electron density, and natural orbitals, among others. We refer the reader to Reference 60 for an overview of wavefunction and electron density methods and to Reference 61 for that of natural orbital-based methods.

In solid-state physics and chemistry, density functional theory (DFT) [62] shows an appropriate equilibrium between efficiency and accuracy. Therefore, this approach has become the most popular, in its Kohn–Sham formulation [63], and will be considered throughout this chapter. Nevertheless, the main drawback of this approach is the lack of a known universal exchange-correlation functional. Consequently, a huge number of approximate functionals have been developed, and, unfortunately, it is not possible to suggest a priori one functional for a given set of calculations. Due to the paramount number of functionals, we will consider only a small amount in this section. Concretely, some are widely used for calculating correct geometries and energetics, to predict accurately new structures and the relative energies between different polymorphs. Even in this case, the appropriate choice of the functional depends on the material under study. In this vein, in further subsections we will detail the functionals that have been used in each case.

Local density approximation (LDA) [64] is one of the simplest functional based on the homogeneous electron gas density. In the calculations considered in this chapter, the local density is far from being homogeneous. Therefore, functionals within the generalized gradient approximation (GGA) are recommended. Functionals such as the revised Perdew–Burke–Ernzerhof (rPBE) [65,66], the Perdew–Wang (PW91) [67], and B3LYP [68–70] have been routinely used for large cluster systems and solid-state calculations. Nevertheless, these functionals do not consider dispersion interactions properly. In the systems considered in this chapter, this is not a big deal, since dispersion effects are small in comparison to others. However, for calculations where these interactions are not negligible, dispersion effects must be considered. This can be done, for instance, including it empirically [71–73]. New families of functionals have been developed in recent years, such as meta-GGA functionals, and long-range functionals. They are not considered further in this chapter, but interested reader may found references in standard computational chemistry software manuals, for instance.

Other aspect that one must consider is that not all electrons of a given atom take part on chemical bonds. Indeed, the core electrons may be replaced by pseudopotentials. This reduces calculation time and expenses, with the prize of neglecting the effect of core electrons in the valence shell. Therefore, a balanced core–valence must be chosen. However, there are methods that consider this effect, like the plane augmented wave method [74], when using plane-wave basis set. Likewise, a large number of pseudopotentials may be found in the literature; for instance, the Troullier–Martins norm-conserving pseudopotentials [75]. For the valence electrons, a basis set to expand the valence electron density must be chosen. In finite systems, like molecules, or cluster of clusters (CA in this chapter), localized functions such as Gaussians have been widely used. In infinite systems, like those solids described with PBC calculations, delocalized functions like plane waves have been routinely used. In any case, both types of basis sets could be used. However, note that when using plane waves in finite systems, one should place the structure inside a large enough box, in order to neglect the intercell interactions. Most of the pseudopotentials include basis sets for the valence electrons, like the relativistic compact effective core potentials (ECP) and shared exponent basis sets of Stevens et al. (SKBJ) [76], Stuttgart/Dresden ECPs [77–79] and references therein, or Los Alamos ECPs plus DZ basis set (LAN-L2DZ) [80–82].

Finally, in the case of PBC calculations, it is also necessary to perform appropriate integrations in reciprocal space, to accurately describe the Brillouin zone. This is done, in general, by defining a Monkhorst–Pack grid [83] of special k-points, which are usually different depending on the system and the properties under study. The general rule is that the larger the unit cell is, the smaller the number of k-points needed, since they are defined in the reciprocal space. Likewise, the symmetry of the unit cell may be used to reduce the number of k-points. Once the structures are optimized, harmonic frequency calculations in the case of finite systems, or phonon calculations in the case of infinite systems, should be performed, in order to ensure that the optimized geometries correspond to a minimum and not to a higher order stationary point in the PES.

There are quite a lot of standard programs that can carry out all these kinds of calculations, for instance, Gaussian 09 package [84], SIESTA computer code [85], and Vienna ab-initio Simulation Program (VASP) [86–88].

4.2.3   Global Minimum Geometry Optimizations and Metastability

By the use of only LMO procedures, it is almost impossible to search for the whole *energy landscape* of a given solid. In other words, only few polymorphs can be studied, depending on the amount of initial structures the researchers may envision. A systematic way of generating a large number of initial structures is provided by GMO techniques. These procedures usually explore the PES or energy landscape in order to search for the lowest-lying polymorphs. However, in some cases the energy function may be substituted by other simpler function, based on bond lengths and coordination numbers, named cost function. Due to the large amount of evaluations that are done along the simulations, usually potential-based methods are used instead of ab initio procedures. Once the energy or cost function is defined, the GMO techniques are applied in order to generate different possible atomic configurations or geometries. These GMO approaches are, among others, genetic algorithm methods [89–92], simulating annealing [93,94], and threshold algorithm [95,96]. Other methods to generate possible structures are topological modeling methods [97–99] and structure models by analogy [100–102]. The candidate structures generated according to these procedures are further locally optimized on the full quantum chemical level using the aforementioned procedures. This multiscale approach has become a routine procedure, and, for more details, the reader is referred to [103].

In spite of the success of the strategy mentioned earlier in locating different minima of the *energy landscape* for a given material, it presents several problems. On one hand, some structures may be favored unrealistically, and on the other hand, they are not valid to check for the metastability of such structures. The metastability of a given arrangement of atoms is connected to the energy barriers that arrangement must gain to move to other minima, and the possibility of gaining these barriers is dependent on the temperature or pressure conditions. Therefore, in order to obtain a more complete picture of the energy landscape, that is, not only local minima but also the maxima connecting these minima, ab initio methods should be used. Moreover, the temperature effect may also be included by means of theoretical thermostats, for instance, that of Nose–Hoover [104], in a number of computational techniques, such as Monte Carlo simulations and molecular dynamics simulations.

A simple first approach to study the metastability of the predicted solid structures is the calculation of energy versus volume diagrams. In these diagrams (see Figure 4.3),

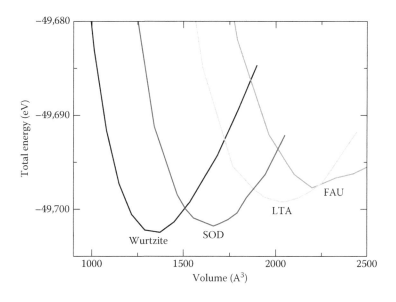

FIGURE 4.3 Energy versus volume for three $(CdS)_{12}$-assembled polymorphs compared to wurtzite. (Taken from Jimenez-Izal, E. et al., *Phys. Chem. Chem. Phys.*, 2012, 14, 9676.)

Chapter 4

the energy of a given polymorph versus the cell volume is depicted. In this way, the minima corresponding to different polymorphs are located with respect to their cell volumes, and in addition, the crossing of such lines gives an indication about the energy barrier necessary to move from one polymorph to other. In the example of Figure 4.3 (taken from Reference 105), it is observed that the $(CdS)_{12}$-assembled sodalite solid polymorph could be a metastable phase. However, a large number of polymorphs should be calculated in order to achieve an accurate picture of the PES. Even in this case, energy to volume calculations are not accurate in the calculation of the energy barriers, and therefore, in order to calculate more accurately the transition from one polymorph to others, other methods are necessary. However, these kinds of calculations give information about the stability of a certain bulk structure toward compression and expansion.

Fortunately, the paramount increase in the computational capabilities has led to the development and implementation of new approaches that make use of ab initio methods combined with GMO techniques like genetic algorithm methods. In this vein, such methods allow, for instance, the prediction of new polymorphs and the study of the metastability of these polymorphs at different temperatures and the connection between different polymorphs, namely, to study how the phase transition between these connected polymorphs occur. In finite systems, like in the aforementioned CA approach, these methods allow, in principle, for locating transition states between different conformations.

Furthermore, a number of different old and new techniques have been implemented quite recently, combined with the ab initio DFT-based energy calculations: Monte Carlo simulated annealing, molecular dynamics (Car–Parrinello algorithm [106] is probably the most known but not the only one), genetic algorithm methods, Basin [107,108] and Minima [109] hopping, metadynamics [110], and others. For more detailed description of these methods and some applications in solid state, we refer the readers to Reference 111 and references therein.

These kinds of approaches are relatively new, and few studies have been carried out concerning or including cluster-assembled materials. For instance, Zagorac et al. studied several polymorphs of ZnO and the PES connection between them by using prescribed path method [112]. Ab initio molecular dynamics have been widely used in studying the metastability of different finite systems, but few studies on cluster-assembled solids have been carried out [57].

4.3 Nanocluster-Assembled Compounds

To characterize cluster-assembled materials, it is convenient to follow some basic steps. The first one is to unveil the structure of the monomer, namely, the bare cluster. Although metastable clusters can be assembled into solids, the ground state of the structure should be taken into account. The first step to evaluate if a cluster could be assembled or not should be to look at dimers, in order to know the nature of the cluster–cluster interaction and to elucidate if the nanoclusters are stable enough to retain their structure or, otherwise, if they coalesce. Last, the bulk must be characterized and the energetics and structural stability with respect to the existing phases should be analyzed. When bringing binary nanoclusters together, it is important to ensure the formation of the favorable X–M interactions between nanoclusters, in opposition to less favorable X–X and M–M interaction. As a consequence, the unit cell should be defined at least by the dimer, instead of using just the monomer. In this way, one will warrant that the monomer will be a low-energy kinetically stable cluster and that the intercluster interaction (being energetically favorable) will allow the bulk formation without losing the structural integrity of the monomer.

4.3.1   Nanoclusters of 1:1 Stoichiometry

In the field of cluster-assembled materials, hitherto the cluster that has been most used as building unit is the $(MX)_{12}$ one, which usually exhibits exceptional stability for many chemical compositions. The ground state for a variety of $(MX)_{12}$ nanoclusters made of different elements has been found to be a spheroidal cage composed of eight hexagons and six squares, having the T_h symmetry, as it is shown in Figure 4.4. The fact that it is so symmetric makes the assembly easier to be thought and modeled, so it has been used as a starting point in many theoretical works for cluster-assembled solids. In this subsection, we will review the assembly of nanoclusters with 1:1 stoichiometry of ionic nature like alkali halides, as well as those made of II–VI and III–V semiconductor elements, among others.

The assembly of different alkali halides was studied by Bromley et al. [113] using $(MX)_{12}$ compounds, with M = Li, Na, K, Rb, and Cs and X = F, Cl, Br, and I, as monomers. The calculations were computed using the plane-wave-based VASP code [86] with the PW91 functional [114,115]. First, the relative energy of different isomers of $(MX)_{12}$ was analyzed. For $(LiX)_{12}$ series (X = F, Cl, Br, I), the cagelike structure is more stable, while for $(NaX)_{12}$, $(KX)_{12}$, $(RbX)_{12}$, and $(CsX)_{12}$ nanoclusters, the slablike structure is the most stable isomer. In Figure 4.5, both structures are depicted. Regarding the periodic compounds, two phases were studied: the rock salt, rs-MX, and the one that is reminiscent of the sodalite zeolite, SOD-MX (see Figure 4.6). The former is the energetically most stable polymorph and can be thought of as arising from the assembly of the slablike $(MX)_{12}$ isomer, where all atoms have octahedral coordination. On the other hand, the later results from the assembly of cagelike $(MX)_{12}$ clusters by square links, in such a way that an equivalent empty cage is created. This polymorph is nanoporous and much less dense than rs-MX. While the ground state of the studied $(MX)_{12}$ nanoclusters is determined by the alkali atom, for all the bulk structures, SOD phase is higher in energy than the corresponding rs-MX, although the energy difference between both phases decreases for the LiX series. The observed trends were explained in terms of

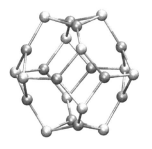

FIGURE 4.4 $(MX)_{12}$ spheroidal cluster.

Chapter 4

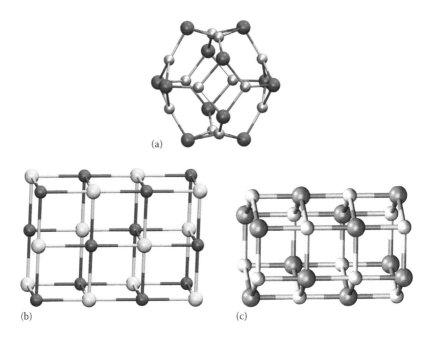

FIGURE 4.5 The ground state of different $(MX)_{12}$ nanoclusters. (a) The cagelike $(ZnO)_{12}$ (Zn and O atoms in blue and red, respectively), (b) the slablike $(NaCl)_{12}$ (Na and Cl atoms in red and green, respectively), and (c) hexagonal tubelike $(MgO)_{12}$ structures (Mg and O atoms in orange and green, respectively).

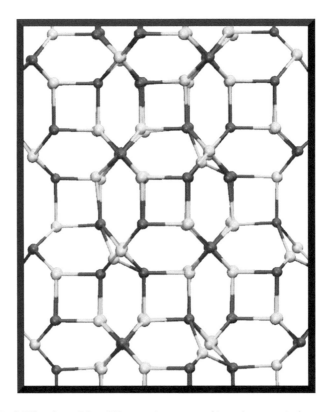

FIGURE 4.6 SOD-$(MX)_{12}$ phase. M and X atoms in green and in red, respectively.

ionic packing, where the ratio of ionic radii between cations and anions plays an important role. Nevertheless, in all alkali halides SOD phase was proven to be stable toward compression and expansion at 0 K by energy versus volume diagrams. Moreover, after performing classical MD simulations at 300 K for the case of SOD-LiF, it was confirmed that this bulk structure is thermally stable, showing that its synthesis, in spite of being challenging, might be a realistic target.

The possible assembly of $(MgO)_{12}$ and $(ZnO)_{12}$ was also theoretically addressed [116] using PW91 exchange-correlation functional [114,115] as implemented in the VASP code [86]. Under ambient conditions, MgO adopts the rock salt dense structure, while ZnS's ground state is the wurtzite one. For $(ZnO)_{12}$ the most stable isomer was predicted to be the cagelike structure shown in Figure 4.5. The lowest energy structure of $(MgO)_{12}$, conversely, was found to be a hexagonal tube, also depicted in Figure 4.5, although it is almost degenerate with the cagelike isomer. It is interesting to note that the 5–5 phase, whose density is between that of rock salt and wurtzite phases, can be viewed as arising from the assembly of the hexagonal tube structure. However, it is known that the most energetically stable cluster might not be the most abundant in experiments but a competing metastable isomer kinetically trapped by a significant energy barrier. And that's it so, as both the cage-to-tube and tube-to-cage transitions were predicted to have a high-energy barrier of ~0.8 eV. So, this pioneer work suggested the possibility of assembling the cagelike $(MgO)_{12}$ and $(ZnO)_{12}$ clusters in three diverse ways, thus yielding three new crystalline magnesium oxide and three new crystalline zinc oxide phases. Since the cagelike $(MX)_{12}$ structure is composed of squares and hexagons, three orientations were taking into account: via edge to edge interactions, by square–square interactions, and by bonding monomers via hexagonal faces. The interaction between clusters forming the dimers has been found to be barrierless and even energetically favorable. The dimers linked in these three ways are shown in Figure 4.7 together with the bulk phases that are yielded by their assembly. The obtained polymorphs resemble sodalite (SOD), linde type A (LTA), and faujasite (FAU) type zeolites, respectively [49]. In FAU phase, each cluster is linked to other four, in LTA solid clusters are bonded to six cages and in SOD solids monomers are attached to other eight building units. The latter has a tetrahedral lattice framework, while the remaining two are cubic phases. In the characterized phases, atoms are four-coordinated unlike in the MgO–rock salt and ZnO–wurtzite ground-state structures. The new solids are low-density nanoporous solids. The calculated total energy versus volume diagrams points out that the new phases are stable toward compression and expansion with respect to other phases (like the most stable rock salt one) as well as with respect to each other. Moreover, conversion between different phases would be difficult due to the predicted high-energy barriers. Interestingly, a later theoretical analysis of H atom in the SOD-ZnO showed that it would be able to transport H atoms with relatively small energy barriers [117].

Woodley et al. [118,119] studied different bulk structures based on $(SiC)_n$ and $(ZnO)_n$ clusters' assembly. They use interatomic potentials within the GULP software [120]. First, the ground-state structure of $(SiC)_n$ clusters was predicted, which coincides with the already established lowest-energy structures of $(ZnO)_n$ compounds. They focus on the assembly of $(SiC)_n$ and $(ZnO)_n$, $n = 12, 16, 24, 28, 36, 48,$ and 64 structures because these are the first eight octahedral nanoclusters. In these clusters, tetragonal and octagonal faces can be seen as octahedral vertices (see Figure 4.8) and the solids were built up by corner sharing the octahedra. They constructed a wide variety of nano- and microporous cubic phases considering a number of possibilities, like monomers linked by cluster–cluster chemical bonds or by merging the atoms in the neighboring cluster. Besides, tetragonal and octagonal rings were inserted between clusters as space fillers, which gave rise to larger pores. It was also taken into consideration constructing different phases by mixing diverse monomers. It should be noted that in several cases, it was needed to resort to the use of different enantiomers of the same stoichiometry to be able to construct some of these phases. These works show the great chances that the use of nanoclusters as building blocks offers yielding bulk materials with different physicochemical properties as the pore size, density, or lattice energy.

ZnO cluster-assembled materials, along with other materials build up from nanoclusters of II–VI elements, have been widely studied. Song et al. [121] studied the possible assembly of $(ZnO)_{12}$ nanoclusters. Calculations were carried out using the GGA-PBE [65] approximation as implemented in the DMOL code [122,123]. First, the lowest-lying structure was determined to be the cagelike one. The study of dimers reveals that the interaction via hexagons was energetically preferred over other coalescences. However, the procedure used to yield the cluster-assembled solid was slightly different: the growth pattern of $(ZnO)_{12}$ was studied by adding clusters one by one on hexagonal sites. Then, the least stable structures were rejected. In this manner, for small $((ZnO)_{12})_n$ assemblies, 2D structure is favored. As the size of n increases, 3D structure becomes more stable. Moreover, it was observed that the more interactions between monomers, the

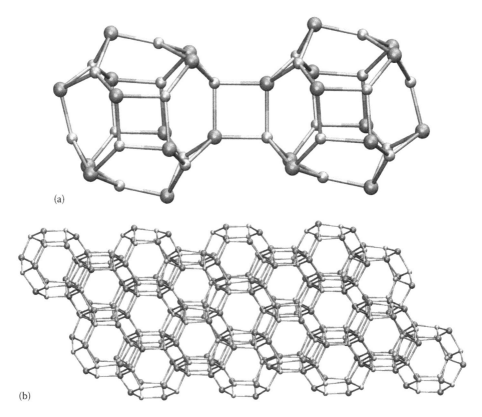

(a)

(b)

FIGURE 4.7 $(MgO)_{12}$ dimers and the corresponding bulk linked via edges (a and b). Lighter, smaller atoms are those of Mg. *(Continued)*

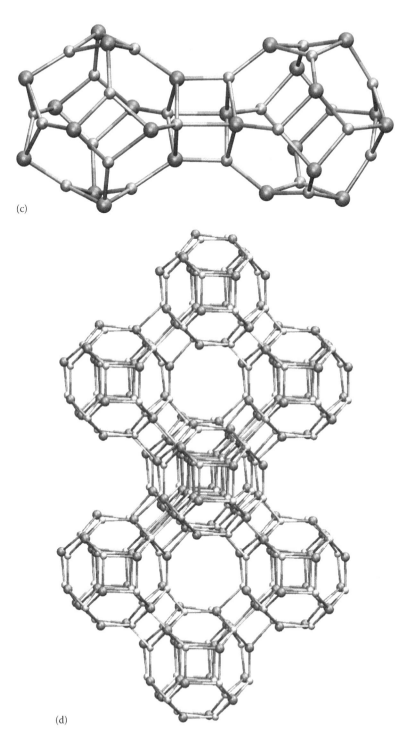

(c)

(d)

FIGURE 4.7 (*Continued*) $(MgO)_{12}$ dimers and the corresponding bulk linked via squares (c and d). Lighter, smaller atoms are those of Mg.   (*Continued*)

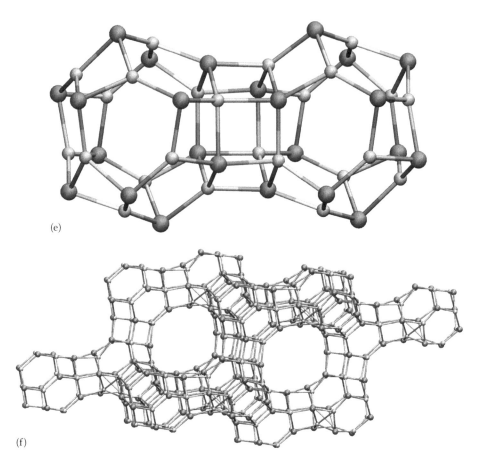

(e)

(f)

FIGURE 4.7 (Continued) $(MgO)_{12}$ dimers and the corresponding bulk linked via hexagons (e and f). Lighter, smaller atoms are those of Mg.

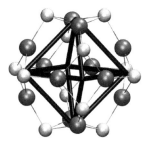

FIGURE 4.8 In black lines the octahedron superimposed on the cagelike $(SiC)_{12}$ nanocluster. Si and C atoms are the largest and smallest ones, respectively.

more stable the final structure is. Finally, a new ZnO phase was achieved, where each cluster is bonded to other eight counterparts on each hexagonal site. The new phase has a rhombohedral lattice and, therefore, it is designed as R-ZnO. Besides, it is predicted to be more stable than the corresponding FAU-ZnO and LTA-ZnO materials.

Recently, Liu et al. [124] focused on the assembly of not only the $(ZnO)_{12}$ nanocluster but also of $(ZnO)_{16}$ nanostructure. $(ZnO)_{16}$ was chosen because it has a hollow cagelike structure too and shows a remarkable stability, as $(ZnO)_{12}$ does [125]. It is composed of 12 hexagons and 6 squares and it has T_d symmetry. Initially, 18 types of links (three types in every one of face to face, face to edge, face to apex, edge to edge, edge to apex, or apex to apex) were considered for constructing dimers. Optimization of those dimers by using PBE [65] and the VASP code [86] lead to four stable low-lying $((ZnO)_{12})_2$ isomers, namely, two linked via hexagonal faces with D_3 and S_6 symmetry, respectively, one bonded by squares and the last one by edges. In the case of $((ZnO)_{16})_2$, three stable low-lying dimers were found: linked by hexagons, by squares, and by edges. The next step was to crack out all the possible multimers, also combining different types of link, but with three constrains: there was no $(ZnO)_{12}$ and $(ZnO)_{16}$ mixing, in every multimer there was a $(ZnO)_n$ cage located at the center, and a periodic system should be obtained through periodic translation of the central cage. After ruling out unstable multimers, the periodic frameworks corresponding to the stable multimers were constructed and fully optimized. Consequently, eight stable phases based on $(ZnO)_{12}$ and other three based on $(ZnO)_{16}$ nanostructures were reached. Some of them have been already mentioned in this text, that is, (SOD-$(ZnO)_{12}$, LTA-$(ZnO)_{12}$, and FAU-$(ZnO)_{12}$), but most of them are new ZnO phases, that is, EMT-$(ZnO)_{12}$ (the name is chosen because it resembles the EMT zeolite), monoclinic Mon-$(ZnO)_{12}$, trigonal Tri-$(ZnO)_{12}$, tetragonal tet-$(ZnO)_{12}$, orthorhombic Ort-$(ZnO)_{12}$, cubic Cub-$(ZnO)_{16}$, tetragonal Tet-$(ZnO)_{16}$, and orthorhombic Ort-$(ZnO)_{16}$. Note that EMT and FAU polymorphs are very similar. In both phases the monomers are linked via hexagonal phases. The difference arises from the different D_3 and S_6 bonds. Indeed, they are almost energetically degenerate phases. Additionally, the authors pointed out that the rhombohedral R-ZnO proposed by Song et al. [121] is not achievable through the bottom-up assembly mode, due to the collapse in the central monomer in the growth. Nevertheless, they propose that this phase can be obtained from other phases, such as SOD-ZnO, by pressure-induced transitions.

Although all of the new nanoporous low-density phases are metastable, through the total energy versus volume diagrams, some of them are found to be more stable than the already synthesized rs-ZnO material. In order to unveil the range of pressures required to stabilize these materials, the corresponding expression of enthalpy was derived from the EOS curves. It was concluded that under achievable negative pressures, most of the characterized solids can be stabilized. Additionally, the phonon density of states of the most stable ZnO phase, namely, SOD-$(ZnO)_{12}$, was computed and no imaginary phonon modes were found in all the Brillouin zone, indicating that the phase is thermodynamically stable. To further check its thermal stability, first-principles MD was performed on a supercell containing 144 atoms at 500 K. After 10 ps the structure remained unaltered, showing a high thermal stability.

Continuing with the solids based on nanoclusters made of II–VI elements, two interesting works have been focused on the assembly of endohedrally doped nanoclusters. Jimenez-Izal et al. [105] studied the possible assembly of $(CdS)_{12}$ and $(CdS)_{16}$ nanoclusters, which have the same spheroidal shape of their $(ZnO)_{12}$ and $(ZnO)_{16}$ counterparts. These two nanostructures were already found to trap both alkali metals and halogen atoms [126]. Moreover, electron affinities (EA) of endohedrally halogen doped clusters and ionization potentials (IE) of endohedrally alkali doped clusters were predicted to be very similar. The novelty of this work lies in the fact that, in addition to the study of the assembly of these bare nanoclusters, the assembly of the doped endohedrally doped nanoclusters was also considered. Both bare Cd_iS_i and endohedral $K@Cd_iS_i$–$X@Cd_iS_i$ (i = 12, 16, X = Cl, Br) clusters were assembled through three different orientations: edge to edge, square to square, and hexagon to hexagon. As a first step, the characterization of the dimers was made, and the endohedral doping was found to be a stabilizing factor, in spite of the monomers retaining their structure. The theoretical characterization of the solids was performed using the rPBE [65,66] functional as implemented in SIESTA code [85,127]. Using the dimers as starting point, six new crystalline bulk structures were characterized: SOD-$(CdS)_{12}$, LTA-$(CdS)_{12}$, FAU-$(CdS)_{12}$, SOD-$(CdS)_{16}$, LTA-$(CdS)_{16}$, and FAU-$(CdS)_{16}$. In contrast to solids made of $(CdS)_{12}$ building blocks, in all phases constructed by $(CdS)_{16}$ nanoclusters, there are square–square contacts due to the geometry of the monomer. All the resulting solids were predicted to be metastable, the most stable ones being the most compact materials. Moreover, it has been seen that the endohedral doping stabilizes those phases besides of having a direct effect on their electronic properties. The energetic ordering of the characterized phases is the same as for those built up by $(MgO)_{12}$ and $(ZnO)_{12}$ building units [116] and it does not change with the endohedral doping. We must emphasize that although solids made of $(CdS)_{16}$ do not resemble sodalite, linde zeolite A, and faujasite zeolites, they are identified with these acronyms for consistency, since they are linked by edges, squares, and hexagons, respectively. Notice that LTA-$(CdS)_{16}$ is called Cub-$(ZnO)_{16}$ in the previously mentioned work by Liu et al. [124].

$Zn_{12}S_{12}$ nanoclusters were also found to encapsulate several alkali metals and halogens [128]. Thus, Matxain et al. [57] focused on the assembling of bare $Zn_{12}S_{12}$ and endohedral $X@Zn_{12}S_{12}$–$Y@Zn_{12}S_{12}$ dimers (X = Na, K; Y = Cl, Br) by considering the square-faces-square orientation of every two adjacent clusters, which leads to an fcc cubic LTA-ZnS crystal structure. However, all attempts to characterize $K@Zn_{12}S_{12}$–$Cl@Zn_{12}S_{12}$ solids failed. The calculations were performed with rPBE [65,66] and SIESTA package [85,127]. The stabilizing effect of the doping, together with the change in the electronic properties, was also observed in this work. Additionally, quantum MD were performed at 298 K for two selected cases, that is, LTA-$(ZnS)_{12}$ and LTA-$Na@(ZnS)_{12}$–$Cl@(ZnS)_{12}$, and their thermal stability was confirmed. It is worth noting that solids based on endohedrally doped nanoclusters of other elements have been theoretically characterized [129,130]. The case of silicon is especially interesting, as doping has emerged as a valuable mechanism to stabilize Si fullerenes. For more information, the reader can go to References 131 and 132 inter alia.

Compounds III–V combine an element from the group III (i.e., B, Al, Ga, In) with an element from the group V (i.e., N, P, As, Sb). Boron nitride is probably the most widely exploited III–V material [133]. It is isoelectronic with carbon compounds and thus displays a similar chemical bond. As a consequence, bulk BN can adopt two phases, cubic and hexagonal, as carbon does (graphite and diamond, respectively). Several works had dealt with the assembly III–V of nanoclusters. A remarkable example is the finding made by Pokropivny [134]. Indeed, he unveiled the structure of the BN E-phase, known from 1965. The comparison between experimental and theoretical data clearly suggested that this material was made of $(BN)_{12}$ cagelike clusters, copolymerized by hexagonal faces. Thus, this framework of cubic lattice can be seen as FAU-$(BN)_{12}$ due to the resemblance with the faujasite zeolite. The existence of these solids is a support to the theoretical predictions based on the assembly of nanostructures, even if its synthesis was not that of bottom-up.

The existence of other periodic materials built up with $(BN)_{12}$ nanoclusters has been hypothesized too [135]. Specifically, a solid where nanoclusters are linked via squares (LTA-$(BN)_{12}$) with covalent and van der Waals type of bonding was studied using the PBE functional [65] within SIESTA code [85,127]. Both phases have face-centered cubic lattice. The main difference between both solids, as expected, is the intercluster distance. Although it was possible to characterize the two distinct solids, the study of the dimers shows that the process to move from the van der Waals to the covalent dimer is barrierless. This fact indicates that the relative stability of the van der Waals solid is small and it will transform into the covalent phase.

In addition, Alexandre et al. [136] analyzed the possibility of assembling the stoichiometric $(BN)_{16}$ and $(BN)_{36}$ and the nonstoichiometric $B_{12}N_{16}$ nanoclusters, with the GGA approximation and SIESTA program [85,127]. The three of them have the cagelike structure, with T_d symmetry, but whereas the two stoichiometric boron nitride clusters are composed of squares and hexagons, $B_{12}N_{16}$ is composed of pentagons and hexagons. In the case of $(BN)_{16}$ and $(BN)_{36}$ nanoclusters, the coalesce is via square covalent bonds, yielding cubic lattice frameworks. Likewise, $B_{12}N_{16}$ compounds are predicted to form a solid with a diamond structure by linking the building units by hexagonal faces.

Research on the assembly of other III–V nanoclusters is scarce. The obtaining of solids using $(MN)_{12}$, where M = Al, Ga, as building blocks was recently proposed [137]. In this work the gas phase calculations were carried out using the PBE [65] functional within DMol3 code [122,123], while the plane-wave pseudopotential DFT VASP [86] code was used for solid-state calculations. First, the stability of these cagelike structures was proven by using ab initio MD at 300 K. Then all the possible dimers were taken into consideration and once again, the lowest-lying isomers were predicted to be those linked by edges, by squares, and by hexagons (one with D_3 symmetry and the other one with S_6). For $(AlN)_{12}$ and $(GaN)_{12}$ dimers, the most stable isomer is the one bonded via hexagons with S_6 symmetry, and not the one with D_3 symmetry, oppositely to what happens with $((ZnO)_{12})_2$ compounds. Nevertheless, in both cases the dimer structures are almost degenerate. After that, multimers were explored with the same constrains as in [116]. At the end eight solid phases were achieved. In four of them, namely, SOD-$(MlN)_{12}$, LTA-$(MlN)_{12}$, FAU-$(MlN)_{12}$, and Tri-$(MlN)_{12}$, there is just one type of interaction between cages. In the other four solids, (EMT-$(MlN)_{12}$, Ort-$(MlN)_{12}$, Mon-$(MlN)_{12}$, and Tet-$(MlN)_{12}$), conversely, there are two or three different type of bonds. The same energetic ordering is found for AlN and GaN phases. Comparing with ZnO phases having the same structures [124], the energetic ordering is the same, with the exception of FAU (EMT) and trigonal polymorphs. The authors assert that most of these bulk nanoporous materials are achievable under triaxial tensile stresses.

Finally, Liu et al. [138] theoretically examine $(InAs)_n$ nanoclusters, where $n = 1$–15 and concluded that the cagelike $(InAs)_{12}$ is the most suitable one for the assembly due to its high stability and highly spheroidal shape. The subsequent study of dimers and trimers yields to the FAU-$(InAs)_{12}$ solid, where every cage is linked by four neighboring cages, with hexagon–hexagon interactions. The energy versus volume diagrams revealed that this phase, although metastable, is more stable than the already synthesized rs-AsIn.

So, all these characterized solids are zeolite-like low-density nanoporous, which might be very useful for applications such as atomic or molecular sieves or heterogeneous catalysis. It is important to note that, oppositely to what happens with the metallic nanoclusters that tend to coalesce, in all the cases mentioned in this subsection the structural integrity of the nanoclusters is predicted to be retained after assembling them into solids. Besides, there is a general trend regarding the stability: monomers are less stable than dimers, and dimers are less stable than solids. Likewise, it is observed that in general the density of the bulk is inversely proportional to its thermodynamic stability. Note that in most of the theoretical predictions of the $(MX)_{12}$ assembly, the SOD phase, although metastable, seems to possess a high stability. In this vein, Bromley et al. [139] calculated the transition pressures under which SOD polymorph may be obtained for a variety of ZnX, CdX, and MgX compounds, where X = O, S, Se, Te. SOD phase was found to be thermodynamically stable under conditions of negative pressure for all compounds. However, negative pressure required to obtain the SOD phase is more achievable for compounds with the biggest anions, that is, CdTe, MgTe, and CdSe.

4.3.2 Other Binary and Ternary Compounds

In this subsection, some examples of cluster-assembled solids of a stoichiometry different to 1:1 will be analyzed. First, we will focus on XO_2 (SiO_2, TiO_2, etc.) and other oxides. Then, brief highlights will be given regarding ionic (Al_{13}^- and As_7^{-3}), intermetallic (Mg_2Al_3 and others), weakly bonded (Ti_8C_{12} and others), and other covalently bonded ($B_{12}X_2$, X = O, As) cluster-assembled materials. In contrast to the structures of the previous subsection, not all the assembled clusters are cage structures.

4.3.2.1 XO_2 and Other Oxides

Among oxide compounds, in addition to those of stoichiometry 1:1 analyzed in previous subsection, the most studied are the XO_2 cluster-assembled solids. In 2004, Astala et al. [140] studied the structures, stabilities, and mechanical properties of a number of zeolite structures, namely, sodalite (SOD), silica LTA, chabalite (CHA), mordenite (MOR), and silicalite (MFT). In addition, they studied the R-and *â*-quartz, R- and *â*-cristobalite, and *â*-tridymite structures, with the sake of benchmarking and comparisons. They performed periodic supercells DFT calculations using the LDA functional, combined with pseudopotentials and plane-wave basis sets. According to their calculations, despite large differences on the geometries of the calculated structures, the cohesive energies of all structures were very similar, almost structure independent. They proposed that their results could be used for the development of models based on empirical potentials and to test their reliability. The paramount number and complexity of the hypothetical silica zeolite structures have led to other similar studies applying top-down methods based on mathematical tiling and graph theory [97,141–144], or other based on quantum theory [145]. At this point, it should be mentioned that the structure patterns generated in these studies are not only for SiO_2 materials, but also for others that are the basic components of zeolites, like $AlPO_4$ or $AlGaO_4$ [145,146]. However, for such materials, no bottom-up studies have been performed and will not be further analyzed throughout this chapter.

In addition to top-down approaches to study new hypothetical silica structures, bottom-up approaches have also been used, where well-defined small cluster structures are used, in order to predict cluster-assembled materials based on $(SiO_2)_n$ clusters. For instance, Wojdel et al. [147] studied the assembling of the magic $(SiO_2)_8$ cluster, characterized previously [148], to predict different super-(Tris)tetrahedral materials. In order to do so, authors characterized different crystal structures based on the different patterns generated after assembling the clusters via siloxane bridge forming reaction: $4[Si=O] \rightarrow 8[Si-O-Si]$.

The $(SiO_2)_8$ cluster may be encased in a tetrahedron or a tristetrahedron that leads to the different ways in which the cluster might be assembled, namely, ST or STT structures by linking from the four or eight vertexes of the tetrahedron and the tristetrahedron, respectively. All these structures were fully optimized based on DFT periodic calculations, using the PW91 functional [67] combined with core potentials and plane-wave basis sets, and suitable k-point meshes generated via the Monkhorst–Pack scheme [83], using the VASP code [86]. Interestingly, all calculated structures were thermodynamically less stable than known synthesized materials, but within an accessible window common to other mesoporous silicas. Specially, the STT1 structure was seen to lie lowest (21 kJ/mol above the R-quartz structure) among the studied frameworks, showing that the one-to-two construction modes for STT structures could lead to almost optimal assemblies of the $(SiO_2)_8$ cluster. Despite the fact that this kind of assembly do not lead to more stable materials, it motivated further theoretical searches for new viable materials employing other types of clusters as building blocks.

Concretely, Bromley [149] studied the viability of new inorganic polymorphs based on the cluster assembly of fully coordinated $(SiO_2)_{12}$ rings and $(SiO_2)_{24}$ cages. In previous studies, the structure [150,151] and stability [152,153], even at high temperatures, of these two silica clusters were examined. These two clusters may be seen as analogous of C_{12} ring and C_{24} fullerene, which may be assembled into cluster-based crystals. Following with this analogy, author showed that these silica clusters might also be assembled to form new silica polymorphs. In order to predict that, potential-based calculations were first performed, with potentials developed to predict correctly the experimental crystal structure of fibrous silica [154]. These potential-based calculations were carried out with the GULP code [155], considering the large amount of possible crystal packing generated with the MOLPAK [156] code. Then, the favored crystal structures were further optimized based on DFT periodic calculations, using the PW91 functional [67] combined with core potentials and plane-wave basis sets, using the VASP code [86]. The calculated lowest-lying crystal structures for the assembling of $(SiO_2)_{12}$ rings and $(SiO_2)_{24}$ cages lie slightly higher in energy compared to the bulk ground state. However, these calculations served to propose the idea of the viability of new crystal polymorphs based on discrete clusters that mimic the topology of carbon nanoclusters.

Related to SiO_2, there are other XY_2 compounds, such as TiO_2, CeO_2, $BeCl_2$, and SiS_2. These other materials could also lead to cluster-assembled materials, similar to SiO_2. In addition, new nanocluster-based materials could be predicted in the future, based on further calculations. As far as we know, the possibility of assembling clusters of these chemical compositions has not been explored yet. However, clusters of these compounds have been studied in the literature (see, for instance, [157,158]). The structure of these clusters, along with the already mentioned assembling patterns for SiO_2 or others, like those based on cubic Si_8O_{12} double four-ring units [146], makes promising the study of cluster-assembled patterns for these valence isoelectronic materials.

The assembly of clusters having different stoichiometries has not been theoretically studied, as far as we know. However, we should point out that recently new polyoxoniobates, with stoichiometries $(NbO_3)_{24}$, $(NbO_3)_{32}$, and $K_{12}(NbO_3)_{96}$, have been

Chapter 4

experimentally synthesized [159]. These and other polyoxometalates could be seen as the combination of smaller units, that is, clusters of clusters. Further theoretical studies of these or other related compounds would be now feasible due to the development of more accurate theoretical methods, as described in previous sections. The interested reader is referred to [159] for more details. In addition to this, it should be pointed out that the structures of other oxides, like Ca_3SiO_5 and WO_3, are very complex and not known yet [51]. Large and complex nanocluster-assembled zeolite-like structures as in the case of some intermetallic clusters [160,161], briefly analyzed in the next subsection, could be the answer. As for the latter, a new approach, based on graphs and tiling, might be used, called *nanocluster approach* by authors [160] and implemented in the TOPOS program package [162].

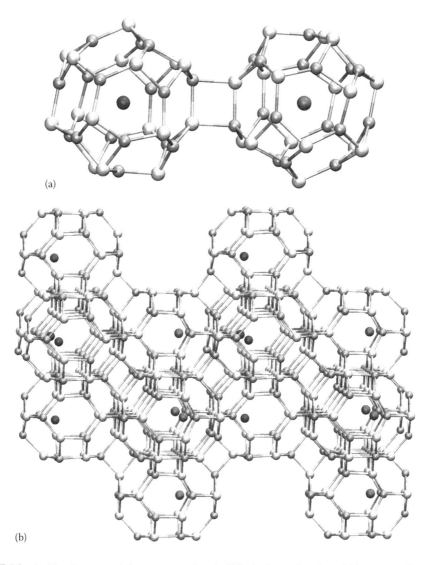

(a)

(b)

FIGURE 4.9 $(CdS)_{16}$ dimers and the corresponding bulk linked via edges (a and b). *(Continued)*

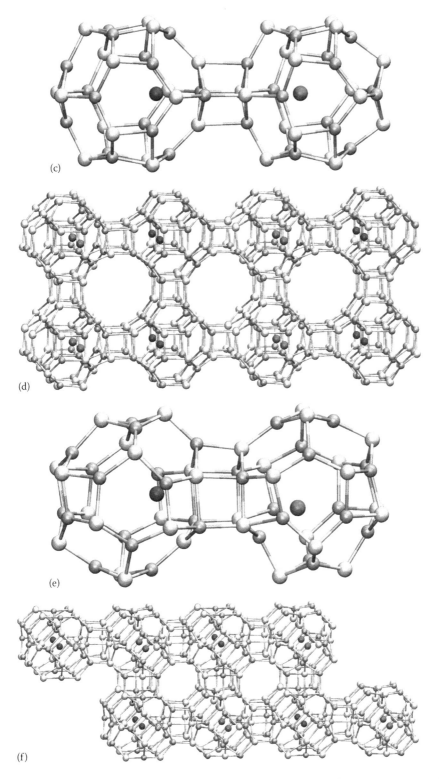

(c)

(d)

(e)

(f)

FIGURE 4.9 (*Continued*) (CdS)$_{16}$ dimers and the corresponding bulk linked via squares (c and d) and hexagons (e and f).

4.3.2.2 Other Interesting Binary and Ternary Cluster-Assembled Solids

We would like to end this chapter highlighting some studies regarding the cluster-assembled structures generated by the assembling of some other binary and ternary materials. Concretely, we will focus on the ionic assemblies based on Al_{13}^- (and related compounds) and As_7^{-3} clusters, the intermetallic assemblies like β, β'-Mg_2Al_3 and related compounds, and some cases that might be somehow related to assembly of doped elementary clusters, but that exhibit some characteristic features: weakly bonded X_8C_{12} and covalently bonded $B_{12}X_2$ compounds.

Many theoretical works have dealt with the assembly of aluminum clusters. Since Al_{13}^- nanocluster acts as a halogen and it bounds ionically to potassium, Liu et al. [8] proposed a CsCl-like crystal made of KAl_{13}. Their calculations suggested that a new metastable polymorph could be formed, where Al_{13} cluster became cuboctahedral instead of remaining icosahedral due to the crystal-field effect, despite it keeps its identity. Ashman et al. [163] observed that $B@Al_{12}^-$, being isoelectronic to Al_{13}^-, was the most stable cluster among different Al_nB_m structures and the most likely to form crystals with $(BAl_{12})Cs$ as monomer. They concluded that a metastable solid with either icosahedral or cuboctahedral $(BAl_{12})Cs$ units could be synthesized (see Figure 4.5). In the same vein, Gong confirmed the stability of both $Al_{12}C$ and $Al_{12}C(Si)$ solids [26]. Zi-zhong et al. [164] found $(Al_{12}B)Li$ solid to be a metallic crystal, where the $Al_{12}B$ cluster has cuboctahedral symmetry. These examples give an idea about how one can *play* with diverse variables, such as the size, composition, or oxidation state, to achieve highly stable structures that may be used to yield cluster solids.

Castleman et al. [165] proposed a route to produce such materials. The proposed pathway combines three steps: (1) identification of potential cluster building units in gas phase experiments, (2) theoretical characterization along with the bonding patterns that could guide the nature of the assembly, and (3) synthetic approach designed for the successful assembly using the information obtained from the previous steps. The reliability of this method was then proved with a zintl phase derivate. Zintl phases are naturally occurring cluster-assembled materials. They contain cations of metals and polyatomic cluster anions of nonmetal atoms, called zintl ions, such as Ge_9^{2-} or Sb_7^{3-}. In this case, the existing As_7K_3 zintl phase was studied following steps (1) and (2). It was revealed that As_7K_3 was a prominent stable cluster. Moreover, theoretical calculations showed that the cluster can be regarded as As_7^{3-} (see Figure 4.11), where the role of the potassium atoms was to donate charge. With this information, it was concluded that it could be possible to stabilize a new phase by changing the nature of the cation. The knowledge gained was then used to stabilize a new polymorph replacing some K atoms by cryptated species. Oppositely to As_7K_3 phase where As_7 units arrange in a compact structure with each As_7 surrounded by potassium atoms, the units in the new solid were linked just by single cation, since the cryptated potassium atoms stabilize the new phase as space fillers and via charge transfer. Moreover, it has been shown with similar systems that the choice of the counterion is another variable that can be used for tuning the properties of the final solids, as the clusters' size and composition are [22].

Some intermetallic binary and ternary compounds, like β,β'-Mg_2Al_3 and related compounds, have very complex crystal structures. Despite the effort of many researchers,

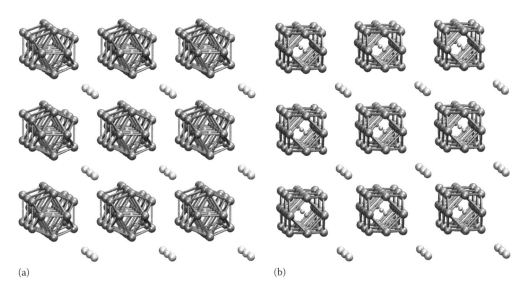

FIGURE 4.10 $Cs(B@Al_{12})$-based cluster-assembled solids. (a) Icosahedral structure and (b) cuboctahedral one.

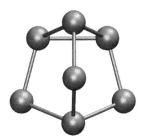

FIGURE 4.11   The global minimum structure As_7^{3-} nanocluster.

a common and well-accepted description of their 3D structures is still lacking [166]. During the decade of the 1960s, Samson [167–171] determined a number of extremely complicated crystal structures of $ZrZn_{22}$, $NaCd_2$, Cu_4Cd_3, and β-Mg_2Al_3, being the last three disordered and with variable composition. The complexity of these structures is related with the high amount of atoms, over 1000, in the unit cell. This amount of atoms is prohibitive for performing ab initio calculations, and, therefore, in order to elucidate their structures, other kind of approaches must be used. The previously mentioned *nanocluster approach* [159], based on graphs and tiling as implemented in the TOPOS [161] package, was used by Blatov et al. to study these structures and compare to the much simpler $MgCu_2$ structure.

They proposed that these complex crystal structures may be seen as nanocluster-assembled materials, where each nanocluster has an onion-like structure. A question arises at this point: are these individual nanoclusters stable at melt? Quantum chemical methods could help in finding the answer to this question. The increase of computational resources in the near future could also make feasible to perform these kinds of

Chapter 4

calculations in these materials, which would be desirable. Unfortunately, it is not possible yet. In addition to this, it was concluded that the *nanocluster approach provides a universal and objective approach to modeling of any crystal structure*. Indeed, this methodology has been applied very recently for the study of a great number of hypothetical and known zeolite structures [160]. This approach may bridge intermetallic structures and nanoalloys [172], cluster-assembled materials [173,174], and icosahedral quasicrystal approximants [175–177]. For more information regarding the structures of these intermetallic compounds and the proposed *nanocluster approach*, see [159,160] and the references therein.

In 1992, Castleman et al. discovered the Ti_8C_{12} [178], a highly stable transition-metal carbide cluster, which led to a new family of clusters called metallocarbohedrenes (Met-Cars). These clusters may be related to the smallest carbon fullerene, C_{20}, by the substitution of eight carbon atoms by eight titanium atoms. In the case of C_{20}, cluster-assembled materials have been already synthesized [179–181]. However, Ti_8C_{12} clusters, although stable, are very reactive [182] and therefore the cluster-assembled materials have not been synthesized yet. In order to search for less reactive materials, Berkdemir et al. [183] studied several metal-substituted MTi_7C_{12} and $M_2Ti_6C_{12}$ clusters, founding that one of the $Sc_2Ti_6C_{12}$ isomers was energetically more favorable, and with a much larger HOMO-LUMO gap than the Ti_8C_{12} species, which suggest its lower reactivity. In order to study the possibility of assembling the mentioned $Sc_2Ti_6C_{12}$, the interaction in the dimers of such compounds was studied and compared to the dimers of pure Ti_8C_{12} species. In order to do so, they perform spin-polarized DFT calculations using the PBE [65] functional combined with Vanderbilt ultrasoft pseudopotentials [184] was performed (with the exception of Mg, for which Troullier Martins [75] pseudopotentials were used) as well as plane-wave basis sets as implemented in the QUANTUM ESPRESSO program package [185]. It was concluded that the $Sc_2Ti_6C_{12}$ dimer binds significantly more weakly than Ti_8C_{12} dimers, and suggested that in order to avoid the reactivity problems encountered in the assembly of pure Ti-based Met-Cars, the $Sc_2Ti_6C_{12}$ species could be used.

The assembly of other Met-Car analogs has also been studied in the literature. For instance, Domingos studied several of the most stable cluster-assembled phases of X_8Y_{12}, X_8C_{12}, and YX_7C12 (X = B, N, Si) [186]. Concretely, the B_8N_{12}, B_8C_{12}, N_8B_{12}, N_8C_{12}, SiB_7C_{12}, SiN_7B_{12}, and BN_7C_{12} clusters assembled in the sc, bcc, fcc orthorhombic and triclinic lattices were analyzed. To study these solid structures, DFT calculations using the PW91 functional [67], combined with ultrasoft pseudopotentials [184] and plane-wave basis sets were performed considering PBCs, using the CASTEP [187] program. These calculations confirmed the high stability of several of the studied solids, showing for the first time that molecular solids of these clusters are possible.

The electron-deficient nature of B allows the formation of 3-centered bonds. Based on this, the assembling of $B_{12}X_2$ structures (X = O, P, As), where the X_2 is bonded exohedrally to the B_{12} icosahedra via one of these atoms, is possible. Beckel et al. [188] studied the lattice vibrations of $B_{12}As_2$ structures, which experimental preparation was described previously by Aselage et al. [189], by using a classical valence-force-field analysis, to account for the observed IR and Raman spectra. They identified the vibration modes that belong primarily to the B_{12} cluster and enlightened the effects of crystal forces and As atom interactions. Notice that at that time, available computational resources did not allow for the use of ab initio procedures to perform such analysis. Even using classical procedures, they were able to predict qualitatively the observed intensities. Recently, Wang et al. performed a similar study for the related $B_{12}O_2$ cluster-assembled structure [190] by means of DFT calculations. The calculated vibrational frequencies agreed well with the experimental data, and, compared to related $B_{12}As_2$ and $B_{12}P_2$ materials, they concluded that in $B_{12}O_2$, the intrachain bonding (O–O, compared to P–P and As–As) is weaker, due to the large O–O distance (3.1 Å). Based on these first-principles calculations, experimental data were properly interpreted.

4.4   General Conclusions

Since the C_{60} buckminsterfullerene was discovered, the nanotechnology and nanoscience have grown at a vibrating speed. First, the attention was focused on the nanostructures. The fact that nanomaterials usually possess properties totally different from those of their corresponding bulk materials revolutionized the chemistry. After the study of these nanocompounds, the next step is likely to be their assembly to yield new solids with distinct properties. The importance of nanoclusters as building blocks can hardly be underestimated, as it will enable us to create new solid phases by using rational design. Where experimental techniques are not yet at hand, computational theory has become an essential tool as a predictive basis for guiding experiment. Indeed, the improvement of computational methods and capabilities allows the prediction and the better understanding of new materials, avoiding, in some cases, the high cost of experimental random trial-and-error search.

In this chapter, a large number of predicted binary and ternary cluster-assembled compounds have been reviewed. Most of the assembled nanoclusters are cagelike structures, but not all of them. As_7 clusters, $(SiO_2)_8$ and $(SiO_2)_{12}$ are examples of this. This fact broadens the idea that cage, spherical structures are necessary to build up cluster-assembled structures. In addition and relation to the nature of the building blocks, ionic, covalent, metallic, or weakly bonded assemblies have been shown. Ionic assemblies include those of alkali halides; covalent, the compounds made of III–V and II–VI as $B_{12}X_2$; metallic ones, the intermetallic Mg_2Al_3 and related compounds; and finally weakly bonded assemblies, those based on $Sc_2Ti_6C_{12}$. Both facts, different cluster shapes and different assembling nature, exponentially increase the amount of possible metastable polymorphs of different materials. Likewise, many works have to be done in this field and a wide range of alternatives to construct to solid phases based on nanoclusters should be analyzed. Learning from the *traditional* atomic and molecular solids, as well as from the large amount of available zeolite patterns, could guide us to new strategies to achieve diverse solid materials. Meanwhile, in order to theoretically address the search for different polymorphs, several theoretical approaches have been developed. The main problems of these calculations are

- The difficulty of obtaining accurate description of the relative energies between large enough local minima (polymorphs) of the PES
- The difficulty of studying accurately the connection (change of phase) between these polymorphs

Chapter 4

The amazing increase of the computational resources occurred in the last years depicts an optimistic picture to overcome these issues. In fact, the possibility of using simulating annealing or genetic algorithm methods combined with ab initio methods would allow for a wider and more accurate knowledge of the PESs of these compounds. In addition, the chance of combining these ab initio methods with methods including temperature, such as molecular dynamics or metadynamics, would allow a more proper study of the metastability of the predicted minima. Moreover, recent promising techniques have been developed to study the connection between different polymorphs in the PES. All these improvements could lead to a more accurate understanding of the whole PES (energy landscape), both local minima (valleys) and transition states (saddle points).

Hopefully, all this theoretical effort will help experimentalists in the synthesis of new compounds. New experimental techniques are being developed in the last years in order to produce metastable polymorphs. However, as has been already pointed out by many authors, new experimental routes should be defined in order to successfully synthesize new solids based on the kinetic control. It is essential that this exciting and challenging journey is walked together, hand in hand, by both theoreticians and experimentalists for this scope to be successful.

Acknowledgments

JMM thanks the Spanish Ministry of Science and Innovation for funding this project through a Ramon y Cajal fellow position (RYC 2008-03216).

References

1. W. D. Knight, K. Clemenger, W. A. de Heer, W. A. Saunders, M. Y. Chou, M. L. Cohen, *Phys. Rev. Lett.* 1984, *52*, 2141.
2. H. W. Kroto, J. R. Heath, S. C. O'Brien, R. F. Curl, R. E. Smalley, *Nature* 1985, *318*, 162–163.
3. R. S. Berry, *J. Phys. Chem. A* 2013, *117*, 9401.
4. W. Kratschmer, L. D. Lamb, K. Fostiropoulos, D. R. Huffman, *Nature* 1990, *347*, 354.
5. S. N. Khanna, P. Jena, *Phys. Rev. Lett.* 1992, *69*, 1664.
6. S. N. Khanna, P. Jena, *Phys. Rev. B* 1995, *51*, 13705.
7. S. N. Khanna, P. Jena, *Chem. Phys. Lett.* 1994, *219*, 479.
8. F. Liu, M. Mostoller, T. Kaplan, S. N. Khanna P. Jena, *J. Chem. Phys. Lett.* 1996, *248*, 213.
9. B. K. Rao, S. N. Khanna, P. Jena, *J. Clust. Sci.* 1999, *10*, 477.
10. P. Jena, *J. Phys. Chem. Lett.* 2013, 4, 1432.
11. R. E. Leuchtner, A. C. Harms, A. W. Castleman, *J. Chem. Phys.* 1989, *91*, 2753.
12. A. W. Castleman Jr., *J. Phys. Chem. Lett.* 2011, *2*, 1062.
13. Y. Han, J. Jung, *J. Am. Chem. Soc.* 2008, *130*, 2.
14. B. Friedrich, *Found. Chem.* 2004, *6*, 117.
15. L. Cheng, J. Yang, *J. Chem. Phys.* 2013, *138*, 141101.
16. W. Ekardt, *Phys. Rev. B* 1984, *29*, 1558.
17. X. Li, H. Wu, X. Wang, L. Wang, *Phys. Rev. Lett.* 1998, *81*, 1909.
18. J. E. Fowler, J. M. Ugalde, *Phys. Rev. A* 1998, *58*, 383.
19. W. J. Zheng, O. C. Thomas, T. P. Lippa, S. J. Xu, K. H. B. Jr., *J. Chem. Phys.* 2006, *124*, 144304.
20. D. E. Bergeron, P. J. Roach, A. W. Castleman Jr., N. O. Jones, J. U. Reveles, S. N. Khanna, *J. Am. Chem. Soc.* 2005, *127*, 16048.
21. A. B. Laursen, K. T. Hojholt, L. F. Lundegaard, S. B. Simonsen, S. Helveg, F. Schuth, M. Paul et al. *Angew. Chem.* 2010, *122*, 3582.
22. S. Mandal, A. C. Reber, M. Qian, P. S. Weiss, S. N. Khanna, Y. Sen, *Acc. Chem. Res.* 2013, 46, 2385.

23. J. Forster, B. Rosner, R. H. Fink, L. C. Nye, I. Ivanovic-Burmazovic, K. Kastner, J. Tuchera, C. Streb, *Chem. Sci.* 2013, *4*, 418.

24. L. Motte, M. P. Pileni, *J. Phys. Chem. B* 1998, *102*, 4104.

25. J. Li, J. Jia, X. Liang, X. Liu, J. Wang, Q. Xue, Z. Li, J. S. Tse, Z. Zhang, S. B. Zhang, *Phys. Rev. Lett.* 2002, *88*, 066101.

26. Y. Pan, M. Gao, L. Huang, F. Liu, H. J. Gao, *Appl. Phys. Lett.* 2009, *95*, 093106.

27. R. Robles, S. N. Khanna, *Phys. Rev. B* 2009, *80*, 115414.

28. P. Jena, S. N. Khanna, B. K. Rao, *J. Cluster Sci.* 2001, *12*, 443.

29. P. Jena, S. N. Khanna, B. K. Rao, *Mater. Sci. Forum* 1996, *232*, 1.

30. F. Duque, A. M. Nanes, L. M. Molina, M. J. Lopez, J. A. Alonso, *Int. J. Quantum Chem.* 2002, *86*, 226.

31. X. G. Gong, *Phys. Rev. B* 1997, *56*, 1091.

32. H. Häkkinen, M. Manninen, *Phys. Rev. Lett.* 1996, 76, 1599.

33. J. A. Alonso, M. J. Lopez, L. M. Molina, F. Duque, A. Mananes, *Nanotechnology* 2002, *13*, 253.

34. W. C. McCrone, in *Polymorphism in Physics and Chemistry of the Organic Solid State*, Vol. 2, Interscience, New York, 1965.

35. M. A. Zwijnenburg, F. Illas, S. T. Bromley, *Phys. Rev. Lett.* 2010, *104*, 175503.

36. M. A. Zwijnenburg, S. T. Bromley, *J. Mater. Chem.* 2011, *21*, 15255.

37. A. Hebard, M. J. Rosseinsky, R. C. Haddon, D. W. Murphy, S. H. Glarum, T. T. M. Palstra, A. P. Ramirez, A. R. Kortan, *Nature* 1991, *350*, 600.

38. G. Hodes, D. Cahen, *Acc. Chem. Res.* 2012, *45*, 705.

39. U. F. Keyser, B. N. Koeleman, S. van Dorp, D. Krapf, R. M. M. Smeets, S. G. Lemay, N. H. Dekker, C. Dekker, *Nat. Phys.* 2006, *2*, 473.

40. D. Fologea, M. Gershow, B. Ledden, D. S. McNabb, J. A. Golovchenco, J. Li, *Nano Lett.* 2005, *5*, 1905.

41. M. L. Pinto, L. Mafra, J. M. Guil, J. Pires, J. Rocha, *Chem. Mater.* 2012, *23*, 1387.

42. T. X. Nguyen, H. Jobic, S. K. Bhatia, *Phys. Rev. Lett.* 2010, *105*, 085901.

43. D. Cohen-Tanugi, J. C. Grossman, *Nano Lett.* 2012, *12*, 3602.

44. Z. Zhang, Y. Wang, X. Wang, *Nanoscale* 2011, *3*, 1663.

45. S. Tang, S. Vongehr, Z. Zheng, H. Ren, X. Meng, *Nanotech.* 2012, *23*, 255606.

46. J. M. Thomas, J. C. Hernandez-Garrido, R. Raja, R. G. Bell, *Phys. Chem. Chem. Phys.* 2009, *11*, 2799.

47. G. Q. Lu, X. S. Zhao (eds.), in *Nanoporous Materials, Science and Technology*, Imperial College Press, London, U.K., 2004.

48. M. E. Davis, *Nature* 2002, *417*, 813.

49. S. M. Auerbach, K. A. Carrado, P. K. Dutta (eds.), in *Handbook of Zeolite Science and Technology*, Marcel Dekker, New York, 2003.

50. C. Feldmann, *Angew. Chem. Int. Ed.* 2013, *52*, 7610.

51. M. O'Keeffe, *Phys. Chem. Chem. Phys.* 2010, *12*, 8580.

52. A. Bach, D. Fischer, M. Jansen, *Z. Anorg. Allg. Chem.* 2009, *635*, 2406.

53. Y. Liebold-Ribeiro, D. Fischer, M. Jansen, *Angew. Chem. Int. Ed.* 2008, *47*, 4428.

54. D. Fischer, A. Muller, M. Jansen, *Z. Anorg. Allg. Chem.* 2004, *630*, 2697.

55. B. Slater, T. Ohsuma, Z. Liu, O. Terasaki, *Faraday Discuss.* 2007, *136*, 125.

56. H. Lee, S. I. Zones, M. E. Davis, *Nature* 2003, *425*, 385.

57. J. M. Matxain, M. Piris, X. Lopez, J. M. Ugalde, *Chem. Eur. J.* 2009, *15*, 5138.

58. M. Born, J. R. Oppenheimer, *Ann. Physik* 1927, *44*, 455.

59. C. J. Cramer, *Essentials of Computational Chemistry. Theories and Models.* John Wiley and Sons, West Sussex, U.K., 2002.

60. J. M. Mercero, J. M. Matxain, X. Lopez, D. M. York, A. Largo, L. A. Eriksson, J. M. Ugalde, *Int. J. Mass Spectrom.* 2005, *240*, 37–99.

61. M. Piris, *Int. J. Quant. Chem.* 2013, *113*, 620.

62. P. Hohenberg, W. Kohn, *Phys. Rev.* 1964, *136*, B864.

63. W. Kohn, L. J. Sham, *Phys. Rev.* 1965, *140*, A1133.

64. S. H. Vosko, L. Wilk, M. Nusair, *Can. J. Phys.* 1980, *58*, 1200.

65. J. P. Perdew, K. Burke, M. Ernzerhof, *Phys. Rev. Lett.* 1996, *77*, 3865–3868.

66. B. Hammer, L. B. Hansen, J. K. Norskov, *Phys. Rev. B* 1999, *59*, 7413.

67. J. P. Perdew, K. Burke, W. Yang, *Phys. Rev. B* 1996, *54*, 16533.

68. A. D. Becke, *Phys. Rev. A* 1988, *38*, 3098.

69. A. D. Becke, *J. Chem. Phys.* 1993, *98*, 1372.

70. C. Lee, W. Yang, R. G. Parr, *Phys. Rev. B* 1988, *31*, 785.

Chapter 4

71. A. J. Austin, G. Petersson, M. J. Frisch, F. J. Dobek, G. Scalmani, K. Throssell, *J. Chem. Theory Comput.* 2012, *8*, 4989.
72. S. Grimme, *J. Comp. Chem.* 2006, *27*, 1787.
73. S. Grimme, J. Antony, S. Ehrlich, H. Krieg, *J. Comp. Chem.* 2006, *27*, 1787.
74. P. E. Blöchl, *Phys. Rev. B* 1994, *50*, 17953.
75. N. Troullier, J. L. Martins, *Phys. Rev. B* 1991, *43*, 1993.
76. W. J. Stevens, M. Krauss, H. Basch, P. G. Jasien, *Can. J. Chem.* 1992, *70*, 612.
77. X. Y. Cao, M. Dolg, *J. Chem. Phys.* 2001, *115*, 7348.
78. X. Cao, M. Dolg, *J. Mol. Struct. Theochem.* 2002, *581*, 139.
79. X. Cao, M. Dolg, H. Stoll, *J. Chem. Phys.* 2003, *118*, 487.
80. P. J. Hay, W. R. Wadt, *J. Chem. Phys.* 1985, *82*, 270.
81. P. J. Hay, W. R. Wadt, *J. Chem. Phys.* 1985, *82*, 299.
82. W. R. Wadt, P. J. Hay, *J. Chem. Phys.* 1985, *82*, 284.
83. H. J. Monkhorst, J. D. Pack, *Phys. Rev. B* 1973, *13*, 5188.
84. M. J. Frisch, G. W. Trucks, H. B. Schlegel, G. E. Scuseria, M. A. Robb, J. R. Cheeseman, G. Scalmani et al., 2009, Gaussian09 Program Package, Gaussian Inc., Wallingford, CT, 2009.
85. J. M. Soler, E. Artacho, J. D. Gale, A. Garcia, J. Junquera, P. Ordejon, D. Sanchez-Portal, *J. Phys. Condens. Matter* 2002, *14*, 2745.
86. G. Kresse, J. Hafner, *Phys. Rev. B* 1993, *47*, 558.
87. G. Kresse, J. Furthmiiller, *Comput. Mater. Sci.* 1996, *6*, 15.
88. G. Kresse, J. Furthmiiller, *Phys. Rev. B* 1996, *54*, 11169.
89. J. H. Holland (ed.), in *Adaptation in Natural and Artificial Systems*, University of Michigan Press, Ann Arbor, MI, 1986.
90. D. A. Coley, in *An Introduction to Genetic Algorithms for Scientists and Engineers*, World Scientific, New York, 1999.
91. H. M. Cartwright, B. Hartke, K. D. M. Harris, R. L. Johnston, S. Habershon, S. M. Woodley, V. J. Gillet, R. Unger, in *Structure and Bonding 110: Applications of Evolutionary Computation in Chemistry*, Springer, New York, 2004.
92. S. M. Woodley, *Phys. Chem. Chem. Phys.* 2007, *9*, 1070.
93. S. Kirkpatrick, J. C. D. Gellat, M. P. Vecchi, *Science* 1983, *20*, 671.
94. V. Czerny, *J. Optim. Theo. Appl.* 1985, *45*, 41.
95. J. C. Schön, H. Putz, M. Jansen, *J. Phys. Condens. Matter* 1996, *8*, 143.
96. J. C. Schön, *Ber. Bunsenges.* 1996, *100*, 1388.
97. M. M. J. Treacy, I. Rivin, E. Balkovsky, K. H. Randall, M. D. Foster, *Microporous Mesoporous Mater.* 2004, *74*, 121.
98. M. D. Foster, A. Simperler, R. G. Bell, O. Delgado-Friedrichs, F. A. Almeida-Paz, J. Klinowski *Nat. Mater.* 2004, *3*, 234.
99. M. O'Keeffe, *Acta Crystallogr.* 2008, *64*, 425.
100. S. Curtarolo, D. Morgan, K. Persson, J. Rodgers, G. Ceder, *Phys. Rev. Lett.* 2003, *91*, 135503.
101. C. C. Fischer, K. J. Tibbetts, D. Morgan, G. Ceder, *Nat. Mater.* 2006, *5*, 641.
102. D. W. M. Hofmann, J. Apostosakis, *J. Mol. Struct.* 2003, *647*, 17.
103. S. M. Woodley, R. Catlow, *Nat. Mater.* 2008, *7*, 937.
104. S. Nosé, *J. Chem. Phys.* 1984, *81*, 511–519.
105. E. Jimenez-Izal, J. M. Matxain, M. Piris, J. M. Ugalde, *Phys. Chem. Chem. Phys.* 2012, *14*, 9676.
106. R. Car, M. Parrinello, *Phys. Rev. Lett.* 1985, *55*, 2471.
107. Z. Li, H. A. Scheraga, *Proc. Natl. Acad. Sci.* 1987, *84*, 6611.
108. D. J. Wales, J. P. K. Doye, *J. Phys. Chem.* 1997, *101*, 5111.
109. S. Goedecker, *J. Chem. Phys.* 2004, *120*, 9911.
110. A. Laio, M. Parrinello, *Proc. Natl. Acad. Sci.* 2002, *99*, 12562.
111. J. C. Schön, K. Doll, M. Jansen, *Phys. Stat. Sol. B* 2010, *247*, 23.
112. D. Zagorac, J. C. Schön, M. Jansen, *J. Phys. Chem. C* 2012, *116*, 16726.
113. W. Sangthong, J. Limtrakul, F. Illas, S. T. Bromley, *J. Mater. Chem.* 2008, *18*, 5871.
114. J. P. Perdew, J. A. Chevary, S. H. Vosko, K. A. Jackson, M. R. Pederson, D. J. Singh, C. Fiolhais, *Phys. Rev. B* 1992, *46*, 6671.
115. J. A. White, D. M. Bird, *Phys. Rev. B* 1994, *50*, 4954.
116. J. Carrasco, F. Illas, S. T. Bromley, *Phys. Rev. Lett.* 2007, *99*, 235502.
117. D. Stradi, F. Illas, S. T.-Bromley, *Phys. Rev. Lett.* 2010, *105*, 045901.

118. S. M. Woodley, M. B. Watkins, A. A. Sokol, S. A. Shevlin, C. R. A. Catlow, *Phys. Chem. Chem. Phys.* 2009, *11*, 3176.
119. M. B. Watkins, S. A. Shevlin, A. A. Sokol, C. R. A. Catlow, S. M. Woodley, *Phys. Chem. Chem. Phys.* 2009, *11*, 3186.
120. J. D. Gale, *J. Chem. Soc. Faraday Trans.* 1997, *93*, 629.
121. Y. Yong, B. Song, P. He, *J. Phys. Chem. C* 2011, *115*, 6455.
122. B. Delley, *J. Chem. Phys.* 1990, *92*, 508.
123. B. Delley, *J. Chem. Phys.* 2000, *113*, 7756.
124. Z. Liu, X. Wang, J. Cai, G. Liu, P. Zhou, K. Wang, H. Zhu, *J. Phys. Chem. C* 2013, *111*, 17633.
125. A. Al-Sunaidi, A. A. Sokol, C. R. A. Catlow, S. M. Woodley, *J. Phys. Chem. C* 2008, *112*, 18860.
126. E. Jimenez-Izal, J. M. Matxain, M. Piris, J. M. Ugalde, *J. Phys. Chem. C* 2010, *114*, 2476.
127. E. Artacho, J. Gale, A. Garcia, J. Junquera, R. M. Martin, P. Ordejon, D. Sanchez-Portal, J. M. Soler, SIESTA program package, http://www.uam.es/departamerLtos/ciencias/fismateriac/siesta.
128. J. M. Matxain, L. A. Eriksson, E. Formoso, M. Piris, J. M. Ugalde, *J. Phys. Chem. C* 2007, *111*, 3560.
129. J. Zhao, R. Xie, *Phys. Rev. B* 2003, *68*, 035401.
130. M. Tanaka, S. Zhang, K. Inumaru, S. Yamanaka, *Inorg. Chem.* 2013, *52*, 6039.
131. L. Guo, X. Zheng, C. Liu, W. Zhou, Z. Zeng, *Comput. Theor. Chem.* 2012, *982*, 17.
132. M. B. Torres, E. M. Fernández, L. C. Balbás, *Int. J. Quant. Chem.* 2011, *111*, 444.
133. R. J. C. Batista, H. Chacham, in *Boron Nitride Fullerenes and Nanocones in Handbook of Nanophysics: Clusters and Fullerenes*, CRC Press, Taylor & Francis Group, Boca Raton, FL, 2011.
134. A. V. Pokropivny, *Diamond Relat. Mat.* 2006, *15*, 1492.
135. J. M. Matxain, L. A. Eriksson, J. M. Mercero, X. Lopez, M. Piris, J. M. Ugalde, J. Poater, E. Matito, M. Sola, *J. Phys. Chem. C* 2007, *111*, 13354.
136. S. S. Alexandre, R. W. Nunes, H. Chacham, *Phys. Rev. B* 2002, *66*, 085406.
137. Z. Liu, X. Wang, G. Liu, P. Zhou, J. Sui, X. Wang, H. Zhu, Z. Hou, *Phys. Chem. Chem. Phys.* 2013, *15*, 8186.
138. Z. Liu, X. Wang, H. Zhu, *RSC Adv.* 2013, *3*, 1450.
139. W. Sangthong, J. Limtrakul, F. Illas, S. T. Bromley, *Phys. Chem. Chem. Phys.* 2010, *12*, 8513.
140. R. Astala, S. M. Auerbach, P. A. Monson, *J. Phys. Chem. B* 2004, *180*, 9208.
141. M. D. Foster, O. Delgado-Friedrichs, R. G. Bell, F. A. Almeida-Paz, J. Klimowski, *J. Am. Chem. Soc.* 2004, *126*, 9769.
142. M. D. Foster, A. Simperler, R. G. Bell, O. Delgado-Friedrichs, F. A. Almeida-Paz, J. Klimowski, *Nat. Mater.* 2004, *3*, 234.
143. M. M. J. Treacy, K. H. Randall, S. Rao, J. A. Perry, D. Chadi, *J. Z. Kristallografiya* 1997, *212*, 768.
144. M. B. Boisen, G. V. Gibbs, M. O'Keeffe, K. L. Bartelmehs, *Microporous Mesoporous Mater.* 1999, *29*, 219.
145. C. Mellot-Draznieks, S. Girard, G. Férey, *J. Am. Chem. Soc.* 2002, *124*, 15326.
146. M. V. Peskov, V. A. Blatov, G. D. Ilyushin, U. Schwingenschlögl, *J. Phys. Chem. C* 2012, *116*, 6734.
147. J. C. Wojdel, M. A. Zwijnenburg, S. T. Bromley, *Chem. Mater.* 2006, *18*, 1464.
148. E. Flikkema, S. T. Bromley, *J. Phys. Chem. B* 2004, *108*, 9638.
149. S. T. Bromley, *Cryst. Eng. Comm.* 2007, *9*, 463.
150. S. T. Bromley, M. A. Zwijnenburg, T. Maschmeyer, *Phys. Rev. Lett.* 1992, *90*, 035502.
151. S. T. Bromley, *Nano Lett.* 2004, *4*, 1427.
152. S. T. Bromley, E. Flikkema, *Comput. Mater. Sci.* 2006, *35*, 382.
153. M. A. Zwijnenburg, S. T. Bromley, E. Flikkema, T. Maschmeyer, *Chem. Phys. Lett.* 2004, *385*, 389.
154. A. Weiss, Z. *Anorg. Allg. Chem.* 1954, *276*, 95.
155. J. D. Gale, A. L. Rom, *Mol. Simul.* 2003, *29*, 291.
156. J. R. Holden, Z. Y. Du, H. L. Ammonn, *J. Comput. Chem.* 1992, *14*, 422.
157. S. T. Bromley, I. P. R. Moreira, K. M. Neyman, F. Illas, *Chem. Soc. Rev.* 2009, *38*, 2657.
158. S. Hamad, C. R. A. Catlow, S. M. Woodley, S. Lago, J. A. Mejias, *J. Phys. Chem. B* 2005, *109*, 15741.
159. P. Huang, C. Qin, Z. M. Su, Y. Xing, X. L. Wang, K. Z. Shao, Y. Q. Lan, E. B. Wang, *J. Am. Chem. Soc.* 2012, *134*, 14004.
160. V. A. Blatov, G. D. Ilyushin, D. M. Proserpio, *Inorg. Chem.* 2010, *49*, 1811.
161. V. A. Blatov, G. D. Ilyushin, D. M. Proserpio, *Chem. Matter.* 2013, *25*, 412.
162. V. A. Blatov, *IUCr. CompComm. Newslett.* 2006, *7*, 4.
163. C. Ashman, S. N. Khanna, F. Liu, P. Jena, T. Kaplan, M. Mostoller, *Phys. Rev. B* 1997, *55*, 15868.
164. Z. Zhi-zhong, T. Bo, *Solid State Comm.* 1998, *108*, 891.

165. A. W. Castleman, S. N. Khanna, A. Sen, A. C. Reber, M. Qian, K. M. Davis, S. J. Peppernick, A. Ugrinov, M. D. Merritt, *Nano Lett.* 2007, *7*, 2734.

166. R. Ferro, A. Saccone, in *Intermetallic Chemistry*, Pergamon, Oxford, U.K., 2008.

167. S. Samson, *Acta Crystallogr.* 1961, *14*, 1229.

168. S. Samson, *Nature* 1962, *195*, 259.

169. S. Samson, *Acta Crystallogr.* 1964, *17*, 491.

170. S. Samson, *Acta Crystallogr.* 1965, *19*, 401.

171. S. Samson, *Acta Crystallogr.* 1967, *23*, 586.

172. R. Ferrando, J. Jellinek, R. L. Johnston, *Chem. Rev.* 2008, *108*, 845.

173. D. V. Talapin, *ACSNano* 2008, *2*, 1097.

174. A. Shelley, S. A. Claridge, A. W. Castleman Jr., S. N. Khanna, C. B. Murray, A. Sen, P. S. Weiss, *ACSNano* 2009, *3*, 244.

175. Q. Lin, J. D. Corbett, *Inorg. Chem.* 2003, *42*, 8762.

176. Q. Lin, J. D. Corbett, *Struct. Bonding (Berlin)* 2009, *133*, 1.

177. S. Alvarez, *Dalton Trans.* 2005, *133*, 2209.

178. B. C. Guo, K. P. Kerns, A. W. Castleman Jr., *Science* 1992, *255*, 1411–1413.

179. Z. Iqbal, Y. Zhang, H. Grebel, S. Vijayalakshmi, A. Lahamer, G. Benedek, M. Bernasconi, J. Cariboni, I. Spagnolatti, R. Sharma, F. J. Owens, M. E. Kozlov, K. V. Rao, M. Muhammed, *Eur. Phys. J.* 2003, *31*, 509.

180. A. Devos, M. Lanoo, *Phys. Rev. B* 1998, *58*, 8236.

181. I. Spagnolatti, M. Bernasconi, G. Benedek, *Europhys. Lett.* 2002, *59*, 572.

182. H. Sakurai, A. W. Castleman Jr., *J. Phys. Chem. A* 1998, *102*, 10486.

183. C. Berkdemir, A. W. Castleman Jr., J. O. Sofo, *Phys. Chem. Chem. Phys.* 2012, *14*, 9642.

184. D. Vanderbilt, *Phys. Rev. B* 1990, *41*, 7892.

185. P. Giannozzi, S. Baroni, N. Bonini, M. Calandra, R. Car, D. Cavazzoni, D. CeresolL et al., *J. Phys. Condens. Matter* 2009, *21*, 395502.

186. H. S. Domingos, *J. Phys. Condens. Matter* 2005, *17*, 2571.

187. S. J. Clark, M. D. Segall, C. J. Pickard, P. J. Hasnip, M. J. Probert, K. Refson, M. C. Payne, "First principles methods using CASTEP", *Zeitschrift fuer Kristallographie*, 2005, 220, 567.

188. C. L. Beckel, N. Lu, B. Abbott, M. Yousaf, *Inorg. Chim. Acta* 1999, *289*, 198.

189. T. L. Aselage, D. R. Tallant, D. Ermin, *Phys. Rev. B* 1997, *56*, 3122.

190. B. Wang, Z. Fan, Q. Zhou, M. Feng, X. Cao, Y. Wang, *Physica B* 2011, *406*, 297.

Properties

5. Melting and Phase Transitions

Florent Calvo

5.1   Introduction 151

5.2   Phenomenology and Experimental Methods............................ 153
     5.2.1   The Lindemann Theory of Melting.............................. 153
     5.2.2   Capillary Theories of Melting in Nanoparticles 153
     5.2.3   Fluctuations 154
     5.2.4   Experimental Approaches 155

5.3   Computational Methods for the Melting Problem...................... 157
     5.3.1   Statistical Ensembles 157
     5.3.2   Energy Landscapes... 159
     5.3.3   Numerical Simulations 160
     5.3.4   Observables from Computer Simulations 162

5.4   Melting of Inorganic Clusters and Nanoparticles: Selected Examples 164
     5.4.1   Van der Waals Clusters....................................... 164
     5.4.2   Metal Clusters and Nanoalloys 166
     5.4.3   Ionic Clusters and Nanoparticles.............................. 169
     5.4.4   Semiconducting and Covalent Nanoparticles 170

5.5   Toward the Liquid–Vapor Transition................................ 171
     5.5.1   Evaporation and the Low-Energy Regime 171
     5.5.2   Multifragmentation 173

5.6   Outlook 174

References 175

5.1   Introduction

Phase transitions refer to qualitative changes in the collective behavior of an atomic or molecular system that occurs when some external condition is varied, usually temperature or pressure. Phase transitions in macroscopic systems are generally dramatic, with clearly visible manifestations that alter many of their properties. In bulk materials, phase transitions are important because applications require the material to be in a specific phase; hence, it is desirable to ensure that it keeps its integrity in the expected range of operating thermodynamical conditions. The same issues apply to nanomaterials, especially in concern with the necessity that they do not change phase upon miniaturization.

At the nanoscale, it is not so obvious that the notions valid in the bulk limit are still relevant, especially in the context of phase transitions that are associated with divergences in some thermodynamic functions. Strictly speaking, there can be no divergence in the thermodynamical properties of a finite system, because for a well-behaved and physically realistic potential energy surface, the partition function varies smoothly with increasing temperature unless any limit $N \to \infty$ is taken. However, even in small systems the concept of phases seems to hold in many situations. For the melting problem, to which much of this chapter will be devoted, it is rather clear that a solid phase can be defined when the internal energy (or temperature) is very low, even though it does not correspond to a periodic crystal. Likewise, it is expected that the same system may undergo some global transformation if a sufficient amount of energy is pumped into it, losing its atomically resolved shape in favor of something more disordered and fluxional. Unfortunately, contrary to the case of the solid phase, it is not straightforward to define a liquid at the nanoscale, because the definition used in the macroscopic limit (a phase in which the system adopts the shape of its container) does not apply. We will show at the end of this chapter that the fluid phases of a nanoscale system are not well defined without knowledge of the time scale of observation.

For a ubiquitous phenomenon, melting surprisingly remains poorly understood, with no satisfactory theory available to date [1] despite some contributions for model systems, such as the classification of the transition from the Yang–Lee theorem [2] based on the distributions of zeroes of the partition function in the complex plane. These issues are mostly ascribable to the difficulties of building a quantitative theory of liquids. Yet, phenomenological approaches exist and can be quite insightful for large nanoparticles lying in the so-called scalable regime, where the properties smoothly depend on size. In contrast, very small systems can be prone to strong nonmonotonic variations in their properties, and this size dependence further impedes any analytical development, while promoting computational modeling. But special care should be taken before embarking into the evaluation of thermodynamic properties by simulation due to the frequent lack of ergodicity suffered from conventional approaches.

This chapter reviews the general aspects of phase transitions in nanoscale systems, with a strong focus on the melting transition. It also attempts to illustrate some of the specific features observed in simulations or real experiments of various inorganic nanomaterials, including metals, and ionic, oxide, or semiconducting systems. After discussing some generalities on phase transitions and the changes occurring at the nanoscale based on phenomenological approaches, we briefly review the existing experimental methods used so far to characterize those transitions. We then focus on the computational methods that are useful to study phase transitions in nanomaterials. Our subsequent survey of insightful examples taken from recent literature on various materials does not claim to be exhaustive but only illustrative of either some general trends or aspects that are more peculiar of the nanoscale.

Before concluding, this chapter considers the gas transition and the stability of nanosystems over long but finite times. In addition to discussing some existing connections between melting and statistical evaporation, we will show that the multifragmentation regime at high excitations can indeed be described from equilibrium theories of the liquid–gas transition.

5.2 Phenomenology and Experimental Methods

In the absence of a general *ab initio* theory describing the corresponding transition in bulk materials, several phenomenological approaches have been proposed [3]. This section emphasizes some of these approaches adapted to the case of nanoparticles.

5.2.1 The Lindemann Theory of Melting

The causes of bulk melting were identified at the beginning of the twentieth century by Lindemann [4] who realized that the amplitude of atomic vibrations in a crystal cannot exceed the lattice constant. He formalized this observation by stating that melting occurs once the relative bond fluctuations reach about 10% of the average bond length itself. This criterion has since been extended by the Berry group [5] and shown to be a useful indicator of melting in numerical simulations.

The Lindemann criterion relates to the amount of energy stored in a bond, which after crossing some threshold value creates a defect, nucleating with other defects to induce melting. In nanoparticles, the main difference with a bulk material lies in the different environments of the outer atoms, which are less coordinated, hence less bound and less stable against thermal motion. Such a property already suggests that, upon approaching the (bulk) melting temperature, a macroscopic material should melt from its surface, a feature that has indeed been reported for various materials [6,7].

5.2.2 Capillary Theories of Melting in Nanoparticles

One immediate consequence of surface melting is that it should be easier to melt nanoparticles (and thin films) than the corresponding bulk material, because of their high surface/volume ratio. However, for the reverse process of freezing, the formation of a critical nucleus is required, with a rate of formation that is proportional to the volume. Hence the liquid tends to become supercooled instead of freezing, a phenomenon experimentally verified by Turnbull and coworkers on micrometer size droplets [8].

The importance of surface energies as a source of destabilization of the solid was first recognized by Pawlow [9]. Since then, various approximate theories [10–13] have been proposed to determine the size variations of the melting temperature T_m as a function of the number N of atoms in the nanoparticle, and similar approaches exist in the context of the liquid–vapor transition [14]. These theories treat the droplets or nanoparticles as spherical for simplicity and consider either a chemical-type equilibrium between entirely solid and entirely liquid particles or a mechanical-type equilibrium between a solid core surrounded by a liquid layer. In any case they rely on the Gibbs–Thomson equation, completed by the Laplace equation for the surface pressure. The outcome of these theories based on capillary considerations can generally be expressed as a formula relating the change in the melting temperature relative to the infinite system to the radius R of the nanoparticle as

$$T_m(R) - T_m(\infty) = -\frac{\alpha}{R},$$
(5.1)

Chapter 5

where α depends on the surface tensions of the various interfaces, on the liquid and solid densities, and possibly on the thickness of the liquid layer covering the solid core [9–13]. The depression in the melting point predicted by the aforementioned formula scales with the number of atoms as $N^{-1/3}$ for a 3D particle, and this behavior has been confirmed by a number of experiments [10–12,15,16] and simulations [17–20] on metal droplets. It should be noted, however, that semiconductors do not seem to behave similarly, as variations scaling *quadratically* with the inverse radius have been reported [21].

Besides the melting point, it is possible to evaluate the variations of the latent heat L_m due to the finite system size, assuming a chemical-type equilibrium model [12,14] between entirely solid and entirely liquid particles. The result can be expressed again as a simple scaling law involving the nanoparticle radius R as [12]

$$L_m(\infty) - L_m(R) = -\frac{\alpha'}{R} + \int_{T_m(R)}^{T_m(\infty)} \Delta C_p(T)dT, \tag{5.2}$$

where α' is again related to the surface tensions between the liquid, solid, and gas phases and $\Delta C_p = C_p^L - C_p^S$ is the difference in specific heats between the solid and liquid phases. The dominating term in the equation given earlier is the first one; hence, the latent heat should also exhibit a depression in $1/R$ with respect to the bulk limit. Measurements of the latent heat of melting of nanoparticles are more scarce than those for the melting temperature; however, the scaling as $N^{-1/3}$ could be verified in the case of tin [12].

5.2.3 Fluctuations

Owing to their statistical origin, phase transitions are subject to fluctuations. Statistical fluctuations apply, for instance, to a set of equivalent systems at the same temperature, but which may not be all in the same phase. Let us consider the case of a set of finite nanoparticles having all N atoms and at the same temperature T. The probabilities p_S and p_L to find any particle in the solid or liquid phase satisfy

$$\frac{p_S(N,T)}{p_L(N,T)} = \exp\left[-\beta(\mu_S - \mu_L)N\right], \tag{5.3}$$

where
$\beta = 1/k_B T$ with k_B the Boltzmann constant
μ_S and μ_L are the chemical potentials of the solid and liquid particles, respectively

In a bulk system, the ratio p_S/p_L is always either 0 or ∞ because of the exponential suppression with N. The only exception is at the melting point, where by definition $\mu_S = \mu_L$ and $p_S = p_L = 1/2$ even for $N \to \infty$.

The situation drastically differs in a finite system: even away from the temperature where $\mu_S = \mu_L$, the ratio p_S/p_L remains finite, and in a statistical ensemble of nanoparticles some of them will be solid, some others liquid. Statistical fluctuations will thus *smear out* the phase transition in the finite system, attenuating the sudden changes exhibited by thermodynamic functions into smoother variations. Imry [22] has generalized this

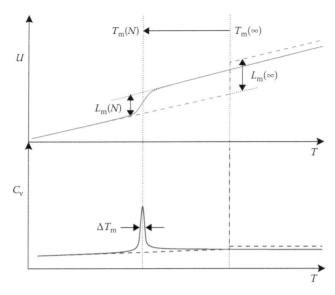

FIGURE 5.1   Finite-size effects on the caloric curves of nanoparticles across the melting transition in the scaling regime. The upper and lower panels depict the internal energy U and the heat capacity $C_v = \partial U/\partial T$, respectively. The three main finite-size effects highlighted are the decrease in the melting point from $T_m(\infty)$ to $T_m(N)$, the broadening of the transition with a width $\Delta T_m(N)$, and the decrease in the latent heat of melting from $L_m(\infty)$ to $L_m(N)$.

idea and estimated the magnitude of this broadening ΔT_m as a function of N. He found a simple relation for ΔT_m involving the entropy difference $\Delta S = N L_m/k_B T_m$ between the solid and liquid phases, namely,

$$\Delta T_m(N) \simeq \frac{k_B T_m^2}{N L_m}.$$

(5.4)

From the previous discussion, the major thermodynamical changes occurring at the nanoscale for the internal energy and the heat capacity near the melting transition are schematically represented in Figure 5.1. The most striking feature for such a first-order transition is the disappearance of the sharp increase in the internal energy and the concomitant divergence in the heat capacity, which at finite size are replaced by a smooth increase and a finite peak, respectively. In the nanoparticle, the melting point and the latent heat both decrease, and the transition itself is smeared out. Strictly speaking, it does not exist anymore in the mathematical sense, and should be better described as a *phase change*. However, for ease and continuity of terminology we will keep referring to melting (and other phase changes) in nanosystems as to phase transitions, implicitly recognizing the presence of rounding size effects.

5.2.4   Experimental Approaches

The earliest measurements of melting points in low-dimensional systems were carried out for deposited particles [10–12,15,16], the substrate playing the role of a thermostat. The particles were rather large in order that their solid phase at sufficiently low temperature could be identified from the Bragg peaks in diffraction measurements. By increasing the substrate temperature by regular increments, melting could be monitored from the disappearance of these diffraction patterns. This method is robust and could in principle apply to any material, but one difficulty in analysing the results is that the particles should have a narrow size distribution.

The same technique was applied by several groups who confirmed from the variations of $T_m(N)$ the general validity of the scaling law of Equation 5.1 for nanoparticles of gold [10] and bismuth [16].

This method cannot be directly applied to gas-phase nanoparticles, but a similar idea was proposed by Martin and coworkers [23] who inferred the melting temperature from the disappearance of geometric magic numbers in the mass spectra over a broad size range. Since magic numbers are identified from the relative intensities between neighboring peaks, the method has some significant uncertainties; however, the authors were able to provide evidence for a size-dependent melting temperature in relatively large nanoparticles containing thousands of atoms [23].

Differential nanocalorimetry experiments performed on supported nanoparticles have given additional insight into the melting behavior [12]. Briefly, the method consists in measuring the heat produced by the system upon increasing the substrate temperature and subtracting the contribution from the sole substrate as determined independently. The results obtained for tin confirmed the general $1/R$ depression law for the melting point and showed a comparable behavior for the latent heat L_m [12]. However, similar measurements for bismuth [24] found that nanoparticles with a radius below than 7 nm had a melting point higher than the bulk material by about 50 K. The authors attributed this discrepancy to a possible size-dependent structural change and to the possible superheating of the solid.

The aforementioned methods were designed for sufficiently large nanoparticles with clear crystallographic signatures or a measureable thermal response to temperature variations. In very small nanoparticles (clusters containing tens to hundreds of atoms only), sensitive methods are needed to circumvent the aforesaid limitations. Early investigations have tried to detect melting in small gas-phase clusters from some measurable change in properties that indirectly depend on the thermodynamical phase through fluctuations in atomic structure. Optical spectroscopy, in particular, has had mitigated successes on van der Waals [25,26] or sodium [27] clusters. Attempts to correlate melting with other experimental signatures have been proposed for the volume inferred either from ion-mobility measurements [28] or from the determination of electronic polarizability [29]. Alternatively, the chemical reactivity has been shown to vary with temperature through the structural disordered induced by melting [30,31]. Diffraction experiments have also been performed on trapped, size-selected ionized clusters, with promising results showing some dependence with temperature [32].

But a breakthrough came with the complete experimental characterization of caloric curves for size-selected clusters using a nanocalorimetry method designed for isolated systems by Haberland and coworkers [33]. Those authors suggested to evaluate the internal energy of the system through its fragmentation pattern, heating being provided either from thermalization through a gas chamber at controlled temperature or from laser heating with a well-defined wavelength in the absorption range. The method could provide the caloric curve of cationic sodium clusters containing between a few tens and a few hundreds of atoms and revealed very intriguing variations in the melting temperature and latent heat of melting as a function of size [34]. In this small sizes regime where each atom counts, the complex variations in the thermodynamical properties originate from the interplay between electronic and geometric shell closings and could be rationalized based on some combinatorial models [35] and first-principles molecular dynamics (MD) simulations [36].

The idea underlying Haberland's experiment has been generalized by the Jarrold group, who suggested to perform the heating by multiple soft collisions with inert species [37], and more recently by L'Hermite and coworkers [38] who used low-energy sticking collisions. These methods are applicable to broader varieties of nanosystems, in particular those that do not absorb laser light, although the excitation energy is not so well controlled. Among the most prominent results of these investigations, the surprisingly high melting points of gallium clusters [37] and complex size effects for aluminum clusters with multiple features in the heat capacity curves [39] were notably demonstrated.

5.3   Computational Methods for the Melting Problem

We address now the practical issues regarding the characterization of melting properties of nanomaterials, and in particular the determination of caloric curves from simulation.

5.3.1   Statistical Ensembles

The most common ensembles of statistical mechanics have different relevance for nanoscale systems:

- *The microcanonical ensemble* $(N, \mathcal{V}, E)$, in which the number of atoms N, the volume $\mathcal{V}$, and the total energy E are conserved, is appropriate for isolated systems. As such, it is the natural ensemble for gas-phase clusters. Its characteristic function is the density of states Ω defined as

$$\Omega(N,\mathcal{V},E) = \frac{1}{\hbar^{3N}} \iint d^{3N}\mathbf{R}\, d^{3N}\mathbf{P}\, \delta\big[H(\mathbf{R},\mathbf{P}) - E\big], \qquad (5.5)$$

from which the microcanonical entropy S and the temperature T_μ are derived as

$$S(N,\mathcal{V},E) = k_B \ln \Omega,$$

$$\frac{1}{T_\mu}(N,\mathcal{V},E) = \frac{\partial S}{\partial E}. \qquad (5.6)$$

- *The canonical ensemble* $(N, \mathcal{V}, T)$, in which the temperature T is fixed by contact to a thermostat, allowing the energy of the system to fluctuate. This ensemble is appropriate for nanoparticles in contact with an environment (substrate, matrix, or solvent). Its properties derive from the partition function Z,

$$Z(N, \mathcal{V}, T) = \frac{1}{\hbar^{3N}} \int \int d^{3N} \mathbf{R} \, d^{3N} \mathbf{P} \exp\left[-\frac{H(\mathbf{R}, \mathbf{P})}{k_B T}\right], \tag{5.7}$$

and the internal energy and heat capacity are defined as $U = -\partial \ln Z / \partial \beta$ and $C_v = \partial U / \partial T$, respectively, with $\beta = 1/k_B T$.

- *The grand-canonical ensemble* $(\mu, \mathcal{V}, T)$ does not constrain the number of particles, which is allowed to fluctuate given the chemical potential μ. This ensemble is mostly relevant in absorption or nucleation problems; it is characterized by the grand-canonical partition function Ξ

$$\Xi(\mu, \mathcal{V}, T) = \sum_{N \geq 0} Z(N, \mathcal{V}, T) \exp(-\beta \mu N). \tag{5.8}$$

This list is, of course, not exhaustive, as other ensembles can be generated from appropriate control and conjugate variables through Legendre transformations. The isothermal–isobaric ensemble (N, p, T), in which the pressure is controlled, is relevant for nanoparticles embedded in an external medium such as a fluid or a different solid. The melting of embedded nanoparticles has been occasionally investigated by computer modeling, and the results generally show a strong role of pressure on stabilizing the solid [40] (thus shifting the melting point to higher temperatures), and sometimes a qualitatively richer behavior involving multistage melting [41]. Experiments on confined bismuth nanoparticles [42] have confirmed the stabilizing role of the substrate, which leads to strong undercooling for a wide range of nanoparticle radius.

Although constant-pressure ensembles have their practical interest, it is simpler to focus on the microcanonical and canonical ensembles, which are usually considered to be equivalent to each other in the thermodynamic limit $N \to \infty$ [43]. This statement implies that the system under study does have a thermodynamic limit, and notably that it is *extensive*: the energy of a double size system is twice the energy of the single system. However, systems with long-range forces such as self-gravitating or Coulomb-bound systems are not extensive. Nanoscale systems, in which the range of the interaction is comparable to their physical extension, can also be considered as nonextensive, suggesting that the two ensembles are not strictly equivalent for a finite set of atoms.

A rather peculiar manifestation of ensemble inequivalence is the existence of negative heat capacities in the microcanonical ensemble of finite nanosystems or equivalently the *decrease* of temperature upon increasing the energy. Negative heat capacities are impossible in the canonical ensemble, because they are proportional to the energy fluctuation [43]

$$C_v(N, \mathcal{V}, T) = \frac{1}{k_B T^2} \left(\langle E^2 \rangle - \langle E \rangle^2\right) \geq 0. \tag{5.9}$$

Expressions for the microcanonical specific heat $C_\mu(N, \mathcal{V}, E)$ are slightly more complicated [44] but do not imply that C_μ should be positive. Negative specific heats are a

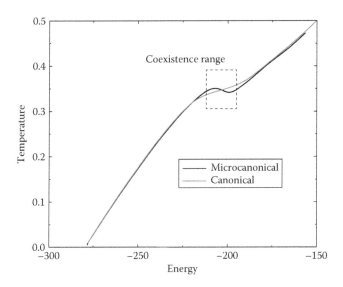

FIGURE 5.2 Microcanonical and canonical caloric curves of the model Lennard-Jones cluster with 55 atoms, as obtained from Monte Carlo simulations. The coexistence range is highlighted.

particular feature of nanoparticles and nonextensive systems that may not occur systematically but pose quite a challenge to intuition when they do.

The first conclusive evidence of negative specific heats in the microcanonical ensemble was found from molecular simulations of model clusters bound by pairwise Lennard–Jones (LJ) interactions [45]. The work by Labastie and Whetten convincingly showed that the microcanonical temperature of the LJ_{55} cluster exhibits nonmonotonic variations with increasing total energy, under the form of a van der Waals loop in the same (energy, temperature) region where the internal energy only displays an inflection in the canonical ensemble. Those variations, depicted in Figure 5.2, are related to the aforementioned coexistence between entirely solid and liquid forms of the cluster that is possible in a range of energy or temperature. In this range, the solid phase has a temperature higher than the liquid phase due to its higher potential energy compensated by a smaller kinetic energy. As total energy increases, the ratio between liquid and solid clusters varies continuously from 0 to 1, quicker than the variations of the temperature in both phases. This leads to the temporary decrease of the microcanonical temperature, as observed in the simulations. Since those theoretical predictions were first made, negative specific heats in finite systems have been indirectly confirmed from experiments on sodium [46] and hydrogen [47] clusters.

5.3.2 Energy Landscapes

Without a complete general and satisfactory theory of melting, numerical simulation techniques must be used to explore the statistical properties of materials at finite temperatures. Before discussing the main categories of available methods, it is important to emphasize the role of the energy landscape or the topography of the multidimensional potential energy function V of the atomic coordinates **R**. The concept of energy landscape is classical by nature, both from the perspective of the electrons (V is the ground-state electronic surface in the Born–Oppenheimer approximation) and with respect to nuclei as well (no delocalization of the vibrational wavefunction). Under such assumptions, the energy landscape can be conveniently partitioned into basins of attractions associated with local minima or stable structures, separated from each other by stationary points. Based on this partitioning, the statistical properties can be expressed from simple sums over the various basins of attraction. In this superposition approach [48], the partition function thus reads

$$Z(N,V,T) = \sum_{\alpha} n_{\alpha} Z_{\alpha}(N,V,T), \tag{5.10}$$

where
  Z_{α} accounts for the contribution of local minimum α
  n_{α} is a constant factor associated with the degeneracy due to permutation–inversion symmetry

A similar expression can be written for the microcanonical density of states $\Omega(N,V,E)$ from the individual densities of states $\Omega_{\alpha}(N,V,E)$ of separate minima. Practical use of Equation 5.10 requires suitable expressions for Z_{α}, which at the lowest level of approximation can be taken as harmonic [48] or can be improved by anharmonic corrections [49]. Perhaps more importantly, a set of minima $\{\alpha\}$ that are representative of the landscape is needed as well. Unless the number of constituents is small, the number of different local minima is huge as it scales approximately exponentially with N [50]. Only a very limited subset can thus be reasonably harvested from computational procedures, and it is all the more important for the accurate evaluation of thermal properties that the sample of minima correctly reflects the low-energy parts of the landscape, even though those parts may form a broad set with multiple funnels.

5.3.3 Numerical Simulations

The superposition approach can be used to evaluate properties other than the thermodynamical functions [51], provided that the observable of interest is computable for individual minima. Alternatively, the properties can be accessed maybe more directly from simulations based on MD or Monte Carlo (MC). Those two classes of methods provide a statistical sample of configuration space, usually at constant energy with MD and at constant temperature with MC. However, it is also possible to sample the canonical ensemble with MD through the use of appropriate thermostats such as Nosé-Hoover or Langevin [52], or the recent scheme of Bussi and coworkers [53]. Similarly, MC methods are not restricted to sample the canonical ensemble, and microcanonical versions have been proposed by several authors [54].

By construction, MC methods evaluate statistical averages by evolving the system randomly in phase space, in such a way that each point carries a desired weight typically proportional to the Boltzmann factor $\exp[-\beta V(\mathbf{R})]$ (in the case of the canonical ensemble with $T = 1/k_B\beta$), the resulting trajectory $\{\mathbf{R}_i, i = 1 \ldots m \to \infty\}$ being referred to as a Markov chain. The statistical average $\langle A \rangle$ of a given property A is the simple arithmetic average of this property along the Markov chain:

$$\langle A \rangle = \frac{1}{Z} \int d^{3N} \mathbf{R} A(\mathbf{R}) \exp[-\beta V(\mathbf{R})] = \lim_{m \to \infty} \frac{1}{m} \sum_{i=1}^{m} A(\mathbf{R}_i). \tag{5.11}$$

In contrast, MD trajectories yield time averages $\overline{A}$ from the continuous time series $A(\mathbf{R}(s))$:

$$\overline{A} = \lim_{t \to \infty} \frac{1}{t} \int_{t=0}^{t} A(\mathbf{R}(s)) ds. \tag{5.12}$$

The equivalence of the two methods, performed within a common statistical ensemble, relies on the hypothesis of ergodicity $\langle A \rangle = \overline{A}$, in the limit of complete sampling. However, although nonergodicity can be detected, strict ergodicity cannot be proven in general, and it is important to recognize factors that impede ergodicity in simulations.

Observables produced by the MD and MC algorithms may seem different, and it is true that only MD can produce time-dependent information that is useful, notably for transport or kinetic properties. However, it is important to stress that phase transitions *at thermal equilibrium* assume that the system has had an infinitely long time to reach this equilibrium. Keeping in mind that the natural time scales of MD do not exceed the micro- to millisecond range, this raises the question of nonergodicity that could occur in MD trajectories if some important, low-lying regions of configuration space are connected through high-energy barriers. In such situations, the time needed to cross those barriers may exceed the time that can be covered by orders of magnitude, and all properties obtained from the time series will be flawed by lack of ergodicity. Conventional MC trajectories also suffer from slow relaxation on such landscapes, and those difficulties have prompted the development of various techniques aimed at accelerating sampling and circumventing broken ergodicity in simulations.

The two most popular strategies consist either of combining information obtained at higher temperature or energy or of biasing the sampling by driving the system artificially toward the less probable regions of the barriers. The two strategies are schematized in Figure 5.3 for a typical 1D energy landscape with two funnels of low-lying minima.

In the first approach, an efficient algorithm is the replica-exchange method of molecular dynamics [55] or its equivalent parallel tempering Monte Carlo (PTMC) version [56], in which several trajectories are propagated simultaneously at different temperatures or total energies. Configurations from different (usually neighboring) trajectories can be exchanged with some probability that preserves the corresponding weights in the appropriate statistical ensembles. The method is particularly suited to multiple-funnel landscapes but does not scale favorably with size because more trajectories must be introduced to cover a given temperature range. This limitation is due to the need that the potential energy distributions of exchanging trajectories overlap with each other.

The second strategy has also been declined in different flavors suitable for MD or MC and works usually with single trajectories but on a modified potential energy surface $\tilde{V}(\mathbf{R}) = V(\mathbf{R}) + \Delta V(\mathbf{R})$, ΔV being a biasing potential chosen *ad hoc* or determined on the fly in order to broaden sampling and guide it toward nonphysical or rare regions,

Chapter 5

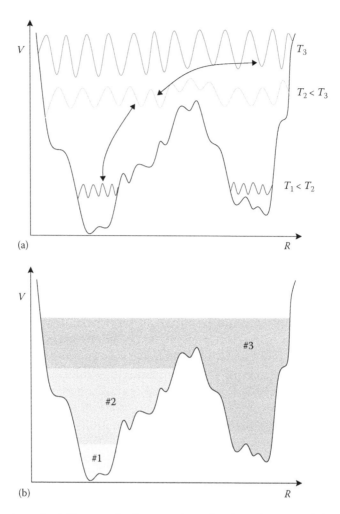

FIGURE 5.3 A schematized 1D energy landscape with two low-lying funnels and the ways it is sampled by some efficient algorithms. (a) In parallel tempering (or replica exchange), different trajectories are performed simultaneously, and configurations between neighboring trajectories are occasionally swapped, allowing the various funnels to be sampled even at low temperature. (b) Multicanonical methods aim to sample the landscape uniformly by penalizing already visited regions, the biasing function eventually obtained after several iterations #1, #2, #3... giving the thermodynamical functions of interest over the uniformly visited configuration space.

according to specific order parameters. MC versions of the method include the so-called multicanonical ensemble sampling [57] or the Wang–Landau approach [58]. In the context of MD, a popular such method called metadynamics [59] proceeds by adding small Gaussian penalizing functions $\delta V(\mathbf{R})$ during the trajectory to the overall bias ΔV.

5.3.4   Observables from Computer Simulations

To analyze phase transitions in nanoscale systems from a given potential energy surface, a primary goal is usually the characterization of the caloric curves, either canonical (internal energy U or heat capacity C_v versus temperature) or microcanonical (temperature T_μ versus energy). Besides direct calculations from time or statistical averages and the use of appropriate formulas, optimized data analyses based on histogram reweighting prove to be particularly efficient to recover the entire thermodynamical functions, starting with the microcanonical density of states Ω, from the potential energy distributions [60].

However, thermodynamical functions may show only weak signals, especially if the transitions are broad as in the case of the smallest systems. It is therefore insightful to monitor other properties in order to assist interpretation of the simulation results. Properties based on pure averages or fluctuations can be inferred from simulations, possibly enhanced with the ergodicity restoring strategies discussed in the previous section, and after unbiasing if necessary. Thermodynamically relevant observables include shape indicators such as the principal momenta of inertia, the radial or pair distribution functions, and more generally properties that are sensitive to changes in shape or coordination, such as the electric polarizability or bond-orientational order parameters [61]. Other observables with a sensitivity to thermodynamical phase through the details of atomic structure are obviously system dependent and can be spectroscopic, electronic, or magnetic in nature.

From single unbiased trajectories, a useful quantity to probe the extent of disorder in the whole system or its subparts is the root-mean-square bond-length fluctuations, which can be considered as a generalization of the aforediscussed ideas of Lindemann [5]:

$$\delta = \frac{2}{N(N-1)} \sum_{i<j} \frac{\sqrt{\langle r_{ij}^2 \rangle - \langle r_{ij} \rangle^2}}{\langle r_{ij} \rangle}, \tag{5.13}$$

where the sum acts on all pairs and not only nearest neighbors. Melting is then conveniently associated with δ exceeding some threshold value near 10%–15%. With MD, time-dependent properties can also be evaluated and tentatively correlated with the thermodynamical phase. For instance, the self-diffusion constant D can be computed by integrating the velocity time autocorrelation function or equivalently by differentiating the long-time mean square displacement. More generally, the density of vibrational states can be rather sensitive to the presence of strong anharmonicities that develop in the liquid state, whereas only discrete phonon-type modes are expected in the vibrating solid particle.

It can also be extremely useful to connect the simulation properties to the underlying topography of the energy landscape and to interpret the observed physical features in terms of the structures effectively sampled. This connection can be made by systematically quenching the trajectories to locate the closest local minima, referred then to as *inherent structures*. In addition to telling which parts of configuration space are visited as a function of temperature, this quenching analysis can be used to monitor the extent of ergodicity in the simulation by comparing the thermal properties evaluated along the trajectories to the predictions of the superposition method.

Chapter 5

5.4   Melting of Inorganic Clusters and Nanoparticles: Selected Examples

The following overview mostly focuses on computational and modeling of thermal effects on inorganic systems but also highlights some of the most prominent results inferred from recent experiments.

5.4.1   Van der Waals Clusters

Clusters bound by weak London-type interactions are usually stable only at low temperatures and as such rarely constitute potential candidates for solid-state materials. There are however exceptions provided, for instance, by molecular compounds exposing large aromatic cycles causing noncovalent dispersive interactions with cohesive energies approaching 1 eV [62]. The thermodynamics of those systems is rather peculiar, because due to the significant geometrical constraints the interactions are short ranged and melting is quickly followed by sublimation [63], as also occurs in the case of clusters of fullerene molecules [64].

Rare-gas clusters are archetypal of nanosize systems bound by van der Waals forces. They can be reasonably well modeled by simple pairwise LJ interactions, and their atomic nature has made them probably the most studied type of clusters for thermodynamical and statistical investigations, dating back to the early 1970s [65]. In particular, they have been repeatedly used for benchmarking computational methods for sampling and global optimization. Despite lacking direct experimental comparison, it therefore remains important to discuss the melting behavior of rare-gas clusters in view of the generic phenomena they shed light on.

The thermodynamics of LJ clusters containing several tens of atoms has been addressed in detail by various authors, and the results generally show various premelting features in the caloric curves [66] or the dynamical indicators [67]. Premelting can be due to structural transitions in which some minima higher in energy and also lower in entropy with respect to the global minimum become more stable at temperatures lower than the melting point, possibly with thermal manifestations. Such phenomena have been investigated specifically for the 38- and 75-atom clusters [68,69]. Surface melting, while also rather general in these systems [67], has limited thermal signatures on the heat capacity. Premelting effects are often associated with peculiar structures of the global minimum, although this is not systematic either. We have represented in Figure 5.4 the heat capacities obtained for argon clusters modeled using an extended LJ potential that describes more accurately the quantum chemical energy curve for the dimer [70]. The simulations were conducted by PTMC on a series of perfect icosahedral clusters containing up to 923 atoms (6 shells) [18]. Only below the size of 309 atoms does the heat capacity exhibit a single peak that becomes narrower and is shifted to higher temperatures as size increases. Premelting features indicative of surface reconstruction are clearly seen in the largest clusters, but they are progressively masked by the main melting peak. Interestingly, representing the melting temperature as a function of size distinctly shows linear scaling with $N^{-1/3}$, as predicted by the phenomenological models discussed previously, which extrapolates as $N \rightarrow \infty$ slightly above the experimental melting temperature

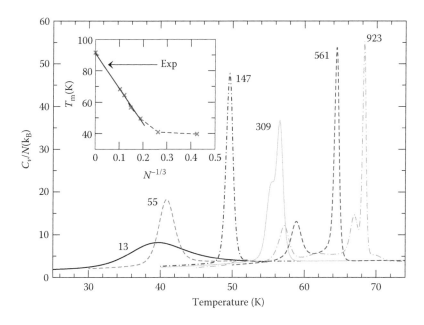

FIGURE 5.4   Heat capacities of increasingly large argon clusters modeled using an accurate pair potential, as a function of temperature. The inset shows the variations of the melting temperature inferred from the highest peak in the heat capacity, as a function of $N^{-1/3}$. (Data taken from Pahl, E. et al., *Angew. Chem. Int. Ed.*, 47, 8207, 2008.)

of argon. Note that upon the addition of three-body interactions, the agreement improves significantly [18]. The generic behavior of rare-gas clusters resembles the one illustrated here for argon, with the exception of neon clusters in which quantum effects become more important across the entire melting range, with the consequence of lowering the melting point by stabilizing the liquid [71]. In the case of Ne_{38}, the classical global minimum is destabilized as well, rendering the heat capacity featureless [72].

Many clusters of small molecules are mainly bound by dispersion forces as well as some multipolar electrostatic interactions. In some cases those multipolar interactions are essential, as is the case of water that is stabilized through a network of hydrogen bonds. Conversely, if those multipolar interactions are minor, the main difference with atomic rare-gas clusters lies in their additional rotational degrees of freedom that can impact the stable structures and their thermodynamics to a large extent, similarly as they are responsible for the much more complex phase diagrams usually displayed by molecular crystals. At the nanoscale, molecular clusters can show particular transitions possibly seen as premelting, originating from the differentiated behaviors of the rotational and translational degrees of freedom. Cluster analogs to plastic transitions, in which the molecules have some rotational freedom while their centers of mass remain close to equilibrium, have notably been reported from simulations for nitrogen [73], methane [74], dimethylnitramine [75], or even C_{60} [64]. However, in larger molecules such as carbon dioxide, sulfur hexafluoride, or the aforementioned polycyclic aromatic hydrocarbons, the rotational and translational variables are released simultaneously upon melting.

Chapter 5

5.4.2 Metal Clusters and Nanoalloys

Whereas the thermodynamics of rare-gas clusters has only been indirectly addressed from the point of view of experiments [25,26], metallic clusters are much more documented owing to their easier production and to the variety of excitations and detections they offer. From electron diffraction or nanocalorimetry experiments, large metal nanoparticles essentially follow the $N^{-1/3}$ scaling law for their melting temperature, with a similar behavior for the latent heat. This notably seems to hold for gold [10], tin [12], or bismuth [16].

Sodium and aluminum clusters have received a sustained attention, with several dedicated experiments on size-selected ions revealing complex variations of the caloric curves with the number of atoms [76]. The case of sodium, with the seminal measurements by the Haberland group [33,34], has motivated numerous computational investigations aimed at interpreting those size variations. Eventually it was shown by Aguado and López [36] that a rather sophisticated level of modeling with explicit account of electronic structure (but within an orbital-free treatment of density-functional theory) was necessary in order to reproduce the experimental data, simpler models based on many-body potentials turning out insufficiently accurate. The quantitative agreement reached by the results from these simulations with the measurements by the Haberland group, reproduced in Figure 5.5, confirmed the subtle interplay between electronic and geometric shell closings.

Neither in the experiment nor in these reference simulations does the heat capacity show any feature except a main, single peak. Conversely, the caloric curves of aluminum clusters display a broad variety of pre- and even postmelting features, as well as nonequilibrium structural transitions [39] that so far have eluded complete theoretical understanding despite an overall satisfactory reproduction of finite-size effects in simulations [77]. Aluminum is also an interesting case where nanocalorimetry experiments have been conducted on larger sizes, bridging the gap between the quantum and scaling

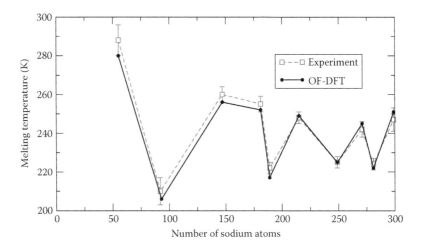

FIGURE 5.5 Melting temperatures obtained for cationic sodium clusters from orbital-free molecular dynamics simulations by Aguado and López [36] and experimental measurements by the Haberland group [34].

regimes. However, the results obtained for such large nanoparticles appear to be some-what scattered, with the convergence to the bulk melting point (933.5 K) varying among authors [78–80]. As discussed by Sun and Simon [78], one reason for this discrepancy may originate from the formation of an alumina shell due to the presence of residual oxygen under experimental conditions.

The melting temperature measured for some aluminum clusters by Jarrold and coworkers [39] happens to exceed the bulk melting point by a small amount. Higher-than-bulk melting points have also been observed in gallium [37] and tin [81] clusters by the same group. In the latter cases, this peculiar phenomenon originates from changes in chemical bonding as the cluster size is lowered and could be essentially interpreted on the basis of dedicated simulations with an explicit account of electronic structure [82,83]. For these small clusters, the electronic energy levels are not numerous enough to form a clear conducting band, and chemical bonding is more akin to covalent than metallic. Covalent bonds being stronger than most metallic bonds, the melting point is higher than its bulk limiting value.

Special attention has been granted to supported metal particles, for their greater potential in applications than free-standing clusters. Experiments on gold nanoparticles have found a significant dependence on the substrate, emphasizing the importance of particle–substrate interactions and the contact angle [84] and confirming earlier simu-lation results [85,86]. Size effects due to the contact angle were also reported by Shibuta and Suzuki [87] who performed MD simulations for reasonably large nanoparticles of various metals. Finally, Hendy [88] showed using a simple thermodynamic model that the melting point depression is related to the difference in contact angles between the solid and liquid phases of the supported nanoparticle.

A special type of metal nanoparticles consists of bimetallic clusters, in which chemi-cal ordering may be altered during the synthesis or relaxed later to some thermal equi-librium [89]. Growing evidence from simulation suggests that already a single atom impurity is able to change the melting temperature of a nanoparticle containing more than a hundred atoms owing to the strain it can release by accommodating to its envi-ronment better than for homogeneous clusters [90].

In general, mixed clusters are better suited to MC investigations owing to a much faster equilibration of chemical order through particle swap moves. Although not for-bidden in MD explorations, those moves involve the crossing of high energy barriers and will therefore constitute rare events that challenge ergodicity. Figure 5.6 illustrates the rich diversity of melting behaviors in bimetallic nanoparticles, as can be interpreted from PTMC simulations with many-body potentials. The examples chosen are Ag–Au, Ag–Pt, and Ag–Ni clusters at the same composition of 75% silver and at the same total size of $N = 309$ atoms [91]. In bulk form, silver readily mixes with gold at all composi-tions but is immiscible with nickel. Platinum is intermediate, with various alloys in the phase diagram. These different behaviors are reflected at the nanoscale.

For the Ag–Au cluster, a solid solution is found in the simulation below the melt-ing point, as shown by the radial distributions (Figure 5.6a). Alloying does not impact the caloric curve, as is also observed in simulations of Pd–Pt clusters [92], and it also becomes increasingly favored in larger clusters.

At 0 K, the other two clusters display core/shell phase separated structures with the element with the lower surface energy (silver) lying outside. The Ag–Ni cluster remains

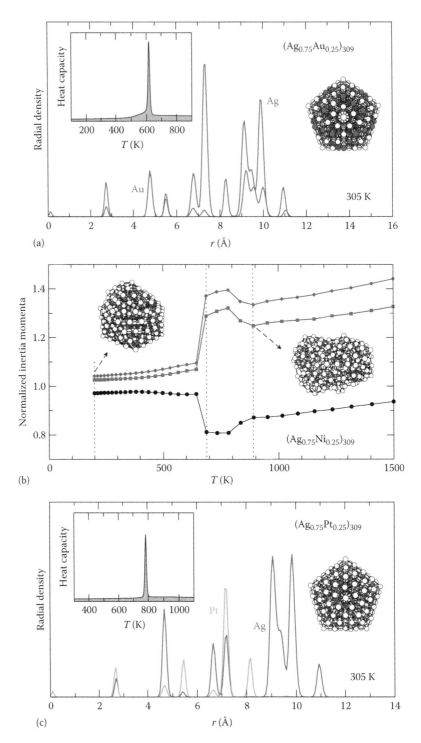

FIGURE 5.6 Melting features of 309-atom, silver-rich nanoclusters modeled by empirical potentials. (a) $Ag_{233}Au_{76}$, radial densities at $T = 305$ K, with the heat capacity as an inset; (b) $Ag_{233}Ni_{76}$, principal momenta of inertia and two representative snapshots at low and high temperatures; (c) $Ag_{233}Pt_{76}$, radial densities at $T = 305$ K, with the heat capacity as an inset. (Data taken from Calvo, F. et al., *Phys. Rev. B*, 77, 121406, 2008.)

phase separated across melting, but the silver shell melts more easily than the nickel core, as expected from the different bulk melting temperatures. The heat capacity peak is then associated with a significant deformation of the overall shape, as revealed by the principal momenta of inertia (Figure 5.6b). Finally, the radial distributions for the Ag–Pt cluster indicate that some mixing takes place prior to melting, but only within the innermost layers. Below the melting point, the equilibrium state of the system is better represented as a core alloy surrounded by a pure silver shell.

5.4.3   Ionic Clusters and Nanoparticles

Ionic bonds caused by electronegativity differences between unlike elements are quite strong due to the Coulomb force acting at molecular distances. An unexpected manifestation of this stronger binding was experimentally observed on the higher melting point in Na_n^+ clusters doped with an oxygen atom, found to be comparable to the melting temperature of Na_{n-2}^+ [93], hence pointing to the formation of a very stable ionic Na_2O impurity sinking to the cluster core.

Due to the presence of long-range and repulsive interactions, the energy landscape of ionic clusters and nanoparticles has much fewer minima than homogeneous clusters of the same size, identical ions tending to avoid each other. Ionic clusters are generally convenient to model (using polarizable force fields). Despite a limited amount of experimental data [94], the thermodynamics of ionic clusters has received a sustained attention from computational theorists, especially in the case of alkali halide clusters [95–98].

Simulations conducted on *magic* (cubic) KCl clusters [96] have confirmed the general phenomenon of dynamical coexistence between solid-like and liquid-like phases originally predicted for LJ clusters. These effects have been indirectly supported by Breaux et al. in their experimental study [94]. In nonmagic clusters, melting proceeds through several stages reflecting a more hierarchical nature of the energy landscape [97]. Such multistep processes can be analyzed using various computational tools, but the Lindemann index turns out to be particularly insightful [98]. We show in Figure 5.7 the variations of both the heat capacity and the rms bond length fluctuation δ for two small sodium fluoride clusters, $(NaF)_7$ and $(NaF)_{13}$, as obtained from MC simulations. For these clusters, the heat capacity exhibits at least one extra peak or shoulder at a temperature lower than the main melting peak, and each feature is associated with a concomitant jump in the Lindemann index δ. Systematic quenching of the MC trajectories shows that new bunches of isomers arise with each of these transitions, which therefore correspond to successive openings of the landscape.

Oxide clusters are a special type of ionic systems, whose binding has a significant polarization contribution making it harder to model with explicit potentials. Among the very few experimental data available, the depression in the melting point of indium oxide nanoparticles has been reported by Singh and Mehta [99]. From the computational point of view, melting in metal oxides has been simulated using first-principle MD simulations by Bromley and coworkers [100], who found for $(MgO)_6$ a premelting stage in which the ions rearrange between two isomers before being able to access higher-lying minima. This process is very akin to the phenomenology reported earlier in the alkali halide clusters. Silica clusters or nanoparticles should also be considered as

Chapter 5

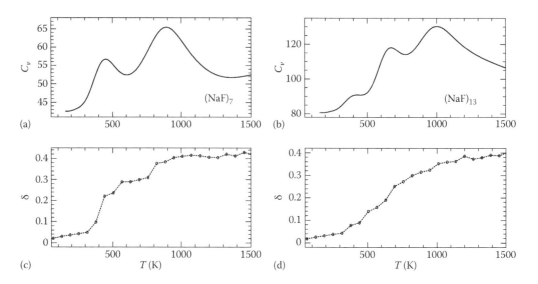

FIGURE 5.7 (a) and (b) Heat capacities; (c) and (d) rms bond length fluctuation indices, as computed for the small ionic clusters $(NaF)_7$ (a and c) and $(NaF)_{13}$ (b and d) from Monte Carlo simulations, as a function of temperature. (Data taken from Calvo, F. and Labastie, P., *J. Phys. Chem. B*, 102, 2051, 1998.)

ionic materials, rather than covalent. *Ab initio* MD of silica nanoparticles has revealed similarities in the thermal behavior of small silica clusters embedded in dielectric media to the experimentally measured properties of bulk silica, including deviation from the Debye law at low temperatures, and a glassy behavior at high temperatures before melting [101].

5.4.4 Semiconducting and Covalent Nanoparticles

The amount of investigations devoted to the thermal behavior of semiconducting and covalent nanomaterials is relatively modest in comparison to other materials, even though they occupy an important role in the context of miniaturized electronics. The directional bonding in covalent interactions stabilizes amorphous structures and glassy phases in the bulk. At the nanoscale, clusters of semiconducting elements also present a broad variety of stable structures that have not yet been fully characterized experimentally.

The apparent variations in the melting point of CdS nanoparticles found by Farrell and coworkers to vary quadratically with inverse radius $1/R$ [21] appear to be similar to the variations in the band gap of quantum dots, which according to the effective mass approximation should scale as $1/R^2$ as well [102]. This led Nanda [103] to speculate that a correlation between the melting temperature and the electronic energy gap exists in semiconductor nanoparticles.

From the computational point of view, different models are available to describe semiconductor and covalent bonding, ranging from explicit potentials to approximate electronic structure theories and more sophisticated density-functional approaches. The thermal behavior of Si nanoparticles containing hundreds to thousands of atoms could be studied by MD simulations with the Stillinger–Weber potential, the results of

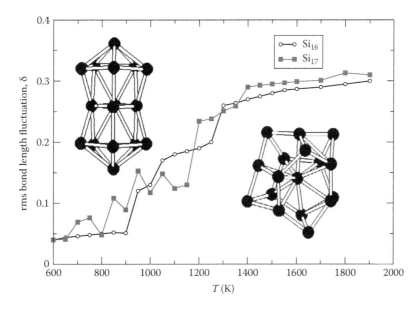

FIGURE 5.8 Variations of the rms bond length fluctuation index δ in molecular dynamics simulations of the melting of Si_{16} (left) and Si_{17} (right) clusters, described with a nonorthogonal tight-binding model. (Data taken from Wang, J. et al., *Chem. Phys. Lett.*, 341, 529, 2001.)

which indicate that the depression in the melting temperature scales inversely as the cluster radius [104]—although the model significantly overestimates the bulk melting point by more than 200 K. Smaller clusters containing tens of atoms only can be modeled using the more demanding tight-binding approach, which yields a more realistic picture of covalent bonding. The simulations performed by Wang and coworkers [105] revealed interesting multistage melting transitions in which the clusters could undergo various shape-changing regimes before fully melting. However, these premelting effects were also found to depend quite significantly on cluster size, as illustrated in Figure 5.8, where the rms bond length fluctuation index exhibits markedly different variations with increasing temperature. The sharper transition exhibited by the 16-atom cluster is connected to its higher symmetry, whereas Si_{17} is already somewhat amorphous and melts more continuously as temperature is increased. Similar mechanisms have also been identified in germanium clusters of comparable size using a related tight-binding model [106], and they have been confirmed on the vibrational spectra obtained from *ab initio* MD [107].

5.5 Toward the Liquid–Vapor Transition

5.5.1 Evaporation and the Low-Energy Regime

Evaporation refers to the thermal dissociation occurring in a system when its internal energy exceeds a certain threshold. In a macroscopic compound the available energy is usually huge with respect to the energy needed to dissociate a constituent, and the probability that a sufficient fraction of this energy will successfully localize into a dissociative mode is vanishingly small. One early formulation of the problem was proposed by Rice,

Chapter 5

Ramsperger [108], and independently by Kassel [109] who incidentally also extended the theory to quantum mechanics. Assuming that both the parent and evaporation product behave as a collection of harmonic oscillators, these authors found a simple relation between the evaporation rate k, the internal energy E, the dissociation energy E_0, and the number of atoms N as

$$k(E) = \nu_0 \left(1 - \frac{E_0}{E} \right)^{\kappa - 1}, \tag{5.14}$$

where ν_0 denotes a constant frequency factor and $\kappa = 3N-6$ is the number of independent degrees of freedom for a nonrotating and nontranslating system.

If we introduce the approximate temperature T of the system such that $E = (3N - 6) k_B T$, the limit $N \to \infty$ in Equation 5.14 leads to an Arrhenius formula $k(T) \simeq \nu_0 \exp(-E_0/k_B T)$ for a bulk material. This expression shows that for a large N-atom system the internal energy has to scale as $3NE_0$ for the rate constant to be of the order of ν_0. Conversely, in a nanoparticle, the energy may not have to be much larger than E_0 for the rate constant to be already significant. This is sufficient to question the existence of the nanoparticle itself if the internal energy makes $1/k$ comparable to the observation time scale. Such considerations have led Klots [110] to suggest that the appropriate statistical ensemble for isolated nanosystems should include knowledge of the experimental time scale and of the dissociation rates.

The harmonic approximation used here is not realistic for a material undergoing melting, and a more accurate treatment was provided by Weisskopf in the context of nuclear physics [111], who obtained general expressions for the rate constant involving the ratio Ω'/Ω between the microcanonical densities of states of the product and parent clusters. Statistical properties at equilibrium are characterized by the densities of states; hence, it is not surprising that phase transitions found as special features in Ω (or Ω') could also have some signature in evaporation properties. Besides the rate constant, the kinetic energy ε of the dissociating fragment is also useful because it can be measured in experiments. Within statistical theories of dissociation, the distribution $p(\varepsilon)$ and its average $\langle \varepsilon \rangle$ only involve the product properties Ω' [111].

This was exploited by Bréchignac and coworkers [112] who could correlate the average kinetic energy released $\langle \varepsilon \rangle$ to the internal energy E in the dissociation of strontium cluster cations. The results obtained were interpreted as a caloric curve giving evidence of the possible phase transition in the product clusters (see Figure 5.9a). Such experiments are difficult, because the simultaneous measurement of the internal energy and the fragment kinetic energies requires dedicated setups. Computational modeling is here significantly easier and appears very complementary to experiments. We show in Figure 5.9b the average kinetic energy released in the evaporation of a Xe atom from the mixed $KrXe_{13}$ LJ cluster [113], as obtained from direct MD simulations at high energies, and extrapolated to low energies using the phase space theory extension [114] of the Weisskopf approach.

The inflections found in $\langle \varepsilon \rangle$ at two energies are caused by concomitant variations in the density of states of the product cluster $KrXe_{12}$, which exhibits near $E \simeq -40$ LJ Xe units a preliminary isomerization transition involving the equilibrium of the krypton atom between the center and the core of the icosahedron, before fully melting near $E \simeq -30$ units.

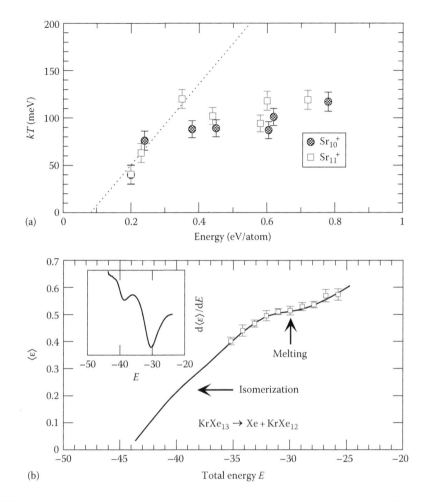

FIGURE 5.9 Caloric curves obtained from dissociation observables. (a) Temperature of Sr_n^+ clusters as a function of internal energy measured by Bréchignac et al. (b) Average kinetic energy $\langle \varepsilon \rangle$ of a Xe atom evaporating from $KrXe_{13}$ computed as a function of internal energy from molecular dynamics simulations (symbols) and extrapolated at low energies by phase space theory. The signatures of isomerization and melting marked by arrows are highlighted in the inset where the derivative $d\langle \varepsilon \rangle/dE$ is shown. (a: Data taken from Bréchignac, Ph. et al., *Phys. Rev. Lett.*, 89, 203401, 2002; b: The data are in LJ xenon units and taken from Parneix, P. et al., *Chem. Phys. Lett.*, 381, 471, 2003.)

5.5.2 Multifragmentation

Evaporation is a thermodynamically driven kinetic process occurring randomly but on a statistically equilibrated ground. If the excitation energy is high and deposited quickly with respect to vibrational redistribution, this extra energy converted into a large amount of kinetic energy may produce the rapid ejection of multiple fragments under short time scales. This so-called multifragmentation scenario is particularly relevant in nuclear physics, which deals with clusters of neutrons and protons, but many computational works have been devoted to simulating the fragmentation of atomic and molecular clusters [115–118].

Chapter 5

In such high-energy regimes, the distribution of fragments is a more appropriate observable than individual properties or even averages, which can sometimes be accessed in experiments equipped with 4π detectors. By varying the excitation energy, different fragmentation regimes can be reached and monitored. Inspired by nuclear physics, the interpretation of mass distributions is usually made in relation with critical phenomena and the liquid–gas transition [119–121]. The Fisher model originally proposed in the context of liquid–vapor nucleation [122] turns out to offer a rather good phenomenology of experiments and simulations. Based on equilibrium thermodynamical arguments, this model predicts the mass distribution to be given by [122]

$$\mathcal{P}(n,T) \propto n^{-\tau} A^{n^{2/3}} B^{n}, \tag{5.15}$$

where the coefficients A and B are functions of temperature. The Fisher model describes different fragmentation regimes depending on these parameters. Above a critical temperature $T > T_c$, $A = 0$ and the distribution is peaked for small fragments and decays rapidly with increasing n. At $T = T_c$, $B = 0$ and the distribution is a power law. Below $T < T_c$, the distribution becomes U-shaped with the occurrence of large fragments.

The Fisher model has been supported by MD simulations of nuclear [119], atomic [115], and molecular [116] clusters, who found evidence of the three fragmentation regimes. In addition to this model, percolation theories [123] have been used to interpret experimental distributions of the fragments of C_{60} excited by highly charged ions [124]. Campi scatter plots [125] are a more refined way to look at the fragment distributions, by representing the fragment multiplicity as a function of the largest fragment.

Computational modeling of the multifragmentation products has also been achieved by MC methods, in which the excitation energy is assumed to distribute statistically into all degrees of freedom (internal, translational, rotational, and interfragment), allowed to occupy a fixed volume [126,127].

5.6 Outlook

This chapter attempted to present the thermodynamical properties of finite-size nanoparticles, with a focus at phase transitions and, in particular, melting. From a theoretically oriented perspective, we have discussed some of the fundamental differences between nanosystems and bulk matter and provided some quantitative estimates of these differences in the large size scaling regime. In general, the melting point of a nanomaterial exhibits a depression that decreases proportionally with the inverse radius. Although exceptions exist for semiconductors and oxide-embedded metals, the relation seems to hold for rare gases and pure metals, and probably for ionic systems as well. The broadening of the transition, which affects all statistical observables, is manifested by the dynamical fluctuation (or coexistence) of the system between its fully solid and fully liquid forms. In the microcanonical ensemble, this phenomenon can produce a backbending in the caloric curve with the temporary decrease in the temperature as total energy increases.

Besides negative heat capacities, atomic and molecular nanoparticles display a broad variety of size-specific features, ranging from surface melting, dynamical isomerization, or entropy-driven structural transitions at temperatures lower than the complete melting point. Although these premelting processes do not necessarily have a thermal signature, they happen to be quite general, and they are particularly prominent in ionic systems. In some cases, clusters melt at temperatures higher than the bulk material. This surprising phenomenon is generally caused by changes in chemical bonding and the partly covalent interactions occurring before the metallic electrons are fully delocalized.

The computational tools required to address melting and finite temperature properties based on molecular simulations imply that a potential energy surface is available, obtained either at the level of a force field or with increasingly refined treatments of electronic structure. Systems with open shells, for which the excited electronic states could have a contribution to the thermodynamics, are usually harder to handle due to the difficulty of evaluating the excited state surfaces [128]. We have emphasized that the two main types of computational methods (MD and MC) are essentially equivalent for the purpose of finite temperature simulations. In particular, the efficient sampling of configuration space can benefit from the same types of ergodicity-enhancing tricks that belong to the parallel tempering or multicanonical biasing families. Practical differences exist though, as only MD can produce time-dependent properties, whereas MC has the ability to bypass the possibly slow kinetics through clever sets of random moves. In practice, the choice of a method over the other should also be dictated by the potential energy surface itself and the associated chemical model. If the energy is a complex many-body function of the coordinates, it may be advantageous to perform MD simulations where all atoms move at once. Conversely, if the gradient is very computer intensive, MC could be far more preferable.

This chapter was also limited to a classical treatment of nuclear motion. The quantum statistical description can be achieved through the use of dedicated computational approaches, and those based on path integrals should be particularly recommended because they are rigorous, controllable, and relatively easy to implement. Quantum effects may have a drastic influence on the thermal properties, as the energy stored in vibrational modes is quantized and always exceeds the zero-point energy even at absolute zero temperature. The heat capacity and internal energy are modified below the Debye temperature, which in most materials is about one-third of the melting point. Nuclear quantum effects are naturally significant in weakly bound clusters, as they are, for instance, responsible for superfluidity in helium droplets. They are also distinctly measurable in metals, and the variations of the low temperature heat capacity could give insight into the vibrations of the nanomaterial [71].

References

1. H. Löwen, Melting, freezing and colloidal suspensions, *Phys. Rep.* **153**, 249–324 (1994).
2. C. N. Yang and T. Lee, Statistical theory of equations of state and phase transitions. I. Theory of condensation, *Phys. Rev.* **87**, 404–409 (1952); Statistical theory of equations of state and phase transitions. II. Lattice gas and Ising model, *Phys. Rev.* **87**, 410 (1952).
3. P. Papon, J. Leblond, and P. H. E. Meijer, *The Physics of Phase Transitions: Concepts and Applications.* Springer-Verlag, Berlin, Germany (2006).
4. F. A. Lindemann, The calculation of molecular natural frequencies, *Phys. Z.* **11**, 609–612 (1910).
5. R. S. Berry, T. L. Beck, H. L. Davis, and J. Jellinek, Solid-liquid phase behavior in microclusters, *Adv. Chem. Phys.* **70B**, 75–138 (1988).
6. J. W. M. Frenken and J. F. van der Veen, Observation of surface melting, *Phys. Rev. Lett.* **54**, 134–137 (1985).

Chapter 5

7. U. Tartaglino, T. Zykova-Timan, F. Ercolessi, and E. Tosatti, Melting and nonmelting of solid surfaces and nanosystems, *Phys. Rep.* **411**, 291–321 (2005).
8. D. Turnbull and R. E. Cech, Microscopic observation of the solidification of small metal droplets, *J. Appl. Phys.* **21**, 804–810 (1950).
9. P. Pawlow, Über die abhängigkeit des schmelzpunktes von der oberflächenergie eines festen körpers, *Z. Phys. Chem.* **65**, 1–35 (1909).
10. Ph. Buffat and J.-P. Borel, Size effect on the melting temperatures of gold particles, *Phys. Rev. A* **13**, 2287–2298 (1976).
11. P. R. Couchman and W. A. Jesser, Thermodynamic theory of size dependence of melting temperature in metals, *Nature* **269**, 481–483 (1977).
12. S. L. Lai, J. Y. Guo, V. Petrova, G. Ramanath, and L. H. Allen, Size-dependent melting properties of small tin particles: Nanocalorimetric measurements, *Phys. Rev. Lett.* **77**, 99–102 (1996).
13. C. E. Bottani, A. Li Bassi, B. K. Tanner, A. Stella, P. Tognini, P. Cheyssac, and R. Kofman, Melting in metallic Sn nanoparticles studied by surface Brillouin scattering and synchrotron-x-ray diffraction, *Phys. Rev. B* **59**, R15601–R15604 (1999).
14. R. Defay and I. Prigogine, *Surface Tension and Adsorption.* John Wiley & Sons, Inc., New York (1967).
15. M. Takagi, Electron-diffraction study of liquid-solid transition of thin metal films, *J. Phys. Soc. Jpn.* **9**, 359–363 (1954).
16. S. J. Peppiatt, Melting of small particles. II. Bismuth, *Proc. R. Soc. Lond. Ser. A* **345**, 401–412 (1975).
17. P. Puri and V. Yang, Effect of particle size on melting of aluminum at nano scales, *J. Phys. Chem. C* **111**, 11776–11783 (2007).
18. E. Pahl, F. Calvo, L. Koči, and P. Schwerdtfeger, Accurate melting temperatures for neon and argon from ab initio Monte Carlo simulations, *Angew. Chem. Int. Ed.* **47**, 8207–8210 (2008).
19. K. Kai, S. Qin, and C. Wang, Size-dependent melting: Numerical calculations of the phonon spectrum, *Physica E* **41**, 817–821 (2009).
20. J. H. Los and R. J. M. Pellenq, Determination of the bulk melting temperature of nickel using Monte Carlo simulations: Inaccuracy of extrapolation from cluster melting temperatures, *Phys. Rev. B* **81**, 064112 (2010).
21. H. H. Farrell and C. D. Van Siclen, Binding energy, vapor pressure, and melting point of semiconductor nanoparticles, *J. Vac. Sci. Technol. B* **25**, 1441–1447 (2007); H. H. Farrell, Surface bonding effects in compound semiconductor nanoparticles: II, *J. Vac. Sci. Technol. B* **26**, 1534–1541 (2008).
22. Y. Imry, Finite-size rounding of a first-order phase transition, *Phys. Rev. B* **21**, 2042–2043 (1980).
23. T. P. Martin, U. Näher, H. Schaber, and U. Zimmermann, Evidence for a size-dependent melting of sodium clusters, *J. Chem. Phys.* **100**, 2322–2324 (1994).
24. E. A. Olson, M. Yu. Efremov, M. Zhang, Z. Zhang, and L. H. Allen, Size-dependent melting of Bi nanoparticles, *J. Appl. Phys.* **97**, 034304 (2005).
25. M. Y. Hahn and R. L. Whetten, Rigid-fluid transition in specific-size argon clusters, *Phys. Rev. Lett.* **61**, 1190–1193 (1988).
26. U. Even, N. Ben-Horin, and J. Jortner, Multistate isomerization of size-selected clusters, *Phys. Rev. Lett.* **62**, 140–143 (1989).
27. M. Schmidt, C. Ellert, W. Kronmüller, and H. Haberland, Temperature dependence of the optical response of sodium cluster ions Na_n^+, with $4 \le n \le 16$, *Phys. Rev. B* **59**, 10970–10979 (1999).
28. D. E. Clemmer and M. F. Jarrold, Ion mobility measurements and their applications to clusters and biomolecules, *J. Mass Spectrom.* **32**, 577–592 (1997).
29. L. Kronik, I. Vasiliev, M. Jain, and J. R. Chelikowsky, Ab initio structures and polarizabilities of sodium clusters, *J. Chem. Phys.* **115**, 4322–4332 (2001).
30. B. Schmidt and R. B. Gerber, Reactive collisions as a signature for meltinglike transitions in clusters, *Phys. Rev. Lett.* **72**, 2490–2493 (1994).
31. B. Cao, A. K. Starace, O. H. Judd, and M. F. Jarrold, Melting dramatically enhances the reactivity of aluminum nanoclusters, *J. Am. Chem. Soc.* **131**, 2446–2447 (2009).
32. M. Maier-Borst, D. B. Cameron, M. Rokni, and J. H. Parks, Electron diffraction of trapped cluster ions, *Phys. Rev. A* **59**, R3162–R3165 (1999).
33. M. Schmidt, R. Kusche, W. Krönmuller, B. von Issendorff, and H. Haberland, Experimental determination of the melting point and heat capacity for a free cluster of 139 sodium atoms, *Phys. Rev. Lett.* **79**, 99–102 (1997).

34. M. Schmidt, R. Kusche, B. von Issendorff, and H. Haberland, Irregular variations in the melting point of size-selected atomic clusters, *Nature* **393**, 238–240 (1998).

35. M. Schmidt, J. Donges, T. Hippler, and H. Haberland, Influence of energy and entropy on the melting of sodium clusters, *Phys. Rev. Lett.* **90**, 103401 (2003).

36. A. Aguado and J. M. López, Anomalous size dependence of the melting temperatures of free sodium clusters: An explanation for the calorimetry experiments, *Phys. Rev. Lett.* **94**, 233401 (2005).

37. G. A. Breaux, R. C. Benirschke, T. Sugai, B. S. Kinnear, and M. F. Jarrold, Hot and solid gallium clusters: Too small to melt, *Phys. Rev. Lett.* **91**, 215508 (2003).

38. F. Chirot, P. Feiden, S. Zamith, P. Labastie, and J.-M. L'Hermite, A novel experimental method for the measurement of the caloric curves of clusters, *J. Chem. Phys.* **129**, 164514 (2008).

39. G. A. Breaux, C. M. Neal, B. Cao, and M. F. Jarrold, Melting, pre-melting, and structural transitions in size-selected aluminum clusters with around 55 atoms, *Phys. Rev. Lett.* **94**, 173401 (2005).

40. F. Calvo and J. P. K. Doye, Pressure effects on the structure of nanoclusters, *Phys. Rev. B* **69**, 125414 (2004).

41. J. A. Pakarinen, M. Backman, F. Djurabekova, and K. Nordlund, Partial melting mechanisms of embedded nanocrystals, *Phys. Rev. B* **79**, 085426 (2009).

42. G. Kellermann and A. F. Craievich, Melting and freezing of spherical bismuth nanoparticles confined in a homogeneous sodium borate glass, *Phys. Rev. B* **78**, 054106 (2008).

43. H. B. Callen, *Thermodynamics and an Introduction to Thermostatistics*, 2nd edn. Wiley, New York (1985).

44. E. M. Pearson, T. Halicioglu, and W. A. Tiller, Laplace-transform technique for deriving thermodynamic equations from the classical microcanonical ensemble, *Phys. Rev. A* **32**, 3030–3039 (1985).

45. P. Labastie and R. L. Whetten, Statistical thermodynamics of the cluster solid-liquid transition, *Phys. Rev. Lett.* **65**, 1567–1570 (1990).

46. M. Schmidt, R. Kusche, T. Hippler, J. Donges, W. Krönmuller, B. von Issendorff, and H. Haberland, Negative heat capacity for a cluster of 147 sodium atoms, *Phys. Rev. Lett.* **86**, 1191–1194 (2001).

47. F. Gobet, B. Farizon, M. Farizon, M. J. Gaillard, J.-P. Buchet, M. Carré, P. Scheier, and T. D. Märk, Direct experimental evidence for a negative heat capacity in the liquid-to-gas phase transition in hydrogen cluster ions: Backbending of the caloric curve, *Phys. Rev. Lett.* **89**, 183403 (2002).

48. D. J. Wales, *Energy Landscapes*. Cambridge University Press, Cambridge, U.K. (2003).

49. F. Calvo, J. P. K. Doye, and D. J. Wales, Characterization of anharmonicities on complex potential energy surfaces: Perturbation theory and simulation, *J. Chem. Phys.* **115**, 9627–9636 (2001).

50. F. H. Stillinger, Exponential multiplicity of inherent structures, *Phys. Rev. E* **59**, 48–51 (1999).

51. F. Calvo, J. P. K. Doye, and D. J. Wales, Equilibrium properties of clusters in the harmonic superposition approximation, *Chem. Phys. Lett.* **366**, 176–183 (2002).

52. D. Frenkel and B. Smit, *Understanding Molecular Simulation*, 2nd edn. Academic Press, San Diego, CA (2002).

53. G. Bussi, D. Donadio, and M. Parrinello, Canonical sampling through velocity rescaling, *J. Chem. Phys.* **126**, 014101 (2007).

54. F. Calvo and P. Labastie, Monte-Carlo simulations of rotating clusters, *Euro. Phys. J. D* **3**, 229–236 (1998).

55. Y. Sugita and Y. Okamoto, Replica-exchange molecular dynamics method for protein folding, *Chem. Phys. Lett.* **314**, 141–151 (1999).

56. G. Geyer, Markov chain Monte carlo maximum likelihood, in *Computing Science and Statistics: Proceedings of the 23rd Symposium on the Interface*, edited by E. K. Keramidas. Interface Foundation, Fairfax Station, VA, p. 156 (1991).

57. B. A. Berg and T. Neuhaus, Multicanonical ensemble: A new approach to simulate first-order phase transitions, *Phys. Rev. Lett.* **68**, 9–12 (1991).

58. F. Wang and D. P. Landau, Efficient, multiple-range random walk algorithm to calculate the density of states, *Phys. Rev. Lett.* **86**, 2050–2053 (2001).

59. A. Laio and M. Parrinello, Escaping free-energy minima, *Proc. Natl. Acad. Sci. USA* **99**, 12562–12566 (2002).

60. A. M. Ferrenberg and R. H. Swendsen, Optimized Monte Carlo data analysis, *Phys. Rev. Lett.* **63**, 1195–1198 (1989).

61. P. J. Steinhardt, D. R. Nelson, and M. Ronchetti, Bond-orientational order in liquids and glasses, *Phys. Rev. B* **28**, 784–805 (1983).

Chapter 5

62. M. Rapacioli, F. Calvo, F. Spiegelman, C. Joblin, and D. J. Wales, Stacked clusters of polycyclic aromatic hydrocarbon molecules, *J. Phys. Chem. A* **109**, 2487–2497 (2005).

63. M. Rapacioli, F. Calvo, C. Joblin, P. Parneix, and F. Spiegelman, Vibrations and thermodynamics of clusters of polycyclic aromatic hydrocarbon molecules: The role of internal modes, *J. Phys. Chem. A* **111**, 2999–3009 (2007).

64. F. Calvo, Thermal stability of the solidlike and liquidlike phases of $(C_{60})_n$ clusters, *J. Phys. Chem. B* **105**, 2183–2190 (2001).

65. C. L. Briant and J. J. Burton, Molecular dynamics study of the structure and thermodynamic properties of argon microclusters, *J. Chem. Phys.* **63**, 2045–2059 (1975).

66. D. D. Frantz, Magic number behavior for heat capacities of medium-sized classical Lennard-Jones clusters, *J. Chem. Phys.* **115**, 6136–3157 (2002).

67. F. Calvo and P. Labastie, Evidence for surface melting in clusters made of double icosahedron units, *Chem. Phys. Lett.* **258**, 233–238 (1996).

68. J. P. K. Doye and D. J. Wales, Thermodynamics of global optimization, *Phys. Rev. Lett.* **80**, 1357–1360 (1998).

69. V. A. Mandelshtam, P. A. Frantsuzov, and F. Calvo, Structural transitions and melting in LJ_{74-78} Lennard-Jones clusters from adaptive exchange Monte Carlo simulations, *J. Phys. Chem. A* **110**, 5326–5322 (2006).

70. P. Schwerdtfeger, N. Gaston, R. P. Krawczyk, R. Tonner, and G. E. Moyano, Extension of the Lennard-Jones potential: Theoretical investigations into rare-gas clusters and crystal lattices of He, Ne, Ar, and Kr using many-body interaction expansions, *Phys. Rev. B* **73**, 064112 (2006).

71. F. Calvo, J. P. K. Doye, and D. J. Wales, Quantum partition functions from classical distributions: Application to rare-gas clusters, *J. Chem. Phys.* **114**, 7312–7329 (2001).

72. C. Predescu, P. A. Frantsuzov, and V. A. Mandelshtam, Thermodynamics and equilibrium structure of Ne_{38} cluster: Quantum mechanics versus classical, *J. Chem. Phys.* **122**, 154305 (2005).

73. J.-B. Maillet, A. Boutin, S. Buttefey, F. Calvo, and A. H. Fuchs, From molecular clusters to bulk matter. I. Structure and thermodynamics of CO_2, N_2 and SF_6 clusters, *J. Chem. Phys.* **109**, 329–337 (1998).

74. F. Calvo, Largest Lyapunov exponent in molecular systems. II: Quaternion coordinates and application to methane clusters, *Phys. Rev. E* **60**, 2771–2778 (1999).

75. L. Zheng, B. M. Rice, and D. L. Thompson, Molecular dynamics simulations of the melting mechanisms of perfect and imperfect crystals of dimethylnitramine, *J. Phys. Chem. B* **111**, 2891–2895 (2007).

76. A. Aguado and M. F. Jarrold, Melting and freezing of metal clusters, *Annu. Rev. Phys. Chem.* **62**, 151–172 (2011).

77. A. K. Starace, C. M. Neal, B. Cao, M. F. Jarrold, A. Aguado, and J. M. López, Electronic effects on melting: Comparison of aluminum cluster anions and cations, *J. Chem. Phys.* **131**, 044307 (2009).

78. J. Sun and S. L. Simon, The melting behavior of aluminum nanoparticles, *Thermochim. Acta* **463**, 32–40 (2007).

79. S. L. Lai, J. R. A. Carlsson, and L. H. Allen, Melting point depression of Al clusters generated during the early stages of film growth: Nanocalorimetry measurements, *Appl. Phys. Lett.* **72**, 1098–1100 (1998).

80. J. Eckert, J. C. Holzer, C. C. Ahn, Z. Fu, and W. L. Johnson, Melting behavior of nanocrystalline aluminum powders, *Nanostruct. Mater.* **2**, 407–413 (1993).

81. A. A. Shvartsburg and M. F. Jarrold, Solid clusters above the bulk melting point, *Phys. Rev. Lett.* **85**, 2530–2532 (2000).

82. K. Joshi, D. G. Kanhere, and S. A. Blundell, Thermodynamics of tin clusters, *Phys. Rev. B* **67**, 235413 (2003).

83. S. Chacko, K. Joshi, D. G. Kanhere, and S. A. Blundell, Why do gallium clusters have a higher melting point than the bulk?, *Phys. Rev. Lett.* **92**, 135506 (2004).

84. W. Luo, K. Su, K. Li, G. Liao, N. Hu, and M. Jia, Substrate effect on the melting temperature of gold nanoparticles, *J. Chem. Phys.* **136**, 234704 (2012).

85. C.-L. Kuo and P. Clancy, Melting and freezing characteristics and structural properties of supported and unsupported gold nanoclusters, *J. Phys. Chem. B* **109**, 13743 (2005).

86. A. Jiang, N. Awasthi, A. N. Kolmogorov, W. Setyawan, A. Borjesson, K. Bolton, A. R. Harutyunyan, and S. Curtarolo, Theoretical study of the thermal behavior of free and alumina-supported Fe-C nanoparticles, *Phys. Rev. B* **75**, 205426 (2007).

87. Y. Shibuta and T. Suzuki, Effect of wettability on phase transition in substrate-supported bcc-metal nanoparticles: A molecular dynamics study, *Chem. Phys. Lett.* **486**, 137–143 (2010); Phase transition in substrate-supported molybdenum nanoparticles: A molecular dynamics study, *Phys. Chem. Chem. Phys.* **12**, 731–739 (2010).

88. S. C. Hendy, A thermodynamic model for the melting of supported metal nanoparticles, *Nanotechnology* **18**, 175703 (2007).

89. R. Ferrando, J. Jellinek, and R. L. Johnston, Nanoalloys: From theory to applications of alloy clusters and nanoparticles, *Chem. Rev.* **108**, 845–910 (2008).

90. C. Mottet, G. Rossi, F. Baletto, and R. Ferrando, Single impurity effect on the melting of nanoclusters, *Phys. Rev. Lett.* **95**, 035501 (2005).

91. F. Calvo, E. Cottancin, and M. Broyer, Segregation, core alloying, and shape transitions in bimetallic nanoclusters: Monte Carlo simulations, *Phys. Rev. B* **77**, 121406 (2008).

92. F. Calvo, Solid-solution precursor to melting in onion-ring Pd-Pt nanoclusters: A case of second-order-like phase change?, *Faraday Discus.* **138**, 75–88 (2008).

93. C. Hock, S. Strassburg, H. Haberland, B. von Issendorff, A. Aguado, and M. Schmidt, Melting-point depression by insoluble impurities: A finite-size effect, *Phys. Rev. Lett.* **101**, 023401 (2008).

94. G. A. Breaux, R. C. Benirschke, and M. F. Jarrold, Melting, freezing, sublimation, and phase coexistence in sodium chloride nanocrystals, *J. Chem. Phys.* **121**, 6502–6507 (2004).

95. A. Heidenreich, I. Oref, and J. Jortner, Isomerization dynamics of Na_4Cl_4 clusters, *J. Phys. Chem.* **96**, 7517–7523 (1992).

96. J. P. Rose and R. S. Berry, Freezing, melting, nonwetting, and coexistence in $(KCl)_{32}$, *J. Chem. Phys.* **98**, 3246–3261 (1992).

97. J. P. K. Doye and D. J. Wales, Structural transitions and global minima of sodium chloride clusters, *Phys. Rev. B* **59**, 2292–2300 (1999).

98. F. Calvo and P. Labastie, Melting and phase space transitions in small ionic clusters, *J. Phys. Chem. B* **102**, 2051–2059 (1998).

99. V. N. Singh and B. R. Mehta, Nanoparticle size-dependent lowering of temperature for phase transition from In(OH)(3) to In_2O_3, *J. Nanosci. Nanotechnol.* **5**, 431–435 (2005).

100. F. Viñes, J. Carrasco, and S. T. Bromley, Nanoscale thermal stabilization via permutational premelting, *Phys. Rev. B* **85**, 195425 (2012).

101. G. Ottonello, M. Vetuschi Zuccolini, and D. Belmonte, The vibrational behavior of silica clusters at the glass transition: Ab initio calculations and thermodynamic implications, *J. Chem. Phys.* **133**, 104508 (2010).

102. L. E. Brus, Electron-electron and electron-hole interactions in small semiconductor crystallites: The size dependence of the lowest excited electronic state, *J. Chem. Phys.* **80**, 4403–4409 (1984).

103. K. K. Nanda, On the paradoxical relation between the melting temperature and forbidden energy gap of nanoparticles, *J. Chem. Phys.* **133**, 054502 (2010).

104. K.-C. Kang and C.-I. Weng, An investigation into the melting of silicon nanoclusters using molecular dynamics simulations, *Nanotechnology* **16**, 250–256 (2005).

105. J. Wang, G. Wang, F. Ding, H. Lee, W. Shen, and J. Zhao, Structural transition of Si clusters and their thermodynamics, *Chem. Phys. Lett.* **341**, 529–534 (2001).

106. J. Wang, J. Zhao, F. Ding, W. Shen, H. Lee, and G. Wang, Thermal properties of medium-sized Ge clusters, *Solid State Commun.* **117**, 593–598 (2001).

107. D. Kang, Y. Hou, J. Dai, and J. Yuan, Temperature-dependent vibrational spectra and melting behavior of small silicon clusters based on ab initio molecular dynamics simulations, *Phys. Rev. A* **79**, 063202 (2009).

108. O. K. Rice and H. C. Ramsperger, Theories of unimolecular gas reactions at low pressures, *J. Am. Chem. Soc.* **50**, 1617–1629 (1928).

109. L. S. Kassel, Studies in homogeneous gas reactions. I, *J. Phys. Chem.* **32**, 225–242 (1928).

110. C. E. Klots, The evaporative ensemble, *Z. Phys. D: At. Mol. Clusters* **5**, 83–89 (1987).

111. V. Weisskopf, Statistics and nuclear reactions, *Phys. Rev.* **52**, 295–303 (1937).

112. C. Bréchignac, Ph. Cahuzac, B. Concina, and J. Leygnier, Caloric curves of small fragmenting clusters, *Phys. Rev. Lett.* **89**, 203401 (2002).

113. P. Parneix, Ph. Bréchignac, and F. Calvo, Evaporation of a mixed cluster: $KrXe_{13}$, *Chem. Phys. Lett.* **381**, 471–478 (2003).

114. W. J. Chesnavich and M. T. Bowers, Statistical phase space theory of polyatomic systems: Rigorous energy and angular momentum conservation in reactions involving symmetric polyatomic species, *J. Chem. Phys.* **66**, 2306–2315 (1977).

115. V. N. Kondratyev and H. O. Lutz, Signatures of critical phenomena in rare atom clusters, *Z. Phys. D: At. Mol. Clusters* **40**, 210–214 (1997).

116. T. A. Beu, C. Steinbach, and U. Buck, Model analysis of the fragmentation of large H_2O and NH_3 clusters based on MD simulations, *Euro. Phys. J. D* **27**, 223–229 (2003).

117. A. M. Mazzone, Dynamical behaviour of Si clusters studied in real time: Fragmentation and melting, *Comp. Mater. Sci.* **39**, 393–401 (2007).

118. F. Calvo, Role of charge localization on the Coulomb fragmentation of large metal clusters: A model study, *Phys. Rev. A* **74**, 043202 (2006).

119. M. Belkacem, V. Latora, and A. Bonasera, Critical evolution of a finite system, *Phys. Rev. C* **52**, 271–285 (1995).

120. J. B. Elliott et al., Statistical signatures of critical behavior in small systems, *Phys. Rev. C* **62**, 064603 (2000).

121. C. O. Dorso, V. C. Latora, and A. Bonasera, Signals of critical behavior in fragmenting finite systems, *Phys. Rev. C* **60**, 034606 (1999).

122. M. E. Fisher, The theory of equilibrium critical phenomena, *Rep. Prog. Phys.* **30**, 615–730 (1967).

123. J.-M. Debierre, Exact scaling law for the fragmentation of percolation clusters: Numerical evidence, *Phys. Rev. Lett.* **78**, 3145–3148 (1997).

124. J. Schulte, Time-resolved differential fragmentation cross sections of $C_{60}^+ + C_{60}$ collisions, *Phys. Rev. B* **51**, 3331–3343 (1995).

125. X. Campi, Signals of a phase-transition in nuclear multifragmentation, *Phys. Lett. B* **208**, 351–354 (1988).

126. J. P. Bondorf, A. S. Botvina, A. S. Iljinov, I. N. Mishustin, and K. Sneppen, Statistical multifragmentation of nuclei, *Phys. Rep.* **257**, 133–221 (1995).

127. D. H. E. Gross and P. A. Hervieux, Statistical fragmentation of hot atomic metal-clusters, *Z. Phys. D: At. Mol. Clusters* **35**, 27–42 (1995).

128. F. Calvo and F. Spiegelman, Exchange Monte Carlo for molecular simulations with monoelectronic Hamiltonians, *Phys. Rev. Lett.* **89**, 266401 (2002).

6. Nanoparticles and Crystallization

Gareth A. Tribello

6.1   Introduction 181

6.2   Nucleation 184
    6.2.1   Classical Nucleation Theory 186
    6.2.2   Simulating Nucleation 189
    6.2.3   Other Theories of Nucleation 195

6.3   Crystal Growth 198
    6.3.1   Crystal Surfaces 200
    6.3.2   Kinetics and Thermodynamics of Attachment 205
    6.3.3   Understanding the Role of Additives 208

6.4   Conclusion 208

References 209

6.1   Introduction

The process by which solids form from either liquid or solution is not concerted. When materials crystallize, the aggregates that are initially formed are small, nanoscale particles. These first formed particles have far greater conformational flexibility than the bulk crystals that form subsequently. It is therefore likely that any technique that will emerge for engineering crystals or controlling polymorphism will work by perturbing these first formed nanoparticles.[1] In addition, to grow nanoparticles from solution, it is necessary to stabilize small aggregates of material so as to prevent them from growing into full scale crystals.[2] To better understand the physical origin for this stabilization, it is necessary to understand the process via which inorganic materials nucleate and grow.

As discussed in Chapter 1, the high surface-to-bulk ratio in nanoparticles is behind many of their anomalous properties.[3] When the surface-to-bulk ratio is large, the surface's contribution to the energy of the system is no longer negligible and must be considered. This can be done within the framework of classical thermodynamics.[4] We can express the work done, dw, during a reversible process in which the surface area of an interface changes by an infinitesimal amount, dA, as

$$dw = \alpha dA \tag{6.1}$$

Chapter 6

where α is a quantity known as surface tension. This quantity must be positive as only then can an interface exist between two phases. If the surface tension were negative, the area of the interface would tend to increase without limit, the phases would coalesce, and the interface would cease to exist.

Equation 6.1 is analogous to the formula $dw = -PdV$ for the work done during a reversible change in the volume of a gas. When a gas is confined in a vessel, the pressure can be obtained from the force acting on the walls of the vessel as shown in Figure 6.1. By analogy, then the surface tension can be obtained by examining the force per unit area at any point on the interface. Furthermore, because the surface tension must be positive, this force will always be directed along the inward normal to the perimeter.

We can combine the first law of thermodynamics ($dE = dw + dq$) and the second law of thermodynamics ($dS = dq_{rev}/T$) to express the energy of a system of two phases (of the same substance) as:

$$dE = TdS - P_1dV_1 - P_2dV_2 + \mu_1dN_1 + \mu_2dN_2 + \alpha dA \qquad (6.2)$$

Clearly, work is performed whenever the number of atoms in each of the two phases (N_1 and N_2), the volumes of the two phases (V_1 and V_2), or the surface area of the interface between the two phases (A) changes. When the two phases are in equilibrium, Equation 6.2 can be simplified as the chemical potentials for the two phases, μ_1 and μ_2, and the pressures, P_1 and P_2, will be equal. We can thus rewrite Equation 6.2 as

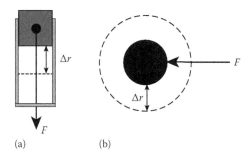

(a)   (b)

FIGURE 6.1 An illustration of the forces acting on the gas inside the piston of an engine (a) and the forces acting on a crystalline nucleus (b). In the piston (a), the heavy piston head applies a downward force, F, on the gas. If the piston head moves up by an amount Δr, the gas must do $-\int_{r}^{r+\Delta r} Fdr$ head. In other words, it must apply a force on this piston head. To quantify the force the gas can apply, we introduce pressure, P, as the force the gas applies on the piston head over the area, A, of the piston head. Replacing F in the previous integral with PA allows us to derive the familiar $dw = -PdV$. The formation of an interface between two phases is always unfavorable. There is thus a *force* on the nucleus encouraging it to redissolve and decrease the area of the interface. As illustrated in (b), this means that work, $\int F(r)dr$, must be done against this force in order to expand the nucleus. The surface tension, α, is introduced by assuming this force, $F(r)$, is a linear function of the surface area.

$$dE = TdS - PdV + \mu dN + \alpha dA \tag{6.3}$$

where $V = V_1 + V_2$ and $N = N_1 + N_2$. Furthermore, if we introduce an appropriate thermodynamic potential, $\Omega_v = E - TS - N\mu$, we can write

$$d\Omega = -SdT - Nd\mu - PdV + \alpha dA \tag{6.4}$$

Thermodynamic quantities such as the energy or entropy can be derived from this potential. We can also introduce a division between the part due to the volume, $d\Omega_v = -SdT - Nd\mu - PdV$, and the part due to the surface, $d\Omega_s = \alpha dA$:

$$\Omega = \Omega_v + \Omega_s \quad \text{where} \quad \Omega_s = \alpha A \tag{6.5}$$

When we write $\Omega_s = \alpha A$, we are assuming that the atoms are divided between the two phases as $n_1 V_1 + n_2 V_2 = N$, where n_1 and n_2 are the numbers of particles per unit volume in the two phases. In other words, we are assuming that the number of atoms at the surface, N_s, is equal to zero. The number of atoms is related to the derivative of the thermodynamic potential with respect to the chemical potential. So this condition is equivalent to assuming that

$$N_s = -\left(\frac{\partial \Omega_s}{\partial \mu}\right)_{T,V} = 0 \quad \rightarrow \quad \left(\frac{\partial \alpha}{\partial \mu}\right)_{T,V} = 0 \tag{6.6}$$

We can use Equation 6.5 to calculate the contribution the surface makes to the total entropy of the system:

$$S = -\left(\frac{\partial \Omega}{\partial T}\right)_{\mu,A,V} \quad \rightarrow \quad S_s = -\left(\frac{d\alpha}{dT}\right)A \tag{6.7}$$

Alternatively, we can calculate the contribution of the surface to the total free energy:

$$F = E - TS = \Omega + N\mu \rightarrow F_s = \alpha A \tag{6.8}$$

or the surface energy:

$$E = F + TS \quad \rightarrow \quad E_s = \left[\alpha - T\left(\frac{d\alpha}{dT}\right)\right]A \tag{6.9}$$

Finally, we can use the second law, to calculate the quantity of heat absorbed during a reversible change in the surface area, ΔA, as long as we assume that during this change the temperature and total volume of the system remain constant:

$$\Delta q_{\text{rev}} = T\Delta S = -T\left(\frac{d\alpha}{dT}\right)\Delta A \qquad (6.10)$$

Equation 6.9 can be recovered by combining this equation with Equation 6.1 for the total work done in agreement with the first law of thermodynamics.

6.2   Nucleation

Nucleation is the first stage in the crystallization of a material. During this initial step, a surface separating the solution/molten phase from the nascent crystal forms. As discussed in the previous section, work has to be done to the system in order to create an interface between the two phases. As such, the phase that is crystallizing must have a lower bulk free energy than the solution/molten phase—we must be in a regime where the solvated/molten phase is metastable. If this is not the case, then there is nothing to drive the reaction—every small nucleus of the new phase that forms due to fluctuations will be unstable and will thus eventually disappear. If, however, we are in a supercooled liquid or a supersaturated solution, crystallization will occur and eventually large nuclei will form. As illustrated in Figure 6.2, in these large nuclei, the gain in free energy

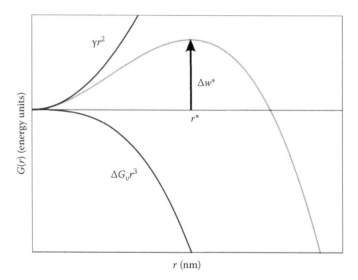

FIGURE 6.2 The free energy according to classical nucleation theory shown as a function of the size of the nucleus. There are two terms in the expression for the free energy. The first is an unfavorable surface term that increases with the square of the cluster radius. This term dominates the free energy when the cluster is small and makes the formation of small nuclei energetically unfavorable. The dependence of this first term on nucleus size is shown in black above the x-axis. The second term is the bulk term that is negative because the solution/molten phase is metastable. The bulk term increases with the cube of the cluster radius and is shown in black below the x-axis in the figure. This second terms ensures that formation of nuclei larger than a certain critical radius r^* is always favorable.

accompanying the formation of the new, more stable phase will easily compensate for the work that has to be done to create the interface. Furthermore, once these crystallites become large, the contribution made by the surface is negligible.

A reasonable question is why is nucleation only observed during condensation/precipitation? Why don't liquid nuclei form during the melting of solids? The reason for this can once again be understood by considering the surface tension and, in particular, by understanding how the value of the surface tension changes when adsorption occurs. To understand this, we have to consider the surface tension between a solid and a solution. We begin by dividing the total volume and the total number of solvent particles between the two phases so that we once again have $n_1V_1 + n_2V_2 = N$ and $V_1 + V_2 = V$. In other words, we once again assume that there are no solvent particles in the interface so that the surface tension is independent of the chemical potential of the solvent. We then consider how solute particles are distributed throughout the system. Much like the thermodynamic quantities discussed in the previous section, we can write the total number of solute particles, n, in the two phases as $n = n_v + n_s$, where n_v is the number of solute atoms in the volumes V_1 and V_2 and n_s is the number of solute atoms at the surface. Clearly, n_s measures whether the concentration of solute atoms in the surface layer is higher or lower than the concentration of solute atoms in the bulk volumes. If solute molecules adsorb on the surface n_s is thus positive.

We can calculate the number of solute atoms at the surface by differentiating the thermodynamic potential for the surface, $\Omega_s = \alpha A$, with respect to the chemical potential of the solute, μ'.

$$n_s = -\left(\frac{\partial \Omega_s}{\partial \mu'}\right)_T = -A\left(\frac{\partial \alpha}{\partial \mu'}\right)_T \quad \rightarrow \quad \gamma = -\left(\frac{\partial \alpha}{\partial \gamma}\right)_T\left(\frac{\partial \gamma}{\partial \mu'}\right)_T \tag{6.11}$$

where in the last step we introduce the surface concentration $\gamma = n_s/A$.

The second law of thermodynamics ensures that an isolated system which has a constant number of atoms, a constant volume, and a constant internal energy, is always in a state where entropy is maximized. This implies in turn that work must always be done to bring about any move away from equilibrium and it is why stable states correspond to minima in the energy landscape. In a minimum derivatives must be zero and the eigenvalues of the matrix of second derivatives must all be positive. This places certain restrictions on the thermodynamic quantities. The particular one that concerns us here being that

$$\left(\frac{\partial \mu'}{\partial \gamma}\right)_T > 0 \tag{6.12}$$

Inserting this inequality into Equation 6.11, we arrive at

$$\left(\frac{\partial \alpha}{\partial \gamma}\right)_T < 0 \tag{6.13}$$

In other words, when solute molecules adsorb on a surface the surface tension is lowered. This is why small particles are often stabilized when additive molecules are present.[5]

Additive molecules bind to the surface of the nanoparticle thereby lowering the surface tension and removing the force that would otherwise drive the particle to redissolve. It also explains why nucleation is only observed for transitions from liquid to solid. When a solid melts a liquid layer can form on any exposed solid surface and thereby lower the surface tension between the solid and gaseous phases of the material. As a result when solids melt nucleation does not take place.

6.2.1 Classical Nucleation Theory

As discussed in the previous section, there is a barrier to nucleation because work has to be done to create a surface of separation between the two phases. At the same time though, when a solution is supersaturated or a liquid is supercooled, there is a thermodynamic force driving the system toward the crystalline/solid state. The interplay between these competing effects is illustrated in Figure 6.2 and is most frequently quantified using the framework of classical nucleation theory.[6-10] According to this theory, the smallest aggregates of a new phase are always unstable because the work done to create the surface is not counterbalanced by the formation of bulk crystal. At a certain critical size, however, this situation changes, the work done to create the interface is equal and opposite to the energy released on creation of the more stable phase. Furthermore, because the volume of bulk material in a nucleus grows much faster than the surface area, it is energetically favorable to create all clusters larger than this critical size.

Classical nucleation theory can be used to calculate the critical size and to calculate the probability of forming a nucleus of critical size. The first step in this procedure is to calculate the critical radius of the cluster. If we assume a system of two phases in which phase one is a sphere surrounded by the second phase, we can write the total thermodynamic potential of the system as

$$\Omega = -P_1 V_1 - P_2 V_2 + \alpha A \tag{6.14}$$

where P_1 and P_2 are the pressures in the two phases and V_1 and V_2 are the volumes of the two phases. In writing this, we assume that the two phases are in equilibrium so the temperatures and chemical potentials in the two phases are equal. Furthermore, $V_1 + V_2$ is constant, which allows us to write:

$$\left(\frac{\partial \Omega}{\partial r}\right)_{T,\mu} = -(P_1 - P_2)\left(\frac{dV_1}{dr}\right) + \alpha\left(\frac{dA}{dr}\right) \tag{6.15}$$

When we form a nucleus with the critical size, the two phases are in equilibrium so the system is at a minimum in the thermodynamic potential. Substituting this condition into the left hand side of Equation 6.15, rearranging and recalling that the surface area and volume of a sphere equal $4\pi r^2$ and $(4/3)\pi r^3$, respectively, we arrive at

$$r^* = \frac{2\alpha}{P_1 - P_2} \tag{6.16}$$

The probability of a fluctuation producing a nucleus of this size is proportional to $\exp(-\Delta w^\star/k_BT)$, where $\Delta w^\star$ is the minimum work required to form the nucleus, which is in turn equal to the change in the thermodynamic potential, Ω (see Equation 6.14), that occurs during the process:

$$\begin{aligned}\Delta w &= (-P_1V_1 - P_2V_2 + \alpha A) - (P_2V_1 - P_2V_2)\\ &= -(P_1 - P_2)V_1 + \alpha A\end{aligned}\tag{6.17}$$

If we substitute the value for the critical nucleus that we obtained from Equation 6.16 and we remember that we have assumed that the nucleus is spherical we can write:

$$\Delta w^\star = \frac{16\pi\alpha^3}{3(P_1 - P_2)^2}\tag{6.18}$$

We can simplify this equation by recalling that the two phases are in equilibrium. As such, the chemical potentials of the two phases satisfy $\mu_1(P_1,T) = \mu_2(P_2,T)$. Furthermore, if we examine Equation 6.16 and imagine that the two phases are separated by a planar interface rather than a curved one, we find that at equilibrium $P_1 = P_2 = P$ because $(P_1 - P_2) \to 0$ as $r \to \infty$. We can therefore write

$$\mu_1(P_1,T) - \mu_1(P,T) = \mu_2(P_2,T) - \mu_2(P,T)\tag{6.19}$$

If we assume that the differences $\delta P_1 = P_1 - P$ and $\delta P_2 = P_2 - P$ are small, we can expand the two sides of this equation in terms of these differences and thus write

$$v_1\delta P_1 = v_2\delta P_2\tag{6.20}$$

where v_1 and v_2 are the volumes of a mole of each phase. We can use this expression to replace the factor of $(P_1 - P_2)$ in Equation 6.18 thereby arriving at

$$\Delta w^\star = \frac{16\pi\alpha^3 v_2^2}{3(v_1 - v_2)^2(\delta P)^2}\tag{6.21}$$

We now need to introduce a measure of the degree of metastability. For a solid forming from a melt, this degree is the difference, δT, between the temperature of the liquid, T, and the melting temperature, T_m. Using the Clapeyron–Clausius formula, we can relate differences in pressure to differences in temperature as

$$\delta P = \frac{q}{T_m(v_1 - v_2)}\delta T\tag{6.22}$$

where q is the latent heat that accompanies the transition from liquid to solid. We can substitute this result into Equation 6.21 to obtain the following expression for the work required to create a solid nucleus of critical size:

$$\Delta w^\star = \frac{16\pi\alpha^3 v_2^2 T_m^2}{3q^2(\delta T)^2}\tag{6.23}$$

Chapter 6

To derive a similar expression for the formation of a solute nucleus from solution,[11] we start from the following chemical potential:

$$\Omega = -P_1 V_1 - P_2 V_2 + \mu_1' n_1 + \mu_2' n_2 + \mu_s' n_s + \alpha A \tag{6.24}$$

where

μ_1' and μ_2' are the chemical potentials of the solute in the solid and solvated systems

n_1 and n_2 are the numbers of solute atoms in the nucleus and in the solution

μ_s' and n_s are the chemical potential for the solute particles in the interface and the number of particles in the interface

P_1, P_2, V_1, V_2, α, and A are as they were in Equation 6.14

Recalling that the total number of solute molecules in the system, $n_1 + n_2 + n_s$, and that the total volume, $V_1 + V_2$, of the system are constant allows us to derive the following expression for the change in the thermodynamic potential that occurs when a nucleus forms:

$$\Delta\Omega = -\left(P_1 - P_2\right)V_1 + \left(\mu_1' - \mu_2'\right)n_1 + \left(\mu_s' - \mu_2'\right)n_s + \alpha A[V_1] \tag{6.25}$$

It is important to note that in deriving this expression we have not assumed that nuclei are spherical. Instead we have asserted that the surface area of the nucleus, A, is a function of its shape, V, and that the surface term is a linear function of the surface area, $A[V_1]$. We do not even have to make this second assumption, however. We can instead write that work involved in building the surface is simply a function of the nucleus shape and the solute chemical potential $\phi(V_1, \mu_2')$.

The critical nucleus is in chemical and mechanical equilibrium with the solution so it must satisfy the following conditions:

$$\left(\frac{\partial \Delta\Omega}{\partial n_1}\right)_{V_1, n_s} = \left(\frac{\partial \Delta\Omega}{\partial n_s}\right)_{V_1, n_1} = \left(\frac{\partial \Delta\Omega}{\partial V_1}\right)_{n_1, n_s} = 0 \tag{6.26}$$

When combined with Equation 6.25, this implies that

$$\mu_1' = \mu_2' = \mu_s' \tag{6.27}$$

$$P_1 = P_2 + \left(\frac{\partial \phi(V_1, \mu_2')}{\partial V_1}\right) \tag{6.28}$$

Substituting the first of these results into Equation 6.25 and recalling that the minimum work, Δw^*, is equal to the change in chemical potential gives

$$\Delta w^* = -\left(P_1 - P_2\right)V_1 + \phi(V_1, \mu_2') \tag{6.29}$$

a rather simple expression for the work required to create a critical nucleus. To establish the relationship between this work and the supersaturation, we differentiate this expression with respect to the chemical potential of the solute, μ_2':

$$
\left(\frac{\partial \Delta w^*}{\partial \mu_2'}\right) = -V_1\left(\frac{\partial(P_1-P_2)}{\partial \mu_2'}\right) - (P_1-P_2)\left(\frac{\partial V_1}{\partial \mu_2'}\right)
$$
$$
+ \left(\frac{\partial \phi(V_1,\mu_2')}{\partial V_1}\right)\left(\frac{\partial V_1}{\partial \mu_2'}\right) + \left(\frac{\partial \phi(V_1,\mu_2')}{\partial \mu_2'}\right)
\tag{6.30}
$$

This is equivalent to differentiating with respect to the supersaturation, or chemical potential difference $\Delta \mu_2'$, as $\Delta \mu_2' = \mu_2' - \mu_2^{coex}$, where μ_2^{coex} is the chemical potential of the solute at coexistence.

Inserting the result in Equation 6.28 into Equation 6.30 allows us to cancel all terms in $(\partial V_1/\partial \mu_2')$. Furthermore, recalling from Equation 6.11 that $\left(\partial \phi(V_1,\mu_2)/\partial \mu_2\right) = -n_s$ and the Gibbs–Duhem relation at constant temperature, $(\partial p/\partial \mu) = n/v$, we can rewrite Equation 6.30 as

$$
\left(\frac{\partial \Delta w^*}{\partial \mu_2'}\right) = -n_1 - n_s + n_2\left(\frac{V_1}{V_2}\right)
\tag{6.31}
$$

This simple equation can be rewritten using number densities, $\rho = n/v$, rather than numbers of atoms.

$$
\left(\frac{\partial \Delta w^*}{\partial \mu_2'}\right) = -\left[\left(1-\frac{\rho_2}{\rho_1}\right)n_1 + n_s\right]
\tag{6.32}
$$

Furthermore, $n_1\rho_2/\rho_1$ is simply the number of solute particles that were in the volume now occupied by the critical nucleus, V_1, before the nucleus formed. In other words, we can rewrite the previous equation as

$$
\left(\frac{\partial \Delta w^*}{\partial \mu_2'}\right) = -\Delta n
\tag{6.33}
$$

where Δn is the difference between the number of solute atoms in the critical nucleus and the number of solute atoms in an equal volume of solution. This is a strikingly simple result and no assumptions concerning the surface energy or the size/shape of the critical nucleus were made in its derivation.

6.2.2 Simulating Nucleation

In previous sections, we have used the tools of classical thermodynamics to understand the process via which crystals form and have glossed over the precise atomistic mechanisms. To understand what is going on at the molecular level we have to use statistical mechanics.[12] One of the most important results in this theory is the fact that thermodynamic potentials are related to the logarithm of an integral of a probability density. This means that for the canonical (NVT) ensemble we can write

$$F(\Gamma) = -k_B T \ln |P(\Gamma)| \tag{6.34}$$

where

$$P(\Gamma) = \frac{\int_\Gamma \mathrm{d}\mathbf{x}\,\mathrm{d}\mathbf{p}\,\exp\left(-\dfrac{H(\mathbf{x},\mathbf{p})}{k_B T}\right)}{\int \mathrm{d}\mathbf{x}\,\mathrm{d}\mathbf{p}\,\exp\left(-\dfrac{H(\mathbf{x},\mathbf{p})}{k_B T}\right)} = \frac{\int_\Gamma \mathrm{d}\mathbf{x}\,\exp\left(-\dfrac{V(\mathbf{x})}{k_B T}\right)}{\int \mathrm{d}\mathbf{x}\,\exp\left(-\dfrac{V(\mathbf{x})}{k_B T}\right)} \tag{6.35}$$

where
  $H(\mathbf{x}, \mathbf{p})$ is the classical Hamiltonian
  $V(\mathbf{x})$ is the potential

The last equality holds because classical Hamiltonians are separable. In the denominator of Equation 6.35 the integral is over the entirety of configuration space. In the numerator, by contrast, the integral is performed only over the particular part of configuration space, Γ, that interests us. These equations can be used to calculate the free energy change (and therefore the equilibrium constants) for chemical reactions such as

$$\mathrm{H} + \mathrm{Cl} \rightarrow \mathrm{HCl} \tag{6.36}$$

To calculate the probability, and thus the free energy, of having the HCl molecule rather than the two separated atoms, we simply have to integrate, using the numerator of Equation 6.35, over all configurations that have a chemical bond between the two atoms. By contrast, to obtain the free energy for the situation where the two atoms are separated, we simply integrate over all configurations where the bond is not present. In this simple case, this can be achieved by separating the Hamiltonian into electronic, translational, rotational, and vibrational parts and integrating over each of these degrees of freedom separately.[13] However, when dealing with more complex chemical systems, the integrals in Equation 6.35 are generally solved by sampling phase space using either Monte Carlo (MC) or molecular dynamics (MD). Using our simple HCl example once more, you would calculate the free energy of HCl from such a simulation by counting the frequency with which configurations in which the bond is formed are visited during the course of the simulation.

There are a number of problems when it comes to applying this sort of logic to examine nucleation. The first is that nucleation is a rare event—as discussed in the previous section, there is a high barrier to the formation of a crystalline nucleus. It is therefore unlikely that we will observe a nucleation event on the relatively short timescales that are accessible to us in an atomistic MD simulation. The second problem is connected to the fact that we generally simulate systems containing relatively few atoms and replicate them throughout the entirety of space using periodic boundary conditions. These conditions ensure that when we simulate a solution and a nucleus of solute atom forms there is an enormous drop in the concentration of the surrounding solution. The same thing does not happen in reality—a liter of a 1 mol dm^{-3} solution will contain 6×10^{23} solute atoms. If a nucleus containing a thousand atoms forms, the solution will still contain 6×10^{23} solute atoms and the concentration will be unchanged. The final problem is a more philosophical one and is connected to the fact that we are uncertain as to what precisely happens to the atoms during the nucleation event. Returning to the example with the HCl molecule described previously, we could use what we know about chemical bonding and reactivity to divide up phase space between configurations where the bond is formed and configurations where the bond is broken. We did not learn to do this by examining Hamiltonians and performing MD simulations. Rather experimentalists at some stage noticed that something strange was happening when hydrogen was combined with chlorine, which was subsequently explained at the atomic level when chemists introduced theories of chemical bonding. If we only look at the Quantum Mechanical Hamiltonian, this division of phase space into a part where the bond is formed and a part where the bond is broken might seem rather arbitrary.

In spite of these difficulties, simulations have been performed in which nucleation events have been observed. One system that has been extensively studied is calcium carbonate in aqueous solution.[14–23] This system is particularly popular because, when the concentrations of the Ca^{2+} and CO_3^{2-} ions are large, clusters of amorphous calcium carbonate form spontaneously on timescales short enough to be investigated with MD. This agglomeration of $CaCO_3$ nanoparticles into an amorphous precursor was first observed in simulation by Martin et al.[24] In their simulations, the initial configuration was composed of a number of small calcite nanoparticles distributed throughout a large box of water. In the very early stage of the simulation, these nanoparticles converted into amorphous clusters. These small clusters then fused together during the remainder of the simulation.

Tribello et al.[25] took Martin's work on calcium carbonate one step further by starting the simulations from single calcium and carbonate ions distributed throughout the solution. Once again all the ions joined together to form a single cluster of amorphous calcium carbonate over the course of their 20 ns simulation. Furthermore, water molecule trapping events were observed, which went some way toward explaining why the amorphous calcium carbonate that forms in experiment is hydrated.[26,27]

In separate simulations, Tribello et al.[25] showed that there is little to no barrier when ions are added to amorphous calcium carbonate clusters but a quite substantial barrier for addition of ions to calcite nanoparticles. This explains why amorphous calcium carbonate grows in preference to calcite but is at odds with what is observed in experiment. Gebauer et al.[28,29] have shown that at high pH small, stable prenucleation clusters of calcium carbonate form. These clusters have a size on the order of 2 nm and are composed of equal numbers of calcium and carbonate ions. In other words, certain sizes of nanoparticle are particularly stable, which is inconsistent with both classical nucleation theory and the diffusion-controlled agglomeration of ions observed by Tribello et al.[25] An explanation for this phenomenon was provided by Raiteri and coworkers.[30–33] In their simulations, they incorporated both the carbonate and bicarbonate ions and found that the behavior of the system depended strongly on the ratio of carbonate to bicarbonate ions. The rapid growth of amorphous calcium carbonate observed by Tribello et al.[25] is only observed when the concentration of carbonate ions is high. When there is a substantial fraction of bicarbonate ions, the growth of amorphous material is much less rapid. Furthermore, when realistic carbonate/bicarbonate ratios are used, an explanation for the prenucleation clusters observed by Gebauer et al.[28] is found. In the simulations at these more realistic conditions, chain-like aggregates of calcium, bicarbonate, and carbonate ions are formed. These short chains (or DOLLOP clusters) are stabilized because they have greater conformational flexibility and thus higher entropies than dense aggregates of amorphous material.[33]

The fact that amorphous calcium carbonate forms on timescales that are accessible in MD simulation is good fortune. When studying other chemical systems, we are not so lucky so, to observe nucleation, we have to perform some form of enhanced sampling. These methods require the user to suggest coordinates, $s(x)$, that measure how far the reaction under study has progressed. A bias potential, $V(s)$, that is a function of this coordinate can then be added to the underlying atomic interaction potential and the system can be forced to undergo some change (see Figure 6.3).[34–36] Alternatively, large numbers of trajectories can be started from the initial state. Inevitably some of these trajectories will move along $s(x)$ toward the product state and new set of trajectories can be started at a new point further along $s(x)$. By repeating this process multiple times, trajectories that connect the reactant state to the product state (see Figure 6.4) can be generated.[37–43]

The main difficulty when using these enhanced sampling methods to examine nucleation is deciding on an appropriate coordinate. When studying nucleation from the melt, this coordinate must measure whether or not the atoms/molecules are ordered as they would be in the bulk solid. An oft used method for measuring this is to use the Steinhardt order parameters[44]:

$$q_{lm}(i) = \frac{\sum_{j \neq i} \sigma(|\mathbf{r}_{ij}|) Y_{lm}(\mathbf{r}_{ij})}{\sum_{j \neq i} \sigma(|\mathbf{r}_{ij}|)} \tag{6.37}$$

where

　　$\mathbf{r}_{ij}$ is the vector connecting atom i to atom j and the sum runs over all the atoms in the system

　　$\sigma(|\mathbf{r}_{ij}|)$ is a switching function that is one when the separation between two atoms is less than a cutoff and zero otherwise

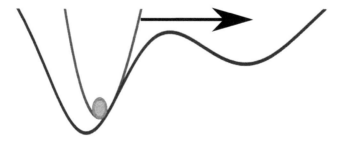

FIGURE 6.3　One method via which collective variables (CV) are used to bias simulations. The darker black line shows the underlying free energy surface as a function of the CV. This surface contains a high barrier, which in a conventional short MD simulation would not be crossed. The lighter gray line shows the underlying free energy surface plus a bias potential that is a simple harmonic function of the collective variable. By adjusting the position of the minimum in the harmonic potential we can force the system to explore the entire collective variable range.

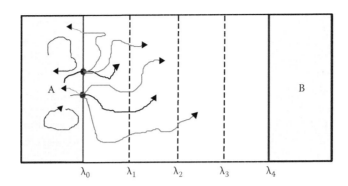

FIGURE 6.4 How methods such as forward-flux sampling, milestoning, and transition interface sampling work. A collective variable is identified that takes the system smoothly from some initial reactant state (A) to some final product state (B). This collective variable is then used to define a set of foliations λ_0, λ_1, λ_3,… between state A and state B. As shown in the figure a large number of trajectories (black lines with arrows) are started from state A. Although none of these trajectories reach the product state B many of these trajectories cross the first foliation line (λ_0). The points where these trajectories cross λ_0 can thus be used as start points for new trajectories (gray lines) and the process can be repeated for λ_1. At the end of the simulations trajectories that connect A to B can be generated by connecting these subtrajectories together. More importantly, the results from all the simulations can be combined and probabilities, $P(\lambda_i \rightarrow \lambda_{i+1}, \Delta t)$, for moving between the ith and $(i + 1)$th foliations can be extracted. Rates of reaction and free energies can be extracted from these probabilities.

The functions $Y_{lm}(\mathbf{r}_{ij})$ are spherical harmonics so l is an integer and m is an integer between $-l$ and $+l$. Generally all the q_{lm} values for a given value of l are combined to give one complex vector per atom in the system, $\mathbf{q}_l(i)$. The magnitude of this vector

$$| \mathbf{q}(i) |= \sqrt{\frac{4\pi}{2l+1} \sum_{m=-l}^{+l} q_{lm}^*(i) q_{lm}(i)} \tag{6.38}$$

gives a measure of the degree of order in the coordination sphere surrounding the atom. It is easy to think of ways to extend this approach so as to measure the degree of order in the coordination spheres around molecules.[45] The difficulty comes when combining the order parameters from all of the individual atoms/molecules so as to get a measure of the global degree of order for the system. The simplest way of doing this—calculating the average Steinhardt parameter[46]—can be problematic. Only the order parameters for the atoms in the nucleus will change significantly when the nucleus forms. The order parameters for the atoms in the surrounding liquid will remain pretty much the same. As such, if one models a small nucleus embedded in a very large amount of solution/melt, any change in the average order parameter will be negligible. In other words, substantial changes in the value of this average are observed in simulations of nucleation but only because the number of atoms is relatively small.

When the average Steinhardt parameter is used to bias the dynamics, problems can occur. These averaged coordinates cannot distinguish between the correct, single-nucleus pathway and a concerted pathway in which all the atoms rearrange themselves into their solid-like configuration simultaneously. This second type of pathway would be impossible in reality because there is a large entropic barrier that prevents concerted processes like this from happening. However, in the finite-sized systems that are commonly simulated, this barrier is reduced substantially. As a result, in simulations where average Steinhardt parameters are biased, there are often quite dramatic system size effects.[47]

If one wants to simulate nucleation using some form of biased dynamics, what is really required is an order parameter that measures

- Whether or not the coordination spheres around atoms are ordered
- Whether or not the atoms that are ordered are clustered together in a crystalline nucleus

One order parameter that measures whether or not these two criteria have been satisfied is the local Steinhardt parameter[48]:

$$\xi = \sum_i \sum_{k \neq i} \sigma(|\mathbf{r}_{ik}|)\mathbf{q}_l(i)\mathbf{q}_l(k) \tag{6.39}$$

In this expression, the sums run over all the atoms in the system and $\sigma(|\mathbf{r}_{kn}|)$ is another switching function. The $\mathbf{q}_l$ vectors are the Steinhardt parameters calculated by Equation 6.37. This coordinate clearly measures whether atoms that are close together have similar orderings of the atoms in their first coordination spheres. It thus satisfies our requirements and is a useful tool for studying nucleation in the melt.[49–53]

The coordinates that were developed for simulating nucleation from the molten state can also be used to study the process via which solute molecules aggregate and crystallize from solution. The great additional difficulty in doing this though is that generally much larger system sizes are required as the simulations must incorporate both the solute particles and the solvent molecules. In these large systems, diffusion is slow, which further slows the rate at which solute particles aggregate and grow. A solution to this problem was suggested by Kawaska et al.[54,55] who, as illustrated in Figure 6.5, proposed generating nuclei by adding atoms one at a time to a single cluster. This is perhaps most useful for getting a qualitative understanding of the way in which additive molecules bind themselves to the earliest formed nuclei. It is difficult to see, however, how this procedure would provide quantitative information on nucleation rates, barriers, or even on the competition between different nucleation pathways.

An alternative strategy for examining nucleation from solution was recently proposed by Giberti et al.[56] In their work, they note that in nuclei the density of solute ions is large and that when no nuclei are present, the solute ions are distributed almost uniformly throughout the solution. In other words, a nucleus is characterized by a fluctuation in the local density, $p(\mathbf{r})$, of solute ions. The presence or absence of nuclei can thus be detected by measuring the gradient of the density of solute ions:

$$S = \frac{1}{2}\int d\mathbf{r}(\nabla\rho(\mathbf{r}))^2 \tag{6.40}$$

where $\quad \rho(\mathbf{r}) = \sum_{i=1}^{N}\frac{1}{(2\pi\sigma^2)^{3/2}}\exp\left(-\frac{(\mathbf{r}-\mathbf{R}_i)^2}{2\sigma^2}\right) \tag{6.41}$

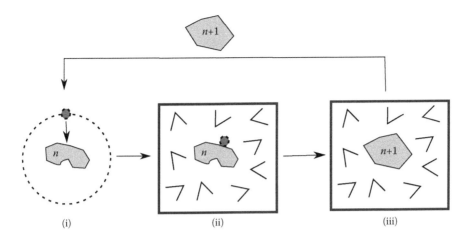

FIGURE 6.5 The operation of Kawaska et al.'s algorithm[54] for simulating nucleation from dilute solution. In this method, there are three steps in the addition of a new ion to the growing nucleus. In step (i) a nucleus containing n ions is simulated in vacuum. The position of the ions in the cluster are held fixed and the $(n + 1)$th ion is added on the surface at the site with the largest binding energy. In step (ii) the cluster is solvated and the ions in the cluster are kept fixed while a short MD simulation is run to equilibrate the positions and orientation of the solvent molecules. Finally, in the last step, the constraints on the ions in the nucleus are removed and the full system is equilibrated. The resulting nucleus containing $n + 1$ ions is extracted from the solution and this process is then repeated to add the $(n + 2)$th ion.

In the second of these expressions, the sum runs over all the solute atoms in the system. Each atom is represented, in this expression for the density, by a normal distribution with finite width, σ, which is centered on its position, $\mathbf{R}_i$. This procedure ensures that the density is differentiable and is standard practice in classical density functional theory.[57]

When a bias potential is added on the gradient coordinate (Equation 6.40), it forces the solute particles to aggregate. It does not force the ions to adopt a particular structure although this could be added by replacing the density in Equation 6.41 with a local average of an order parameter. However, even without this modification, Giberti et al.[56] were able to observe formation of solid nuclei from solutions of sodium chloride. Surprisingly, these ordered aggregates have both the familiar rock salt structure and a structure that resembles wurtzite.

6.2.3 Other Theories of Nucleation

The simulations highlighted in the previous section showed a number of results that cannot be accounted for within the framework of classical nucleation theory. It is increasingly recognized that these peculiarities are important as the results obtained from experiments are now regularly showing that there are many things occurring in solution that are not explained by the classical theory.[29,58–60] One particularly disturbing result is that classical nucleation theory often overestimates the nucleation rate by 10 orders of magnitude.[58]

The problems with classical nucleation theory arise because of the numerous approximations in the derivation of the theory. Three common assumptions that are often highlighted as being particularly problematic are as follows:

1. The number of atoms in the critical nucleus is independent of the supercooling/supersaturation.
2. Nuclei grow by the one-by-one addition of further atoms and these atoms immediately align themselves within the structure of the growing crystal.
3. Nuclei have the same structure as the bulk crystal.

Testing the validity of the first of these assumptions in experiment is straightforward. Equation 6.33 suggests that if the number of atoms in the critical nucleus is constant, the size of the nucleation barrier will decrease linearly as the supersaturation increases.* Consequently, the nucleation rate should increase exponentially with supersaturation, which is what is observed when experiments are performed at low supersaturations. However, when the supersaturation passes a certain threshold value, there is a change in the nucleation rate that does not fit the expected pattern. In addition, analyzing this result with an equation akin to Equation 6.33 gives an estimate of one for the number of atoms in the critical nucleus, which in turn implies that there is no barrier to nucleation.

Similar results have been seen in the results in simulations of the formation of Lennard–Jones solid from the melt. Trudu et al.[61] observed a classical-nucleation-style mechanism when the supercooling was low and a transition to a spinodal decomposition mechanism with large supercoolings. Furthermore, the calculated free energy barriers fit quite well to a linear function of the supercooling. The spinodal mechanism set in for conditions where this linear model predicted a less-than-zero barrier to nucleation.

The fact that the third of the assumptions described earlier is inappropriate is well established. In 1897, Ostwald[62] proposed his step rule and asserted that in general the least stable polymorph of the material crystallizes before the most stable form. This formation of less stable polymorphs during the early stages of crystal growth and subsequent conversion to a more stable form has been observed for a variety of different materials. It must happen because the surface tension for the less stable polymorph is lower than the surface tension of the more stable form.[63] The bulk free energy of the most stable polymorph is always lower than the bulk free energy of the less stable form by definition.

Rapid aggregation of material at high supersaturation and the formation of less stable polymorphs during the early stages of nucleation and growth of crystals appear at first hand to not be consistent with classical picture of critical nuclei and the interplay of surface tension and bulk free energy. However, as discussed in the previous paragraphs, these phenomena can at least be reconciled with the classical picture.

---

* Using Equation 6.33 in this way further assumes that the number of solute atoms in a volume equivalent to that of a critical nucleus is a constant. We are therefore further assuming that the number of solute atoms in a critical nucleus, n_{nucl}, will always be far larger than the number of solute atoms in an equal volume of solution, n_{sol} and hence that $n_{nucl} - n_{sol} \approx n_{nucl}$.

Equation 6.33 predicts that the barrier to nucleation should disappear when the supersaturation/supercooling is large. The spinodal, rapid growth of crystals is observed in this regime. Meanwhile, if the unstable polymorphs have lower surface tensions than the most stable form of the crystal, then of course classical nucleation theory tells us that these unstable polymorphs might form during the early stages of crystallization. There are, however, phenomena that occur during nucleation that are not at all consistent with classical nucleation theory. One that is particularly simple to explain is the prenucleation clusters[29] that were discussed in the previous section. For clusters of a particular size to persist for long periods of time in solution, the free energy as a function of nuclei size must have a shape like that shown in Figure 6.6. There must be a minimum in the free energy for clusters with a size smaller than the critical radius. Thankfully, the prenucleation clusters that have been observed in experiments are small and are thus amenable to study with molecular simulation.[33]

Another example of a nonclassical nucleation pathway is the two-step nucleation mechanism. As discussed previously, classical nucleation theory assumes that atoms are added one at a time to the growing nucleus and that when an atom is added it immediately aligns itself to the crystal structure (the second assumption in the list on Section 6.2.3). The two-step mechanism moves beyond this assumption by imagining nucleation to take place in a space of two order parameters. These order parameters come about because a two-component system can be in one of three states the dilute solution, a dense liquid containing only solute atoms, and the crystal. As illustrated in Figure 6.7, local concentration will distinguish between the first two of these states while a structural parameter (e.g., the Steinhardt parameters) will distinguish the second two. Figure 6.7 shows that the assumed pathway for growth in classical nucleation

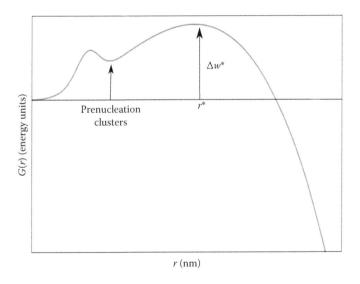

FIGURE 6.6 The postulated free energy as a function of size for systems where the formation of prenucleation clusters is observed. This figure should be compared with Figure 6.2. Now the free energy is not simply the sum of a favorable bulk term and an unfavorable surface term. There is some additional factor that acts to stabilize small nuclei. For the well-studied case of calcium carbonate, simulations suggest that this stabilization is due to the formation of DOLLOP. (From Demichelis, R. et al., *Nat. Commun.*, 2, 590, 2011.)

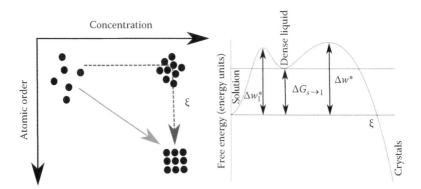

FIGURE 6.7 The reaction coordinate for the two-step mechanism for nucleation and a free energy surface shown as a function of this coordinate. Two order parameters are used to describe crystallization. The first measures local concentration and if there is a sufficient large concentration of ions in any part of the box for there to be a nucleus present. The second coordinate then measures whether or not the close together ions are ordered as they would be in a bulk crystal. The solid arrow shows the nucleation pathway that is assumed in classical nucleation theory. Crystalline nuclei form directly from the solution. The dashed line shows the pathway suggested for the two-step mechanism. Dense liquid-like clusters form in which there is little to no solid structure. Crystalline nuclei then grow within these regions of heightened density. The right panel shows a sketch of the expected free energy profile for this mechanism. Note that this free energy is no longer simply a function of the size of the nucleus but is rather a function of the reaction coordinate ξ. Furthermore, there are now three thermodynamic quantities that are required: the barrier to formation of the clusters of dense liquid Δw_1^*, the difference in free energy between the solution and dense liquid phases $\Delta G_{s \to 1}$, and the barrier to formation of the critical nucleus Δw^*.

theory is along the diagonal in this plane. Meanwhile, in the two-step mechanism, it is assumed that a dense liquid phase forms first and that ordered nuclei then grow in the region where the concentration of solute is enhanced.[64] Hence, in the language of the previous section, we should use two separate order parameters when studying nucleation with simulation. The first should measure whether or not atoms are clustered together, while the second should measure whether or not atoms that are clustered together are ordered.

Two-step nucleation pathways were first proposed based on theoretical considerations and from the results of simulation.[65–67] This mechanism has been observed in experiments on the crystallization of proteins.[68–71] In some cases, the dense liquid state is stable with respect to the dilute solution. In the more intriguing and common case, however, the dense liquid is less stable than the dilute solution.

6.3 Crystal Growth

Crystals have well-defined, regular shapes because the atoms in solids lie in an ordered and repeating pattern. Because of this ordered arrangement, the number of chemical bonds broken when the structure is cleaved depends on the orientation of the cleavage plane. In other words, different crystal faces have different surface tensions and as a result, the minimum energy shape for a solid is not simply the structure that minimizes the exposed surface area—a sphere. Instead, solid nuclei have shapes that minimize the area of any high-surface-tension facets while maximizing the area of the low-surface-tension facets.

As discussed in Section 6.2.1, we can allow for nonspherical nuclei in classical thermodynamics by replacing the γdA terms with an expression for the work, $\phi(V)$, that is a function of the shape, V, of the crystalline nucleus.[72] We can express this function as

$$\phi(V) = \int d\mathbf{n} |\mathbf{A}(\mathbf{n})| \gamma(\mathbf{n}) = \int d\mathbf{n} |\mathbf{A}(\mathbf{n})| \gamma \left(\frac{\mathbf{A}(\mathbf{n})}{|\mathbf{A}(\mathbf{n})|} \right) \tag{6.42}$$

where $\mathbf{A}(\mathbf{n})$ is a unit vector that represents a surface. This vector is directed along the outward normal, $\mathbf{n}$, to the surface and has a magnitude, $|\mathbf{A}(\mathbf{n})|$, that is proportional to the surface area. To find the most stable shape for a given crystal, a set of values for the surface areas of the different facets of the crystal (the $|\mathbf{A}(\mathbf{n})|$ values in Equation 6.42) must be found that minimize $\phi(V)$.[72,73] This minimization problem is actually reasonably straightforward because $\phi(V)$ is a convex function.[72,73] This assertion can be proved by considering what happens when we change the shape of the crystal by removing a particular facet. To do this, at least two new facets have to be added to the crystal shape. If these two new facets have area vectors $\mathbf{a}$ and $\mathbf{b}$, then the area vector for the original facet is $\mathbf{a} + \mathbf{b}$. As shown in Figure 6.8, $\mathbf{a} + \mathbf{b}$ lies in the plane spanned by $\mathbf{a}$ and $\mathbf{b}$. Furthermore, because the triangle inequality holds for vectors, we know that

$$|\mathbf{a} + \mathbf{b}| \le |\mathbf{a}| + |\mathbf{b}| \tag{6.43}$$

In other words, and as is clear from Figure 6.8, the surface area for the combined vector $\mathbf{a} + \mathbf{b}$ is less than the combined surface areas of $\mathbf{a}$ and $\mathbf{b}$. As a result, changing the shape of the crystal by adding new facets to the crystal always increases the surface area.

Equation 6.43 also explains why nuclei do not have facets if γ is independent of $\mathbf{n}$. The crystal can lower the exposed surface area and hence the work done in creating

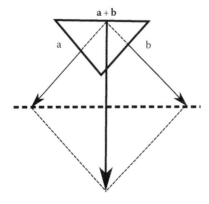

FIGURE 6.8   The orientation of a plane is specified by a normal to it. The surface normals of two planes $\mathbf{a}$ and $\mathbf{b}$, when added together, give rise to a vector that is normal to a third plane. This new plane's orientation, relative to the planes $\mathbf{a}$ and $\mathbf{b}$, is shown in the figure.

the surface, by combining the **a** and **b** facets into a single **a** + **b** facet. In other words, to expose the **a** and **b** facets, the surface tensions of the **a**, **b**, and **a** + **b** faces must make it so that

$$|\mathbf{a}+\mathbf{b}|\gamma\left(\frac{\mathbf{a}+\mathbf{b}}{|\mathbf{a}+\mathbf{b}|}\right)\geq|\mathbf{a}|\gamma\left(\frac{\mathbf{a}}{|\mathbf{a}|}\right)+|\mathbf{b}|\gamma\left(\frac{\mathbf{b}}{|\mathbf{b}|}\right) \tag{6.44}$$

in spite of the inequality in Equation 6.43.

Equation 6.42 can be minimized using a theorem due to Wulff[74] that states that the least work is done to create the nuclei when the lengths of the vectors connecting the center of the crystal to the facets, $|\mathbf{A}(\mathbf{n})|$, are proportional to the surface tensions, $\gamma(\mathbf{n})$. This theorem is proved by first noting that at the minimum in $\phi(\mathcal{V})$

$$\frac{d\phi(\mathcal{V})}{d\mathcal{V}}=0 \quad \rightarrow \quad \int d\mathbf{n}\gamma(\mathbf{n})\frac{d|\mathbf{A}(\mathbf{n})|}{d\mathcal{V}}=0 \tag{6.45}$$

if $d\gamma(\mathbf{n})/d|\mathbf{A}(\mathbf{n})| = 0$. We then note that we are only interested in changes to the shape that do not change the total volume, V, of the nucleus. This is why the magnitude of the vectors, $|\mathbf{A}(\mathbf{n})|$, in Equation 6.42 are proportional to the surface areas of the corresponding facets. If the length of the vector connecting the surface to the center of the crystal is changed and the surface area of the facet is held constant, the total volume of the nucleus **must** increase. This constraint on the total volume allows us to write

$$\int d\mathbf{n}h(\mathbf{n})\frac{d|\mathbf{A}(\mathbf{n})|}{d\mathcal{V}}=0 \tag{6.46}$$

where $h(\mathbf{n})$ is the length of the vector that connects the center of the nucleus to the facet. We can combine this equation with Equation 6.45 and a constant of proportionality, λ, and thus obtain

$$\int d\mathbf{n}\left[h(\mathbf{n})-\lambda\gamma(\mathbf{n})\right]\frac{d|\mathbf{A}(\mathbf{n})|}{d\mathcal{V}}=0 \tag{6.47}$$

The surface areas of the facets will change when the shape of the crystal is changed so $d|\mathbf{A}(\mathbf{n})|/d\mathcal{V} \neq 0$, which means that $h(\mathbf{n})$ must equal $\lambda\gamma(\mathbf{n})$ and the Wulff theorem must hold.

6.3.1 Crystal Surfaces

Figure 6.9 shows how the Wulff theorem can be used to determine the equilibrium morphology of a crystal. A plane perpendicular to the surface normal, **n**, is drawn at a distance $\gamma(\mathbf{n})$ from the center of the plot. This process is repeated for all surfaces for which information on the surface tension is available. The inner envelope of all these planes is then the equilibrium morphology for the crystal. To determine the equilibrium

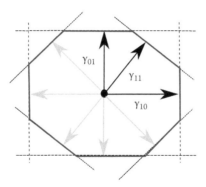

FIGURE 6.9   A two-dimensional Wulff construction that illustrates how this technique can be used to determine the equilibrium morphology for a crystal. Lines parallel to the various surface normals and with lengths equal to the surface tensions are drawn radiating out from a central point. Planes perpendicular to each of these surface normals are drawn at the end points of these vectors. The equilibrium morphology, is the inner envelope of these planes. The structure shown above has a fourfold, rotational symmetry axis at its center so the vectors shown in gray are symmetrically equivalent to those shown in black.

shape of the crystal, we thus need to only determine the surface tensions for the various facets of the crystal. In addition, as shown in Figure 6.9, we are helped in this task by the symmetry of the solid as this ensures that many surfaces are equivalent.

The Wulff construction means that we do not need to sample over all possible nuclei shapes to determine the equilibrium morphology of the crystal. Instead, because we are assuming that the surface tension is independent of the area of the surface, we can study an infinite periodic slab of material as shown in Figure 6.10. Furthermore, we only need to perform simulations for the low index surface as (1) high index surfaces will only make a small contribution to the surface term and (2) in many simulations the system will fluctuate between a number of closely related surfaces so the surface tensions entering Equation 6.42 will actually be an average. In fact, there are even ways of incorporating the dependence of the surface tension on the surface area and for properly

FIGURE 6.10   A snapshot from a simulation of the interface between sodium chloride crystals and water illustrating how the slab geometry is used to study interfaces. Half the cell is taken up with crystal while the other half is filled with liquid water. This cell is replicated in all three directions so care must be taken to ensure that the replicas do not interact with each other.

incorporating the contributions from edges and corners,[75] all of which should probably be applied when nanoscale clusters are being examined.

A number of methods exist for calculating the surface tensions from simulations of slabs such as that shown in Figure 6.10. However, before discussing these methods, it is important to remember that we always simulate finite-sized systems and that as such finite size effects must be controlled. In these types of simulations, controlling finite size effects means ensuring that

- The surface tension does not change when the thickness of the slab of material is increased
- The surface tension does not change when the distance between adjacent slabs is increased

The simplest method for calculating the surface tension from an atomistic simulation is to use a relation derived from the Kirkwood–Buff equation[76]:

$$\gamma = \int_{-\infty}^{+\infty} dz[P_N(z) - P_T(z)] \tag{6.48}$$

where $P_N(z)$ and $P_T(z)$ are the components of the pressure tensor that are normal and tangential to an interface that lies in the xy plane. These pressures depend on the z coordinate because the pressure tensor is a microscopic function that varies from point to point in a fluid. Macroscopic values for the pressure tensor can only be obtained by taking ensemble averages. To calculate a surface tension using this expression, we therefore need to calculate the following ensemble average:

$$\gamma = \frac{1}{A} \left\langle \sum_i \sum_{j>i} \frac{1}{2} \left(1 - \frac{3z_{ij}}{r_{ij}^2} \right) r_{ij} \frac{du(r_{ij})}{dr_{ij}} \right\rangle = L_z (\overline{P_N} - \overline{P_T}) \tag{6.49}$$

where the sum runs over all the pairs of atoms in the system and $u(r_{ij})$ is the potential acting between atoms i and j. Code to calculate the average of this quantity does not need to be added to the MD engine. Instead, the expression after the final equals sign can be used with $\overline{P_N} = \langle P_{zz} \rangle$ and $\overline{P_T} = \left\langle \dfrac{P_{xx} + P_{yy}}{2} \right\rangle$ with P_{xx}, P_{yy}, and P_{zz} being the xx, yy, and zz components of the macroscopic pressure tensor. These quantities are output by all of the standard MD packages. Therefore, calculating the surface tension only involves setting up a simulation cell that contains an interface between the two phases of interest and taking time averages. It can, however, be difficult to converge these ensemble averages as oftentimes the fluctuations in the components of the pressure tensor are quite large.[77,78]

Alternatives to the Kirkwood–Buff-equation-based methods for calculating the surface tension generally involve calculating the free energy of the interface between the solid and liquid phases using methods based on thermodynamic integration.[79–81] In the most commonly used form of thermodynamic integration,[34,82] we start from a reference potential for which the free energy is known—oftentimes the Einstein crystal.

We obtain the free energy difference between the system of interest and the reference system, $\Delta F(A \rightarrow B)$, by performing the following integral:

$$\Delta F(A \rightarrow B) = \int_0^1 d\lambda \left\langle \frac{\partial E(\lambda)}{\partial \lambda} \right\rangle_\lambda \qquad (6.50)$$

where $E(\lambda) = \lambda E_A + (1-\lambda)E_B$ \qquad (6.51)

In this expression, E_A and E_B are the internal energies of the system of interest and the reference system, respectively. $E(\lambda)$ is thus a potential for a coupled system and λ is a coupling parameter. The previous integral can be solved by calculating the ensemble average, $\langle E(\lambda) \rangle$, for a number of values of λ. Derivatives of $\langle E(\lambda) \rangle$ with respect to λ can be calculated and the previous integral can then be calculated numerically.

To calculate a surface tension using thermodynamic integration requires one to integrate between multiple potentials as shown in Figure 6.11.[80] Initially, simulations of the bulk solid and bulk liquid are performed and the potential is changed so as to cleave these two systems into slabs. In this first step, the initial potential is the potential for the bulk liquid/solid. Once $\gamma = 1$ the final potential has the periodic boundary conditions in the z-direction removed. In addition, constraints are added to (1) keep the atoms/molecules confined in the slab and to (2) ensure that the atoms/molecules at the surfaces of the liquid slab are arranged like the atoms/molecules in the solid. In the second step, the liquid and solid systems are stuck together. The end point of the integration has the liquid and solid side by side but still has the restraints introduced in the previous step. Finally, in the third and final step, the restraints introduced in step 1 are removed so

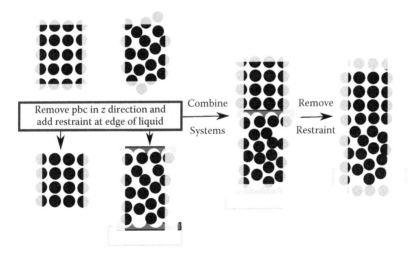

FIGURE 6.11   The steps involved when using thermodynamic integration and the cleaving method to determine interfacial free energies. In the first step, the solid and liquid phases are cleaved along the z axis and a restraining potential is added at the liquid–vacuum interface. During the second step, the two systems are merged with the restraint in place. The restraint is then removed. This entire process starts from two separate liquid and solid phases and finishes with system where an interface is present. Consequently, the sum of the integrals obtained from the three steps is equal to the interfacial free energy.

that the end point in the integration contains an unrestrained interface between the solid and liquid. The sum of the integrals obtained using these three steps is the surface free energy as the only difference between the initial reference state and the final state is the introduction of the interface between the two phases.

Recently, Angioletti-Uberti et al.[83,84] proposed a particularly appealing way of determining the surface free energy between solid and liquid phases at the coexistence temperature. Their method is considerably simpler to implement than the cleaving method described in the previous paragraph and is thus easier to apply to complex chemical systems. In it the metadynamics method[85] is used to generate a bias potential that drives the system between configurations where an interface between the two phases is present and configurations where the interface is absent. The bias potential generated in the metadynamics simulation is a function of two collective variables, which, as shown in Figure 6.12, measure the average values of the atomic order parameters (quantities much like the Steinhardt parameters; Equation 6.37) for the atoms in the two halves of the box. As such, the system is forced to undergo transitions between states in which the system is fully liquid, fully solid, and in which half the box is liquid and half the box is solid. Furthermore, when the free energy obtained from the metadynamics simulation is projected on these order parameters, it is very straightforward to interpret the result. As shown in Figure 6.12, features in the

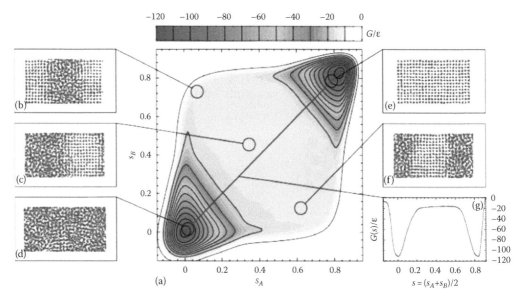

FIGURE 6.12   The free energy surface obtained in Angioletti-Uberti et al.'s work on the surface tension of Lennard-Jones. In panel (a) the free energy is shown as a function of the average order parameter calculated separately for the two halves of the box. The atoms in these two separate regions are colored red and blue in the snapshots of the trajectory shown in the panels labelled (b) through (f) above. The configurations mapped in the top right corner of panel (a) have all the atoms arranged as they would be in a solid. Configurations that are mapped in the bottom left hand corner of panel (a) have all the atoms arranged as they would be in a liquid. Configurations mapped between these two extremes contain an interface between the solid and liquid phases. Panel (g) shows how the surface free energy can be estimated from this free energy surface as this panel shows the free energy projected on the red line shown in the main figure. The difference between the free energy at the bottoms of the basins in panel (g) and the free energy in the plateau region between these two minima is thus the surface excess free energy. (Reproduced from Angioletti-Uberti, S. et al., *Phys. Rev. B*, 81, 125416, 2010.)

bottom left of the free energy surface correspond to configurations in which the system is entirely liquid, while features in the top right correspond to configurations where the system has solidified. All the configurations that are projected outside these two regions contain an interface between the solid and liquid phases. The free energy of the interface can be determined by projecting the free energy surface shown in Figure 6.12 along the line $(s_A + s_B)/2$ and measuring the difference between the basins and the plateau region that corresponds to the presence of the interface. Furthermore, much like the cleaving methods described in the previous paragraph and unlike methods based on the Kirkwood formula, the interface that forms between the two surfaces is rough like a real interface. This work has recently been extended by Cheng et al. and can now be used to calculate the excess surface free energy in supercooled conditions.[86]

The methods described earlier are expensive and are still quite difficult to run so they are not yet routine. It is therefore much more common to assume that entropy makes only a small contribution to the surface tension and to thus equate the surface tension with the surface energy.[87] In these sort of calculations, it is also often assumed that the solution or liquid phase does not contribute to the surface tension of the solid as one can then simulate the slab in contact with a vacuum. Surface energies are obtained from these sorts of calculations by minimizing the energy of the slab and then minimizing the energy of a similar number of atoms/molecules in the bulk crystal. The surface energy is the difference between the final energy of the slab and the final energy of the bulk crystal. Some more refined versions of this technique include a single layer of water molecules above the solid surface in order to incorporate the effect of the solvent at least in part.[88] Clearly, any such technique could be extended by adding more than a single monolayer of water.

6.3.2 Kinetics and Thermodynamics of Attachment

Figure 6.2 shows how the free energy of a nucleus changes as it grows. As discussed at length in previous sections, it is unfavorable to form small clusters of material because of the surface term. However, beyond a certain critical size, the favorable bulk term takes over and makes it always favorable to increase the size of the cluster. Furthermore, Figure 6.2 suggests that small clusters of material will always dissolve and transfer material to the larger clusters as one large cluster will always have a smaller surface area than two small clusters. This is a thermodynamic limit though in reality when you crystallize salts, you often find that multiple single crystals drop out of solution. For multiple crystals to be present, kinetics must be playing a role.

To equilibrate and form the lowest free energy structures available, the high-free-energy crystallites must redissolve. The problem, however, is that once a nucleus of critical size has formed, redissolution involves the crossing of a substantial free energy barrier, which will be a rare occurrence. In addition, the high-free-energy crystallite will continue to grow because, as discussed in the previous paragraph, addition of more material is always energetically favorable. This will further decrease the chance of it redissolving and equilibrating. Apart from causing multiple crystals to form from the solution, this also means that we often find that crystals do not have the equilibrium Wulff morphology. They have morphologies that expose high-surface-tension faces. This is especially prevalent when the low-surface-tension facets grow more rapidly than

Chapter 6

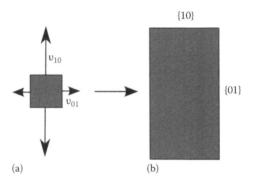

FIGURE 6.13 In panel (a) the arrows illustrate the amount of material that will be added to each of the phases in a unit time. This figure thus illustrates why crystals grown under kinetic control end up with the morphology, shown in panel (b), that is dominated by the slowest growing facets.

the high-surface-tension facets because, as illustrated in Figure 6.13, when crystals grow under kinetic control, the slowest growing facet dominates the final morphology.

If rates of growth for the various surfaces in the crystal can be estimated the final morphology of the crystal can be obtained using a procedure that closely resembles the procedure that was used to extract the equilibrium morphology in the previous section.[89–91] Lines with lengths proportional to the rate of growth, v_n, of the various faces are drawn radiating out from a central points. The kinetic morphology of the crystal is the shape contained inside the planes perpendicular to these vectors.

Extracting the rate at which a surface grows from an MD simulation of an interface between two phases is not straightforward as there is a limit on the number of solid layers that can be added to the solid part of the simulation cell. In simulations of a liquid–solid interface, we do not want to grow the solid slab too large as then the two interfaces will start to interact with each other. Meanwhile, for a solid–solution interface, the problem is that there are a limited number of solute atoms inside the solution phase. Piana and Gale[92,93] developed a particularly attractive two-step solution to this problem and used it to examine the formation of urea crystals from water and methanol solutions. In the first part of their work, MD simulations of a solid urea slab in contact with solution were performed.[92] In these simulations, urea molecules were seen to add themselves and subtract themselves from the solid substrate. By classifying the various sites on the surface at which urea molecules attached and detached and by measuring the frequency with which the urea molecules changed state, Piana and Gale[92] were thus able to develop a kinetic model for the growth of the urea surfaces. This kinetic model for the growth of surfaces is considerably cheaper to run than a MD simulation. As such, it could be used to model how the surfaces of the crystal changed over much longer length and timescales. In fact, Piana and Gale[93] used the kinetic model they obtained to model the formation of a macroscopic-sized crystal. Remarkably, as shown in Figure 6.14, the morphologies they predict for the crystal from these simulations match those obtained in the experiment.

A nice addition to the work performed by Piana and Gale was recently performed by Salvalaglio et al.[94] They did further MD simulations of the interface between a urea crystal and a solution of urea. However, in their work, they analyzed the data that emerged from their simulation in a different manner. This allowed them to explain why

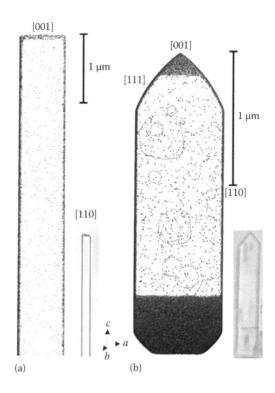

FIGURE 6.14 The results from Piana et al.'s kinetic MC simulations of the growth of urea crystals from solution: (a) shows the shape obtained when the crystallization is from a solution of urea in water while in (b) the solvent is methanol. The insets show the morphologies obtained in experiment. Remarkably, the crystals *grown* in simulation have the same morphology as those grown in experiments. (Reprinted by permission from Macmillan Publishers Ltd., *Nature*, Piana, S., Reyhani, M., Gale, J.D., 438, 70–73, copyright 2005.)

the different faces grow at different rates. It would seem that the growth mechanism is different on different crystal faces. For urea, when molecules attach themselves to the fast-growing {001} face, they almost immediately adopt an orientation like that in the bulk crystal. By contrast on the slow-growing {110} face, the individual layers grow via birth and spread and secondary nuclei of material must form on the exposed face. As a result, if urea molecules add themselves to the surface in locations not near to a secondary nucleus, they quickly drop off again. The formation of these secondary nuclei is an activated process and this two-stage layer-nucleation process is similar to the two-step bulk-nucleation process that was described in previous sections. Clearly, the fact that there is a barrier to the growth of a new layer on the {110} face and no barrier to the addition of a new layer on the {001} face explains why the {110} face grows more quickly.

The problem when it comes to applying the methods developed by Piana and Gale and by Salvalaglio et al.[94] to other chemical systems is that urea is a bit of a special case. Urea molecules add themselves to the solid substrate and dissolve on timescales that are accessible in unbiased MD simulations.[92] This is the exception rather than the rule—to see addition and dissolutions events in the majority of chemical systems we are forced to use enhanced sampling techniques. One problem here, as always, is thinking of appropriate collective variables. These variables must reflect the underlying mechanism,

Chapter 6

which as discussed in the previous paragraph often depends on the particular chemical under study and even on the surface of the crystal that is exposed. In addition, if the CV is used to add a bias potential, extracting rates is far from straightforward.

6.3.3   Understanding the Role of Additives

Increasingly, experiments are showing that crystallization can be controlled through the addition of additive molecules.[95] This is particularly relevant for many nanomaterials as oftentimes the only reason these small clusters persist rather than growing to macroscopic size is the presence of additive molecules in the solution. With many of these techniques, there is very little understanding, at the atomistic level, of how the additive molecules are perturbing the crystallization process. There is thus a great deal of interest in using simulation methodologies to model the process of crystallization when there are additive molecules present.

The majority of simulations that have tried to examine what effect the additives have on the crystallization process have done so by examining how the additive binds to various surfaces of the crystal or to ions in solution. Furthermore, in many of these calculations, finite temperature effects and the effect of the solution are ignored and the energy of a single additive molecule on the surface is minimized in vacuum. It is now no longer necessary to make these kinds of assumptions if the simulations are done with classical potentials. Even if the additive does not bind and unbind to the surface on the simulation timescale, it is straightforward to think of collective variables and to add bias potentials to increase the frequency of the binding and unbinding events.[96–101] These techniques can be used to extract free energies of attachment. From these quantities, conclusions can be drawn as to how the additive affects the rate at which the various surfaces grow.

The paper by Salvalaglio et al.[94] that was mentioned in the previous section has a nice example of how these sorts of calculations can be done in practice. It is known experimentally that adding biureate to solutions of urea changes the shape of the crystals that precipitate. Salvalaglio et al.[94] were able to show that this happens because the biureate molecules bind more strongly to the fast growing {001} than they do to the slower growing {110} face. Additional layers of urea molecules cannot be added to the {001} when biureate molecules are bound it. Consequently, the addition of the biureate slows the rate at which the {001} face grows making it more comparable with the rate of growth of the {110}. This changes the morphology of the final crystal.

6.4   Conclusion

A great deal of interesting work in which simulations have been used to understand the process via which crystalline material forms from solution or from melt has been done. There has even been interesting work performed, which there has not been space in this chapter to discuss, on the nucleation of solids during solid–solid phase transitions.[102] Even so, the field is still in its infancy and there are a large number of unresolved problems. One in particular concerns the considerable amount of evidence that has been accumulated from both experiment and simulation, which shows that classical nucleation theory does not give a complete description of these phenomena. The failure of this theory is perhaps not that surprising given the number of approximations that are made in the derivation. However, finding a suitable replacement for it, which makes fewer approximations, is difficult. For instance, alternate theories of nucleation, which have been used to explain the data from experiments, often work by introducing additional, fittable parameters in the equation for the free energy of a nucleus. The way these parameters should be physically interpreted, and hence calculated from a simulation, is often not clear. Furthermore, these theories still do not contain a mathematical framework through which the effect of the solvent or the presence of additives can be quantified.

Increasingly, researchers are using a combination of simulation and experiment to understand how crystals form. In these projects, the simulators and experimentalists work on the same chemical system so results from experiments can be explained through the simulations and simulations can suggest new experiments to perform. This sort of project clearly allows one to understand a great deal about the crystallization of a particular system. Furthermore, reusable tools that can be applied to other systems often emerge. Hopefully, a new phenomenological theory, which improves on classical nucleation theory by explaining the solvent's role or the role of chemical additives, can be obtained using the information emerging from these collaborations and from simulations performed on simple model systems such as Lennard–Jones.

References

1. Desiraju, G. R., Vittal, J. J., Ramanan, A. 2011. Crystal Engineering: A textbook. World Scientific Publishing Co.
2. Yin, Y., Alivisatos, A. P. *Nature* 2005, *437*, 664.
3. Edelstein, A., Cammaratra, R. *Nanomaterials: Synthesis, Properties and Applications.* Taylor & Francis Group: New York, 1996.
4. Landau, L. D. *Statistical Physics: Part 1.* Butterworth-Heinemann: Oxford, U.K., 1980.
5. Sharma, V., Park, K., Srinivasarao, M. *Mater Sci Eng R: Rep* 2009, *65*, 1–38.
6. Becker, R., Doring, W. *Ann Phys (Leipsiz)* 1935, *24*, 719.
7. Frenkel, J. *Kinetic Theory of Liquids.* Dover: New York, 1955.
8. Abraham, R. *Homogeneous Nucleation.* Academic Press: New York, 1974.
9. Oxtoby, D. W. *J Phys Condens Matter* 1992, *4*, 7627.
10. Navrotzky, A. *Proc Natl Acad Sci USA* 2004, *101*, 12096.
11. Oxtoby, D. W., Kashchiev, D. *J Chem Phys* 1994, *100*, 7665.
12. Mcquarrie, D. A. *Statistical Mechanics.* University Science Books: Sausalito, CA, 2000.
13. Atkins, P., De Paula, J. *Physical Chemistry*, 8th edn. Oxford University Press: Oxford, U.K., 2006.
14. Brunenal, F., Donadio, D., Parrinello, M. *J Phys Chem B* 2007, *111*, 12219.
15. Di Tommaso, D., de Leeuw, N. H. *J Phys Chem B* 2008, *112*, 6965.
16. Kerisit, S., Cooke, D. J., Spagnoli, D., Parker, C. *J Mater Chem* 2005, *15*, 1454.
17. de Leeuw, N. H., Parker, S. C. *J Chem Soc Faraday Trans* 1997, *93*, 467.
18. Kerisit, S., Parker, S. C. *J Am Chem Soc* 2004, *126*, 10152.
19. Spagnoli, D., Kerisit, S., Parker, S. C. *J Cryst Growth* 2006, *294*, 103.
20. Quigley, D., Rodger, P. M. *J Chem Phys* 2008, *128*, 221101.
21. Quigley, D., Freeman, C., Harding, J., Rodger, P. *J Chem Phys* 2011, *134*, 044703.
22. Archer, T. D., Birse, S. E. A., Doye, M. T., Redfern, S. A. T., Gale, J. D., Cygan, R. T. *Phys Chem Miner* 2003, *30*, 416.
23. de Leeuw, N. H., Parker, S. C. *Phys Chem Chem Phys* 2001, *3*, 3217.
24. Martin, P., Spagnoli, D., Marmier, A., Parker, S. C., Sayle, D. C., Watson, G. *Mol Simul* 2006, *32*, 1079.
25. Tribello, G. A., Bruneval, F., Liew, C., Parrinello, M. *J Phys Chem B* 2009, *113*, 11680–11687.

26. Rieger, J., Frechen, T., Cox, G., Heckmann, W., Schmidt, C., Thieme, J. *Faraday Discuss* 2007, *136*, 265–277.
27. Addadi, L., Raz, S., Weiner, S. *Adv Mater* 2003, *15*, 959–970.
28. Gebauer, D., Völkel, A., Cölfen, H. *Science* 2008, *322*, 1819.
29. Gebauer, D., Cölfen, H. *Nano Today* 2011, *6*, 564–584.
30. Raiteri, P., Gale, J. D., Quigley, D., Rodger, P. M. *J Phys Chem C* 2010, *114*, 5997–6010.
31. Raiteri, P., Gale, J. D. *J Am Chem Soc* 2010, *132*, 17623–17634.
32. Gale, J. D., Raiteri, P., van Duin, A. C. T. *Phys Chem Chem Phys* 2011, *13*, 16666–16679.
33. Demichelis, R., Raiteri, P., Gale, J. D., Quigley, D., Gebauer, D. *Nat Commun* 2011, *2*, 590.
34. Frenkel, D., Smit, B. *Understanding Molecular Simulation: From Algorithms to Applications*, 2nd edn. Academic Press: Orlando, FL, 2002.
35. Chipot, C., Pohorille, A. *Free Energy Calculations: Theory and Applications in Chemistry and Biology*. Springer: Berlin, Germany, 2007.
36. Bonomi, M., Branduardi, D., Bussi, G., Camilloni, C., Provasi, D., Raiteri, P., Donadio, D. et al. *Comput Phys Commun* 2009, *180*, 1961–1972.
37. Allen, R. J., Valeriani, C., ten Wolde, P. R. *J Phys Condens Matter* 2009, *21*, 463102.
38. Faradjian, A. K., Elber, R. *J Chem Phys* 2004, *120*, 10880–10889.
39. Bolhuis, P. G., Chandler, D., Dellago, C., Geissler, P. L. *Ann Rev Phys Chem* 2002, *54*, 20.
40. Dellago, C., Bolhuis, P. G., Geissler, P. L. *Adv Chem Phys* 2002, *123*, 1–78.
41. Dellago, C., Bolhuis, P. G. *Atomistic Approaches in Modern Biology: From Quantum Chemistry to Molecular Simulations*. Springer: Berlin, Germany, 2007, pp. 291–317.
42. van Erp, T. S., Bolhuis, P. G. *J Comput Phys* 2005, *205*, 157–181.
43. Dellago, C., Bolhuis, P. G. *Advanced Computer Simulation Approaches for Soft Matter Sciences III*. Springer: Berlin, Germany, 2008, pp. 167–233.
44. Steinhardt, P. J., Nelson, D. R., Ronchetti, M. *Phys Rev B* 1983, *28*, 784–805.
45. Santiso, E. E., Trout, B. L. *J Chem Phys* 2011, *134*, 064109.
46. Quigley, D., Rodger, P. M. *Mol Simul* 2009, *35*, 613–623.
47. Desgranges, C., Delhommelle, J. *Phys Rev B* 2008, *77*, 054201.
48. Moroni, D., ten Wolde, P. R., Bolhuis, P. G. *Phys Rev Lett* 2005, *94*, 235703.
49. Coasne, B., Jain, S. K., Naamar, L., Gubbins, K. E. *Phys Rev B* 2007, *76*, 085416.
50. ten Wolde, P. R., Ruiz-Montero, M. J., Frenkel, D. *J Chem Phys* 1996, *104*, 9932–9947.
51. ten Wolde, P. R., Frenkel, D. *Phys Chem Chem Phys* 1999, *1*, 2191.
52. Volkov, I., Cieplak, M., Koplik, J., Banavar, J. R. *Phys Rev E* 2002, *66*, 061401.
53. Lechner, W., Dellago, C. *J Chem Phys* 2008, *129*, 114707.
54. Kawska, A., Brickmann, J., Kniep, R., Hochrein, O., Zahn, D. *J Chem Phys* 2006, *124*, 024513.
55. Schepers, T., Brickmann, J., Hochrein, O., Zahn, D. *Z Anorg Allg Chem* 2007, *633*, 411–414.
56. Giberti, F., Tribello, G. A., Parrinello, M. *J Chem Theory Comput* 2013, *9*, 2526–2530.
57. Ginzburg, V., Landau, L. *Zh Eksp Teor Fiz* 1950, *20*, 1064–1082.
58. Vekilov, P. G. *Nanoscale* 2010, *2*, 2346–2357.
59. Sear, R. P. *Int Mater Rev* 2012, *57*, 328–356.
60. De Yoreo, J. J., Vekilov, P. G. *Rev Miner Geochem* 2003, *54*, 57–93.
61. Trudu, F., Donadio, D., Parrinello, M. *Phys Rev Lett* 2006, *97*, 105701.
62. Ostwald, W. *Z Phys Chem* 1897, *22*, 289–330.
63. Threlfall, T. *Org Process Res Dev* 2003, *7*, 1017–1027.
64. Peters, B. *J Chem Phys* 2011, *135*, 044107.
65. ten Wolde, P. R., Frenkel, D. *Science* 1997, *277*, 1975–1978.
66. Talanquer, V., Oxtoby, D. W. *J Chem Phys* 1998, *109*, 223–227.
67. Soga, K. G., Melrose, J. R., Ball, R. C. *J Chem Phys* 1999, *110*, 2280–2288.
68. Oxtoby, D. *Nature* 2002, *420*, 277–278.
69. Anderson, V., Lekkerkerker, H. *Nature* 2002, *416*, 811–815.
70. Vekilov, P. G. *Cryst Growth Des* 2004, *4*, 671–685.
71. Galkin, O., Vekilov, P. G. *Proc Natl Acad Sci* 2000, *97*, 6277–6281.
72. Cahn, J. W., Carter, W. *Met Trans A* 1996, *27*, 1431.
73. Herring, C. *Phys Rev* 1951, *82*, 87.
74. Wulff, G. *Z Kryst Mineral* 1901, *34*, 449–530.
75. Barnard, A. S., Zapol, P. *J Chem Phys* 2004, *121*, 4276.
76. Kirkwood, J., Buff, F. *J Chem Phys* 1949, *17*, 338–343.

77. Hoyt, J. J., Asta, M., Karma, A. *Phys Rev Lett* 2001, *86*, 5530.
78. Altmann, S. L., Cracknell, A. P. *Rev Mod Phys* 1965, *37*, 19.
79. Broughton, J. Q., Gilmer, G. H. *J Chem Phys* 1986, *86*, 5759.
80. Davidchack, R. L., Laird, B. B. *Phys Rev Lett* 2001, *86*, 5530.
81. Handel, R., Davidchack, R. L., Amwar, J., Brukhno, A. *Phys Rev Lett* 2008, *100*, 036104.
82. Kirkwood, J. G. *J Chem Phys* 1935, *3*, 300–313.
83. Angioletti-Uberti, S., Ceriotti, M., Lee, P. D., Finnis, M. W. *Phys Rev B* 2010, *81*, 125416.
84. Angioletti-Uberti, S. *J Phys Condens Matter* 2011, *23*, 435008.
85. Laio, A., Parrinello, M. *Proc Natl Acad Sci USA* 2002, *99*, 12562.
86. Cheng, B., Tribello, G. A., Ceriotti, C. *Phys Rev B* 2015, *92*, 180102.
87. Gay, D. H., Rohl, A. L. *J Chem Soc Faraday Trans* 1995, *91*, 925–936.
88. de Leeuw, N. H., Parker, S. C. *J Phys Chem B* 1998, *102*, 2914.
89. Hartman, P., Perdok, W. G. *Acta Crystallogr* 1955, *8*, 48–52.
90. Pina, C. M., Becker, U., Risthous, P., Bosbach, D., Putnis, A. *Nature* 1998, *395*, 483–486.
91. Liu, X. Y., Boek, E. S., Briels, W. J., Bennema, P. *Nature* 1995, *374*, 342–345.
92. Piana, S., Gale, J. D. *J Am Chem Soc* 2005, *127*, 1975–1982.
93. Piana, S., Reyhani, M., Gale, J. D. *Nature* 2005, *438*, 70–73.
94. Salvalaglio, M., Vetter, T., Giberti, F., Mazzotti, M., Parrinello, M. *J Am Chem Soc* 2012, *134*, 17221–17233.
95. Song, R.-Q., Colfen, H. *CrystEngComm* 2011, *13*, 1249–1276.
96. Tribello, G. A., Liew, C., Parrinello, M. *J Phys Chem B* 2009, *113*, 7081–7085.
97. Raiteri, P., Demichelis, R., Gale, J. D., Kellermeier, M., Gebauer, D., Quigley, D., Wright, L. B., Walsh, T. R. *Faraday Discuss* 2012, *159*, 61–85.
98. Freeman, C. L., Harding, J. H., Quigley, D., Rodger, P. M. *Angew Chem Int Ed* 2010, *49*, 5135–5137.
99. Freeman, C. L., Harding, J. H., Quigley, D., Rodger, P. M. *J Phys Chem C* 2011, *115*, 8175–8183.
100. Freeman, C. L., Harding, J. H., Quigley, D., Rodger, P. M. *Phys Chem Chem Phys* 2012, *14*, 7287–7295.
101. Quigley, D., Rodger, P. M., Freeman, C., Harding, J., Duffy, D. *J Chem Phys* 2009, *131*, 094703.
102. Leoni, S. *Chem A Eur J* 2007, *13*, 10022.

7. Mechanical Properties of Inorganic Nanostructures

Eduardo R. Hernández, Yukihiro Takada, and Takahiro Yamamoto

7.1 Introduction 213

7.2 Simulation Techniques and Models 214

7.3 Mechanical Properties 222
    7.3.1 Mechanical Characterization of Materials at the Nanoscale 222
    7.3.2 Experimental Approaches to the Mechanical Characterization
        of Nanoscaled Objects. 227
    7.3.3 Atomistic Simulations of the Mechanical Properties
        of ZnO Nanowires 237

7.4 Conclusions, Challenges, and Perspectives 242

Acknowledgments 244

References 244

7.1 Introduction

Over the last two decades, research on nanotubes and nanowires (NTs and NWs) of various compositions has increased. Advances in synthesis techniques, combined with the availability of increasingly powerful nanoscale probes, has resulted in a very rapid pace of development in this field. This interest has been fuelled in equal measure by scientific curiosity and by the potential for technological applications made possible by these systems. It has been recognized for some time that the physical properties of nanoscaled objects (NSOs) can differ, sometimes substantially, from those of the parent bulk material. Understanding the physical origins of this variability has been one of the motives behind the interest in NSOs in general and NTs and NWs in particular. But furthermore, NTs and NWs are seen as one of the building blocks of future nanotechnologies, and in this context, it is important to characterize their physical and chemical properties and, if possible, to learn how to control them. This process will no doubt result in the finding of new ways of exploiting NTs and NWs in practical applications.

There is little doubt that a significant number of future technological applications of NTs[1,2] and NWs[3] will critically depend on their mechanical properties and on their coupling to other characteristics of the material, for example, as occurs in piezoelectricity.[4]

Chapter 7

Many of those foreseeable applications, such as nanomass sensors[5,6] and energy-harvesting devices,[7,8] are already being explored at the level of laboratory prototypes. Undoubtedly others, as yet unforeseen, will follow in the future. This fact motivates the study and understanding of the mechanical properties of NSOs.

In this chapter, our aim is to illustrate how atomistic simulation techniques have been used in the study of the mechanical properties of inorganic nanostructures and in particular NTs and NWs. We start by providing a brief description of the simulation techniques and atomistic models most frequently used for this purpose; this is done in Section 7.2. Section 7.3 is devoted to the mechanical properties of nanostructures. We start by briefly reviewing how these are quantified in terms of a number of elastic constants and discuss the difficulties that may be encountered in the case of certain NSOs. We then describe some of the most common experimental techniques used to measure not only the mechanical properties of nanostructures, particularly NTs and NWs, but also 2D systems such as graphene and similar layered materials. We complete this part with a discussion of some examples of simulation studies that have focused on the mechanical properties of NWs, focusing on the particular case of ZnO NWs. This chapter concludes with a discussion in Section 7.4 of the challenges and perspectives for atomistic simulations in this topic.

7.2 Simulation Techniques and Models

Atomistic simulation studies require two main ingredients: (1) a *model* to represent the physical/chemical interactions between the atoms in the system, interactions that are ultimately responsible for its mechanical properties, and (2) a *simulation technique*, to be used in conjunction with the physical model, that allows to simulate and analyze the response of the model system under a specified distortion. The models used in this context range from classical force field potentials to first-principles electronic structure methods (usually based on the density functional theory [DFT]), going through semiempirical electronic structure models such as tight-binding (TB).[9] As for the simulation strategies, the most frequently used are *structural relaxation* (SR) techniques and *molecular dynamics* (MD) methods. We will cover succinctly these different aspects, trying to illustrate each one of them with specific examples drawn from the literature.

An atomistic model, be it a force field, a TB model, or an ab initio calculation, is ultimately a recipe for calculating the total energy of the system once its atomic configuration and, for the case of periodic systems, cell parameters have been specified. A given model helps to calculate not only the total energy but also its derivatives with respect to the atomic positions, that is, the forces on the atoms, and with respect to deformations of the cell (strains), that is, the stress components (see Section 7.3.1). If we regard the total energy as a function of the atomic positions $\mathbf{r}_i$ and, if needed, lattice parameters $\mathbf{a}_\alpha$,

$$E_{tot} = E(\{\mathbf{r}\},\{\mathbf{a}\}), \tag{7.1}$$

where $\{\mathbf{r}\}$ represents the set of all atomic positions in the system, and likewise, $\{\mathbf{a}\}$ are all the lattice vectors, then the force on atom i is $\mathbf{f}_i = -\nabla_{r_i} E_{tot}$, and the stress components are $\sigma_{\alpha\beta} = (1/V_0)\partial E_{tot}/\partial \in_{\alpha\beta}$, α, $\beta = 1, 2, 3$. Here, V_0 is the equilibrium volume and $\in_{\alpha\beta}$ is the $\alpha\beta$ component of the strain tensor. The strain tensor defines the deformation of the cell with respect to the equilibrium cell as follows:

$$\mathbf{a}'_\alpha = \mathbf{a}_\alpha + \sum_\beta \in_{\alpha\beta} \mathbf{a}_\beta, \tag{7.2}$$

where
α, $\beta = 1, 2, 3$
$\mathbf{a}_\alpha$ is the αth cell vector (the prime indicates the deformed cell vector, while unprimed vectors are undeformed)

Given the ability to calculate the total energy of the system, together with the forces and stresses acting on it, it is possible to ask the question, which are the stable configurations of the system? That is, which are the atomic arrangements and cell shapes that make the forces and stresses zero? One way to address the issue is by performing a so-called SR simulation, in which the total energy (Equation 7.1) is regarded as a function to be optimized by finding the atomic positions and, if required, the lattice vectors that make the energy a minimum. Optimization problems like this constitute an important area of applied mathematics, and numerous methods have been developed to solve them. Perhaps the most widely used among them, at least in the field of atomistic simulations, is the so-called *conjugate gradient* method; this and other similar methods are described in detail in standard sources on numerical methods.[10] A nuance to be aware of in SR problems is the fact that it is not usually possible to ascertain that the optimal configuration found will be the absolute minimum, that is, the one having the lowest energy. Except in very simple systems having relatively few degrees of freedom, a thorough exploration of the accessible configuration space is usually not possible, and therefore, one can never rule out completely the possible existence of a structure with lower energy than those already found.

The second simulation strategy that finds frequent use in the computational study of NSOs is MD. In a nutshell, MD consists of numerically solving Newton's equations of motion for the collection of atoms in the system under study. This can be done once one has a model, Equation 7.1, to describe the energetics of the system and from which the forces on the atoms and, if needed, the stress components on the container walls, can be evaluated. This simple description pertains solely to the most basic form of MD, the so-called *microcanonical* or NVE MD, in which the number of atoms, N; the volume, V; and the total energy, E, remain constant. Under these conditions, the system is assumed to be isolated from the rest of the universe. However, it is frequently desirable to simulate the system under more general conditions, such as isothermal, in which the system is in contact with a thermal bath, thus fixing its average temperature, or isobaric conditions, in which the external pressure, rather than the system's volume, is fixed. These conditions are frequently closer to the experimental situation that one wishes to reproduce, and thus, generalizations of the basic MD method have been developed to incorporate them. MD methods are widely used simulation tools, as, in contrast to SR, they make it possible to model the system of interest at finite temperatures, and study its time evolution. More details about MD techniques can be found in a series of standard references.[11-14]

As discussed earlier, a key issue in the atomistic modeling of a system is an adequate description of the interactions taking place between the atoms that constitute it. These interactions are the result of the subtle interplay between electrons and nuclei, an interplay that can give rise to a variety of bonding situations. These are typically classified as metallic, covalent, or ionic bonding, although it should be recognized that these are extreme cases and that real systems often display an intermediate behavior (e.g., partly ionic, partly covalent). There are generally two different approaches that can be adopted to model the energetics of an atomic system. The first one is to employ a parametrized potential form, that is, a physically plausible function, depending on the relative interatomic positions (distances, perhaps angles) and a series of parameters. These parameters are fitted so as to reproduce as faithfully as possible a number of properties of the system of interest, such as structural, elastic, or vibrational properties, which must be known either from experiment or from higher levels of theory. In this approach, the electronic structure of the system is ignored, its effects on the energetics of the system approximately accounted for by the chosen potential energy function. This first approach is frequently called the *empirical potential* or *force-field* approach. There is generally no fundamental principle to guide the form of the potential energy, beyond the fact that it should be repulsive at short interatomic distances, attractive at intermediate ones, and should decay to zero for large separations. The second approach consists of retaining the image of the system that is composed of electrons and nuclei and obtaining its energy, forces, and stress, from a quantum mechanical treatment of its electronic structure, either at a semiempirical level or from a first-principles treatment. This approach is theoretically more sound, but also more expensive numerically.

Each of these different approaches to model the energetics of the system has its advantages and disadvantages: empirical potentials are computationally cheap, which allows for the study of large systems over long simulation time scales. However, fitting a potential model for a new system can be a difficult task, particularly for systems with multiple chemical species, and the transferability of the resulting model to physical conditions outside of those considered in the fitting procedure is always open to question. At the other end of the spectrum, first-principles calculations are in general more reliable than empirical potential models, but they are also orders of magnitude more computationally demanding, a circumstance that hampers their application to large systems or to problems that require long simulation time scales. Sometimes semiempirical electronic structure methods, being less demanding than first-principles approaches, can offer a compromise between these extremes, although they are not themselves without disadvantages (fitting, transferability problems, etc.).

Traditionally, empirical potentials have been tailored to the nature of the particular system of interest. Specific model families have been developed and used for ionic, covalent, or metallic systems. For example, in ionic systems, a key ingredient is the long-range electrostatic interaction between ions, which must be supplemented by a short-range potential that is repulsive at short distances, to prevent ions of opposite charge from collapsing onto each other. Such ionic system models frequently include a polarizable ion description, which can be particularly important for the anions.[15] A rigid-ion model of this kind (i.e., without polarizable ions) was used by Kulkarni et al.[16,17] and by Agrawal and coworkers[18] to perform their simulations on the mechanical properties of ZnO NWs and nanobelts. These simulations will be discussed in more detail in the following sections. On occasion, for partly covalent materials, they can also include bond-angle three-body terms.

Covalent systems, on the other hand, are characterized by their directional bonding, and in order to adequately account for this, empirical potentials that aim to describe such systems must depend on the angles formed by a given atom and its neighbors, as well as on the interatomic distances. Many different models have been proposed for covalent systems, but here let us just cite two examples that are frequently used: the first one is the Stillinger–Weber potential[19] for Si, and the second one is the Tersoff *many-body* potential,[20] parametrized for both Si and C, as well as for SiC. In particular, the Tersoff potential has the following expression:

$$U_{tot} = \frac{1}{2}\sum_i \sum_{j \neq i} f_C(r_{ij})[f_R(r_{ij}) + b_{ij}f_A(r_{ij})], \tag{7.3}$$

where

$f_R(r)$ and $f_A(r)$ are repulsive and attractive pairwise interactions, respectively
$f_C(r)$ is a cutoff function that smoothly approaches zero between the first and second nearest neighbor distance

The many-body nature of this potential is contained in the b_{ij} term, which is a function of the bond angles formed by atoms i, j and every other possible neighbor of i. This factor modulates the strength of the attractive term between atoms i and j, making it dependent on the local environment of atom i, reflecting the concept of bond order in covalent systems. This is why the Tersoff potential and other models derived from it are sometimes called *bond-order potentials*. An example of the use of the Tersoff potential in the context of this chapter is provided by the work of Makeev et al.[21]; these authors used it to model $\langle 111 \rangle$-oriented cubic SiC NWs of diameters in the range of 0.89–3.56 nm. They found that, contrary to what happens in the case of ZnO NWs (see Sections 7.3.2 and 7.3.3), Young's modulus was largely insensitive to the NW diameter, although only relatively small diameters were considered. In contrast, both the bending and torsion moduli were found to increase with diameter, as expected from the theory of elasticity.[22]

Potentials have also been developed for metallic materials, such as the embedded atom model (EAM),[23] the Cleri and Rosato[24] model, or the Finnis–Sinclair[25] potential, to name a few examples. In the EAM model, it is assumed that the cohesive energy of a metallic system can be written as a sum of atom energy terms, where the energy of atom i results from the embedding of that atom into the electron density obtained from the superposition of the spherically averaged atomic electron densities due to all other atoms, evaluated at the position of atom i, supplemented by a short-range repulsive term. The total potential energy has the form

$$U_{EAM} = \sum_i G_i(\rho_i) + \sum_{j \neq i} U_{rep}(r_{ij}),$$

$$\rho_i = \sum_{j \neq i} \rho_j(r_{ij})$$

(7.4)

where

$G_i(\rho)$ is the energy that results when embedding an atom of the chemical species of atom i in an electron density ρ

U_{rep} is a pairwise short-range potential

An example of the application of this model to the structural and mechanical properties of NWs is provided by the work of Diao and coworkers.[26] These authors used a revised version of the EAM model (the *modified*, or MEAM[27]) to study the structural stability of fcc Au NWs. In so doing, they uncovered a surface-stress-induced structural transformation from the fcc structure adopted by bulk Au to a bct structure in the case of NWs with lateral dimensions of 1.83 × 1.83 nm and smaller. Surface effects gain increasing relevance as the dimensions are reduced, and eventually, for sufficiently small systems, they can overcome the role of the bulk energetics, which obviously govern the structure in the case of large systems. Diao et al. simulated gold NWs that initially had the fcc bulk structure, with ⟨100⟩-orientation, having a squared cross section and [100], [010], and [001] free surfaces (i.e., nonperiodic boundary conditions were employed); the axial length was 32 nm, and the lateral dimensions were smaller than 4 nm. When these fcc NW structures were relaxed using the conjugate gradient technique (see earlier), it was observed that wires with sides greater than 1.83 nm contracted axially by less than 4%. However, wires with sides of 1.83 nm or less contracted upon relaxation by more than 30%, see Figure 7.1. In these wires, inspection of the relaxed configurations showed that they had transformed to a bct structure oriented along the c axis. The c lattice parameter was found to be 2.824 Å according to the MEAM model used, which, when compared to the bulk fcc lattice parameter of 4.07 Å predicted within the same model, accounts for the large 30% distortion upon relaxation. The same transformation was observed to occur in MD simulations at low temperatures (100 K). The fact that the transformation was seen is probably due to the fact that nonperiodic boundary conditions were employed, as the free surfaces at the ends of the NW serve as nucleation sites for the transformation, which then propagates inward from both ends.

More recently, there have been some attempts to construct models that in principle need not incorporate any previous assumption or knowledge on the physicochemical properties of the system.[28,29] In such schemes, one does not attempt to parametrize the energy in terms of an ad hoc function, as done in the aforementioned examples; the aim is rather to employ strategies that are flexible enough to represent any kind of physical behavior with equal accuracy and without imposing preconceived ideas about the system. In the scheme of Behler and Parrinello,[28] this is achieved using neural network methods.

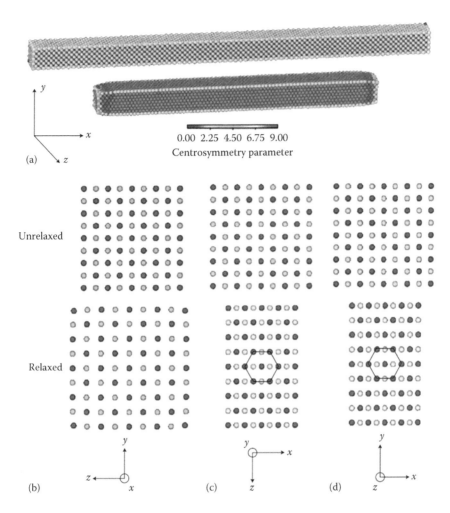

FIGURE 7.1   Phase transformation in a 1.83 nm × 1.83 nm Au [100] wire. (a) shows the unrelaxed and relaxed configurations of the wire. Atoms are colored according to a centrosymmetry parameter, which is zero for a perfect fcc crystal and nonzero at defects and surfaces. (b)–(d) show cross sections at the centre regions of the unrelaxed and relaxed wires, observed from x, y, and z directions, respectively, only two adjacent lattice planes of atoms are shown, and atoms in different lattice planes are shown in different colors. In (c) and (d), only partial regions of the wires are shown in the x direction. (Reproduced from Diao, J. et al., *Nat. Mater.*, 2, 656, 2003. With permission.)

A neural network is trained to predict the energy and forces of the system of interest, using extensive data sets from first-principles calculations. Once the neural network has been adequately trained, it can be used to predict the energy and forces for arbitrary configurations, provided these do not deviate significantly from those considered within the training data set. The novelty of the scheme of Behler and Parrinello[28] with respect to earlier attempts to use neural networks in this context is that rather than attempting to predict directly the energy of the whole system, it is assumed that the total energy can be partitioned into a sum of individual atom terms, each one of which will be dependent on the local environment of the relevant atom; the neural network is trained to predict the value of this atomic energy term, given the local environment of each atom. The Gaussian approximation potential method of Bartók and coworkers[29] is similar in spirit to the neural network approach in that the total energy of the system is partitioned as a sum of atom terms, each one dependent on the local environment. However, in this method, instead of using neural network techniques, the atomic energy dependence on the local environment is quantified by Gaussian process regression. A common feature of these methods is the need to find adequate descriptors of the local atomic environment, in order to make the atomic energy terms sufficiently sensitive and flexible to correctly represent the total energy of the system. Both methods show great promise, and it is very likely that the future of empirical force fields will see further developments in this direction.

Compared to empirical potential models, semiempirical electronic structure methods, such as TB models,[9] constitute a step toward a more complete description of the material, given that they do not totally obviate the electronic structure of the material, even if they do treat it at a rather simplistic and approximate level compared to first-principles methods. In TB models, the matrix elements of the electronic Hamiltonian are not evaluated rigorously from the Hamiltonian operator and basis set of choice; rather, they are treated in a manner akin to that in which the total energy itself is treated in the empirical potential models: they are assumed to have a certain parametrized dependence on the interatomic distances. This makes the cost of constructing the matrix representation of the TB Hamiltonian small compared to that of the rigorous calculation used in first-principles methods. However, this is at the cost of assuming a given form for these matrix elements, which, as with empirical potentials, may be physically sound, but is ultimately ad hoc. TB models have become extremely popular in materials modeling,[9] due to their combination of methodological simplicity and reasonable accuracy and due to the fact that they can provide a qualitative description of the electronic structure of materials, something beyond the capabilities of empirical potentials. Interested readers may find more details about TB models in several review articles[9,30] and books.[31,32]

One of the interesting features of TB models is that they can be tailored to the particular problem at hand. A TB model can be constructed to be more or less sophisticated, depending on the nature of the questions to be addressed. For example, the electronic properties of graphene close to the Fermi energy can be largely understood on the basis of a minimal TB model consisting of a single π-electron, that is, a singly occupied p_z orbital per carbon atom.[33] At the other extreme, TB models can be made to approach the limit of first-principles methods (see the following text) by gradually eliminating the approximations incurred in the TB model while still remaining comparatively cheap. An example of the latter case is the *density-functional-based tight-binding* (DFTB) model.[34] This model has been successful at accounting for the mechanical properties of carbon-based and BN NTs,[35,36] phosphorus[37] and silicide-based NTs,[38] and even MoS_2[39,40] NTs. It has recently been used in combination with an *objective* structural description[41,42] to study chiral MoS_2 NTs.[40] The description of chiral NTs within conventional periodic-boundary conditions can be costly, because they can have very long unit cells along the axial direction. A more effective way to describe these systems is as objective structures, structures that have a symmetry that consists of a translational component combined with a rotational one; in other words, they have a helical symmetry. The corresponding unit cell within this objective description is typically much smaller than the corresponding purely translational unit cell. Using this methodology, Zhang et al.[40] have analyzed the dependence of the band gap on the chiral angle for a series of

$(n + k, n - k)$ MoS$_2$, where $k \in [0, n]$ and $n = 10, 12, 14, 16, 18$. The size of the band gap was found to decrease slightly with decreasing k, or equivalently, increasing chiral angle within a given value of n. They also found that, like reported earlier for (n, n) MoS$_2$ NTs[39] using the same model, chiral NTs have an indirect band gap, while $(n, 0)$ ones have a direct one. These gaps are all of similar size and increase with NT diameter, tending toward the bulk value. Zhang et al. also found that chiral NTs of narrow diameters (below ~6 nm) have an intrinsic twist, meaning that the tubular structure that conceptually results from the usual cutting and folding of the flat MoS$_2$ sheet[43] has a residual stress, indicating that the structural parameters predicted by the folding procedure are inaccurate and the structure that results is strained. When the tubular structure thus created is relaxed, a chiral angle results that is different (slightly larger) from the one imposed in the folding construction procedure. This is illustrated in Figure 7.2 for the (10,10), … (20,0) family of tubes.

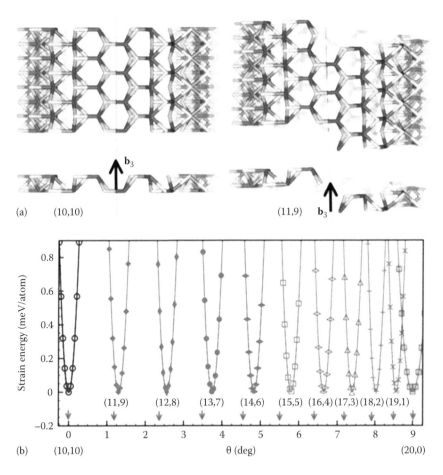

FIGURE 7.2 (a) Objective computational cells for a (10,10) (left) and (11,9) MoS$_2$ nanotubes. The nanotube structures displayed are obtained by helical replication of the objective cells shown in the following, by helical replication along the vector **b**$_3$. (b) displays the strain energy versus chiral angle θ for the (10,10), …, (20,0) family, having nearly equal diameters. The energy minima correspond to the stress-free (relaxed) nanotubes with indices indicated under the curve. (Reproduced from Zhang, D.-B. et al., *Phys. Rev. Lett.*, 104, 065502, 2010. With permission.)

Chapter 7

The formulation of DFT by Kohn and collaborators[44,45] during the 1960s is certainly one of the milestones in the history of computational materials science. In subsequent decades, DFT was further developed into a mature methodology that made possible the undertaking of accurate total energy electronic structure calculations for complex materials, and it also brought about the possibility of performing first-principles MD calculations, first demonstrated by Car and Parrinello,[46] which is itself another landmark in the development of computational materials science. DFT is certainly not the only methodology available to study the electronic properties of materials, nor is it the most accurate one; quantum chemistry methods and quantum Monte Carlo techniques are superior in terms of accuracy, but their computational cost is much higher, and for the time being, this precludes them from being used in combination with MD, while DFT-based MD simulations have been relatively standard for nearly two decades now. There are many technicalities that form part of DFT calculations, and we cannot cover them here in any detail; interested readers should consult the excellent book by R. M. Martin[31] and the references therein for a full description.

7.3   Mechanical Properties

In this section, we discuss the mechanical properties of NTs and NWs, focusing more particularly on the latter. We will first consider the question of how to quantify the mechanical properties of NSOs in Section 7.3.1. We will then describe succinctly in Section 7.3.2 the different experimental techniques that have been employed to measure the mechanical properties of NTs and NWs, and finally, we will describe some examples of simulation studies applied to the particular case of ZnO NWs in Section 7.3.3. Our intention here is not to provide an exhaustive review on the mechanical properties of these systems, but simply to introduce and motivate the topic. More extensive reviews on the mechanics of NTs[47–49] and NWs[50] can be found elsewhere.

7.3.1   Mechanical Characterization of Materials at the Nanoscale

The classical theory of elasticity[22] shows that an isotropic medium is mechanically characterized by two material-dependent constants, which can be chosen as Young's modulus and Poisson's ratio, or alternatively as the Lamé elastic constants, μ, the shear modulus, and λ. Thus, within the elastic regime of small distortions, the mechanical response of a deformed isotropic material is fully defined in terms of these two constants, or combinations thereof. Since at the macroscopic scale it is not infrequent to find materials that are effectively isotropic (e.g., polycrystalline materials), the theory of elasticity of isotropic media is of great practical use and has been frequently employed in fields as widely varied as engineering or geophysics. However, not even at the macroscopic level can all materials be regarded as isotropic, for example, single crystals and oriented polymeric materials. When this is the case, elasticity theory tells us that more constants are required to characterize the material. The number of elastic constants necessary to fully specify the mechanical behavior of a nonisotropic system is larger when the number of symmetries that are present in the system is fewer. Thus, for a cubic crystal, only three elastic constants are required (C_{xxxx}, C_{xxyy}, and C_{xyxy}), while a total of 21 is needed for a triclinic crystal.[51]

In some sense, the meso-/nanoscale is a transition zone between the macroscopic world, in which the concept of an isotropic system makes sense and frequently constitutes a useful approximation, and the microscopic world of atoms, where an isotropic system is a physical impossibility. This dichotomy reflects in the fact that in many practical situations, it is still possible to make use of macroscopic models such as rods and shells to analyze the mechanical properties of NWs and NTs, while in other situations, forcing the analogy between macro-objects and nano-objects too far can lead to inconsistencies and absurd results, as will be shown in the following.

As noted earlier, the mechanical response to deformation of an isotropic body is fully specified by two constants, typically chosen as Young's modulus and Poisson's ratio (although other choices are possible). Young's modulus, Y, is defined as

$$Y = \frac{1}{V_0} \left(\frac{\partial^2 E}{\partial \varepsilon^2} \right)_{\varepsilon=0}. \tag{7.5}$$

where
 V_0 is the equilibrium volume
 E is the total energy
 ε is the *strain*, or fractional deformation of the system

The strain is taken to be uniaxial (i.e., the system is deformed, stretched, or compressed, along a given direction), but since the material is assumed to be isotropic, Y has no directional dependence. If the equilibrium length of the piece of material being strained along the axis of deformation is L_0, then ε is given as $\varepsilon = (L-L_0)/L_0$. The general definition given in Equation 7.2 reduces to this in the case of uniaxial strain. Following this convention, compressive strains are negative and dilating ones are positive. Strictly speaking, Equation 7.5 is only valid at zero temperature; at finite temperatures, E should be replaced by F, the free energy. Another important observation to be made on the definition of Young's modulus is the following: since the (free) energy is an extensive property, the factor of V_0^{-1} on the rhs of Equation 7.5 is required in order to make Y an intensive property (i.e., independent of the system's dimensions). Alternative factors that would also result in an intensive Y would be possible, but perhaps for historical reasons, this is the one that is used most frequently. However, it is worth noticing that there are good reasons for reconsidering the definition of Young's modulus, particularly when contemplating its application to atomically thin NSOs. The definition via Equation 7.5 poses no difficulty at the macroscopic scale; however, at the nanoscale, it is not always obvious to define V_0. Indeed, for what is the volume of a single layer of atoms, as one encounters in graphene, or in single-walled NTs? We will return to this issue later.

The physical interpretation of Equation 7.5 is simple: materials for which the (free) energy changes rapidly with applied strain are less easily deformable and have, correspondingly, a high Young's modulus. Conversely, softer, easily deformable materials have lower Young's modulus values. The dimensions of Y, definition given in Equation 7.5, are those of an energy density, or equivalently, those of a pressure. Thus, Y values are frequently quoted in units of GPa. The highest values of Young's modulus have been measured for carbon nanotubes (CNTs), and as we shall see in the following, they are in the range of the TPa.

The second constant that characterizes the mechanical behavior of an isotropic material is its Poisson's ratio[52], σ. It is a well-known fact that straining a piece of material along a particular direction will usually also cause deformation along the perpendicular directions. Hence, Poisson's ratio is defined by

$$\varepsilon_\perp = -\sigma\varepsilon_\parallel, \tag{7.6}$$

where

σ is Poisson's ratio
$\varepsilon_\parallel, \varepsilon_\perp$ are the strains parallel and perpendicular to the axis along which the deforming load (stress) is applied

The negative sign on the rhs of Equation 7.6 accounts for the fact that in the immense majority of cases, a material that is stretched (compressed) along a given direction is compressed (stretched) along the perpendicular ones. In other words, the usual behavior is that $\varepsilon_\parallel$ and $\varepsilon_\perp$ have opposite signs. Thus, with the definition embodied in Equation 7.6, conventional materials have positive values of σ. However, there are exceptions to this rule: some materials, such as cork, have practically zero Poisson's ratio (cork hardly deforms at all along perpendicular directions when stretched, at least for small deformations). There are also cases of what is known as auxetic behavior, displayed by materials with a negative Poisson's ratio. Though such materials are relatively rare, auxetic behavior has important potential for technological applications, and there is a growing interest in them.[52,53] Mechanical stability arguments require that Poisson's ratio of any material be confined in the range of −1 to 0.5.[22,54]

As argued earlier, nonisotropic materials require a larger number of elastic constants to characterize their mechanical response to an imposed stress. Although the concept of Young's modulus is still a useful one in this context, it is no longer the case that one can speak of a unique value; rather, Young's modulus becomes direction dependent.

Due to obvious structural similarities, NTs and NWs are frequently conceptualized as nanoscopic cylinders or shells and rods, respectively. Thus, it is not surprising that their mechanical properties are rationalized in terms of Young's modulus and Poisson's ratio. Even if the material constituting the NW or NT cannot be regarded as isotropic, the uniaxial character of these nanostructures makes it reasonable to speak about their Young's modulus, which refers to that along the main axial direction of the nanostructure. Applying Equation 7.5 to an NW poses no major conceptual difficulty; however, the case of a single-walled nanotube (SWNT), or an atomically thin membrane such as graphene, is more problematic. The difficulty arises because of the V_0^{-1} factor: defining the equilibrium volume in the case of a wall that is only one atom thick is simply not possible without adopting some convention. The literature on CNTs is plagued by calculations of Young's modulus employing different criteria to define V_0, thus leading to large discrepancies that are, in most cases, artificial. This somewhat embarrassing situation is illustrated in Table 7.1. As can be seen, some choices of the shell thickness h result

Table 7.1   Selection of Calculated Values for Young's Modulus of Single-Walled Carbon Nanotubes

Method	Wall Thickness h (nm)	Young's Modulus Y (TPa)	Scaled Modulus $\tilde{Y} = \dfrac{h}{0.34} Y$ (TPa)	Reference
Structural mechanics	0.34	~1	~1	Li and Chou[89]
Continuum model	0.075	4.7	1.04	Tu et al.[90]
Continuum model	0.0617	4.88	0.88	Vodenitcharova et al.[91]
Continuum model	0.075	4.84	1.07	Pantano et al.[92]
Force field	0.066	5.5	1.07	Yakobson et al.[93]
Force field	0.34	0.974	0.974	Lu[94]
Force field	0.34	1.21–1.37	1.21–1.37	Jin and Yuan[95]
Force field/DFT	0.0665	~5	~1	Wang et al.[96]
Force field	0.0734	3.21	0.69	Huang et al.[97]
Tight-binding model	0.34	1.24	1.24	Hernández et al.[35,36]
Tight-binding model	0.07	5	1.03	Xin et al.[98]
DFT	0.34	~1	~1	Sánchez-Portal et al.[99]
DFT	0.0894	3.859	1.01	Kudin et al.[100]
DFT	~3.3	~1	~1	Wagner et al.[56]

Note: Different authors have used different conventions in specifying the wall thickness h, and the resulting Young's modulus value is quoted directly; also given is the value of the scaled modulus, that is, the value obtained by taking the convention $h = 0.34$ nm.

in SWCNTs having extraordinarily large values of Young's modulus, up to five times larger than the value of the C_{xxxx} elastic constant associated with the in-plane deformation of graphite along the basal plane. But clearly, if we accept that elastic constants ultimately reflect the resistance of chemical bonds to deformation, and we consider that the bonding situation in a SWCNT wall is not too different from that in a graphene layer of graphite, then it follows that such large disparities do not make any physical sense. This view is confirmed by the last column in Table 7.1, in which we list rescaled values of Young's modulus, obtained using a common wall thickness equal to the interlayer spacing in graphite. In Table 7.1, when this is done, the different results reported in the literature appear to be much more consistent. This shows that the apparent dispersion of Young's modulus values of SWCNTs found in the literature is artificial.

The difficulty in defining V_0 in this case is a genuine one, and it can only be resolved by adopting a general convention, or else by taking an alternative definition of Young's

Chapter 7

modulus that is more appropriate at the nanoscale. One alternative definition was put forward by Robertson et al.,[55] who modified Equation 7.5 as

$$\hat{Y} = \frac{1}{N} \left(\frac{\partial^2 E}{\partial \varepsilon^2} \right)_{\varepsilon=0}. \tag{7.7}$$

Here N is the number of atoms in the periodic unit of the structure in question. This definition has the advantage of not depending on any cell volume definition; the number of atoms is always a well-defined quantity, independent of the actual geometry or dimensionality of the system. In cases where the cell volume can be unambiguously defined, the conventional expression for Young's modulus can be recovered by multiplying $\hat{Y}$ with the equilibrium particle density, $\rho_{eq} = N/V_{eq}$. More recently, Wagner et al.[56] have suggested that the solution to this ambiguity is by retaining the traditional definition of Young's modulus embodied in Equation 7.5, if the volume is chosen based on an electron density criterion. Their proposal consists of defining an *active volume*, which would be the volume in which the numerical value of the electron density is larger than a certain cutoff value, c. Wagner and coworkers suggest taking c as the value at which the NSO (say an SWNT or a graphene sheet) has the same average electron density as the parent bulk material (graphite). With such a definition, the resulting volume that encloses the NSO is found to contain more than 99% of the electron density, testifying to the physical soundness of the criterion. It also results in consistent values for Young's modulus for one-atom-thick nano-objects and the parent bulk material (see Table 7.2). In the particular case of graphene/graphite, and that of other layered materials considered by the authors (BN, MoS_2, WS_2, $MoSe_2$, $MoTe_2$), the active volume found is very similar to the one obtained by simply dividing the volume of the unit cell of the parent material by the number of layers contained in it, thus lending some justification to the standard criterion adopted in most experimental studies. As we have seen earlier, though, this criterion is not always the one adopted by theory and simulation studies, as can be seen in Table 7.1.

Young's modulus and related elastic constants discussed earlier characterize the mechanical response of a material to a distorting load (a stress) within the linear, also called *elastic*, regime. Within this regime, the material is able to recover its original undistorted configuration once the stress is removed. However, if the material is tensile loaded beyond the limits of the elastic regime, it either fails by fracture, in the case of brittle materials, or it yields, and plastic deformation sets in, in the case of ductile materials. In the latter case, plastic deformation under tension leads to progressive necking and eventually to (ductile) fracture. The stress at which fracture occurs is known as the material's *strength*. The fracture dynamics of materials is important from both an applied and a fundamental perspective; therefore, it is not surprising that this is an active area of research. An example of how MD simulations can help in this field is provided by the work of Kang and Cai.[57] These authors used MD simulations of [110]-oriented Si NWs with diameters in the range of 2–7 nm, employing the MEAM potential. The NWs were subject to a strain rate of 5×10^8 s^{-1}, and simulated at various temperatures (from 300 to 1200 K) until failure was observed. These simulations resulted in the observation of two distinct failure mechanisms: it was observed to occur either by cleavage

Table 7.2   Calculated In-Plane Young's Modulus for Various Layered Materials Following the Method Proposed by Wagner et al.[56]

System	$Y(c)$ (TPa)	$c(e^-/a_0^3)$	h (Å)	N_q (%)
Monolayer graphene	1.059	0.00240	3.31	99.64
Bilayer graphene	1.059	0.00247	3.32	99.81
Trilayer graphene	1.058	0.00237	3.32	99.88
Four-layer graphene	1.055	0.00226	3.32	99.91
Graphite (bulk)	1.055	—	3.32	100.0
Monolayer BN	0.898	0.00268	3.19	99.60
Bilayer BN	0.891	0.00288	3.19	99.78
Trilayer BN	0.886	0.00277	3.19	99.86
h–BN (bulk)	0.880	—	3.19	100.0
Monolayer WS_2	0.251	0.00290	6.14	99.89
WS_2 (bulk)	0.242	—	6.17	100.0
Monolayer MoS_2	0.222	0.00293	6.12	99.85
MoS_2 (bulk)	0.219	—	6.14	100.0
Monolayer $MoSe_2$	0.188	0.00335	6.35	99.87
$MoSe_2$ (bulk)	0.188	—	6.36	100.0
Monolayer $MoTe_2$	0.132	0.00329	6.87	99.87
$MoTe_2$ (bulk)	0.132	—	6.91	100.0

Source: Data taken from Wagner, P. et al., *J. Phys. Condens. Matter*, 25, 155302, 2013.
Note: c is the cutoff density for which the resulting volume of the nano-object is such that its average electron density is equal to that of the bulk, that is, $Q(c)/V(c) = \rho_{bulk}$, where $Q(c)$ is the integrated electron charge over volume $V(c)$; h is the corresponding shell thickness, and $Nq = Q(c)/Q_{total}$ is the fraction of electrons actually enclosed in the volume $V(c)$.

along a (110) plane (brittle fracture), or by shear on {111} planes (ductile fracture), or a mixture of both, depending on the temperature and the NW diameter. An example of both mechanisms is provided in Figure 7.3, where it is seen that, at low temperatures ($T = 300$ K), a 7 nm Si NW undergoes brittle fracture; the same NW undergoes ductile fracture by shear at $T = 1000$ K. The transition from brittle to ductile fracture was seen to depend on the NW's diameter; indeed, NWs with $D = 2$ nm were observed to undergo ductile failure at temperatures as low as 100 K.

7.3.2   Experimental Approaches to the Mechanical Characterization of Nanoscaled Objects

The mechanical characterization of NSOs such as NWs, NTs, or atomic layers is immensely challenging, and the different experimental techniques that have been set up over the last decade and a half to achieve this goal are examples of nanotechnological feats. The need to perform a controlled manipulation of the NSO that has to be mechanically tested is difficult. An even greater challenge is the performance of the

Chapter 7

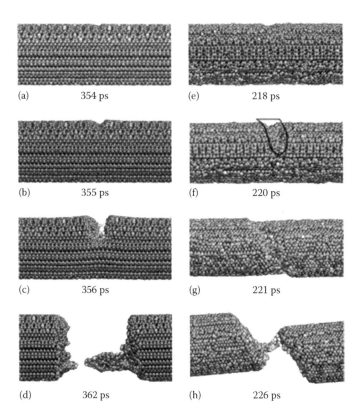

FIGURE 7.3 Fracture of a $D = 7$ nm [110] Si nanowire under tensile loading. (a)–(d) show the NW at $T = 300$ K, which results in failure by cleavage along a (110) plane, that is, perpendicular to the NW axis. (e)–(h) show the NW at $T = 1000$ K; the lines in (f) show the surface area that undergoes slippage. (Reproduced from Kang, K. and Cai, W., *Int. J. Plasticity*, 26, 1387, 2010. With permission.)

testing itself; not only is it necessary to apply appropriate distortions to the NSO, but its response to the testing has to be recorded as well, and its dimensions have to be accurately determined, in order to attain the sought mechanical characterization. In this section, we will review some of the experimental techniques that have been developed to achieve this aim. Our intention is not to provide an exhaustive review, but rather to give a bird's-eye view of the most important experimental techniques used in this context. Nor do we intend to provide a discussion of all the practical applications of these techniques to specific cases of NSOs; given the volume of published results, that would be a daunting task. Our aim is rather to illustrate each technique with appropriately chosen examples of its use. As we shall see in the following text, many of these techniques were first developed and applied in the field of CNTs during the 1990s and were later used directly or adapted for use with other NSOs.

We will see that a common denominator of many of these mechanical testing techniques is their reliance on the classical theory of elasticity as applied to rods, cylinders, or plates, since these can be assimilated to NWs, NTs, and atomic shells, respectively.

Ever since the first observations of CNTs, there had been speculations about the possibility that these NSOs had exceptional mechanical properties, since it was known that the CC bond along the basal plane of graphite is the strongest chemical bond in any extended system, and this should ultimately be reflected in a large value of Young's modulus. Treacy, Ebbesen, and Gibson[58] reported the first experimental determination of Young's modulus of multiwalled carbon nanotubes (MWCNTs), and their results confirmed the expectation. These authors analyzed the amplitude of the thermally induced oscillations of anchored NTs observed under a transmission electron microscope (TEM). They noticed that while it was possible to obtain a sharply focused image of clamped NTs close to their anchoring point, the free tip was always comparatively blurred. This blurring was not due to a mismatch between the focal plane and the plane containing the axis of the NT; had that been the case, simple refocusing would have brought the NT tip into focus. Furthermore, they observed that the blurring of the tip area increased with temperature, and thus concluded that the blurring was caused by the thermally induced oscillations at the tip. By using a careful image-processing procedure, they were able to extract the vibrational amplitude of the tip of a series of NTs as a function of the imposed temperature. By assimilating the NT to a cantilevered cylinder of length L, outer and inner radii a and b, respectively, and Young's modulus Y, it is possible to show[59] that the squared amplitude of the thermally induced vibrational profile at the NT tip, Σ, is linearly proportional to the temperature:

$$\Sigma^2 = 0.4243 \frac{L^3 k_B T}{Y\left(a^4 - b^4\right)}, \tag{7.8}$$

where k_B is Boltzmann's constant. Hence, by measuring the length, the radii, and the amplitude of the thermal oscillations at a series of temperatures, Treacy et al. were able to determine Young's modulus of a series of 11 NTs. Their measured values ranged between 0.4 and 4.15 TPa, with an average value of 1.8 TPa. The large variations in the data for individual NTs probably reflect the inherent difficulties in quantifying the vibrational amplitudes and the exposed length of the tubes (it was not always possible to determine their exact anchoring point). Nevertheless, their results clearly showed that MWCNTs indeed have a Young's modulus in the TPa range and constituted the first experimental proof of the exceptional mechanical properties of these nanostructures.

In subsequent work, Krishnan and coworkers[59] applied the same experimental procedure to a set of SWCNTs. This study confronted the issue of assigning values to the outer and inner radii of a cylindrical shell that was only one atom thick, a question that the authors resolved by setting $a - b = 3.4$ Å, equal to the interlayer spacing in graphite. From a set containing a total of 27 cantilevered SWCNTs analyzed, they extracted an average value of Young's modulus of 1.25 TPa, which is similar to the C_{xxxx} elastic constant of graphite. The same technique was used by Chopra and Zettl[60] in a multiwalled BN NT. They reported a Young's modulus of 1.22 TPa, not too different to that obtained for CNTs. This was the first study to report a measurement of Young's modulus for a non-CNT.

The technique of Treacy et al.[58] for determining Young's modulus of NSOs consisted of exploiting an intrinsic property such as the thermally induced tip oscillations of adequately cantilevered tubes. The only means of controlling the tubes were exerted indirectly through the ambient temperature. Subsequent studies, however, introduced an additional element of control on the analyzed nanostructure. For example, Wong et al.[61]

used the tip of an atomic force microscope (AFM) to push and bend clamped SiC NWs and MWCNTs while recording the restoring force exerted by the bent NSO on the AFM tip. Poncharal et al.[5] used an electrostatic approach to similarly distort clamped NTs. Let us look at these techniques in more detail.

Wong, Sheehan, and Lieber's[61] approach consisted of depositing the NT/NWs to be mechanically tested onto a substrate and subsequently clamping them by depositing a regular array of squared SiO pads through a mask. Some of the deposited nanorods would thus become trapped at one end by a SiO pad while remaining unclamped at the other end. This was done on a MoS_2 substrate, as this material can be easily cleaved, producing extended atomically flat surfaces of low friction. The exposed ends of clamped NT/NWs were then located with the help of an AFM, and the tip of the microscope was then used to measure the exposed length of the NT/NW, and to bend it laterally while at the same time recording the restoring force exerted by the bending NSO onto the AFM tip. By recording this restoring force at different points along the exposed length of the cantilevered NSO as a function of the lateral deflection, it was possible to extract Young's modulus of the tested NT/NW. In order to test the exper-imental procedure, data were first accumulated for SiC NWs of various diameters; these yielded a Young's modulus in the range of 610–660 GPa, consistent with earlier measurements performed on micrometer-sized SiC whiskers. Then the same proce-dure was applied to a series of six MWCNTs, with diameters in the range of 26–76 nm. They obtained an average Young's modulus of 1.28 ± 0.59 TPa, a value that is close to the in-plane modulus of graphite (1.06 TPa). Another interesting observation to result from this work was the fact that while SiC NWs would eventually suffer brittle fracture, MWCNTs were capable of sustaining large bending distortions without any detectable sign of failure.

The original approach of Wong et al.[61] has been adapted and applied to work with vertically aligned ZnO NWs by Song and coworkers.[62] From a sample of 15 vertically aligned ZnO NWs, they determined an average value of Young's modulus of 29 ± 8 GPa, with individual values distributed in the range of 15–47 GPa. Aside from the scatter of the data for individual wires, the average value of Y found by this team seems rather low, when compared to Young's modulus of bulk ZnO crystals along the [0001] direction, which is estimated to be ~140 GPa. Later in this section we will discuss in more detail the case of ZnO NWs and the experimental efforts that have been reported to character-ize their mechanical properties, putting this work in the context of other studies.

A similar approach to that of Wong et al.[61] was used by Salvetat and coworkers.[63] In this work, an AFM was also used, but rather than clamping the CNTs as done by Wong et al.,[61] they dispersed them onto an ultrafiltration alumina membrane; such membranes contained pores with diameters of the order of a few hundred nanometers, and the deposition of CNTs on the membrane resulted in frequent occurrences of the CNTs bridging across one of the pores. An AFM would then be used to locate favorable situations in which a CNT is held, partly suspended, across a pore. The AFM tip was then used to bend laterally the suspended CNT and measure the restoring force against the displacement. In particular, Salvetat et al.[63] used this technique to determine Young's modulus of ropes of SWCNTs, finding values in the range of 1 TPa. This approach relies on the assumption that the CNT remains clamped on either side of the pore and does not slip when distorted with the AFM tip.

Mechanical testing of suspended NSOs using an AFM tip has also been reported for the case of 2D laminar materials. For example, Lee et al.[64] used this technique to extract Young's modulus of a single layer of graphene, obtaining a value of 1 ± 0.1 TPa, and a third-order elastic stiffness of −2 ± 0.4 TPa, values that were obtained assuming a graphene thickness of 3.35 Å. By indenting the suspended graphene layer until failure was observed, they were able to extract an intrinsic tensile strength of 130 ± 10 GPa, occurring at a strain of 25% (see Figure 7.4). Similar experiments have been reported on thin graphite flakes consisting of a small number of graphene layers by Poot and van der Zant[65] and by Castellanos-Gomez et al.[66] on sheets of MoS_2 containing between 5 and 25 layers. Poot and van der Zant[65] were able to measure the drum frequencies of the suspended graphite flakes, which were found to depend both on the number of graphene layers and on the diameter of the hole over which the flake was suspended, finding values in the range of 1–10 GHz. Castellanos-Gomez et al.[66] reported an in-plane Young's modulus of MoS_2 as 330 ± 70 GPa, which, though far from the values reported for graphene and CNTs, is still considerably high.

Rather than using the tip of an AFM to mechanically distort NSOs, Poncharal and coworkers[5] employed an electrostatic method. This approach consisted of depositing a mass

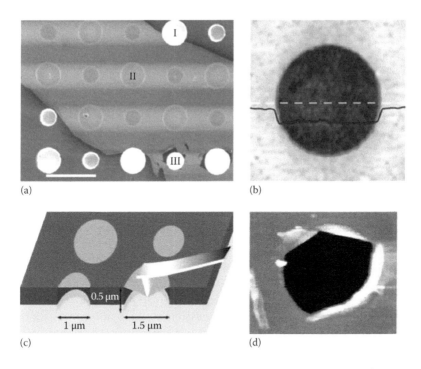

(a)　　　　　　　　　　　　　　　　(b)

0.5 μm

1 μm　　1.5 μm

(c)　　　　　　　　　　　　　　　　(d)

FIGURE 7.4　Mechanical properties of a single graphene layer. (a) shows a scanning electron microscopy image of a graphene sheet deposited on an array of holes with alternating diameters of 1 and 1.5 μm; the hole labeled as I is seen to be only partially covered, while II is fully covered; the result of an indentation experiment resulting in the puncturing of the graphene membrane is seen in III; (d) shows this area in more detail. (b) shows a noncontact AFM image of graphene stretched over a hole of 1.5 μm in diameter, with the solid line giving the height profile along the dashed line in the image; the step at the edge of the membrane is ~2.5 nm. (c) shows a schematic representation of the experimental indentation of suspended graphene, and (d) shows an atomic force microscope image of an indented and fractured area. (Reproduced from Lee, C. et al., *Science*, 321, 385, 2008. With permission.)

Chapter 7

of MWCNTs on a fine metal wire to which a potential bias could be applied. This was then mounted onto a special sample holder that included a grounded counterelectrode, such that the CNTs protruding from the wire were located at a distance of a few micrometers away from it. The whole assembly was then introduced into a TEM, for simultaneous observation of the CNTs. When a potential was applied to the supporting wire, the MWCNTs attached to it became electrically charged and were seen to bend toward the counterelectrode. Furthermore, if the applied bias was tuned to depend on time in a sinusoidal fashion, it was possible, by tuning the frequency of the bias, to excite different fundamental vibrations of the exposed NTs. By assimilating a MWCNT to a cantilevered elastic beam, it is possible to extract an estimation of Young's modulus of the MWCNT from the observation of these resonant frequencies. Elasticity theory applied to the problem of a cantilevered beam states that this system has vibrational resonances at frequencies given by

$$\nu = \frac{\beta_i^2}{8\pi L^2}\sqrt{D_{out}^2 + D_{inn}^2}\sqrt{\frac{E}{\rho}} \tag{7.9}$$

where

β_i is a constant for each mode

L is the exposed length of the tube

$D_{inn/out}$ are the inner and outer diameters of the CNT, respectively

ρ is the density

E is an effective bending modulus, the value of which can be extracted from the previous expression once the resonant frequency and the structural parameters L and $D_{inn/out}$ have been determined for a given tube

For relatively narrow tubes ($D_{out} \leq 0$ nm), E can be identified as Young's modulus, because in such tubes, bending occurs by alternate stretching of the outer arc and compression of the inner arc of the tube. Indeed, for tubes in this range of diameters, Poncharal et al.[5] reported values of $E \approx 1$ TPa, which were in line with the expected value on the basis of previous experiments.[58,59,61] For wider NTs, however, the effective bending modulus dropped dramatically to about 0.1 TPa, signaling the onset of a different bending mechanism in these tubes. Indeed, the authors found that, in the wider tubes, bending resulted in the formation of a wave-like distortion along the inner arc. The tube walls along the inner arc were seen to buckle sideways in response to the initial compression, a kind of distortion that provides a more accessible route to accommodate the bending in the wider tubes. But Poncharal et al.[5] did not limit themselves to demonstrating a new way of determining the mechanical properties of NSOs. They also provided proof of principle for potential practical applications of cantilevered NSOs as nanobalances and Kelvin probes. Indeed, by determining the resonant fundamental frequency of a cantilevered tube loaded with a spherical carbon nanoparticle and comparing it to that predicted for an unloaded tube of the same dimensions on the basis of Equation 7.9, they were able to deduce a mass of 22 ± 6 fg for the carbon nanoparticle, a value that compared favorably with that estimated from the nanoparticle's diameter, with the further assumption of perfect sphericity and a density equal to that of bulk amorphous carbon. Likewise, by providing an offset constant bias to the applied time-dependent voltage,

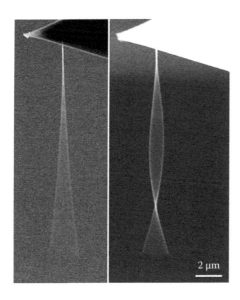

FIGURE 7.5 Scanning electron microscopy image of a vibrationally excited cantilevered boron nanowire. The first two resonance modes are shown. (Reproduced from Ding, W. et al., *Compos. Sci. Technol.*, 66, 1112, 2005. With permission.)

they were able to demonstrate the possible use as a Kelvin probe to determine differences in the work function of the anchored tube serving as probe and a sample.

The resonance method of Poncharal and coworkers[5] has been used to determine Young's modulus of CNTs,[67] Si NWs,[68] SiC NWs,[69] amorphous SiO_2 NWs,[70] B NWs,[71] BN NTs,[72] GaN NWs,[73] ZnO NWs,[74] and nanobelts,[75] among other systems. The particular case of ZnO NWs and nanobelts[74,75] will be discussed in more detail later. Figure 7.5 shows two SEM images of the first and second resonance modes of a tethered boron NW, as obtained by Ding and coworkers.[71]

Yet another approach was used by Yu et al.,[76] who constructed a nanoscale tensile testing device consisting of two opposing AFM tips; these were welded to an MWCNT using the electron beam of a TEM. Once the NT was securely attached to the tips, pulling them apart helped to measure the imposed stress and the response (strain) of the tested NT, that is, the stress–strain relation of the NT. The measured stress–strain relations were compatible with a Young's modulus ranging between 270 and 950 GPa. The welding between NT and AFM tips was strong enough that failure of the tested NT could be observed. This was seen to occur in a *sword-in-sheath* fashion, that is, the outer wall of the NT would fail under the imposed stress, and the inner tubes would then be pulled out of the resulting fragments. The measured tensile strength, that is, the stress required to cause failure, ranged between 11 and 63 GPa. The same approach has been used by Ding et al.[71] to characterize mechanically B NWs.

The final experimental approach for the mechanical characterization of NT/NWs that we will cover here is one that relies on the use of a microelectromechanical system (MEMS), such as the device developed by Zhu and Espinosa[77] or by Desai and Haque.[78] Contrary to most of the previously discussed experimental techniques (the only exception is that of Yu et al.[76]), which usually involve the lateral distortion (bending) of the tested NT/NW, with the use of MEMS-based testing devices, it is possible to apply direct

Chapter 7

axial strain and measure the resulting stress, which considerably simplifies the analysis of the results. However, these are sophisticated devices that are difficult to build and require careful calibration. Key components in these devices are a displacement actuator, responsible for pulling the tested object, which can be a thermal or electrostatic actuator as employed by Zhu and Espinosa,[77] or a piezoactuator, as used by Desai and Haque,[78] and a load sensor to measure the resulting stress. An example of this kind of device, used in the experiments determining Young's modulus of a series of ZnO NWs,[18] is shown in Figure 7.6.

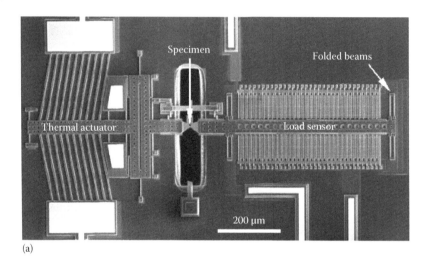

(a)

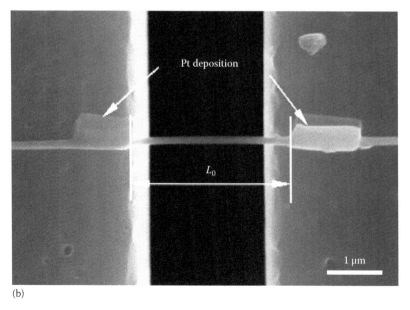

(b)

FIGURE 7.6 Scanning electron micrograph of the nanoscale mechanical testing device used by Agrawal and coauthors[18] to determine Young's modulus and its diameter dependence, for a series of ZnO [0001] nanowires. (a) shows general view of the device, while in (b) a detail of the specimen in shown. The nanowire to be tested is perfectly welded to the testing system by platinum deposition. (Reproduced from Agrawal, R. et al., *Nano Lett.*, 8, 3668, 2008. With permission.)

The different experimental methodologies that have been reviewed earlier illustrate the technical and practical challenges involved in the experimental mechanical characterization of NSOs. In view of these difficulties, it is not surprising that there has been some scatter in the measured values of, for example, Young's modulus reported in the literature cited in this chapter. Let us take, for the sake of illustration, the particular case of ZnO nanobelts and NWs. These systems, as noted earlier, are particularly interesting from the point of view of technological applications, and thus, their mechanical properties have been the focus of considerable attention. One of the first studies to attempt a mechanical characterization of ZnO nanobelts was that of Bai et al.[75] These authors employed the resonant electric-field approach developed by Poncharal et al.[5] to determine the bending mode of ZnO nanobelts with the crystalline [0001] direction oriented along the main axis of the nanobelts, and with facets $\pm\left(2\bar{1}\bar{1}0\right)$ and $\pm\left(01\bar{1}0\right)$. In these experiments, the bending modulus was determined for four different nanobelts with lengths in the 4–9 μm range and lateral dimensions between 19 and 55 nm; the values found varied between ~38 and ~65 GPa, without any obvious correlation with the structural parameters of the nanobelts. These values are to be compared with that of the C_{33} elastic constant (210 GPa) or Young's modulus (~140 GPa) reported for bulk crystals of ZnO along the [0001] direction. Similar experiments employing the same methodology by Huang et al.[79] reported bending modulus values in the range of 48.9–69.8 GPa for ZnO NWs with lengths between 4.77 and 11.4 μm and diameters 43–110 nm, again without any apparent correlation of the bending modulus with the size of the NMs.

However, in subsequent work employing the same experimental procedure, Chen and coworkers[74] analyzed the bending modulus of ZnO [0001] NWs with diameters ranging from 17 to 550 nm. These authors revealed for the first time a clear dependence of Young's modulus on the diameter of the NWs; they observed that in the limit of large diameters, Young's modulus tended toward the bulk value of 140 GPa, but for smaller diameters, it gradually increased to values above 220 GPa. They explained this size dependence as resulting from a surface modification of the NWs, the influence of which would be larger for small diameters, with their larger surface to volume ratio; they argued that ample evidence existed to show that the $\left(10\bar{1}0\right)$ surfaces forming the facets of the NW suffered considerable relaxation; they thus proposed a composite wire model consisting of a stiffer outer shell and a bulk ZnO core, which would account for the reported observations.

A different approach to determine the mechanical properties of ZnO [0001]-oriented NWs was employed by Song et al.[62] They employed the tip of an AFM to bend vertically aligned NWs, recording the topography and lateral force when displacing an NW. For ZnO NWs with an average diameter of 45 nm, a modulus of 29 ± 8 GPa was reported. This value is again surprisingly low compared to the bulk Young's modulus along the [0001] direction. Ni and Li[80] obtained similar values of 38.2 ± 1.8 GPa from bending experiments of partially suspended ZnO nanobelts, and 31 ± 1.3 GPa from indentation experiments. On the other hand, Hoffmann et al.[81] reported values of Young's modulus averaging to 97 ± 18 GPa, obtained from tensile experiments of vertically aligned ZnO [0001] NWs; their experiments consisted of pulling NWs that had been welded to the tip of an AFM microscope by focusing the electron beam of the SEM in which the experiment was carried out. Although they did not detect any noticeable dependence of Young's modulus on the NW diameter, possibly because of uncertainties in

Chapter 7

their diameter estimations, nevertheless, their values are closer to the expected bulk value. Stan and coworkers,[82] using contact resonance AFM and friction measurements, reported Young's modulus values in the range of 198–100 GPa for ZnO [0001] NWs with diameters between 25.5 and 134.4 nm. These values, though they still fall somewhat short of the limiting bulk value for wide NWs (≈140 GPa), are much closer to it, and in contrast to the values reported by Song et al.[62] and Ni and Li,[80] they do find a dependence of Young's modulus on the diameter that is in line with the report of Chen and coworkers.[74]

Using a microfabricated mechanical test stage capable of imposing direct axial strain, Desai and Haque[78] measured Young's modulus of three ZnO [0001] NWs with diameters in the range of 217–314 nm, obtaining an average value of ~21 GPa. These results are comparable to, though smaller than, those obtained earlier by Song et al.[62] but significantly smaller than the bulk Young's modulus: (29 ± 8 GPa) and Ni and Li[80] (38.2 ± 1.8 GPa). No dependence on the diameter was reported by these authors. Yet these results are in contrast to those of the combined experimental/theoretical work of Agrawal et al.,[18] who, using a nanoscale materials testing system,[77] measured Young's modulus of ZnO [0001] NWs with diameters in the range of 20.4–412.9 nm under uniaxial strain. These authors found very clear evidence of a diameter dependence of Young's modulus, corroborating the earlier results of Chen et al.[74] They also performed atomistic simulations that confirmed and clarified the core-shell model that Chen et al. had proposed to account for the structural dependence of Young's modulus found in their work. We will discuss these simulation results in more details in the next section.

As evident from the previous discussion, there is a large scatter of measured values of Young's modulus for ZnO NWs; what are the reasons for this? Clearly, these widely different values reflect the difficulties and challenges in performing accurate experimental mechanical tests on NSOs. There are many key factors in performing an accurate determination of Young's modulus of an NW, and to adequately control all these factors, sufficient precision is impossible. For example, in AFM-based experiments, the contact geometry between AFM tip and NW is difficult to characterize; in the electric-field-induced resonance method,[74,75,79] one seeks to identify the natural resonance frequency of a cantilevered NW and, from this, to extract Young's modulus. However, as shown by Shi et al.,[83] it is possible to observe super- and subresonances, which, if wrongly interpreted as the natural resonance frequency, would result in erroneous values of Young's modulus. Finally, experiments in designed test stages such as those used by Desai and Haque[78] and Agrawal and coworkers[18] rely on adequate positioning and welding of the sample to the test stage in order to avoid slippage during loading. The small values found by Desai and Haque[78] in the three samples that they examined are probably indicative of inadequate fixing of the NWs to the test stage elements.

We have seen in the aforementioned text that it is generally challenging to adequately control the various factors that intervene in an experimental characterization of NSOs. This is a drawback that simulation methods do not have: simulations allow for a minute control of every aspect in what ultimately is nothing but a *numerical* experiment. This, however, should not be taken to mean that simulations are to supersede experiments in any way; far from it: it only means that experiments and simulations have different strengths and weaknesses. While simulation methods have full control over the numerical experiment conditions, frequently, these conditions are too idealized, and not truly representative of the relevant experimental situation. Also there are limitations with respect to the accuracy of the models employed to represent the system under study; the accessible time scale or the system size constitutes important shortcomings of simulations. Nevertheless, there is a degree of complementarity between experimental techniques and modeling, that is, at least in part, the reason why a growing number of combined experimental/simulation studies of the physical–chemical properties of materials are being published in the literature. An overview of the simulation techniques and models relevant to the computational mechanical characterization of materials is provided next.

7.3.3 Atomistic Simulations of the Mechanical Properties of ZnO Nanowires

In the same way as was done in Section 7.3.2, we will take ZnO NWs as an illustrative example of the ways in which atomistic simulation techniques have been helpful in the study of the structural and mechanical properties of NSOs. As we saw there, the particular case of ZnO NWs is of great interest, in view of their relevance to technological applications on the one hand, but also because of their intriguing mechanical behavior, on the other. Again, our objective here is not to provide an exhaustive review of ZnO NW simulations but rather to describe in some detail a number of relevant examples taken from the literature.

Kulkarni et al.[16] reported MD simulations of ZnO NWs and nanobelts under tensile strain employing an empirical rigid-ion potential. Specifically, they considered NWs with [0001], $\left[00\bar{1}0\right]$, and $\left[2\bar{1}\bar{1}0\right]$ orientation, with different squared cross-sectional dimensions, ranging between 1×1 and 4×4 nm². To model the interactions between the ions, they used a rigid-ion model due to Binks,[84] having the following expression:

$$U = \sum_{i \neq j} \left(A_{ij} e^{-r_{ij}/r_{ij}^c} - \frac{C_{ij}}{r_{ij}^6} + \frac{q_i q_j}{r_{ij}} \right). \tag{7.10}$$

where
  the sum is over all distinct pairs of ions, r_{ij} being their interatomic distance
  q_i is the ionic charge on atom i (with its corresponding sign)
  A_{ij}, r_{ij}^c, and C_{ij} are suitably adjusted parameters, chosen so as to ensure that the model correctly reproduces the equilibrium lattice energies, cell parameters, and elastic and dielectric constants of the bulk material

The parametrization of Binks has been shown to correctly reproduce defect and surface energies. The latter are particularly important in order to get a reasonable description of NSOs. In Equation 7.10 the first two terms are known as a pairwise Buckingham potential, consisting of a short-range exponential repulsion and a dispersion-type r^{-6} attractive term; the last term in the equation accounts for the electrostatic interaction between ions. The slow-decaying r^{-1} nature of this last term makes it necessary to use special summation techniques such as the Ewald sum[11] in order to properly and efficiently account for this term's contribution to the total energy.

Initial structures with appropriate orientation and lateral dimensions were generated by cutting from the bulk crystal and subsequently equilibrated during 10 ps of MD simulation at 300 K. During this process Kulkarni et al.[16] found two different relaxation modes, depending on the lateral dimensions of the NW. NWs with sides of 20 Å and larger were seen to retain their rectangular cross sections during the equilibration period, but smaller NWs underwent a transformation to concentric tubular shells, similar in structure to that of a double-walled NT, exposing a low energy (0001) surface. Subsequently, stress–strain numerical experiments were performed in a quasi-static fashion, consisting of two steps. During the first one, the boundary ions were gradually pulled along the axial direction until a maximum displacement of ~0.6 Å was reached; in the second step, the stretched NW was equilibrated during 3 ps of MD at 300 K. This process of sequential stretching followed by finite-temperature relaxation was repeated until the onset of failure was observed. These simulations uncovered a direction and size dependence of Young's modulus and strength of the modeled NWs. Both were found to be very similar for $\left[01\bar{1}0\right]$ and $\left[2\bar{1}\bar{1}0\right]$ NWs and somewhat smaller for the [0001]-oriented NWs (values are given in Table 7.3), in parallel to what is observed in the bulk material. However, the values found were considerably larger than those obtained for the bulk crystal. For example, for $\left[01\bar{1}0\right]$ and $\left[2\bar{1}\bar{1}0\right]$ NWs with lateral dimension of 20 Å, Young's modulus was found to be ~255 GPa, compared to 156 GPa for the bulk along both directions; for the [0001] NW, the result was 172.6 GPa, compared to the bulk value of 119.7 GPa. NWs with larger cross section were found to have smaller Young's modulus and breaking strength, with Young's modulus tending toward the bulk value in the corresponding direction. This size dependence was explained in terms of surface relaxation effects, which causes compressive stress along the axial direction; the influence of such surface-induced stresses is more noticeable in NWs of smaller lateral dimensions, where the ratio of surface to bulk-like atoms is correspondingly larger. For sufficiently wide NWs, the effect of surface-induced compressive stresses becomes negligible, and thus, Young's modulus approaches the bulk value. Tensile strengths were also computed, and it was found that these also decreased with lateral dimension. In $\left[01\bar{1}0\right]$ and $\left[2\bar{1}\bar{1}0\right]$ NWs, failure was found to occur by shear along $\left(\bar{1}2\bar{1}0\right)$ planes, while in [0001] NWs, failure had a brittle character. It should be considered that though the rigid-ion model is probably sufficiently flexible to account for the elastic behavior of these systems, its ability to describe correctly the plastic behavior is probably more questionable, given that the plastic regime involves significant distortion of the material with respect to the conditions for which the model potential was fitted. In a subsequent

Table 7.3 Orientation and Size Dependence of the Calculated Young's Modulus and Strength of ZnO Nanowires as Reported by Kulkarni et al.[16]

Lateral Dimension (Å)	Young's Modulus (GPa)			Strength (GPa)		
	$\left[2\bar{1}\bar{1}0\right]$	$\left[01\bar{1}0\right]$	[0001]	$\left[2\bar{1}\bar{1}0\right]$	$\left[01\bar{1}0\right]$	[0001]
10	307.4	325.75	339.8	15.0	23.4	36.3
20	256.5	254.2	172.65	13.8	12.7	10.9
30	210.3	219.9	140.4	9.2	9.5	8.6
Bulk	156.2	156.2	119.7	—	—	—

study,[17] these authors considered [0001]-oriented NWs of hexagonal cross section (as opposed to rectangular[16]), with lateral dimensions of 19.5, 26.0, 32.5, 39.0, and 45.5 Å. Like in their previous study, they found that Young's modulus and strength reduced with increasing NW diameter, although the values found for the hexagonal cross section NWs were larger than those obtained for rectangular ones of comparable dimensions. Specifically, the reported values for hexagonal cross section NWs were 299.5 GPa (for the NW of 19.5 Å in diameter), 271 GPa (26.0 Å), 250 GPa (32.5 Å), 238.2 GPa (39.0 Å), and 227.5 GPa (45.5 Å). While considering the dependence of the mechanical properties on the NW diameter, the effect of temperature was also investigated; numerical tensile experiments were carried out at temperatures in the range of 300–1500 K at intervals of 300 K. Perhaps not unexpectedly, Young's modulus was found to decrease slightly with increasing temperature, as a result of the larger temperature-induced atomic vibrations and defect formation. In the particular case of an NW of 32.5 Å in diameter, Young's modulus was found to decrease from ~250 to 204 GPa when raising the temperature from 300 to 1500 K. These authors also uncovered a stress-induced transformation from the hexagonal wurtzite structure to a body-centered tetragonal one; we will discuss this later.

One limitation of the simulations reported by Kulkarni and collaborators[16,17] is the fact that the range of NW diameters considered is relatively small; indeed, the largest diameter they were able to consider was 4.5 nm, which should be compared to the experimentally relevant diameters, starting at ~5 nm and going up to over 500 nm. It must be borne in mind that the number of atoms to be considered in the simulations grows very rapidly with NW diameter (in fact as d^2), and the cost of the simulations grows at least linearly with the number of atoms. Thus, to perform long time scale MD simulations at various strain levels and temperatures, for a range of NW diameters, as was done by these authors, is no small challenge. Nevertheless, simulations like these were taken a step further by Agrawal et al.,[18] who complemented their experimental studies of the mechanical properties of ZnO [0001] NWs (see Section 7.3.2) with MD simulations employing the same model[84] as used by Kulkarni and collaborators;[16,17] Agrawal et al. were able to simulate NWs with diameters between 5 and 20 nm, which, though still comparatively small, begin to overlap with the range of experimentally relevant diameters (see Figure 7.7). Similar to the work of Kulkarni and collaborators,[16,17] Agrawal et al. found a size dependence in Young's modulus that follows the same trend as seen in the experimental results, although there is a small mismatch (of about 10 GPa) between the computed value and the experimental one for diameters of ~2 nm, the range accessible to both experiments and simulations. This mismatch could be attributable to a slight overestimation of the elastic constants by the model of Binks.[84] There is also a mismatch between the simulation results of Wang et al.[17] and Agrawal et al. (see inset in Figure 7.7), probably due to the different implementations used by these two groups. In order to uncover the reasons behind the observed stiffening of ZnO [0001] NWs with decreasing diameter, Agrawal et al. calculated the contribution to Young's modulus coming from surface and core atoms in the NW, finding that the former made a contribution that was substantially larger than the bulk value along the same direction, an effect that was attributed to surface reconstruction. Conversely, the core of NWs was found to have a slightly smaller value than the bulk crystal along the same direction (see Figure 7.8). They also calculated

Chapter 7

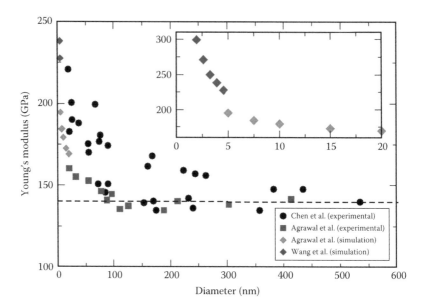

FIGURE 7.7 Young's modulus of ZnO [0001]-oriented nanowires as a function of the nanowire diameter. Black dots are experimental results from Chen et al.,[74] obtained using the resonant electrostatic field approach of Poncharal and coworkers.[5] Red squares are experimental data obtained by Agrawal et al.,[18] employing their nanoscale materials testing device, and the blue diamonds are results from molecular dynamics simulations obtained by the same authors; red diamonds are simulation results obtained by Wang et al.[17] The inset shows the range of simulated nanowire diameters in more detail, and the horizontal dashed line marks the value of Young's modulus obtained for bulk ZnO along the [0001] direction.

Young's modulus for individual layers of atoms, normalized to the value of the wire, for a series of diameters. The result is reproduced in Figure 7.8b, where it can be seen that Young's modulus for the surface shell can be considerably larger (for the 5 nm wire nearly twice as large) than that of the whole wire. It can also be seen that the surface stiffening effect decays rapidly with the diameter of the atomic layer; this allowed the authors to define an effective shell thickness, taken as the region in which the elastic modulus was larger than the combined value. This was found to be ~15% of the wire diameter. Thus, these authors were able to confirm and characterize, on the basis of their detailed atomistic simulations, the core-shell model that had been proposed by Chen and coworkers[74] to explain the experimentally observed Young's modulus dependence on the wire diameter.

Thus far, we have discussed the use of atomistic simulations to explore the mechanical properties of ZnO NWs within the elastic regime, but they have also been employed to study the behavior of these systems in the plastic limit. For example, Wang et al.[17] reported the observation of a stress-induced transformation from the wurtzite structure to a body-centered tetragonal phase, taking place in ZnO NWs at strain levels above 7%. The transformation occurred at critical stress levels that decreased slightly with NW diameter (~22 GPa for a diameter of 1.95 nm, 16.5 GPa for a diameter of 4.55 nm). Agrawal et al.[18] also found this transition in their atomistic simulations, but they could not confirm its occurrence in their experiments. In the latter, they observed brittle fracture to occur along (0001) planes at strain levels in the range

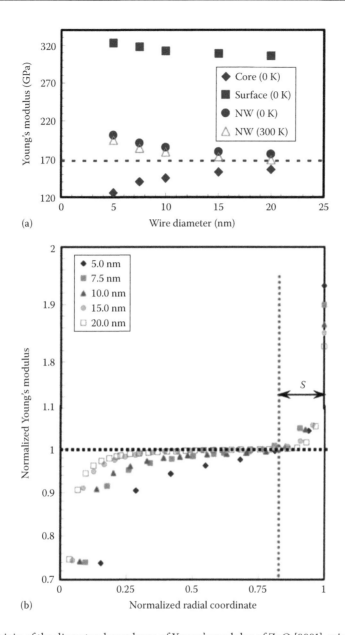

FIGURE 7.8 Origin of the diameter dependence of Young's modulus of ZnO [0001]-oriented nanowires. (a) shows the contribution of surface atoms and the remaining (core) atoms, together with the total elastic modulus, as a function of nanowire diameter. (b) plots Young's modulus calculated for concentric shells of atoms normalized to that of the whole wire, computed for wires of various diameters. The region labeled S is the deduced shell thickness. (Reproduced from Agrawal, R. et al., *Nano Lett.*, 8, 3668, 2008. With permission.)

of 2.4%–6%. In their simulations, which actually used the same model[84] as used by Wang and collaborators, the transformation occurred at strain levels of ~6.5%, that is, larger than the failure strain observed in the experiments. A number of reasons could account for these differences between experimental results and simulations. For example, the experimental samples were seen to contain surface defects (not present in the simulations) that could lead to stress concentration and result in early failure; thus, simulations were also conducted on model NWs containing various concentrations of defects. However, the presence of these did not preclude the phase transition from occurring.

The examples cited illustrate both the power and the current limitations of atomistic simulations in this field. Simulations provide a detailed picture of the studied process at the atomic scale, but that picture is only trustworthy if the model employed is sufficiently accurate, and if the simulated system is truly representative of the experimental one. We have seen that models are not always reliable, particularly when considered under conditions for which they were not fitted. Furthermore, computational costs frequently preclude the simulation of systems with sizes comparable to those on which experiments are actually carried out. Finally, the accessible simulation times are usually much shorter than the experimentally relevant time scales. We will revisit these shortcomings in Section 7.4.

7.4   Conclusions, Challenges, and Perspectives

As illustrated in Section 7.3, advances in the study of the mechanical properties of NTs and NWs have been considerable in recent years. Experimentally, a number of mechanical testing procedures have been developed and employed. A measure of how challenging these experiments are is provided by the scatter in the measured values that have been reported in, for example, the case of ZnO NWs (see Section 7.3.2). Nevertheless, it can be expected that these differences, where they exist, will be reduced as techniques are further refined and improved. At the same time, simulation methods have contributed to this advance by providing a detailed atomistic picture of the response to deformation (both within the elastic and plastic regimes) of NTs and NWs; this has been instrumental in the rationalization and understanding of experimental results. However, a considerable gap remains between the systems on which experiments are performed and those that can be simulated: First, computational costs usually preclude the simulation of systems of experimentally relevant sizes, over experimentally relevant time scales; second, defects and imperfections are always present in experimental samples, while it is not easy to account for them in simulations; finally, it must be remembered that the model (force field, etc.; see Section 7.2) is only an approximation (and frequently a crude one) to the true interactions that govern the mechanical properties of the real system. In this respect, it can be expected that improvements in algorithms and models will contribute to reduction in the size of this gap.

We have seen (Section 7.2) that a hierarchy of models exists, ranging from empirical potentials up to first-principles electronic structure methods, that allows us to model materials at different levels of accuracy and fidelity. To some extent, it is possible to tailor the model to the particular system under study and to the specific questions being investigated. We have also seen that materials modeling techniques, combined with suitable models, are gradually establishing themselves as a powerful research tool that can frequently complement experimental studies, assisting in the interpretation and explanation of measurements. Nevertheless, a number of limitations must be overcome before computer simulations of materials can unfold their true potential in the study of nanoscaled systems. In our opinion, the most pressing limitations of present simulation techniques are the following:

- Poor reliability and transferability of empirical potentials and semiempirical electronic structure models. The ad hoc nature of these models confines their range of applicability to the environmental conditions at which they were fitted; using them outside those conditions can lead to unexpected, unphysical results. We saw an example of this in Section 7.3.3, where we discussed the simulations of Wang et al.,[17] who reported the observation of a strain-induced structure transformation in ZnO NWs. Specifically, they reported the transformation from the wurtzite structure to a body-centered tetragonal one in [0001]-oriented ZnO NWs under ~7% strain; in experiments, however, it was found that the NWs suffer from brittle fracture before reaching such strain levels.
- The sizes of experimentally relevant nanoscaled systems are frequently too large for atomistic simulations. This is so even at the level of empirical potentials, but much more so for first-principles methods. The computational cost of these methods grows superlinearly with the number of atoms in the system, N [typically as $O(N^3)$ or higher], which at present limits the size of the systems that can be reasonably considered to a few hundreds, or a few thousands at most. A number of linear-scaling methods (methods for which the computational cost grows linearly with the system size) have been proposed in the literature (see References 85 and 86), but to date, these have not as yet acquired a level of maturity that makes them sufficiently robust and generally applicable.
- The accessible simulation times are frequently many orders of magnitude shorter than the experimentally relevant time scales. This is so even for simulations at the level of empirical potentials. In the particular case of mechanical properties, this implies that the simulated strain rates are usually much higher than the experimentally attainable ones, and this raises questions as to whether the response of the model at such high strain rates is really representative of that of the real system.

It is to be expected that atomistic simulation methods will evolve to overcome, at least partially, some of the limitations discussed. For example, there will be a pressing need for accurate and transferable, yet simple and cost-effective, empirical potentials. As discussed in Section 7.2, empirical potentials have been traditionally designed for specific classes of materials, such as the EAM potential[23] for metallic systems and the Tersoff potential[20] for covalent systems. The strategy behind such an approach is to design a specific functional form that captures the essence of a particular type of material, dependent on a relatively small number of adjustable parameters that can then be fitted for each particular system. This approach has been reasonably successful in the past but is ultimately limited in that the functional forms that are adopted to represent the potential energy surface do have restricted flexibility. They cannot in general be expected to reproduce the complex dependence of the potential energy surface on atomic positions that would be given, for example, by a DFT calculation.

Recently, however, there have been two interesting developments, also mentioned in Section 7.2, that may signal the start of a new trend in the design of empirical potentials. These are the so-called neural network potentials[28] and the Gaussian approximation potentials.[29] Although these methods are different, they are so similar that they give up any attempt to find a physically motivated potential function form and opt for a less physically motivated, but mathematically more flexible, description. Either approach has the capacity of providing a more faithful representation of the potential energy surface than the traditional schemes, with the added benefit of providing a universal procedure, that can be applied to any kind of system. It can also be expected that there will be a continuing effort to develop more efficient algorithms in the implementation of first-principles electronic structure calculations, as well as improvements in the accuracy of these methods (better exchange-correlation functionals in DFT, etc.).

Perhaps the biggest challenge lying ahead for computer simulations will be, however, the development of new methodologies that can successfully encompass the varying length and time scales that are becoming increasingly relevant in the study of materials. Through evolution, nature has uncovered a variety of materials, such as nacre, bone, and silk, having a high strength and fault tolerance.[87] Invariably, such materials are nanocomposites, displaying a series of hierarchical structures on several length scales, with basic building blocks that are nanoscopic in size. As we learn to emulate nature to obtain new synthetic materials with improved properties (see, e.g., Reference 88), modeling techniques that can bridge over the relevant length and time scales are needed, if simulation is to continue to assist experimental efforts in this respect.

Acknowledgments

YT and TY are partly supported by Grants-in-Aid (Nos. 24681021, 24860057, 25104722) from the Ministry of Education, Culture, Sports, Science and Technology of Japan. The work of ERH is supported by the Spanish DGCYT through project FIS2012-31713.

References

1. M. S. Dresselhaus, G. Dresselhaus, and P. Avouris, eds., *Carbon Nanotubes, Synthesis, Structure, Properties and Applications*, vol. 80 of *Topics in Applied Physics* (Springer, Berlin, Germany, 2001).
2. A. Jorio, M. S. Dresselhaus, and G. Dresselhaus, eds., *Carbon Nanotubes, Advanced Topics in the Synthesis, Structure, Properties and Applications*, vol. 111 of *Topics in Applied Physics* (Springer, Berlin, Germany, 2008).
3. C. M. Lieber and Z. L. Wang, *MRS Bull.* **32**, 99 (2007).
4. Z. L. Wang, *Adv. Funct. Mater.* **18**, 3553 (2008).
5. P. Poncharal, Z. L. Wang, D. Ugarte, and W. A. de de Heer, *Science* **283**, 1513 (1999).
6. J. Chaste, A. Eichler, J. Moser, G. Ceballos, R. Rurali, and A. Bachtold, *Nat. Nanotechnol.* **7**, 301 (2012).
7. Z. L. Wang and J. Song, *Science* **312**, 242 (2006).
8. Y. Qin, X. Wang, and Z. L. Wang, *Nature* **451**, 809 (2008).
9. C. M. Goringe, D. R. Bowler, and E. R. Hernández, *Rep. Prog. Phys.* **60**, 1447 (1997).
10. W. H. Press, S. A. Teukolsky, W. T. Vetterling, and B. P. Flannery, *Numerical Recipes in Fortran: The Art of Scientific Computing* (Cambridge University Press, Cambridge, U.K., 1992).
11. M. P. Allen and D. J. Tildesley, *Computer Simulation of Liquids* (Clarendon Press, Oxford, U.K., 1987).
12. D. Frenkel and B. Smit, *Understanding Molecular Simulation* (Academic Press, San Diego, CA, 1996).
13. J. M. Thijssen, *Computational Physics* (Cambridge University Press, Cambridge, U.K., 1999).
14. E. R. Hernández, *AIP Conf. Proc.* **1077**, 95 (2008).

15. G. V. Lewis and C. R. A. Catlow, *J. Phys. C: Solid State Phys.* **18**, 1149 (1984).
16. A. J. Kulkarni, M. Zhou, and F. J. Ke, *Nanotechnology* **16**, 2749 (2005).
17. J. Wang, A. J. Kulkarni, F. Ke, Y. Bai, and M. Zhou, *Comput. Methods Appl. Mech. Eng.* **197**, 3182 (2008).
18. R. Agrawal, B. Peng, E. E. Gdoutos, and H. D. Espinosa, *Nano Lett.* **8**, 3668 (2008).
19. F. H. Stillinger and T. A. Weber, *Phys. Rev. B* **31**, 5262 (1985).
20. J. Terso, *Phys. Rev. B* **37**, 6991 (1988).
21. M. Makeev, D. Srivastava, and M. Menon, *Phys. Rev. B* **74**, 165303 (2006).
22. L. D. Landau and E. M. Lifshitz, *Theory of Elasticity*, vol. 7 of *Course of Theoretical Physics*, 2nd edn. (Pergamon Press, Oxford, U.K., 1970).
23. M. S. Daw, S. M. Foiles, and M. I. Baskes, *Mater. Sci. Rep.* **9**, 251 (1993).
24. F. Cleri and V. Rosato, *Phys. Rev. B* **48**, 22 (1993).
25. M. W. Finnis and J. E. Sinclair, *Philos. Mag. A* **50**, 45 (1984).
26. J. Diao, K. Gall, and M. L. Dunn, *Nat. Mater.* **2**, 656 (2003).
27. M. I. Baskes, *Phys. Rev. B (Condens. Matter)* **46**, 2727 (1992).
28. J. Behler and M. Parrinello, *Phys. Rev. Lett.* **98**, 1 (2007).
29. A. P. Bartók, M. C. Payne, R. Kondor, and G. Csanyi, *Phys. Rev. Lett.* **104**, 136403 (2010).
30. L. Colombo, *Nuovo Cimento* **28**, 1 (2005).
31. R. M. Martin, *Electronic Structure: Basic Theory and Practical Methods* (Cambridge University Press, Cambridge, U.K., 2004).
32. A. P. Sutton, *Electronic Structure of Materials* (Clarendon Press, Oxford, U.K., 1993).
33. A. H. C. Neto, F. Guinea, N. M. R. Peres, K. S. Novoselov, and A. K. Geim, *Rev. Mod. Phys.* **81**, 109 (2009).
34. D. Porezag, T. Frauenheim, T. Koehler, G. Seifert, and R. Kaschner, *Phys. Rev. B* **51**, 12947 (1995).
35. E. Hernández, C. Goze, P. Bernier, and A. Rubio, *Phys. Rev. Lett.* **80**, 4502 (1998).
36. E. Hernández, C. Goze, P. Bernier, and A. Rubio, *Appl. Phys. A* **68**, 287 (1999).
37. G. Seifert and E. Hernández, *Chem. Phys. Lett.* **318**, 355 (2000).
38. G. Seifert, T. Koehler, H. M. Urbassek, E. Hernández, and T. Frauenheim, *Phys. Rev. B* **63**, 193409 (2001).
39. G. Seifert, H. Terrones, M. Terrones, G. Jungnickel, and T. Frauenheim, *Phys. Rev. Lett.* **85**, 146 (2000).
40. D.-B. Zhang, T. Dumitrică, and G. Seifert, *Phys. Rev. Lett.* **104**, 065502 (2010).
41. R. D. James, *J. Mech. Phys. Solids* **54**, 2354 (2006).
42. T. Dumitrică and R. D. James, *J. Mech. Phys. Solids* **55**, 2206 (2007).
43. R. Saito, G. Dresselhaus, and M. S. Dresselhaus, *Physical Properties of Carbon Nanotubes* (Imperial College Press, London, U.K., 1998).
44. P. Hohenberg and W. Kohn, *Phys. Rev.* **136**, 864 (1964).
45. W. Kohn and J. J. Sham, *Phys. Rev.* **140**, 1133 (1965).
46. R. Car and M. Parrinello, *Phys. Rev. Lett.* **55**, 2471 (1985).
47. B. I. Yakobson and P. Avouris, in *Carbon Nanotubes, Synthesis, Structure, Properties and Applications*, vol. 80 of *Topics in Applied Physics*, eds. M. Dresselhaus, G. E. Dresselhaus, and P. Avouris (Springer-Verlag, Berlin, Germany, 2001), pp. 287–328.
48. M. S. Dresselhaus, G. Dresselhaus, J. C. Charlier, and E. Hernández, *Philos. Trans. R. Soc. Lond. A* **362**, 2065 (2004).
49. T. Yamamoto, K. Watanabe, and E. R. Hernández, in *Carbon Nanotubes, Advanced Topics in the Synthesis, Structure, Properties and Applications*, vol. 111 of *Topics in Applied Physics*, eds. A. Jorio, M. S. Dresselhaus, and G. Dresselhaus (Springer, Berlin, Germany, 2008), chap. 5, pp. 165–194.
50. H. S. Park, W. Cai, H. D. Espinosa, and H. Huang, *MRS Bull.* **34**, 178 (2009).
51. J. F. Nye, *Physical Properties of Crystals* (Oxford Science Publications, Oxford, U.K., 1985).
52. G. N. Greaves, A. L. Greer, R. S. Lakes, and T. Rouxel, *Nat. Mater.* **10**, 823 (2011).
53. H. Mitschke, J. Schwerdtfeger, F. Schury, M. Stingl, C. Körner, R. F. Singer, K. Robins, K. Mecke, and G. E. Schröder-Turk, *Adv. Mater.* **23**, 2669 (2011).
54. R. P. Feynman, R. B. Leighton, and M. I. Sands, *The Feynman Lectures on Physics*, vol. 2 (Addison-Wesley, Reading, MA, 1989).
55. D. H. Robertson, D. W. Brenner, and J. W. Mintmire, *Phys. Rev. B* **45**, 12592 (1992).
56. P. Wagner, V. V. Ivanovskaya, M. J. Rayson, P. R. Briddon, and C. P. Ewels, *J. Phys. Condens. Matter* **25**, 155302 (2013).

Chapter 7

57. K. Kang and W. Cai, *Int. J. Plasticity* **26**, 1387 (2010).
58. M. M. Treacy, T. W. Ebbesen, and J. M. Gibson, *Nature (Lond.)* **381**, 678 (1996).
59. A. Krishnan, E. Dujardin, T. W. Ebbesen, P. N. Yianilos, and M. M. J. Treacy, *Phys. Rev. B* **58**, 14013 (1998).
60. N. G. Chopra and A. Zettl, *Solid State Commun.* **105**, 297 (1998).
61. E. W. Wong, P. E. Sheehan, and C. M. Lieber, *Science* **277**, 1971 (1997).
62. J. Song, X. Wang, E. Riedo, and Z. L. Wang, *Nano Lett.* **5**, 1954 (2005).
63. J.-P. Salvetat, G. A. D. Briggs, J.-M. Bonard, R. R. Bacsa, A. J. Kulik, T. Stöckli, N. A. Burnham, and L. Forró, *Phys. Rev. Lett.* **82**, 944 (1999).
64. C. Lee, X. Wei, J. W. Kysar, and J. Hone, *Science* **321**, 385 (2008).
65. M. Poot and H. S. J. van der Zant, *Appl. Phys. Lett.* **92**, 3111 (2008).
66. A. Castellanos-Gomez, M. Poot, G. A. Steele, H. S. J. V. D. Zant, N. Agraït, and G. Rubio-Bollinger, *Adv. Mater.* **24**, 772 (2012).
67. R. Ciocan, J. Gaillard, M. J. Skove, and A. M. Rao, *Nano Lett.* **5**, 2389 (2005).
68. Z. L. Wang, *Adv. Mater.* **12**, 1295 (2000).
69. S. Perisanu, V. Gouttenoire, P. Vincent, A. Ayari, M. Choueib, M. Bechelany, D. Cornu, and S. Purcell, *Phys. Rev. B* **77**, 165434 (2008).
70. D. A. Dikin, X. Chen, W. Ding, G. Wagner, and R. S. Ruo, *J. Appl. Phys.* **93**, 226 (2003).
71. W. Ding, L. Calabri, X. Chen, K. M. Kohlhaas, and R. S. Ruoff, *Compos. Sci. Technol.* **66**, 1112 (2005).
72. A. P. Suryavanshi, M.-F. Yu, J. Wen, C. Tang, and Y. Bando, *Appl. Phys. Lett.* **84**, 2527 (2004).
73. C.-Y. Nam, P. Jaroenapibal, D. Tham, D. E. Luzzi, S. Evoy, and J. E. Fischer, *Nano Lett.* **6**, 153 (2006).
74. C. Q. Chen, Y. Shi, Y. Zhang, J. Zhu, and Y. Yan, *Phys. Rev. Lett.* **96**, 075505 (2006).
75. X. Bai, P. X. Gao, Z. L. Wang, and E. G. Wang, *Appl. Phys. Lett.* **82**, 4806 (2003).
76. M.-F. Yu, O. Lourie, M. J. Dyer, K. Moloni, T. F. Kelly, and R. S. Ruo, *Science* **287**, 637 (2000).
77. Y. Zhu and H. D. Espinosa, *Proc. Natl. Acad. Sci.* **102**, 14503 (2005).
78. A. V. Desai and M. A. Haque, *Sensors Actuators A* **134**, 169 (2007).
79. Y. Huang, X. Bai, and Y. Zhang, *J. Phys. Condens. Matter* **18**, L179 (2006).
80. H. Ni and X. Li, *Nanotechnology* **17**, 3591 (2006).
81. S. Ho mann, F. Östlund, J. Michler, H. J. Fan, M. Zacharias, S. H. Christiansen, and C. Ballif, *Nanotechnology* **18**, 5503 (2007).
82. G. Stan, C. V. Ciobanu, P. M. Parthangal, and R. F. Cook, *Nano Lett.* **7**, 3691 (2007).
83. Y. Shi, C. Q. Chen, Y. S. Zhang, J. Zhu, and Y. J. Yan, *Nanotechnology* **18**, 5709 (2007).
84. J. D. Binks, Computational modelling of zinc oxide and related oxide ceramics, PhD thesis, University of Surrey, Surrey, U.K. (1994).
85. S. Goedecker, *Rev. Mod. Phys.* **71**, 1085 (1999).
86. D. R. Bowler and T. Miyazaki, *Rep. Prog. Phys.* **75**, 036503 (2012).
87. H. Gao, B. Ji, I. L. Jäger, E. Arzt, and P. Fratzl, *Proc. Natl. Acad. Sci.* **100**, 5597 (2003).
88. H. D. Espinosa, T. Filleter, and M. Naraghi, *Adv. Mater. Weinheim* **24**, 2805 (2012).
89. C. Li and T. W. Chou, *Int. J. Solids Struct.* **40**, 2487 (2003).
90. Z.-C. Tu and Z.-C. Ou-Yang, *Phys. Rev. B* **65**, 233407 (2002).
91. T. Vodenitcharova and L. Zhang, *Phys. Rev. B* **68**, 165401 (2003).
92. A. Pantano, D. M. Parks, and M. C. Boyce, *J. Mech. Phys. Solids* **52**, 789 (2004).
93. B. I. Yakobson, C. J. Brabec, and J. Bernholc, *Phys. Rev. Lett.* **76**, 2511 (1996).
94. J. P. Lu, *Phys. Rev. Lett.* **79**, 1297 (1997).
95. Y. Jin and F. G. Yuan, *Compos. Sci. Technol.* **63**, 1507 (2003).
96. L. Wang, Q. Zheng, J. Liu, and Q. Jiang, *Phys. Rev. Lett.* **95**, 105501 (2005).
97. Y. Huang, J. Wu, and K. Hwang, *Phys. Rev. B* **74**, 245413 (2006).
98. Z. Xin, Z. Jianjun, and O.-Y. Zhong-Can, *Phys. Rev. B (Condens. Matter Mater. Phys.)* **62**, 79506 (2000).
99. D. Sánchez-Portal, E. Artacho, J. M. Soler, A. Rubio, and P. Ordejón, *Phys. Rev. B* **59**, 12678 (1999).
100. K. Kudin, G. Scuseria, and B. Yakobson, *Phys. Rev. B* **64**, 235406 (2001).

8. Thermal Properties of Inorganic Nanostructures

Yukihiro Takada, Eduardo R. Hernández, and Takahiro Yamamoto

8.1   Introduction 247

8.2   Thermal Transport in Inorganic Nanomaterials 248
    8.2.1   Thermal Conductivity and Thermal Conductance 248
    8.2.2   Quantized Thermal Conductance. 249
    8.2.3   High-Thermal-Conductivity Materials. 250
        8.2.3.1   Boron Nitride Nanotubes 251
        8.2.3.2   Theoretical and Experimental Studies 251
        8.2.3.3   Toward Higher Thermal Conductivity 254
        8.2.3.4   Boron Nitride Nanoribbons 255
    8.2.4   Low-Thermal-Conductivity Materials. 256
        8.2.4.1   Bismuth Telluride Nanowires 259
        8.2.4.2   Silicon Nanowires 261
        8.2.4.3   Core–Shell Nanowires 263

8.3   Conclusions, Challenges, and Perspectives 265

Acknowledgments. 266

References 266

8.1   Introduction

Thermal properties offer abundant scope for future technological applications of nanotubes (NTs)[1,2] and nanowires (NWs).[3] There are two complementary road maps to achieve this, involving the use of either high- or low-thermal-conductivity materials. Since, in the field of electronic devices, removal of excess heat constrains further device miniaturization, nanomaterials with high thermal conductivity are expected to be good candidates for the conducting channels in future device generations. At the other extreme, that is, the low-thermal-conductivity materials are present: these also have a great potential for technological applications. A typical example is the fabrication of thermoelectric devices, whose power generation efficiency is inversely proportional to the thermal conductivity. Many of these foreseeable applications are already explored at the level of laboratory prototypes and will be introduced in this chapter. Undoubtedly others, as yet unforeseen, will follow in the future. This fact motivates the study and understanding of the thermal transport properties of nanoscaled objects.

Chapter 8

In this chapter, we focus on the issue of thermal transport properties. The concepts of thermal conductivity and conductance are introduced; the conductance is shown to be quantized at the nanoscale. Technological applications of the thermal properties of nanoscaled materials may result from materials with either a high or low thermal conductance. Research into each of these cases is outlined in this section. We conclude this chapter with a discussion in Section 8.3 on the challenges and perspectives for atomistic simulations in this topic.

8.2 Thermal Transport in Inorganic Nanomaterials

8.2.1 Thermal Conductivity and Thermal Conductance

For a conventional bulk material, the thermal conductance κ is directly proportional to the cross-sectional area S of the sample and inversely proportional to its length L, namely,

$$\kappa = \frac{S}{L}\lambda. \tag{8.1}$$

where λ is the thermal conductivity, the property that characterizes the ability to conduct heat by a material; it is independent of the sample's dimensions. λ is defined as the ratio of the thermal current density, J, to the temperature gradient, dT/dx, via Fourier's law:

$$J = -\lambda \frac{dT}{dx}. \tag{8.2}$$

Fourier's law correctly describes thermal transport phenomena in bulk materials, where the thermal current is diffusive.[4,5] The same is the case with Ohm's law for electrical current in bulk materials, $I = \sigma E$, where I is the current density, E is the electrical field, and σ is the electrical conductivity. However, Fourier's law in Equation 8.2 can not be applied to a sample whose dimension is much smaller than the phonon mean free path, because the thermal conductivity approaches infinity, $\lambda = |J/(dT/dx)| \rightarrow \infty$, as the temperature gradient tends to zero in this limit. In this case, the phonons in the sample behave ballistically, rather than diffusively. The issue of ballistic phonon transport has been one of long-standing theoretical interest, going back to Peierls' early work in the 1920s.[6] Understanding ballistic and/or coherent phonon transport at the nanoscale is therefore of fundamental scientific interest and is essential for exploiting novel technology that relies on the phononic wave nature.

In a nanoscale system whose dimensions are much smaller than the phonon mean free path, phonons propagate ballistically, conserving their momentum. As mentioned earlier, the thermal conductivity λ is no longer a suitable quantity to represent the ability to conduct heat in nanoscale materials. Thus, instead of λ, we introduce the thermal conductance κ, which is defined by

$$\kappa = \frac{I}{T_{hot} - T_{cold}}, \tag{8.3}$$

where $I = JS$ is the thermal current flowing from the hot to cold heat reservoirs. Because the thermal conductance is defined by the temperature difference $\Delta T = T_{hot} - T_{cold}$ between the hot and cold heat reservoirs, and not by the temperature gradient, it is not divergent, even when the temperature gradient (dT/dx) is zero in the material, and therefore, it is applicable to describe the ability to conduct heat in the nanoscale materials.

8.2.2   Quantized Thermal Conductance

Phonon thermal transport at the nanoscale has been actively studied since the pioneering work by Rego and Kirczenow in 1998.[7] They predicted on the basis of the Landauer theory[7] that in the low-temperature region, where the optical phonons are not excited, the thermal conductance of ballistic phonon transport in NWs is quantized in multiples of

$$\kappa_0 = \frac{\pi^2 k_B^2 T}{3h} \equiv g_0 T, \tag{8.4}$$

where
  k_B is Boltzmann's constant
  h is Plank's constant giving $g_0 = 9.4 \times 10^{-13}$ W/K^2 irrespective of the material species[7-12]

The quantized thermal conductance κ_0 gives an upper limit of the thermal conductance that a single acoustic phonon mode can carry. This fundamental upper limit comes from quantum mechanics and can be regarded as the thermal conductance quantum. The quantization of thermal conductance is very similar to the quantization of electrical conductance $G_0 = e^2/h$ in ballistic electron transport in NWs.[13] It is noted that the thermal conductance quantum, κ_0, contains the thermal energy $k_B T$, just as the electrical conductance quantum contains the electrical charge e.

It is extremely challenging to experimentally observe the quantized thermal conductance κ_0, because its value is extremely small (see Equation 8.4). In 2000, Schwab and colleagues[8] succeeded in detecting it in a NW consisting of silicon nitride (SiN) at temperatures below 0.6 K using a sophisticated fabrication technique.[8] Figure 8.1a shows the experimental setup used to measure the quantum κ_0. The small square at the center is a SiN membrane. The membrane is heated by two C-shaped heaters made of a gold thin film and acts as a heat reservoir. Since this membrane is suspended by four SiN bridges and is in vacuum, the heat reservoir and the four bridges are thermally isolated from the environment and the heat flows from the membrane into 4 SiN NWs without heat dissipation. The perfect phonon transmission was realized by constructing smooth connections between the membrane and the SiN NWs, as shown in Figure 8.1b. Figure 8.1c shows the temperature dependence of the phonon thermal conductance, normalized to a universal value of $16\kappa_0$, of a SiN NW. The total thermal conductance of the device in Figure 8.1a is expected to be $16\kappa_0$, because each of the four acoustic modes in each of the four bridges carries the quantum κ_0. At extremely low temperature, below a threshold temperature ~0.6 K, the normalized thermal conductance is one or lower, as we expected. Above the threshold temperature, the thermal conductance curve increases, because the optical phonon modes begin to contribute to the thermal transport.

Chapter 8

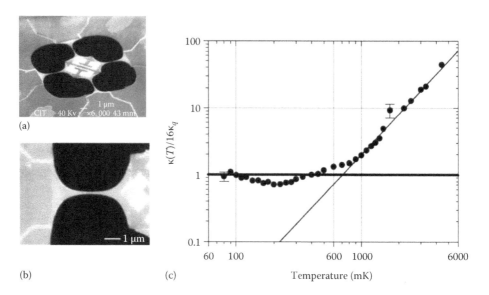

FIGURE 8.1 (a) Experimental setup used to measure the quantum of thermal conductance κ_0. (b) The enlarged view of a NW consisting of SiN in (a). (c) The temperature dependence of thermal conductance, normalized to a universal value of $16\kappa_0$, of the NW. (From Schwab, K. et al., *Nature*, 404, 974, 2000.)

As well as in SiN NWs, the quantization of thermal conductance has also been observed in carbon NTs[12] (CNTs) shortly after its theoretical prediction.[11] Owing to the small diameter of CNTs (~1 nm), the threshold temperature is a few tens of kelvins, which is higher than that of SiN NWs with wider width.[11]

8.2.3   High-Thermal-Conductivity Materials

Electronic devices have been evolving into the nanoscale in accordance with Moore's law thus far, and they now play a fundamental role in society.[14] The progressive shrinking down of electronic devices, however, has reached a limit, in spite of continuing social and technological requirements. Since power densities in integrated circuits have risen exponentially over the last few decades,[15] one of the key limiting factors to further device miniaturization is the excess heat removal. There is, consequently, a strongly growing demand for nanoscale structures that can transfer heat efficiently. In other words, high-thermal-conductivity nanomaterials are essential for future technological developments.

Heat conduits constructed with nanoscale materials for future nanoscale devices require not only high thermal conductivity but also mechanical strength and flexibility. CNTs and boron nitride NTs (BNNTs) are two possible candidates, owing to their promising mechanical and thermal properties. Many studies have reported that BNNTs and CNTs both reveal exceptionally high thermal conductivity.[16–18] Electronically, BNNTs, however, are semiconducting, in contrast to CNTs, whose electronic properties strongly depend on their chirality. A semiconducting feature seems to be suitable for heat removal applications, because it will not interfere with electronic functions. We will thus briefly introduce the basics and describe recent attempts to improve the thermal conductivity of BNNTs and CNTs in this section.

8.2.3.1   Boron Nitride Nanotubes

BNNTs were initially proposed theoretically by A. Rubio et al. in 1994,[19] by analogy to CNTs, since hexagonal BN is also a laminar material similar to graphite. The first successful synthesis of multiwalled BNNTs, with inner diameter of 1–3 nm and with length up to 200 nm, was reported.[20] Similar to a CNT, a BNNT has a one-dimensional cylindrical structure produced by rolling up a hexagonal boron nitride sheet.[21] BNNTs, with a similar phonon dispersion to that of CNTs, can be expected to have a thermal conductivity value similar that of CNTs.[22–24] Furthermore, as we have seen earlier, the Young's modulus of BNNTs is also similar in magnitudes to that of CNTs; both do have extremely high stiffness.[25] But in contrast to CNTs, BNNTs are partly ionic; as each boron (nitrogen) atom must form chemical bonds with nitrogen (boron) atoms, some of the charge on boron will transfer to nitrogen and produce an asymmetric charge distribution between them. The bonds thus possess some ionic character and are not purely covalent, like in CNT, which gives rise to an energy bandgap between valence and conduction bands in BNNT. The reported value of energy bandgap is around 5 eV in BNNT and it can be expected thus to behave like a wide-gap semiconductor.[19,26]

Since boron has large natural isotope abundance ratio (19.9% ^{10}B and 80.1% ^{11}B), it can also be expected that naturally synesized BNNTs contain approximately 20% of light boron isotopes.[18] These isotopes significantly affect the thermal conductivity, and details of such isotope effects will be discussed in the following text. Furthermore, the biocompatibility of BNNTs has also been reported in literature. Owing to their chemically inert and structurally stable properties, BNNTs showed no cytotoxicity in experimental studies that investigated the interaction between nanomaterials and biosystems.[27] CNTs and other nanocarbon materials, in contrast, have been shown to be toxic for living cells.[28,29] This has led to the suggestion that BNNTs might have a potential for use as a biologically friendly material.[27]

8.2.3.2   Theoretical and Experimental Studies

In general, there are two contributions to thermal conductivity: heat conduction by electrons and the phonon contribution. Since BNNTs are electronically insulating, only phonons make a significant contribution to the thermal conductivity, and the influence of conduction electrons can be neglected. The lattice thermal conductivity basically incorporates the effect of three processes, such as phonon-isotope scattering, phonon-boundary scattering, and phonon–phonon scattering. The lattice thermal conductivity of both single-walled BNNTs and CNTs has been theoretically calculated using the exact solution to the phonon Boltzmann equation.[30] Figure 8.2 shows the numerical results of lattice thermal conductivity for a (10, 10) BNNT (black symbols; lower curves) and a (10, 10) CNT (red symbols; upper curves) with length $L = 3$ μm as a function of temperature, T. In this figure, the dashed curves correspond to an isotopically pure BNNT and CNT composed only ^{11}B and ^{12}C, and the solid curves correspond to a BNNT and CNT contained isotopes with natural abundance, 19.9% ^{10}B and 1.1% ^{13}C, respectively. First of all, it can be seen that the thermal conductivities of the BNNT and the CNT are substantially higher than that of the typical bulk materials, such as silicon, whose thermal conductivity is 150 W/mK at room temperature.[31] Second, comparing with the naturally synthesized BNNT and CNT, the thermal conductivity of BNNT is lower than that of the CNT. We can observe the same relationship between the isotopically

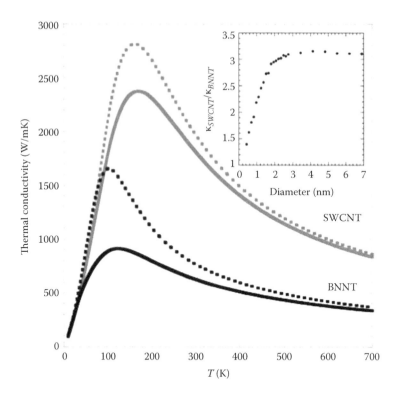

FIGURE 8.2 Calculated values of the thermal conductivities for a (10, 10) boron nitride nanotube (BNNT) (black symbols; lower curves) and a (10, 10) carbon nanotube (CNT) (gray symbols; upper curves) with length, $L = 3$ μm, as a function of temperature, T. The dashed curves correspond to isotopically pure NTs and the solid curves to NTs with naturally occurring boron and carbon abundances. The inset shows the ratio between the thermal conductivities of the isotopically pure CNT and BNNT as a function of diameter with $T = 300$ K. (Reproduced from Lindsay, L. and Broido, D.A., *Phys. Rev. B*, 85, 035436, 2012. With permission.)

pure BNNT and CNT. These are largely due to an harmonic phonon–phonon scattering in the BNNT, which are invoked by the mass difference between boron and nitrogen atoms. Furthermore, because of differences between chemical bonds formed in BNNT (covalent + ionic) and CNT (covalent), the thermal conductivity of the BNNT is three times lower than that of the CNT, due to phonon–phonon scatterings at room temperature, as shown in the inset of Figure 8.2.

As shown in Figure 8.2, the thermal conductivities have maximum values at around 100 K in BNNTs and 180 K in CNTs. Paying attention to the individual curves, the thermal conductivity falls with decreasing T from the peak temperature, because phonon-boundary scattering provides the dominant effect to thermal transport. We also see that the thermal conductivity drops with increasing T from the peak value. This is due to the fact that phonon–phonon scattering becomes more dominant, which is mainly due to the increase of Umklapp processes, rather than the phonon-isotope and phonon-boundary scattering. These characteristics seem to be common irrespective of the conditions of both BNNTs and CNTs. We can further clarify the phonon-isotope scattering effects by comparing the isotopically pure NTs with the naturally synthesized ones.

We can see that the thermal conductivity of the naturally synthesized NTs is smaller than that of the isotopically pure ones throughout the entire temperature range by virtue of the abundant supply of natural isotope. A detailed study of the isotope effects of the thermal transport has been reported using molecular dynamics simulations.[32] In this study, BNNTs with varying concentrations of randomly distributed [11]B or [10]B isotopes were constructed, and the calculated thermal conductivities have shown that the isotopically pure BNNTs exhibit the largest value of thermal conductivity, while the maximum reduction from isotope scattering was found in the NTs with an equal ratio of [11]B or [10]B. The calculated thermal conductivities are in the range of 340–500 W/mK above 200 K, in close agreement with the experimental observations.[18] Furthermore, this study has also investigated the effect of unstable isotopes [8]B and [9]B on the thermal conductivity. The calculated thermal conductivity decreases linearly with decreasing mass of the isotopes and indicates that lowering the mass difference between boron and nitride may improve the thermal conductivity, due to the suppression of the anharmonic phonon–phonon scattering. Other studies by molecular dynamics simulations[33] and kinetic theory[34] have also reported consistent results of the thermal conductivity in BNNTs with or without isotopes. Consequently, phonon-isotope scattering effects contribute to the reduction of thermal conductivity, suggesting that this effect can be reduced by the fabrication of isotopically pure NTs.

Experimental studies have also measured isotope effects on thermal conductivity in BNNTs, showing it to be sensitive to isotopic concentrations.[18,35] Figure 8.3 displays the thermal conductivity for a naturally synthesized BNNT, an isotopically pure multi-walled BNNT of [11]B, and a CNT. Since boron has larger natural isotopic disorder than does carbon, the phonon-isotope scattering will strongly affect the thermal conductivity of BNNTs, but not that of CNTs. It is seen that isotopic enrichment is shown to

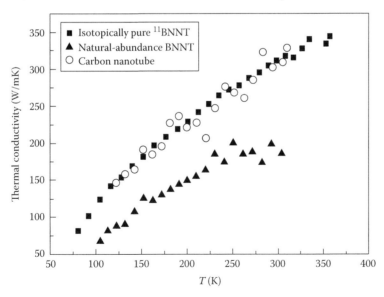

FIGURE 8.3 The thermal conductivity of a carbon nanotube (open circles), a naturally synthesized boron nitride nanotube (BNNT; solid triangles) that contains the natural abundance of isotopes, and an isotopically pure BNNT (solid squares) with similar outer diameters. (Reproduced from Chang, C.W. et al., *Phys. Rev. Lett.*, 97, 085901, 2006. With permission.)

Chapter 8

have a dramatic effect on the enhancement of the thermal conductivity of BNNTs, with approximately 50% increase throughout the measured temperature range, as shown in Figure 8.3, and this is consistent with the theoretical predictions.[23] Moreover, overlap of the data sets for the isotopically pure BNNT and the CNT reflects similarities of their intrinsic phonon dispersion relationship. For all materials, thermal conductivity increases with increasing temperature with no sign of saturation or decrease up to room temperature. Such a behavior of the thermal conductivity, however, deviates from the theoretical predictions above room temperature and indicates the necessity of further investigations in the theoretical aspects of thermal transport.

8.2.3.3 Toward Higher Thermal Conductivity

In order to achieve a higher thermal conductivity, it is important to gain an understanding of the microscopic mechanisms of thermal transport. As stated earlier, the phonon contribution is dominant in thermal conduction in BNNTs and CNTs. Thus, a detailed analysis of phonon transport plays an important role in the understanding of these microscopic mechanisms. Coherent phonon transport can be classified into three regimes: ballistic, diffusive, and localized.[13] It has been speculated that localized phonon transport could be the reason for the reduction of the thermal conductivity in naturally synthesized BNNTs, which contain both [11]B and [10]B isotopes. However, Savić et al.[36] have proposed that the dominant contribution in the observed reduction is diffusive phonon scattering, as no sign of localization effects was observed in their work,[36] in contrast to the case of CNTs.[37] They performed an *ab initio* study of atomistic Green's function formalism to calculate the localization length, mean free path, and phase relaxation length of coherent phonon transport in a (7, 0) CNT with 10.7% of ^{14}C, as shown in Figure 8.4. According to the transport theory, the transition to the localized

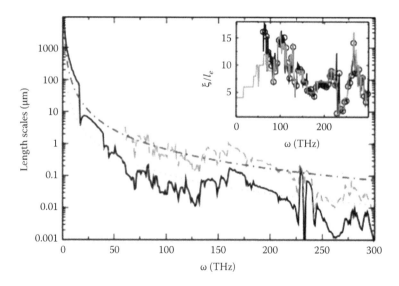

FIGURE 8.4 Transport characteristic lengths as a function of frequency for a (7, 0) carbon nanotubes with 10.7% of ^{14}C isotopes. Solid and dashed lines represent the mean free paths and localization lengths of phonons, respectively. Dash-dotted and dotted lines correspond to estimated phase relaxation lengths for temperatures of 50 and 300 K. (Reproduced from Savić, I. et al., *Phys. Rev. B*, 78, 235434, 2008. With permission.)

transport regime for a certain phonon frequency will happen only if incoherent effects are weak enough, so that phonons preserve their phase information up to the localization length. As indicated in Figure 8.4, however, the phase relaxation length of phonons is comparable to the localization length for most frequencies not only at 300 K but also at 50 K. It was eventually suggested that localization effects do not occur in the system and the thermal conductivity is dominated by the ballistic and diffusive contributions of phonons. Further analysis of the diffusive transport also showed that the characteristics of thermal conduction depend on isotope concentrations. In the case of high isotope concentration (up to 10%), multiple phonon scattering takes place and has a large influence on the thermal conductivity. On the other hand, phonon transport was well described by the additive contribution of single scatterers in the case of smaller isotope concentrations.

The thermal conduction path has attracted much attention. In multiwalled BNNTs, an atomistic Green's function approach has shown that isotope effects were much greater for heat flow carried by a few NT shells than by all NT shells.[38] It was suggested that thermal conduction in multiwalled BNNTs takes place mostly through a few shells, rather than being an all-shell homogeneous conduction process. This has been confirmed by the good agreement obtained between their first-principles calculated thermal conductivity for single-walled BNNTs and the experimental measurements for multiwalled BNNTs obtained in the following year by Stewart et al.,[39] thus suggesting that indeed thermal conduction in these systems occurs mainly through the outer NT shell.

The influence of mechanical deformation on thermal transport has also generated considerable interest. The thermal conductivities of continuously bent multiwalled BNNTs, with outer diameters in the range 30–40 nm, and of CNTs, with diameters in the range 10–33 nm, were experimentally examined by Chang and coauthors.[40] These experiments revealed that thermal transport remained unaffected by the mechanical deformation inflicted on the NTs. Subsequent molecular dynamics simulations have confirmed this robustness of the thermal conductivity against bending deformation in a single-walled CNT, qualitatively agreeing with the experimental results.[41,42] However, plastic deformation and fracture of the NT result in a considerable decrease in the thermal conductivity, as expected (Figure 8.5).

8.2.3.4 Boron Nitride Nanoribbons

Similar to NTs, the thermal conductivity of boron nitride nanoribbons (BNNRs) has also been widely investigated.[43] Similar to a graphene nanoribbon (GNR), a BNNR is made from a hexagonal boron nitride sheet cut off to a finite width (Figure 8.6). The calculated thermal conductance of zigzag BNNRs is shown in Figure 8.7 as a function of the ribbon width for various temperatures. It can be seen that at the same temperature and width, the thermal conductance of a zigzag BNNR is comparable to that of the corresponding graphene NR. However, the thermal conductance of zigzag BNNR is larger than that of the GNR below room temperature, indicating a qualitative difference between these materials with regard to the low-temperature thermal transport mechanisms. Above room temperature, this difference is inverted. The room temperature thermal conductivity is approximately 1700 W/mK and is very similar to that found for BNNTs and CNTs.

Chapter 8

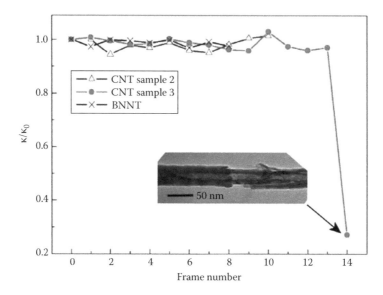

FIGURE 8.5 Normalized thermal conductivity of two carbon nanotube (CNT) samples (open triangles and solid circles) and a boron nitride nanotube (BNNT; crosses) undergoing continuous bending deformation. The inset shows the TEM image of a destroyed CNT sample after an irreversible mechanical damage had been inflicted. (Reproduced from Chang, C.W. et al., *Phys. Rev. Lett.*, 99, 045901, 2007. With permission.)

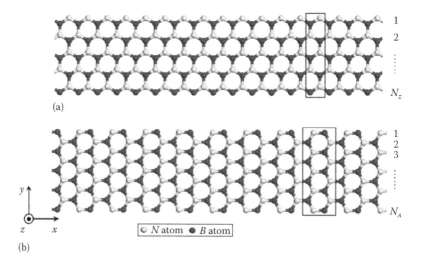

FIGURE 8.6 Schematic image of (a) a zigzag boron nitride nanoribbon of width N_z and (b) an armchair BNNR of width N_A. The black boxes enclose a unit cell of each ribbon. (Reproduced from Ouyang, T. et al., *Nanotechnology*, 21, 245701, 2010. With permission.)

8.2.4   Low-Thermal-Conductivity Materials

As in the case of high-thermal-conductivity materials, materials at the other extreme, that is, low-thermal-conductivity materials, have also great potential from the point of view of technological applications. One of the promising fields for these materials is in the fabrication of *thermoelectrics,* that is, materials that facilitate the conversion of

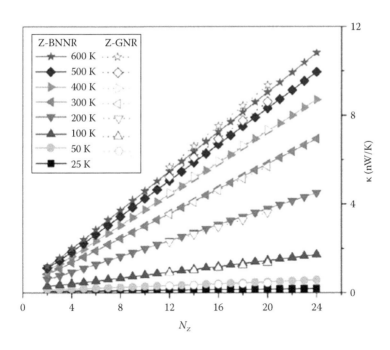

FIGURE 8.7 Thermal conductance κ of zigzag boron nitride nanoribbon and graphene nanoribbon as a function of width at different temperature. (Quoted from Ouyang, T. et al., *Nanotechnology*, 21, 245701, 2010. With permission.)

thermal energy into electricity or vice versa. Their potential has been demonstrated for power generation and refrigeration; however, the low efficiency of currently available thermoelectric materials has so far limited their exploitation.

The thermoelectric performance of a material is characterized by the dimensionless figure of merit, ZT.[44] This is defined as

$$ZT = \frac{\sigma S^2}{\lambda} T,$$

(8.5)

where

T is the temperature

σ is the electrical conductivity

S is the Seebeck coefficient (or thermopower)

λ is the thermal conductivity

ZT is directly related to the ideal efficiency η_{max} of a thermoelectric device for electricity generation, which is given as the Carnot efficiency:

$$\eta_{max} = \frac{T_H - T_L}{T_H} \times \left(1 - \frac{1 + \dfrac{T_L}{T_H}}{\sqrt{1 + ZT} + \dfrac{T_L}{T_H}} \right)$$

(8.6)

Chapter 8

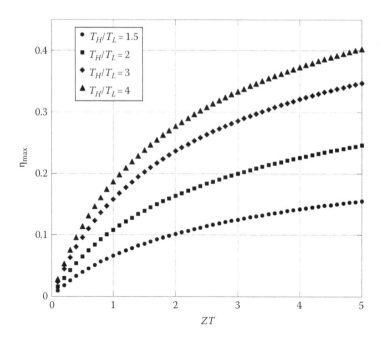

FIGURE 8.8　Ideal efficiency, η_{max}, as a function of the dimensionless figure of merit, ZT, with various ratios between high and low temperatures, T_H/T_L.

where $T_{H/L}$ are the high/low temperatures imposed on the thermoelectric device. This formula tells us that high ZT results in a large value of η_{max}, as shown in Figure 8.8. Materials with $ZT \sim 1$ are regarded as good thermoelectrics; however, a thermoelectric that wants to be as competitive with a conventional refrigerator or power generator needs to have a value equal to or greater than $ZT \sim 3$. Moreover, the thermoelectric efficiency η_{max} is saturated at the ideal Carnot efficiency when $ZT \gg T_L/T_H$. There is, therefore, a significant technological drive toward the design and synthesis of new thermoelectric materials with high ZT values.

According to Equation 8.6 for ZT, an ideal thermoelectric material should possess a high Seebeck coefficient, high electrical conductivity, and low thermal conductivity. This combination of features corresponds to a so-called phonon-glass electron crystal (PGEC), in which phonons are scattered by a disordered glass-like structure, thus giving a low λ, while electrons do not undergo any scattering, resulting in a high S value and σ value. Hicks and Dresselhaus[45] proposed in 1993 that the thermoelectric performance of materials can be significantly improved by preparing them in the form of one-dimensional NWs.[45] There are two main reasons why this would result in an increase of ZT.[46] First, nanoscale constituents introduce quantum-confinement effects that change the electronic density of states, resulting in an enhanced electronic conduction. Second, the use of nanostructures leads inevitably to the presence of many interfaces, barriers, and surfaces, which would result in frequent scattering of phonons, thus reducing thermal conduction.

Since high ZT materials are desirable for highly efficient thermoelectric applications, many nanomaterials constructed from NWs have been investigated.[47] Here, we will introduce and discuss bismuth tellurides, which have long been recognized as materials with the highest figure of merit, with $ZT \approx 1$.[44] We will also discuss silicon NWs as possible candidates for future thermoelectric nanomaterials.

8.2.4.1   Bismuth Telluride Nanowires

Bi_2Te_3, as well as its alloys, has long been proved to be the best thermoelectric material at and around room temperature conditions, with a ZT value about unity. So far, significant enhancement of ZT has been obtained in Bi_2Te_3-based nanostructures, mainly due to the reduction of thermal conductivity.[48–51] Among various nanostructures, Bi_2Te_3 NWs are expected to possess reduced thermal conductivity and consequently enhanced ZT.[50]

The influence of NW geometry on the thermal conductivity in bismuth has been calculated by Hasegawa and coauthors,[52] focusing on the dependence on NW diameter, and is shown in Figure 8.9. These results have revealed that the thermal conductivity reduces in diameters of less than 1 µm and have shown that the NW geometry affects the thermal conductivity. In Bi_2Te_3 NWs, the thermal conductivity calculated by molecular dynamics simulation displays a significant reduction, compared to the bulk value, as the NW diameter is reduced from 30 down to 3 nm,[53] as shown in Figure 8.10. A NW with sawtooth rough surface (STNW) reveals smaller thermal conductivity than one with a smooth surface (SMNW), indicating that surface roughness contributes to further reducing the thermal conductivity. The difference between these two, however, diminished with decreasing NW diameters. Moreover, the thermal conductivity of a Bi_2Te_3 of a 30 nm diameter NW shows good agreement with experimental measurements in a 52 nm diameter Bi_2Te_3 NW with rough surface.[50] The temperature dependence gradually weakens from bulk to SMNW and to STNW and deviates from the T^{-1} trend, as shown in Figure 8.11, consistently with experimental observation.[50] This indicates that Umklapp scattering taking place in the bulk becomes less important in a NW geometry, as boundary scattering plays a more dominant role. This has been supported in an experimental study.[54]

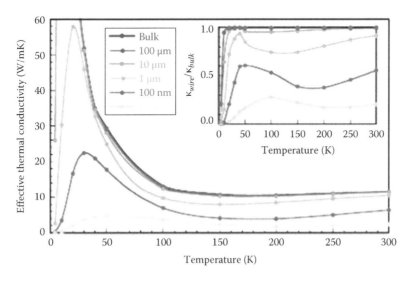

FIGURE 8.9   Temperature dependence of the effective thermal conductivity for various NW diameters. The inset shows the thermal conductivity of NW normalized by that of the bulk. (Reproduced from Hasegawa, Y. et al., *J. Appl. Phys.*, 106, 063703, 2009. With permission.)

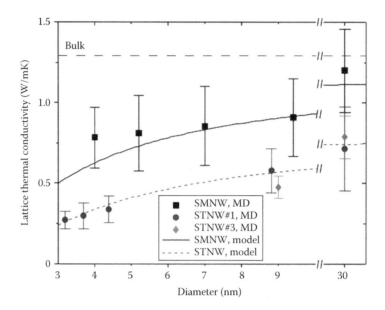

FIGURE 8.10 Calculated lattice thermal conductivity of Bi_2Te_3 NWs with smooth surface (SMNW) and sawtooth rough surface (STNW) at room temperature as a function of diameter. (Reproduced from Qiu, B. et al., *Phys. Rev. B*, 83, 035312, 2012. With permission.)

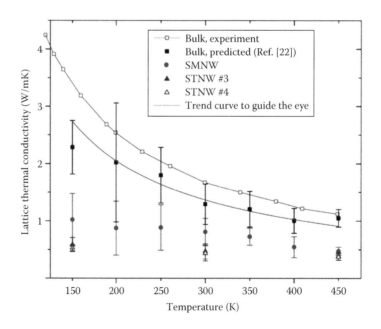

FIGURE 8.11 Lattice thermal conductivities of Bi_2Te_3 NWs with smooth surface (SMNW) and sawtooth rough surface (STNW) as a function of temperature. (Reproduced from Qiu, B. et al., *Phys. Rev. B*, 83, 035312, 2012. With permission.)

8.2.4.2 Silicon Nanowires

Although bismuth telluride NWs have sufficiently low thermal conductivity to make them suitable candidates for thermoelectric applications, there is a significant demand to find alternative thermoelectric materials. This is because bismuth and tellurium are both costly rare metals but also because of their high toxicity. Ideally, such alternatives should be found among the compounds formed by common elements, such as carbon, silicon, and oxygen, that is, elements that are ubiquitous in our environment.

Here, we are going to present and discuss the thermal properties of silicon NWs and some related materials. Silicon is one of the ubiquitous elements and is used in a wide variety of engineering fields and technological applications, such as integrated circuits. Recent progress in nanotechnology makes it now possible to fabricate Si NWs,[55–57] which are expected to have reduced thermal conductivity, due to the reduction of low-frequency phonons and the significant impact of boundary scattering. Furthermore, the compatibility with present engineering process of Si NWs may also be a considerable advantage for future applications of these systems.

In 2008, two critical studies simultaneously reported novel thermoelectric properties of Si NWs.[58,59] Both experimental studies reported the enhancement of ZT by introducing surface roughness in Si NWs. The surface roughness results in an increase of the phonon-boundary scattering and hence a concomitant reduction in thermal conductivity. Hochbaum et al.[58] prepared Si NWs using two different fabrication methods, the aqueous electroless etching (EE) method and the vapor–liquid–solid (VLS) method, and measured the thermal conductivities of NWs with various diameters, as is shown in Figure 8.12. These two synthesis methods provide different characteristics

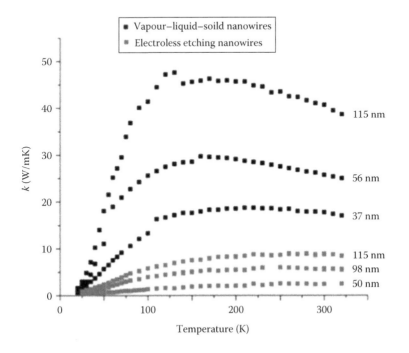

FIGURE 8.12 The temperature dependence of thermal conductivity, κ, of VLS (black squares) and EE (gray squares) of EE NWs. (Reproduced from Hochbaum, A.I. et al., *Nature*, 451, 163, 2008. With permission.)

of NW, with EE-grown Si nanowires having rougher surface than VLS-grown ones. In Figure 8.12, it can be seen that the thermal conductivity of Si NWs strongly depends on their diameter, a dependence that is attributed to the enhancement of phonon-boundary scattering. The magnitude of the thermal conductivity of EE-grown NWs, however, is significantly lower than that of VLS-grown NWs of comparable diameter. This may be caused by the roughness on the NW surface, which contributes to higher rates of phonon scattering.

The instrument made by Boukai et al.[59] has also shown efficient thermoelectric performance due to reduction of thermal conductivity by changing the diameter of Si NW and also by controlling impurity doping levels. The measured thermal conductivity of thin NWs is in the order of unity at room temperature, in contrast to that of bulk Si, which has a value that is 100 times smaller. These results are consistent with previous thermal conductivity measurements.[60,61] The thermal conductance of Si NWs has been shown to display a T^3-dependence in the very low temperature limit, below 3 K, which shows evidence of phonon contributions to thermal transport.[61] In the range of 20–100 K, λ has a T-linear dependence, which deviates from the ordinary T^3 behavior,[62] suggesting that there may be a frequency dependence of the phonon-boundary scattering.[63]

In order to explore the possibility of gaining further reduction in the thermal conductivity, many theoretical studies have been performed focusing on surface roughness,[64] isotopes, vacancies and atomic substitution effect,[46,65–67] and surface decoration.[68] Martin et al.[64] have theoretically demonstrated a remarkable effect of surface roughness on thermal conductivity in Si NWs. As shown in Figure 8.13, the theoretical results have revealed that a significant decrease of the thermal conductivity is observed in Si NWs with a rough surface and provided a good agreement with the experimental results for a wide range of temperatures.[59] They also found that the thermal conductivity depends on the square of the root mean square surface roughness, suggesting that the surface roughness strongly contributes to the reduction in thermal conductivity.

Similar to what is observed in BNNTs, the presence of different isotopes also provides a promising path for the control of the thermal conductivity in Si NWs. The thermal conductivity of isotope-doped Si NWs was investigated by substituting ^{28}Si atoms into ^{29}Si and ^{42}Si, and the influence of their concentration on the thermal conductivity was calculated using molecular dynamics simulations.[65] Because of the mass difference of ^{28}Si and isotopes, the calculated thermal conductivity is reduced by isotope doping and reaches minimum value, which is about 27% of that of pure ^{28}Si NW, when the isotope concentration is 50%. Similar to the isotope effects, atomic vacancies also result in the reduction of thermal conductivity due to the suppression of phonon transport.[66] Furthermore, SiGe NWs, which can be formed by atomic substitution of Si into Ge, have also been investigated.[46,67] The thermal conductivity of SiGe NWs decreased with increasing concentration of Si in NWs of sufficiently small diameter, as shown in Figure 8.14. Finally, we would like to mention a study considering the effects of surface decoration of silicon NWs.[68] These structural changes reduce the thermal conductivity but do not influence the electrical conductivity. It was thus claimed that these materials are more suitable for thermoelectric applications than NWs with rough surface.

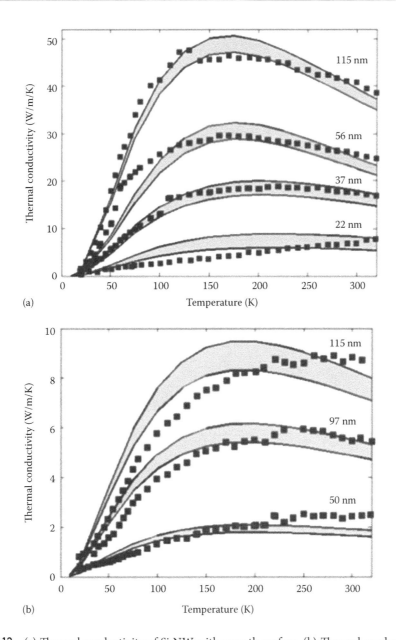

FIGURE 8.13 (a) Thermal conductivity of Si NW with smooth surface. (b) Thermal conductivity of Si NW with rough surface. In both figures, shaded areas are theoretical results with the roughness in different root mean square. Square symbols denote the experimental results. (Reproduced from Martin, P. et al., *Phys. Rev. Lett.*, 102, 125503, 2009. With permission.)

8.2.4.3 Core–Shell Nanowires

Another possible way to control the thermal conductivity of NWs is through their structural design. Core–shell Si and Ge NW structures have been explored in this context, both experimentally and theoretically, as possible candidates for thermoelectric applications.[69–72] Si/Ge core–shell NWs can be synthesized by the VLS method with

Chapter 8

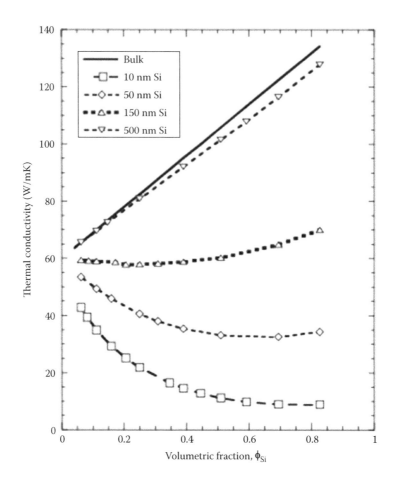

FIGURE 8.14   Calculated thermal conductivity as a function of Si fraction for Si NWs in a Ge host material to constitute a model for the thermal conductivity of Si_xGe_{1-x} nanocomposite samples. (Reproduced from Dresselhaus, M.S. et al., *Adv. Mater.*, 19, 1043, 2007. With permission.)

either a Ge core and Si shell or vice versa, as shown schematically in Figure 8.15a, and can be obtained with a high degree of control on diameter, length, and doping profile.[73,74] In Ge-core, Si-shell structures, the electronic properties revealed that the band offset naturally leads to hole accumulation in the core region, without any impurity doping in Ge–Si core–shell NW,[75] and ballistic hole transport has been reported, indicating the existence of a smooth interface between the Ge core and Si shell that does not scatter the holes. Moreover, the highest valence band states are spatially confined in the Ge-core region, suggesting that electronic conduction may not be influenced by the roughness of the Si shell.[69,76] In contrast to electronic transport, phonon transport can be modified by the surface roughness. Figure 8.15b shows the calculated thermal conductance of pristine and surface-disordered Ge–Si core–shell NWs as a function of temperature. The effect of surface roughness on the thermal conductivity reduction is clearly seen in a wide range of temperatures, in good agreement with an experimental study.[77] Molecular dynamics studies have also reported the same behavior of thermal conductivity and claimed that anharmonic phonon–phonon scattering could be important for core–shell NWs.[70]

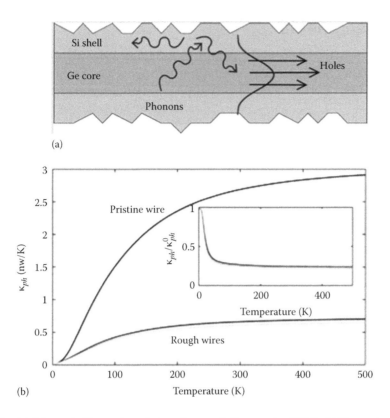

FIGURE 8.15 (a) Schematic illustration of a Ge–Si core–shell nanowire (NW). (b) Thermal conductance of pristine and surface-disordered Ge–Si core–shell NWs. (Quoted from Markussen, T., *Nano Lett.*, 12, 4698, 2012. With permission.)

8.3 Conclusions, Challenges, and Perspectives

In this chapter, we have seen that studies in the field of the thermal properties of NTs and NWs have advanced along two major fronts: the search nanomaterials with either a high or a low thermal conductance. On the one hand, there is a high technological demand for materials that can transfer heat efficiently, and in this respect, both boron nitride and carbon NTs are promising candidates. Simulations have been able to accurately calculate the thermal conductance and/or conductivity in pristine systems and also to account for isotopic effects, the influence of mechanical deformation, multilayers, etc. Nevertheless, theoretical predictions deviate from experimental measurements above room temperature, an indication that there are still advances to be made in the theory of thermal transport. Low-thermal-conductivity materials are also of interest, due to their possible use in the fabrication of thermoelectrics. The use of nanostructures interesting here, because thermoelectric performance can be improved through the reduction of thermal conduction. In this context, the thermal properties of bismuth telluride, silicon, and silicon–germanium NWs have been theoretically calculated, and the influence of factors such as the presence of impurities, surface roughness, and core–shell structures has been explored. The objective here is to assist in the material design of PGEC systems for thermoelectric applications.

Chapter 8

Acknowledgments

YT and TY are partly supported by Grants-in-Aid (Nos. 24681021, 24860057, 25104722) from the Ministry of Education, Culture, Sports, Science and Technology of Japan. The work of ERH is supported by the Spanish DGCYT through project FIS2012-31713.

References

1. M. S. Dresselhaus, G. Dresselhaus, and P. Avouris, eds., *Carbon Nanotubes, Synthesis, Structure, Properties and Applications*, vol. 80 of *Topics in Applied Physics*. Springer, Berlin, Germany (2001).
2. A. Jorio, M. S. Dresselhaus, and G. Dresselhaus, eds., *Carbon Nanotubes, Advanced Topics in the Synthesis, Structure, Properties and Applications*, vol. 111 of *Topics in Applied Physics*. Springer, Berlin, Germany (2008).
3. C. M. Lieber and Z. L. Wang, *MRS Bull.* **32**, 99 (2007).
4. R. E. Peierls, *Quantum Theory of Solids*. Oxford University Press, New York (1955).
5. J. M. Zimann, *Electron and Phonons*. Clarendon Press, Oxford, U.K. (1960).
6. R. E. Peierls, *Ann. Physik.* **3**, 1055 (1929).
7. L. G. C. Rego and G. Kirczenow, *Phys. Rev. Lett.* **81**, 232 (1998).
8. K. Schwab, E. A. Henrlksen, J. M. Worlock, and M. L. Roukes, *Nature* **404**, 974 (2000).
9. D. E. Angelescu, M. C. Cross, and M. L. Roukes, *Superlattices Microstruct.* **23**, 673 (1998).
10. M. P. Blencowe, *Phys. Rev. B* **59**, 4992 (1999).
11. T. Yamamoto, S. Watanabe, and K. Watanabe, *Phys. Rev. Lett.* **92**, 075502 (2004).
12. H. Y. Chiu, V. V. Deshpande, H. W. Postma, C. N. Lau, C. Miko, L. Forro, and M. Bockrath, *Phys. Rev. Lett.* **95**, 226101 (2005).
13. S. Datta, *Electronic Transport in Mesoscopic Systems*. Cambridge University Press, Cambridge, U.K. (1995).
14. G. E. Moore, *Electronics* **38**, 114 (1965).
15. E. Pop and K. E. Goldson, *J. Electron. Packag.* **128**, 102 (2006).
16. B. G. Demczyk, Y. M. Wang, J. Cumings, M. Hetman, W. Han, A. Zettl, and R. O. Ritchie, *Mater. Sci. Eng.* **A334**, 173 (2002).
17. P. Kim, L. S. A. Mujundar, and P. L. McEuen, *Phys. Rev. Lett.* **87**, 215592 (2001).
18. C. W. Chang, A. M. F. A. Afanasiev, D. Okawa, T. Ikuno, H. Garcia, D. Li, A. Majumdar, and A. Zettl, *Phys. Rev. Lett.* **97**, 085901 (2006).
19. A. Rubio, J. L. Corkill, and M. L. Cohen, *Phys. Rev. B* **49**, 5081 (1994).
20. N. G. Chopra, R. J. Luyken, K. Cherrey, V. H. Crespi, M. L. Cohen, S. G. Louie, and A. Zettl, *Science* **269**, 966 (1995).
21. M. L. Cohen and A. Zettl, *Phys. Today* **63**, 63 (2010).
22. D. Sánchez-Portal and E. R. Hernández, *Phys. Rev. B* **66**, 235415 (2002).
23. L. Wirtz, A. Rubio, R. A. de la Concha, and A. Loiseau, *Phys. Rev. B* **68**, 045425 (2003).
24. Y. Xiao, X. H. Yan, J. X. Cao, J. W. Ding, Y. L. Mao, and J. Xiang, *Phys. Rev. B* **69**, 205415 (2003).
25. E. Hernández, C. Goze, P. Bernier, and A. Rubio, *Phys. Rev. Lett.* **80**, 4502 (1998).
26. X. Blase, A. Rubio, S. G. Louie, and M. L. Cohen, *Euro. Phys. Lett.* **28**, 335 (1994).
27. X. Chen, P. Wu, M. Rousseas, D. Okawa, Z. Gartner, A. Zettl, and C. R. Bertozzi, *J. Am. Chem. Soc.* **131**, 890 (2009).
28. J. Muller, F. Huaux, N. Moreau, P. Misson, J. F. Heilier, M. Delos, M. Arras, A. Fonseca, J. B. Nagy, and D. Lison, *Toxicol. Appl. Pharmacol.* **207**, 221 (2005).
29. L. Tabet, C. Bussy, N. Amara, A. Setyan, A. Grodet, M. J. Rossi, J. C. Pairon, J. Boczkowski, and S. Lanone, *J. Toxic. Environ. Health* **72**, 60 (2009).
30. L. Lindsay and D. A. Broido, *Phys. Rev. B* **85**, 035436 (2012).
31. C. J. Glassbrenner and G. A. Slack, *Phys. Rev.* **134**, A1058 (1963).
32. C. Sevik, A. Kinaci, J. B. Haskins, and T. çağin, *Phys. Rev. B* **86**, 075403 (2012).
33. G. Zhang and B. Li, *J. Chem. Phys.* **123**, 114714 (2005).
34. J. W. Jiang and J. S. Wang, *Phys. Rev. B* **84**, 085439 (2011).
35. C. W. Chang, W. Q. Han, and A. Zettl, *Appl. Phys. Lett.* **86**, 173102 (2005).
36. I. Savić, N. Mingo, and D. A. Stewart, *Phys. Rev. Lett.* **101**, 166502 (2008).

37. T. Yamamoto, K. Sasaoka, and S. Watanabe, *Phys. Rev. Lett.* **106**, 215503 (2011).
38. I. Savić, D. A. Stewart, and N. Mingo, *Phys. Rev. B* **78**, 235434 (2008).
39. D. A. Stewart, I. Savić, and N. Ming, *Nano Lett.* **9**, 81 (2009).
40. C. W. Chang, D. Okawa, H. Garcia, A. Majumdar, and A. Zettl, *Phys. Rev. Lett.* **99**, 045901 (2007).
41. F. Nishimura, T. Takahashi, K. Watanabe, and T. Yamamoto, *Appl. Phys. Express* **2**, 035003 (2009).
42. F. Nishimura, T. Shiga, S. Maruyama, K. Watanabe, and J. Shiomi, *Jpn. J. Appl. Phys.* **51**, 015102 (2012).
43. T. Ouyang, Y. Chen, Y. Xie, K. Yang, Z. Bao, and J. Zhong, *Nanotechnology* **21**, 245701 (2010).
44. A. Majumdar, *Science* **303**, 777 (2004).
45. L. D. Hicks and M. S. Dresselhaus, *Phys. Rev. B* **47**, 16631 (1993).
46. M. S. Dresselhaus, G. Chen, M. Y. Tang, R. Yang, H. Lee, D. Wang, Z. Ren, J. P. Fleurial, and P. Gogna, *Adv. Mater.* **19**, 1043 (2007).
47. J. W. Jiang, J. S. Wang, and B. Li, *J. Appl. Phys.* **109**, 014326 (2011).
48. R. Venkatasubramanian, E. Siivola, T. Colpitts, and B. O'Quinn, *Nature* **413**, 597 (2001).
49. B. Poudel, Q. Hao, Y. Ma, Y. Lan, A. Minnich, B. Yu, D. W. X. Yan et al., *Science* **320**, 634 (2008).
50. A. Mavrokefalos, A. L. Moore, M. T. Pettes, L. Shi, W. Wang, and X. Li, *J. Appl. Phys.* **105**, 104318 (2009).
51. C. L. Chen, Y. Y. Chen, S. J. Lin, J. C. Ho, P. C. Lee, C. D. Chen, and S. R. Harutyunyan, *J. Phys. Chem. C* **114**, 3385 (2010).
52. Y. Hasegawa, M. Murata, D. Nakamura, and T. Komine, *J. Appl. Phys.* **106**, 063703 (2009).
53. B. Qiu, L. Sin, and X. Ruan, *Phys. Rev. B* **83**, 035312 (2012).
54. J. Zhou, C. Jin, J. H. Seol, X. Li, and L. Shi, *Appl. Phys. Lett.* **87**, 133109 (2005).
55. A. M. Morales and C. M. Lieber, *Science* **279**, 208 (1998).
56. J. D. Holmes, K. P. Johnston, R. C. Doty, and B. A. Korgel, *Science* **287**, 1471 (2000).
57. Y. Wu, Y. Cui, L. Huynh, C. J. Barrelet, D. C. Bell, and C. M. Lieber, *Nano Lett.* **4**, 433 (2004).
58. A. I. Hochbaum, R. Chen, R. D. Delgado, W. Liang, E. C. Garnett, M. Najarian, A. Majumdar, and P. Yang, *Nature* **451**, 163 (2008).
59. A. I. Boukai, Y. Bunimovich, J. Tahir-Kheli, J. K. Yu, W. A. G. Iii, and J. R. Heath, *Nature* **451**, 168 (2008).
60. D. Li, Y. Wu, P. Kim, L. Shi, P. Yang, and A. Majumdar, *Appl. Phys. Lett.* **83**, 2934 (2003).
61. O. Bourgeois, T. Fournier, and J. Chaussy, *J. Appl. Phys.* **101**, 016104 (2007).
62. R. Chen, A. I. Hochbaum, P. Murphy, J. Moore, P. Yang, and A. Majumdar, *Phys. Rev. Lett.* **101**, 105501 (2008).
63. P. G. Murphy and J. E. Moore, *Phys. Rev. B* **76**, 133313 (2007).
64. P. Martin, Z. Aksamija, E. Pop, and U. Ravaioli, *Phys. Rev. Lett.* **102**, 125503 (2009).
65. N. Yang, G. Zhang, and B. Li, *Nano Lett.* **8**, 276 (2008).
66. T. Markussen, A. P. Jauho, and M. Brandbyge, *Phys. Rev. B* **79**, 035315 (2009).
67. J. Li, T. A. Yeung, and C. H. Kam, *J. Appl. Phys.* **111**, 094308 (2012).
68. T. Markussen, A. P. Jauho, and M. Brandbyge, *Phys. Rev. Lett.* **103**, 055502 (2009).
69. T. Markussen, *Nano Lett.* **12**, 4698 (2012).
70. M. Hu, K. P. Giapis, J. V. Goicochea, X. Zhang, and D. Poulikakos, *Nano Lett.* **11**, 618 (2011).
71. M. Hu, X. Zhang, K. P. Giapis, and D. Poulikakos, *Phys. Rev. B* **84**, 085442 (2011).
72. J. Chen, G. Zhang, and B. Li, *J. Chem. Phys.* **135**, 104508 (2011).
73. L. Lauhon, M. Gudiksen, C. Wang, and C. Lieber, *Nature* **420**, 57 (2002).
74. Y. Zhao, J. T. Smith, J. Appenzeller, and C. Yang, *Nano Lett.* **11**, 1406 (2011).
75. W. Lu, J. Xiang, B. P. Timko, Y. Wu, and C. M. Lieber, *PNAS* **102**, 10046 (2005).
76. M. Amato, S. Ossicini, and R. Rurali, *Nano Lett.* **11**, 594 (2011).
77. M. C. Wingert, Z. Y. Chen, E. Dechaumphai, J. Moon, J. H. Kim, J. Xiang, and R. Chen, *Nano Lett.* **11**, 5507 (2011).

Chapter 8

9. Modeling Optical and Excited-State Properties

Enrico Berardo and Martijn A. Zwijnenburg

9.1 Introduction 269

9.2 Challenges 270
    9.2.1 Excited-State Processes: A Potential Energy Surface Perspective ... 270
    9.2.2 Where Can Modeling Play a Role? 272

9.3 Methods 273
    9.3.1 Desired Properties of the Ideal Excited State Method. 273
    9.3.2 Density Functional Theory 274
    9.3.3 Time-Dependent DFT 275
    9.3.4 Correlated Wavefunction-Based Methods 276
    9.3.5 Many-Body Perturbation Theory Methods: Green's
           Function and Bethe–Salpeter. 277

9.4 Applications 277
    9.4.1 Quasiparticles Spectra and Band Structure. 277
    9.4.2 Vertical Absorption Spectra 278
    9.4.3 Photoluminescence and Excited State Relaxation. 281
    9.4.4 Charge Transfer between Nanoparticles and Adsorbed Molecules ... 284

9.5 Perspectives 285

Acknowledgments 285

References 285

9.1 Introduction

Inorganic nanostructures interact strongly with light. After excitation by light of a certain wavelength, they might re-emit, with some delay, higher wavelength light (i.e., fluorescence and phosphorescence), or, for example, transfer the excited electrons to an external electric circuit where they can be exploited to do useful electrical work (photovoltaics). The excited electrons and the holes formed in the top of the valence band can also be used to drive chemical reactions, for example, the splitting of water in photocatalysis. Understanding the excited-state properties of inorganic nanostructures in general and the fate of excited electron–hole pairs, in particular, is hence of both fundamental and technological importance. As the transient nature of excited states and the inability of obtaining atomically resolved structural information for relevant nanostructures make it difficult to obtain such insight by experiment alone, only computational nanoscience can play a crucial role in unraveling the puzzles.

Chapter 9

In this chapter, we will first review possible computational methods and then their applications in modeling the photochemistry and physics of nanostructures.

9.2 Challenges

9.2.1 Excited-State Processes: A Potential Energy Surface Perspective

The study of the excited-state properties of inorganic nanostructures in general and the fate of excited electron–hole pairs in particular requires the careful consideration of a number of features and processes occurring on and between the ground and excited-state potential energy surfaces (referred to in the remainder of this chapter as energy surfaces). Figure 9.1 displays the most prominent features and processes for a cartoon-like one-dimensional set of energy surfaces. Relevant features are labeled with bold upper case letters (**A–G**) and the pertinent vertical excitations between the different surfaces are labeled by dashed arrows. Figure 9.1 and the remainder of the discussion in this section assume that the initial ground state of the nanostructure is a closed-shell singlet, but a similar analysis can be made for nanostructures with an open-shell ground state.

The first relevant process is the absorption of light by the nanostructure in the ground-state minimum energy geometry **A** and the electronic excitation of the system from S_0 to higher excited singlet states S_n (where n > 0). An electron gets promoted to the conduction band (or lowest unoccupied orbital [LUO]) and at the same time, a hole in the top of the valence band (or highest occupied orbital [HOO]) is formed (Figure 9.2a). This excited electron and hole can either be essentially free charge carriers, where there is negligible interaction between them (Figure 9.2b), or form electron–hole pairs bound by a mutual Coulomb interaction (Figure 9.2c). The latter bound state is referred to as exciton and is stabilized relative to free carrier states by the exciton binding energy (EBE). The size of the EBE in different systems can vary widely, ranging from millielectronvolts to electronvolts. As a result, it is not always clear-cut experimentally if one is dealing

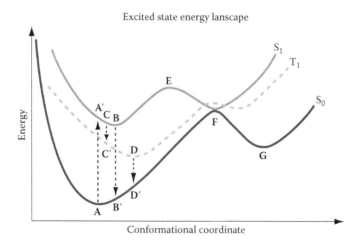

FIGURE 9.1 Cartoon-based description of the ground-state (S_0 dark gray curve) and lowest excited-state (S_1 solid and T_1 dashed light-gray curve) energy surfaces, schematically showing processes involved with light absorption and fluorescence.

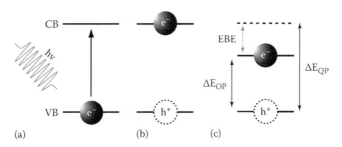

FIGURE 9.2 Excitation process in a material, where the absorption of light (a) causes the promotion of an electron (black sphere from the valence band (VB) to the conduction band (CB), and the creation of a hole (white sphere) in the VB. (b) The electron and the hole are free charge carriers, with negligible interaction between them both. (c) Exciton pair, where the electron and hole are bound together by a reciprocal Coulomb interaction. EBE stands for exciton binding energy, while ΔE_{OP} and ΔE_{QP} represent the optical gap and the quasiparticle gap, respectively.

with either essentially free or excitonic states, especially in the case of nanostructures. A measure of the EBE can be obtained by a comparison between the quasiparticle gap (ΔE_{QP}, the energy required to make free charge carriers, experimentally measured indirectly through a combination of normal and inverse photoemission spectroscopies, also referred to as the band, transport or fundamental gap) and the energy of the lowest optical excitation (the optical gap, ΔE_{OP}).

The experimentally measured UV–visible (UV–Vis) absorption spectrum contains contributions from both vertical excitations (i.e., excitations that involve the same geometry for both the ground and excited states) at **A** and nonvertical adiabatic excitations that involve a vibrationally excited version of, either or both, the ground and excited states. The nonvertical adiabatic excitations give rise to vibronic bands or progression, which however, for nanostructures, are unlikely to be individually resolvable (except for small clusters, e.g., monomers, in the gas phase) and will contribute to vibrational broadening of the main peak instead.

After absorption, the excited system is generally not in a stationary state and will start to relax and lower its energy. If the nanoparticle was initially excited to a higher excited state than the lowest excited state, the system will de-excite radiationlessly through *internal conversion* (IC) to the lowest excited state (e.g., from S_n to S_1, for simplicity, this is not shown in Figure 9.1). The system can also lower its energy by nuclear relaxation along the excited-state surface in the direction of a nearby excited-state minimum (e.g., in the case of S_1 in Figure 9.1, relax from **A′** toward **B**). The electronic relaxation from S_n to S_1 through IC is generally very fast compared to nuclear motion so that one can assume that all nuclear relaxations will take place on the S_1 surface. This observation is canonized in Kasha's rule,[1] which states that fluorescence or phosphorescence (see below) occurs in appreciable amounts only from the lowest excited state of a given multiplicity (i.e., S_1 or T_1). The vast majority of systems appear to follow Kasha's rule though selected experimental examples of non-Kasha behavior have been reported for molecules[2] and, more recently, nanoparticles.[3]

Concentrating on the majority of systems that follow Kasha's rule, nuclear relaxation on the S_1 surface after IC will lead the system to an S_1 excited-state minimum (point **B** in Figure 9.1), from which the system can undergo *fluorescence* back to ground state (**B** → **B′**). Alternatively, the system can experience *intersystem crossing* (ISC) from the singlet S_1 to the T_1 triplet surface (**C** → **C′**), relax to a nearby T_1 minimum (**D**), and

Chapter 9

display *phosphorescence,* where the system relaxes back to the ground state (**D** → **D'**). In both cases, the system after de-excitation to the ground-state surface (at point **B'** or **D'** respectively) can relax back to the ground-state minimum **A**. If there is an energy barrier on the ground-state surface between **B'/D'** and **A** (not shown in Figure 9.1), the system might also relax to an alternative ground-state minimum geometry.

An excited-state barrier (point **E** in Figure 9.1) will separate the excited-state minimum downhill from the ground-state minimum energy structure from other excited-state minima or, alternatively, conical intersections (CX, point **F** in Figure 9.1). The latter are points where two excited-state surfaces or an excited and ground-state surface are degenerate. CXs were once thought to be theoretical rarities but are now believed to be ubiquitous (at least in molecular systems).[4] At a CX, the system can undergo a radiationless transition from one excited state to the other, or, as is the case on point **F** in Figure 9.1, from the lowest excited state to the ground state. In some cases, the barrier **E** might be so small or nonexistent that the system effectively relaxes on S_1 straight from the ground-state minimum energy geometry to the CX. In this scenario, a barrierless route between ground-state minimum energy geometry and CX is expected to result in a dark pathway back to the ground state with no or only limited fluorescence observed experimentally. Finally, as CXs correspond to maxima on the ground-state surface, the system can potentially relax to a different minima when transitioning from excited to ground state (either to the new ground-state minimum **G** or back to the original ground-state minimum **A** in Figure 9.1).

Beyond these processes that only involve the nanostructure, there are also processes that involve molecules adsorbed on the nanoparticle. Transfer of excited electrons and holes to adsorbed molecules, or the other way around, is the key step in both photovoltaics and photocatalysis. A very pertinent question in this context is the nature of the relevant structure for electron and/or hole transfer: the ground-state minimum energy geometry or a relaxed excited-state geometry? In other words, what is the time scale of electron–hole transfer relative to that of nuclear relaxation? Such questions are of fundamental relevance as the ability of the nanostructure to donate and/or accept electrons might be substantially different for different nuclear geometries.

Finally, it should be pointed out that the excited-state surfaces of inorganic nanostructures are potentially more complicated than those of organic molecules. The latter typically have a very small number of well-defined chromophores (specific chemical groups that interact with light), where large parts of the molecule are effectively innocent spectators. In nanostructures, by contrast, it is generally much more complicated to split up the structure into isolated chromophores and the whole nanostructure could potentially be one giant chromophore.

9.2.2   Where Can Modeling Play a Role?

Based on what has been said in the previous section, one could consider a number of possible calculations. For instance:

- Calculation of absorption spectra, not only obtaining excitation energies but also their intensities (UV–Vis). Such spectra could be based on only vertical excitations or could include also vibronic transitions.
- Calculation of the quasiparticle gap and thus the EBE.

- Calculation of the energy of the photoluminescence and phosphorescence peaks, which are calculated on excited state minimum energy geometries, obtained through excited-state optimizations and hence require excited-state gradients.
- Finding the degree of localization of the excited-state density (i.e., the localization of the excited electron and hole) calculated at the ground-state or excited-state minima.
- Locating excited-state energy barriers and calculating their height, the calculation of which requires at least two evaluations of excited-state frequencies (one to start the transition state search from and one to verify that a proper transition state is found).
- Locating other excited-state surface features (e.g., conical intersections).
- Full-excited-state molecular dynamics (MD).

9.3 Methods

In this section, we will review the desired properties of the ideal excited-state method and important characteristics of the main excited-state methods.

9.3.1 Desired Properties of the Ideal Excited State Method

Based on the discussion and general considerations, we can define a number of properties that an ideal excited-state method should preferably have:

1. To yield results that lie numerically close to those that would be obtained by solving the exact Schrödinger equation
2. The ability to treat excited states of any multiplicity on an equal footing
3. The ability to treat all points on the excited-state surface (i.e., conical intersections) similarly well
4. The presence and implementation of excited-state analytical gradients, allowing for an efficient exploration of the excited-state energy surfaces away from the ground-state minimum energy geometry
5. The presence and implementation of second derivatives, allowing for the verification of excited-state minima
6. For the computational effort of calculating excited-state energies, spectra and analytical gradients to scale sufficiently favorably with system size so that true nanostructures can be studied

No practical method in reality combines all these desired properties and the choice of a method generally involves a trade-off.

Desired property (1) is specifically raised with reference to the theoretical ideal of a solution to the exact many-electron Schrödinger equation and not to experimental data, as in practice, there might be uncertainty about the structure responsible for the observed optical signature. Also, unresolved vibrational broadening, in the case of larger nanostructures, or vibronic peaks, when comparing to molecular species in the gas phase, makes a direct comparison between theory and experiment difficult.

Chapter 9

Property (1) might also be system dependent where the fact that a method gives accurate results for one class of system does not necessarily mean that it should be expected to yield good results for all possible classes of systems.

9.3.2 Density Functional Theory

Density functional theory (DFT), the stalwart of contemporary computational chemistry and nanoscience, is essentially a ground-state theory. However, even so, DFT can be used in three ways to model specific aspects of excited states.

First, the energy gaps between the highest energy occupied orbital (i.e., the top of the valence band) and unoccupied virtual orbitals can be thought of as an approximation to the true excitation energies. This approximation can be justified by the fact that, in contrast to Hartree–Fock theory, in DFT, when using a purely density-dependent exchange-correlation functional (e.g., LDA or GGA), the unoccupied orbitals are optimized in the same field of $N - 1$ electrons (where N is the number of electrons in a system) as the occupied orbitals. However, in contrast to the case of LDA/GGA the situation for the commonly used hybrid exchange-correlation functionals is less clear. Please see a recent article by Baerends and coworkers[5] for a more in-depth discussion. The modeling of excited states as orbital energy differences is probably the most computationally efficient method imaginable as the (unoccupied) orbitals are essentially the by-products of the ground-state total energy DFT calculation. However, as excited states are never explicitly calculated, there are no excited-state gradients or second derivatives, ruling out excited-state optimizations. Also, it is impossible to distinguish excited states of different multiplicities in this approximation.

A second approach, delta SCF (ΔSCF), exploits the fact that for excited states with a different multiplicity than the ground state, the relevant lowest excited state (e.g., the lowest triplet excited state, T_1, for a system with a singlet ground state) can be obtained by a simple DFT total energy calculation for that multiplicity. Here the energy difference between that excited state and the ground state at the same geometry is the excitation energy. Similarly, when using symmetry, also the lowest excited states of the same multiplicity as the ground state, but belonging to another irreducible representation, can be obtained. Excited-state optimizations and frequency calculations are within ΔSCF no different from their ground-state equivalents and thus possible with most standard quantum chemistry and solid-state physics codes, and at computational cost not dissimilar to that of the ground state. Calculations on excited states with the same multiplicity (and irreducible representation) as the ground state are, however, inherently not feasible, and the same is true for higher excited states of any symmetry and multiplicity.

Finally, a combination of orbital energy differences and ΔSCF can be used to estimate the quasiparticle gap (ΔE_{QP}). By definition, ΔE_{QP} is equal to the difference between the ionization potential and the electron affinity: $\Delta E_{QP} = IP - EA = E_{N-1} + E_{N+1} - 2 E_N$ (where E_N is the DFT energy of a system with N electrons and E_{N+1} and E_{N-1} that of systems with one electron added and one removed, respectively). In practice, the application of the standard approximate exchange-correlation functional, in contrast to the unreachable ideal of the exact functional, will inherently introduce an error in the calculated ΔE_{QP}. This is especially true as there is also a technical issue with the calculation of IP values for nanostructures where the highest occupied orbital is degenerate.

Here the application of currently known exchange-correlation functionals might result in nonphysical symmetry breaking. An alternative approximation that circumvents the explicit calculation of IP is $\Delta E_{QP} = \varepsilon_{HOO(N+1)} - \varepsilon_{HOO(N)}$ (where $\varepsilon_{HOO(N+1)}$ and $\varepsilon_{HOO(N)}$ are the Kohn–Sham energies of the highest occupied orbital of a systems with N + 1 and N electrons, respectively). Finally, Kroonik and coworkers have recently reported a computational scheme in which the Kohn-Sham gap, the energy difference between the highest occupied and lowest unoccupied orbitals, is formally equivalent to the quasiparticle gap instead of the optical gap.[6]

9.3.3 Time–Dependent DFT

Time-dependent DFT (TD-DFT) is the excited-state version of conventional ground-state (or static) DFT. Based on the Runge–Gross theorem,[7] the time-dependent extension of the Hohenberg–Kohn theorem that underlies DFT, TD-DFT would be correct if one knew the exact exchange-correlation functional. However, just like DFT, in practice, the exact exchange-correlation functional is not known and approximate exchange-correlation functionals have to be used. Within TD-DFT, the exchange-correlation functional should formally be frequency dependent, but in reality, most practical calculations make the adiabatic approximation and standard DFT exchange-correlation functionals without frequency dependence are used instead.

Adiabatic TD-DFT using such standard functionals is found to work generally very well (though the majority of studies into TD-DFT performance focus on organic molecules[8]) but struggles with double-excitations, charge-transfer excitations, and Rydberg states. The latter two problems are inherently linked to the fact that the functional derivative of most standard functionals (the exchange-correlation potential) does not display the correct 1/r asymptote, something that can be (partly) compensated by using either range-separated functionals or by adding a correction term that enforces the correct asymptote. A useful check for the potential of charge-transfer-related problems for a particular system is the ∧ indicator of Peach and coworkers,[9] although one should be careful with the reported boundary values, as they were determined for organic molecules.[9] Generally in inorganic nanostructures, especially those made out of relatively more ionic materials, an excited electron will always transfer from one chemical sublattice to another (e.g., from sulfur to zinc, in zinc sulfide). Even local excitations in inorganic materials will, when compared with local excitation in organic molecules, always have partial charge-transfer character and hence relative low ∧ values.[10–12] Finally, it is known that standard adiabatic TD-DFT cannot accurately reproduce the precise topology of the energy surfaces around a CX,[13] although that problem does not necessarily extend to locating CXs, predicting their approximate geometry and the qualitative dynamics around them.[14,15]

The TD-DFT problem can be solved either via explicit time propagation of the time-dependent Kohn–Sham equations or through a linear-response approach. The explicit time-propagation method allows one to easily calculate spectra over wide energy ranges and to perform simulations for systems in intense laser fields but is impossible to consider one specific excited state. The strong point of the linear response formalism, solving the Casida equation,[16] is that it is incorporated in most standard quantum chemistry codes and that one can consider specific excited states and thus calculate excited-state gradients and frequencies. Spectra calculations within the linear response formalism

Chapter 9

are perfectly feasible but the fact that all excited states up to a certain lid energy have to be included in the calculations can make them rather memory intensive. Analytical TD-DFT gradients have been implemented in a number of quantum chemistry codes, while the derivation of analytical second derivatives for TD-DFT has also recently been reported.[17] Finally, there is also a tight-binding version of TD-DFT that trades in accuracy for system-size and/or speed.[18]

9.3.4   Correlated Wavefunction–Based Methods

Correlated wavefunction methods are approximations to the exact nonrelativistic many-electron Schrödinger equation that add electron correlation on top of the Hartree–Fock wavefunction, where the Full Configuration Interaction (FCI) method in the basis set limit yields formally exact results. FCI, especially with large basis sets, is, in practice, numerically intractable for everything but two to three light atom systems, so methods that make further approximations are required. These methods can be based on a single reference (e.g., equation-of-motion-coupled cluster singles and doubles [EOM-CCSD] and equation-of-motion-coupled cluster single doubles and triples [EOM-CCSDT] methods) or a multireference wavefunction (e.g., complete active space second order perturbation theory [CASPT2], multireference configuration interaction [MRCI], and spectroscopy-oriented configuration interaction [SORCI] methods).

Multireference methods include full correlation for a limited number of electrons and orbitals (the active space), while single reference methods include limited correlation for all electron and orbitals. While for organic systems the choice of the active space is usually based on chemical intuition, multireference methods can be troublesome in the case of inorganic nanostructures. In the latter case, it is not necessarily clear-cut what to include in the active space and the safe option of including all valence orbitals and electrons is often numerically intractable. Single reference methods are by comparison much more black box, with the excitation level as the only (main) option. Here the excitation level refers to the types of electron excitations included in the electron correlation calculation, for example, singles and doubles (SD) and singles, doubles, and triples (SDT), where including all possible excitations would correspond to a FCI calculation. In the case of coupled cluster methods, (EOM-)CCSD normally yields very accurate results that are, barring some spurious error cancelation for selected systems, only improved upon the addition of triples in (EOM-)CCSDT. Multireference methods with a well-chosen active space can treat all points on the excited energy surface on an equal footing. Single-reference methods, however, struggle with multiconfigurational situations such as CXs, an issue that can be (partially) compensated by going to higher excitation levels. Ground-state CCSDT, for example, gives results very close to FCI result for multiconfigurational model systems.[19] Both approaches thus have their advantages and disadvantages, but importantly are generally numerically too expensive for real (>1 nm) nanosystems. The use of correlated wavefunction methods therefore is generally limited to their use in generating benchmark values to compare TD-DFT with systems containing just a few atoms (e.g., up to 10 atoms). Possible exceptions are approximate coupled cluster (like) methods such as CC2, CC3, ADC(2), and ADC(3).

9.3.5   Many-Body Perturbation Theory Methods: Green's Function and Bethe–Salpeter

Many-body perturbation theory (MBPT)[20–28] is generally considered as the physics equivalent of correlated wavefunction methods from computational chemistry. Here typically first the free electron and hole state are calculated using a method that employs one-particle Green's functions (GW) and in the second step, excitonic effects are introduced by solving the Bethe–Salpeter Equation (BSE) on top of the Green's function result. GW calculations normally start from a converged DFT calculation and different levels of approximations exist in terms of what parts of the equations are solved self-consistently,[23–26,28] where the most commonly used GW implementation G_0W_0 is the so-called one-shot method without any self-consistency. Analytical gradients are available for both GW and BSE and recently also BSE calculations that involved a CX (in the case of an organic molecule) have been reported.[29]

9.4   Applications

We will now review the applications of the computational methods, outlined earlier, to the excited-state properties of nanostructured systems. This review is by no means meant to be exhaustive but rather focuses on illustrating what can be done and what are the (remaining) challenges.

9.4.1   Quasiparticles Spectra and Band Structure

Quasiparticle spectra and the ΔE_{QP} (the lowest excitation in the quasiparticle spectra) correspond, as outlined in Section 9.2.1, to excitations that would create free electron and holes, in the absence of excitonic effects. Such spectra are only an approximation to the true vertical absorption spectra discussed in the next section, in which excitonic effects are presumed to be negligible. Calculations of the quasiparticle spectra are still very useful for a number of reasons. First, even though quasiparticle spectra do not correspond to the result of a simple optical absorption experiment, they can still be measured experimentally, for instance, by means of the combination of photoelectron and inverse photoelectron spectroscopy or scanning tunnel spectroscopy. Second, quasiparticle spectra can be used as a theoretical starting point for further calculations including excitonic effects (e.g., when solving the BSE, see the next section). Finally, as discussed in Section 9.2.1, the ΔE_{QP} can be used to estimate the size of the EBE.

The most theoretically well-defined method for calculating quasiparticle spectra is the Green's function–based GW method (see Section 9.3.5). Systems studied with this many-body approach include nanoclusters (0D), nanoribbons and nanotubes (1D), and thin films (2D), including truly nanosized systems (e.g., in the 0D case 1.2 nm $Si_{41}H_{60}$,[30] 1.2 nm $Ge_{35}H_{36}$,[31,32] and 1.6 nm $Ge_{87}H_{76}$[32]). Most of these calculations use the simplest approximation to GW (G_0W_0), which is known for bulk materials to lead to an underestimation of the ΔE_{QP}.[23] An estimate of just the ΔE_{QP} can also be obtained within the DFT framework through the ΔSCF procedure discussed in Section 9.3.2. ΔE_{QP} values obtained from ΔSCF generally follow the same trends as those obtained from GW, but their absolute magnitudes have been reported to be slightly larger[32,33] or smaller[32,34] than their GW counterparts (it is difficult to make general observations as the ΔSCF result will be functional and system dependent).

9.4.2 Vertical Absorption Spectra

Vertical absorption spectra (or optical absorption spectra) are, as discussed in Section 9.2, a key optical property of inorganic nanostructures. As such, spectra are simple to obtain experimentally (compared to, e.g., more complicated luminescence spectra); they are measured for almost all relevant nanostructured systems. They are also the starting point for understanding more complicated phenomena involving excited-state relaxation. Figure 9.3 shows an example of the absorption spectrum of a $Zn_{12}S_{12}$ nanoparticle, predicted with TD-B3LYP.

Because of its low computational cost, there is an absolute multitude of published studies that use the DFT orbital energy approximation to obtain estimates of the lowest excitation energy (ΔE_{OP}) for inorganic nanostructures. In those studies, the energy difference between the DFT HOO and LUO is commonly referred to as the nanostructure's band gap. The term band gap, however, historically has also been used for the ΔE_{QP}, discussed earlier, and so has to be carefully used in the context of DFT orbital energy differences. Because orbital energies are an inherent by-product of a ground-state DFT calculation, as outlined earlier, these estimates can be obtained for any nanostructure for which ground-state DFT calculations are currently tractable (i.e., for systems comprising of hundreds or even thousands of atoms at the moment). Examples from the literature include large hydrated (e.g., $(TiO_2)_{122}(H_2O)_{58}$[35] and $(TiO_2)_{449}(H_2O)_8$[36]) and hydrogen-terminated TiO_2 clusters (e.g., $(TiO_2)_{411}H_{16}$[37] and CdS nanorods (e.g., $(CdS)_{91}$[38]). Besides the HOO-LUO

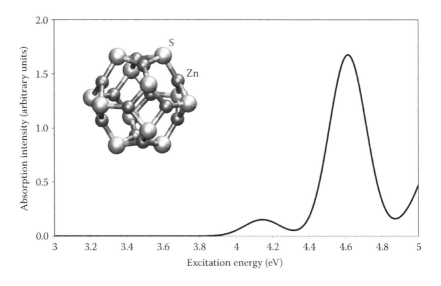

FIGURE 9.3 TD-B3LYP/DZ(D)P calculated optical absorption spectrum up to 5.0 eV for the $Zn_{12}S_{12}$ nanoparticle (structure shown in the top left corner; dark and light gray spheres represent Zn and S atoms, respectively). All excitations plotted are represented as Gaussians with a standard deviation of 0.10 eV.

gap, one can, in principle, also obtain estimates of the vertical absorption spectra by combing the orbital energy differences with intensities approximated through Fermi's golden rule,[39] but this is rarely done in practice.

True absorption vertical spectra can be obtained using TD-DFT, although at a slightly higher cost than orbital energy differences. For example, a linear response TD-DFT calculation including the 12 lowest excitations for $Zn_{12}S_{12}$ takes approximately seven times as long as a ground-state calculation for the same clusters, which only yields the orbital energy differences. The calculation of the spectrum in Figure 9.3 with 48 excitations took 55 times as long as the ground-state calculation. As a result of the higher computational cost, the maximum size of systems for which the absorption spectra is calculated using TD-DFT is typically smaller than those studied by means of orbital energy differences, but calculations on true nanosystems are perfectly feasible. Examples of inorganic nanostructured systems studied with TD-DFT in the literature include BN (e.g., fullerene $B_{36}N_{36}$[40]), CdS (e.g., $(CdS)_{12}$ nanocluster[41]), SiO_2 (e.g., the $Si_{23}O_{46}$ nanoparticle[42]), TiO_2 (e.g., the uncapped $Ti_{10}O_{20}$[10,43] and hydrated $Ti_{23}O_{46}(H_2O)_{34}$[44] nanoparticles), ZnO (e.g., unsaturated $(ZnO)_{6-72}$ nanorods[45] and hydrated $ZnO_{111}(H_2O)_{12}$[46,47] nanoparticles[23,29]), and ZnS (e.g., uncapped $(ZnS)_{26}$[48] and hydrated $(ZnS)_{111}(H_2O)_{12}$[47] nanoparticles). For the linear response version of TD-DFT (see Section 9.3.2), the memory requirements increase with both the size of a system and the number of excitations included. In practice, for larger systems, there is often a trade-off between the system size and the number of excitations in a spectrum that one can calculate. As with ground-state DFT, one has to make a choice for a specific functional. All TD-DFT calculated absorption spectra will be, to a smaller or larger degree, functionally dependent. Validation of TD-DFT results, where possible, with correlated wavefunction-based results (see the following) for small model clusters is therefore always advisable.

Tight-binding TD-DFT (TB-TD-DFT)[18] is computationally cheaper than full TD-DFT but introduces a further approximation. As a result of its lower computational cost, calculations on large systems are tractable (e.g., $Si_{199}H_{140}$)[49] with TB-TD-DFT. Most work seems to have focused for the moment on organic or main group systems, probably because every element requires a new parameterization. The ΔSCF method, finally, is also relatively computationally cheap compared to TD-DFT. As ΔSCF, however, can only be used for states with a multiplicity different from the ground state, states that are inherently dark as excitation into them is spin forbidden, it is not very useful for optical spectra calculations.

Correlated wavefunction-based methods are generally too computationally expensive to calculate vertical excitations for true nanosized systems, with the possible exception of the approximate coupled cluster (like) methods discussed in Section 9.3.4 (e.g., CIS(D) and CC2). Calculation of spectra with correlated wavefunction-based methods, rather than mere excitation energies, is further complicated by the fact that the calculation of oscillator strengths often requires an additional effort. Correlated wavefunction-based methods are, however, very useful to validate TD-DFT results by comparing their results for the lowest excitation energies of small model clusters with TD-DFT results. This has been done, for example, for ZnS (MRCI, CASPT2, CC2)[50-54] and TiO_2 (CASPT2, EOM-CCSD, EOM-CCSDT).[55-58] It is best to focus in such a comparison on the relative energetic ordering of the different excited states (both within a cluster and in between different clusters) as the absolute magnitude of excitation energies calculated with popular correlated wavefunction methods (e.g., CC2, EOM-CCSD,

Chapter 9

CASPT2) do not necessarily fully converge with respect to FCI values. For example, the lowest excitation energies of the TiO_2 monomer decrease by ~0.25 eV when improving the correlated wavefunction description from EOM-CCSD to EOM-CCSDT (see Figure 9.4).[58] Approximate coupled cluster (like) methods have been found to give accurate predictions for some systems[28,59,60] but are known to fail dramatically for other systems.[52,56]

Solving the BSE (BSE-GW) on top of GW is an alternative method for calculating the vertical absorption spectrum. The BSE-GW many-body method has its origin in the solid-state physics community, where it is routinely used to predict the optical absorption spectrum of crystalline solids, including excitonic effects. Examples of the applications of the BSE-GW methods to nanostructured materials are limited but include studies of 0D systems, such as hydrogenated silicon clusters (e.g., $Si_{41}H_{60}$[30], $Si_{35}H_{36}$[61]) and 1D systems, such as silicon nanowires,[62] SiC,[63] and GaN[64] nanotubes.

As discussed in Section 9.2.1, the true absorption spectrum of a nanostructure will contain, besides the peaks arising from vertical excitations, also vibronic peaks, arising from nonadiabatic excitations that involve a vibrationally excited version of, either or both, the ground and excited states. Methods to calculate spectra including these vibronic peaks exist[65] and have been routinely applied to molecules. However, we are not aware of any calculations on nanoparticles that attempt to include vibronic progression, excluding the organic dyes adsorbed on and/or embedded in inorganic nanoparticles.[66–69] This is most likely related to the additional cost of including vibrational peaks in the calculated spectrum (e.g., ground and excited-state gradients need to be calculated) and the fact that they are unlikely to be individually resolvable for nanostructures in experiment and thus will only show up as vibrational broadening of the main peak instead.

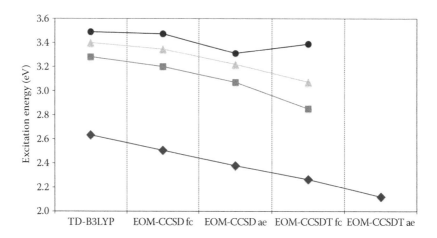

FIGURE 9.4 Comparison in the trend for the four lowest excitations of the TiO_2 monomer as calculated with different method combinations (1^1B_2 diamond markers, 1^1A_2 square markers, 2^1B_2 triangle markers, and 2^1A_1 circle markers). All the methods used a triple-ζ basis set. Fc and ae are the two different approximations used in the EOM-CC calculations and stand for frozen core and all electrons, respectively.

9.4.3   Photoluminescence and Excited State Relaxation

As discussed in Section 9.2.1, after absorption of light, the excited system is generally not in a stationary state and will start to relax and lower its energy. During this relaxation, the nanostructure can, for example, relax to an excited-state minimum from where it can de-excite back to the ground state (luminescence). Alternatively, it could also cross energetic barriers on the excited-state surface and end up in a CX, from where it can cross without emitting any radiation to the ground state or another excited state. What all these processes have in common is that the nanostructure geometry is not fixed but moves on the excited-state energy surface along a path determined by the excited-state nuclear gradients. Hence, the computational effort involved in the calculation of these excited-state gradients effectively determines the size of systems for which excited-state relaxation can be studied computationally. The availability of analytical gradients in this context is an important advantage, as even for moderately sized systems, the effort of calculating $2*(3N - 6)$ finite differences to obtain the gradients numerically is much larger than calculating them analytically. Because of the additional computational cost of describing excited-state relaxation, the number of computational papers studying phenomena that involve nuclear movement on excited-state energy surfaces is much smaller than the amount of papers that merely calculate vertical spectra. Successful examples include the study of MgO nanoparticles with ΔSCF,[70-72] ZnS nanoparticles with TD-DFT,[59,60] silica nanoparticles with TD-DFT,[42] and silicon nanostructures with TB-TD-DFT[49,73-75] and TD-DFT.[76,77]

A specific relevant weak point of TD-DFT in the context of excited-state relaxation are charge-transfer excitations, where there is a small or negligible overlap between the orbital where an electron gets excited from and the orbital, which it gets excited to. Local functionals (e.g., generalized gradient approximation functionals) and hybrid functionals with relatively low percentages of Hartree–Fock-like exchange (HFLE) will underestimate the excitation energies of charge-transfer excitations and may lead to an overall wrong ordering of excited states. Since these charge-transfer states by definition have low or negligible oscillator strengths, as, for example, observed in our recent work on $(TiO_2)_n$ nanoparticles,[10] the underestimation of charge-transfer states is not necessarily an issue for the calculation of optical absorption spectra. For excited-state relaxation, however, where, as canonized in Kasha's rule, most dynamics occurs on the lowest energy excited-state surface, the underestimation of charge-transfer state can become problematic if this leads to the wrong prediction of the electronic nature of the lowest excited state, as recently observed by us for (hydrated) TiO_2 nanoparticles.[44,78] As outlined in Section 9.3.3, a useful check for such problems is the Λ indicator of Peach and coworkers. Range-separated functionals (e.g., CAM-B3LYP) and hybrid functionals with a high percentage of HFLE minimize problems with underestimation of charge-transfer states but at the expense of yielding very high excitation energies. The ΔSCF method does not suffer from charge-transfer problems, but has a weak point: it cannot be used to study the excited-state energy surface of states with the same electronic multiplicity as that of the ground state. For example, in the case of nanostructures with a closed-shell singlet ground state, the ΔSCF method can formally only be used to describe the relaxation of dark triplet states relevant for phosphorescence. Alternatively, one can assume that the singlet and triplet excited states behave alike and that the singlet–triplet splitting for all relevant geometries is small and that the ΔSCF result is thus one-to-one predictive for the lowest singlet excited state, but this assumption is not necessarily warranted.

The basic tool to explore the excited-state energy landscape is energy minimization. Starting from the ground-state minimum energy geometry, this allows one to find the excited-state minima that are connected via an energetic downhill path to the ground-state minimum and hence are likely fluorescence (singlet) or phosphorescence (triplet) sources (e.g., see Figure 9.5 for a comparison between the ground and excited-state minimum energy geometries of $Zn_{12}S_{12}$). Excited-state frequency calculations can be used to verify that proper excited-state minima have been found. However, as the required second derivatives, at present, are normally obtained through numerical differentiation, excited-state frequency calculations are rather costly: this makes it perhaps a nonroutine tool for larger nanostructures. One thing to be careful of during excited state optimisations is that the ground state minimum energy geometry is typically an n-dimensional hyperpoint on the (lowest) excited state energy surface,[52,79] and that hence typically different excited state minima can be reached in an energetic downhill fashion from the ground state minimum energy geometry.[52,60,79] There might be a strong dependence on the direction of the initial geometry step of the energy minimization. One potential

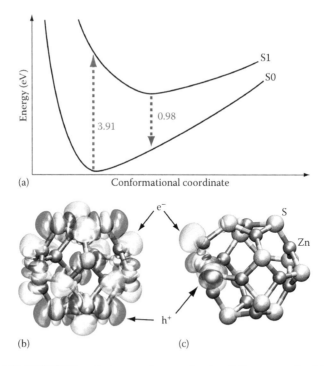

(a)   Conformational coordinate

(b)   (c)

FIGURE 9.5 TD-B3LYP/DZ(D)P excited-state electron density difference at the (b) ground-state and (c) excited-state minimum geometries for the $Zn_{12}S_{12}$ nanoparticle. A cartoon-like description of the S0 and S1 energy surfaces (a) and the TD-B3LYP/DZ(D)P excited-state ground-state electron density difference at the ground-state (b) and excited-state (c) minimum energy geometries for the $Zn_{12}S_{12}$ nanoparticle. Panel (a) illustrates the processes taking place at (b) vertical absorption from the ground-state minimum (3.91 eV) and (c) photoluminescent radiative emission from the excited-state minimum (0.98 eV). The green lobes illustrate the regions where the excited electron localizes, while the blue ones show where the hole is predicted to be. Silver and gold spheres represent Zn and S atoms, respectively.

way of sampling this is to perform a number of minimizations runs for every structure, where in every run, the structure is either distorted along a different random combination of ground-state normal modes[52,79] or corresponds to a snapshot of ground-state MD. In a way, these procedures mirror the fact that in experiment, the initial ground-state structure is likely to be slightly distorted away from the ideal minimum energy geometry due to vibrations. Most studies, irrespective of the chemical composition, seem to find rather large Stokes shift (differences between the lowest vertical excitation energy and the luminescence energy) of up to 2–3 eV for small (~1 nm) nanoparticles. The TB-TD-DFT studies on hydrogen-terminated silicon nanoparticles[46,73] find that upon increasing the nanoparticle size, the Stokes shift gets smaller and effectively appears to become negligible for particles that are larger than 2 nm. This reduction in Stokes shift might be a more fundamental property of larger nanoparticles. However, except for silicon nanoparticles, the >1 nm size range is effectively unchartered with respect to excited-state relaxation. For the moment, TB-TD-DFT appears to be the only method that scales sufficiently well with system size to probe excited-state nuclear dynamics for particles with more than 200 atoms (e.g., $Si_{199}H_{92}$).

Excited-state relaxation in inorganic nanostructures is typically associated with large changes in the excited-state electron density and localization of the excited state. Figure 9.5 shows, for example, the charge density difference of the ground and excited-state density at the ground and excited-state geometries of $Zn_{12}S_{12}$, where green represents an excess of electron density in the excited state (the excited electron) and blue a lack of electron density in the excited state (the hole). Clearly, while at the ground state geometry the excited state (panel b in Figure 9.5) is delocalized over the whole particle, it is only localized on one Zn_2S_2 four-membered ring in the excited state energy minimum (panel c in Figure 9.5). The excited electron and hole can localize together in close proximity, as is the case for $Zn_{12}S_{12}$ and ZnS nanoparticles in general, or in different parts of the cluster.

Beyond the excited state minima that are connected by an energetic downhill path to the ground-state geometry, there might be other excited-state minima or CXs that require crossing of excited-state barriers. For example, in a recent work on ZnS clusters, we found cascades of excited-state minima, separated by low barriers, that end up in a CX with the ground state.[60] Barrier heights can be obtained through conventional transition state searches on the excited-state surface, starting from an initial guess obtained through linear interpolation between the two adjacent minima. A practical bottleneck for these transition state searches is that two excited-state frequency calculations are required: one to start the search and another at the end to verify that the search has indeed converged to a transition state. A more important problem is how to systematically find the relevant excited-state minima beyond those directly connected to the ground-state minimum structure by an energetic downhill path and the barriers between them. In the work on ZnS, the excited-state minima cascade was essentially discovered by a combination of serendipity and analogy between different particle sizes. A possible alternative would be to explore the excited-state energy surface using MD. There, one can use either conventional Bohr–Oppenheimer MD (BO-MD) on the excited-state energy surface or the surface-hopping approach to nonadiabatic *ab initio* molecular dynamics (SH-MD), which explicitly considers hopping between different (excited state) energy surfaces during the MD run. Both of these approaches have been implemented in the context of TD-DFT[15,80] but their use is for the moment only numerically tractable for just the smallest particles

Chapter 9

(also as a large number of independent runs will be required). Moreover, careful thought is required on how to properly include the dissipation of heat to the environment in MD calculations. A rare example of the application of (BO-)MD/TD-DFT to inorganic nanostructures can be found in Reference 60. More advanced excited-state processes such as the photoinduced modification of inorganic nanoparticles (e.g., emission of neutral metal emission after light absorption) can also be modeled using TD-DFT and/or the ΔSCF method but might need a combination of different calculations (see, e.g., References 70, 71, and 72 in the case of MgO).

9.4.4   Charge Transfer between Nanoparticles and Adsorbed Molecules

Charge transfer of excited electrons and/or holes between an inorganic nanostructure and an organic molecule adsorbed on the nanostructure's surface is a crucial physical process in applications like dye-sensitized solar cells (DSSCs) or photocatalysis. We will illustrate the role of modeling charge transfer by focusing on DSSCs in the remainder of this section (for a discussion in the context of photocatalysis see References 44 and 81). DSSC technology combines a nanostructured semiconductor electrode (e.g., TiO_2, ZnO) with a fundamental band gap that is too large for direct excitation by visible light, with a dye that absorbs light in the visible region and transfers the excited electron to the inorganic nanostructure. The electron transfer can be direct or indirect in nature. In the case of the direct mechanism, the electron is excited from the HOO of the dye into the LUO of the semiconductor. In the indirect mechanism, in contrast, the excitation is completely localized on the dye (i.e., an electron is excited from the HOO of the dye in the LUO of the dye), and the excited electron is only subsequently injected to the conduction state of the inorganic material.

The modeling of charge transfer and other relevant processes in DSSC is a very challenging task for a number of reasons. Fundamentally, there are electronic differences between the two components of the interface. The dyes generally exhibit localized electronic states, while the inorganic semiconductors typically have highly delocalized electronic states.[82,83] More practically, as discussed earlier, the properties of inorganic nanoparticles can depend on their size (i.e., on quantum confinement effects); therefore, the study of large systems (~2 nm) is fundamental for an accurate comparison with experimental values. Furthermore, a realistic model of a DSSC would also have to take into account the role of the solvent and the redox species (e.g., I^-/I_3) on the alignment of the dye's level with the semiconductor band edges. For all these reasons, theoretical investigations of DSSCs generally focus on individual aspects of the problem, for example, the modification of the rates of interfacial charge transfer and charge separation, simple processes that will finally affect the efficiency of the devices.

Due to the complex nature of DSSC, the inorganic part of the systems is generally approximated by simple structures (bulk cuts of restricted size, generally nonrelaxed) and the solvent effects (if included) are considered implicitly through a polarizable solvation model (e.g., C-PCM). One of the earliest examples of the study of charge transfer in DSSC was an investigation of the photoinjection mechanism in a catechol-sensitized $(TiO_2)_{38}$ nanoparticle.[84] This study was performed using CIS calculations on top of a semiempirical Hamiltonian (INDO). More recent studies used (TD-)DFT to predict the optical properties and the character of the charge-transfer mechanism in the $Ti_9O_{18}H\text{-}O\text{-}TCNQ^-$ system,[85] the HCat-H$(TiO_2)_6$ and

HCat-H(TiO$_2$)$_{15}$ systems,[86] alizarin supported on (TiO$_2$)$_n$ (with n = 1, 2, 3, 6, 9, 15, 38),[87] and a series of different N749 dyes (prototypical form of Ru(II) complexes) anchored to the (TiO$_2$)$_{28}$ nanoparticle.[88] De Angelis and coworkers have developed an approach for the prediction of the efficiency of dye-sensitized systems, which is based on the calculation of the alignment of the dye's molecular levels and the semiconductor band edges, and the ground-state and excited-state oxidation potential (GSOP and ESOP respectively).[89-94] To date, the largest computational investigations on dye-sensitized materials consist of a molecular dynamic simulation of the squaraine dye absorbed onto a (TiO$_2$)$_{16}$ nanoparticle in the presence of 90 explicit water molecules,[95] and a TD-DFT study of a perylene dye absorbed on a (ZnO)$_{222}$ nanoparticle (~2–5 nm).[96] All the studies mentioned earlier, and nearly all work in the literature, employ a combination of (TD-)DFT with conventional semilocal and hybrid functionals. However, as discussed in Sections 9.3.1 and 9.3.2, this approach suffers from a number of potential deficiencies. A rare example of a study that goes beyond (TD-)DFT is the application of MBPT (G$_0$W$_0$) to simple TiO$_2$ systems. The systems, Ti$_2$cat$_2$ and Ti$_3$INA$_3$, even if much smaller than the particles used in actual DSSCs, employ all the essential physics principles (bar the solvent) of the local interaction between the dye and the TiO$_2$ clusters.[97]

9.5 Perspectives

As sketched in this chapter, the excited-state properties of true inorganic nanostructures can nowadays be successfully modeled. A number of important challenges, however, remain. Most systems studied computationally are still small relative to the experimental systems they aim to mirror (~1–2 nm vs. ~10–100 nm). Also, the accuracy of commonly used methods such as TD-DFT can sometimes be insufficient or, perhaps worse, unknown, while the computational scaling of inherently more accurate methods is generally not favorable enough to routinely study true nanostructures on state-of-the-art supercomputers. Finally, simulation methods for more complex excited-state processes (as discussed in Sections 9.4.3 and 9.4.4) are only just starting to be developed and we are thus entering a promising but largely unknown territory. The future will have to tell us what is possible and/or tractable.

Acknowledgments

We kindly acknowledge Dr. M. Calatayud, Prof. S.T. Bromley, Prof. C.R.A. Catlow, Prof. F. Illas, Dr. A. Kerridge, Dr. K. Kowalski, Dr. S. Shevlin, Prof. A. Shluger, Dr. A.A. Sokol, Dr. C. Sousa, Dr. M. van Setten, M. Wobbe and Dr. S.M. Woodley for stimulating discussion. M.A.Z. acknowledges the UK Engineering and Physical Sciences Research Council (EPSRC) for a Career Acceleration Fellowship (Grant EP/I004424/1), while the preparation of this chapter has further been supported by a UCL Impact studentship award to EB.

References

1. Kasha, M. *Discuss. Faraday Soc.* 1950, *9*, 14.
2. Beer, M., Longuethiggins, H. C. *J. Chem. Phys.* 1955, *23*, 1390.
3. Choi, C. L., Li, H., Olson, A. C. K., Jain, P. K., Sivasankar, S., Alivisatos, A. P. *Nano Lett.* 2011, *11*, 2358.
4. Bernardi, F., Olivucci, M., Robb, M. A. *Chem. Soc. Rev.* 1996, *25*, 321.

5. Baerends, E. J., Gritsenko, O. V., van Meer, R. *Phys. Chem. Chem. Phys.* 2013, *15*, 16408.
6. Kronik, L., Stein, T., Refaely-Abramson, S., Baer, R. *J. Chem. Theory Comput.* 2012, *8*, 1515.
7. Runge, E., Gross, E. K. U. *Phys. Rev. Lett.* 1984, *52*, 997.
8. Jacquemin, D., Wathelet, V., Perpete, E. A., Adamo, C. *J. Chem. Theory Comput.* 2009, *5*, 2420.
9. Peach, M. J. G., Benfield, P., Helgaker, T., Tozer, D. J. *J. Chem. Phys.* 2008, *128*, 044118.
10. Berardo, E., Hu, H.-S., Shevlin, S. A., Woodley, S. M., Kowalski, K., Zwijnenburg, M. A. *J. Chem. Theory Comput.* 2014, *10*, 1189.
11. Wobbe, M. C. C., Kerridge, A., Zwijnenburg, M. A. *Phys. Chem. Chem. Phys.* 2014, *16*, 22052.
12. Wobbe, M. C. C., Zwijnenburg, M. A. *Phys. Chem. Chem. Phys.* 2015, 28892.
13. Levine, B. G., Ko, C., Quenneville, J., Martinez, T. J. *Mol. Phys.* 2006, *104*, 1039.
14. Tapavicza, E., Tavernelli, I., Rothlisberger, U., Filippi, C., Casida, M. E. *J. Chem. Phys.* 2008, *129*, 124108.
15. Tapavicza, E., Meyer, A. M., Furche, F. *Phys. Chem. Chem. Phys.* 2011, *13*, 20986.
16. Jamorski, C., Casida, M. E., Salahub, D. R. *J. Chem. Phys.* 1996, *104*, 5134.
17. Liu, J., Liang, W. *J. Chem. Phys.* 2011, *135*, 014113.
18. Niehaus, T. A., Suhai, S., Della Sala, F., Lugli, P., Elstner, M., Seifert, G., Frauenheim, T. *Phys. Rev. B* 2001, *63*, 085108.
19. Yang, K. R., Jalan, A., Green, W. H., Truhlar, D. G. *J. Chem. Theory Comput.* 2013, *9*, 418.
20. Aryasetiawan, F., Gunnarsson, O. *Rep. Prog. Phys.* 1998, *61*, 237.
21. Lars, H. *J. Phys. Condens. Matter* 1999, *11*, R489.
22. Onida, G., Reining, L., Rubio, A. *Rev. Mod. Phys.* 2002, *74*, 601.
23. van Schilfgaarde, M., Kotani, T., Faleev, S. *Phys. Rev. Lett.* 2006, *96*, 226402.
24. Shishkin, M., Kresse, G. *Phys. Rev. B* 2007, *75*, 235102.
25. Shishkin, M., Marsman, M., Kresse, G. *Phys. Rev. Lett.* 2007, *99*, 246403.
26. Rostgaard, C., Jacobsen, K. W., Thygesen, K. S. *Phys. Rev. B* 2010, *81*, 085103.
27. Faber, C., Boulanger, P., Attaccalite, C., Duchemin, I., Blase, X. *Philos. Trans. R. Soc. A* 2011, *372*, 20130271.
28. Caruso, F., Rinke, P., Ren, X., Scheffler, M., Rubio, A. *Phys. Rev. B* 2012, *86*, 081102.
29. Conte, A. M., Guidoni, L., Del Sole, R., Pulci, O. *Chem. Phys. Lett.* 2011, *515*, 290.
30. Ramos, L. E., Paier, J., Kresse, G., Bechstedt, F. *Phys. Rev. B* 2008, *78*, 195423.
31. Pulci, O., Degoli, E., Iori, F., Marsili, M., Palummo, M., Del Sole, R., Ossicini, S. *Superlattice Microstruct.* 2010, *47*, 178.
32. Marsili, M., Botti, S., Palummo, M., Degoli, E., Pulci, O., Weissker, H.-C., Marques, M. A. L., Ossicini, S., Del Sole, R. *J. Phys. Chem. C* 2013, *117*, 14229.
33. Tiago, M. L., Idrobo, J. C., Oguet, S., Jellinek, J., Chelikowsky, J. R. *Phys. Rev. B* 2009, *79*, 155419.
34. Tiago, M. L., Chelikowsky, J. R. *Phys. Rev. B* 2006, *73*, 205334.
35. Zhang, J., Hughes, T. F., Steigerwald, M., Brus, L., Friesner, R. A. *J. Am. Chem. Soc.* 2012, *134*, 12028.
36. Li, Y.-F., Liu, Z.-P. *J. Am. Chem. Soc.* 2011, *133*, 15743.
37. Nunzi, F., Mosconi, E., Storchi, L., Ronca, E., Selloni, A., Graetzel, M., De Angelis, F. *Energy Environ. Sci.* 2013, *6*, 1221.
38. Sangthong, W., Limtrakul, J., Illas, F., Bromley, S. T. *Nanoscale* 2010, *2*, 72.
39. Ratcliff, L. E., Hine, N. D. M., Haynes, P. D. *Phys. Rev. B* 2011, *84*, 165131.
40. Koponen, L., Tunturivuori, L., Puska, M. J., Nieminen, R. M. *J. Chem. Phys.* 2007, *126*, 214306.
41. Gutsev, G. L., O'Neal, R. H., Jr., Belay, K. G., Weatherford, C. A. *Chem. Phys.* 2010, *368*, 113.
42. Zwijnenburg, M. A., Sokol, A. A., Sousa, C., Bromley, S. T. *J. Chem. Phys.* 2009, *131*, 34705.
43. Shevlin, S. A., Woodley, S. M. *J. Phys. Chem. C* 2010, *114*, 17333.
44. Berardo, E., Zwijnenburg, M. A. *J. Phys. Chem. C* 2015, *119*, 13384.
45. Malloci, G., Chiodo, L., Rubio, A., Mattoni, A. *J. Phys. Chem. C* 2012, *116*, 8741.
46. De Angelis, F., Armelao, L. *Phys. Chem. Chem. Phys.* 2011, *13*, 467.
47. Azpiroz, J. M., Mosconi, E., De Angelis, F. *J. Phys. Chem. C* 2011, *115*, 25219.
48. Zwijnenburg, M. A. *Nanoscale* 2011, *3*, 3780.
49. Wang, X., Zhang, R. Q., Lee, S. T., Niehaus, T. A., Frauenheim, T. *Appl. Phys. Lett.* 2007, *90*, 123116.
50. Matxain, J. M., Fowler, J. E., Ugalde, J. M. *Phys. Rev. A* 2000, *62*, 053201.
51. Matxain, J. M., Irigoras, A., Fowler, J. E., Ugalde, J. M. *Phys. Rev. A* 2001, *64*, 053201.
52. Zwijnenburg, M. A., Sousa, C., Illas, F., Bromley, S. T. *J. Chem. Phys.* 2011, *134*, 064511.
53. Azpiroz, J. M., Ugalde, J. M., Infante, I. *J. Chem. Theory Comput.* 2013, *10*, 76.
54. Nguyen, K. A., Pachter, R., Day, P. N. *J. Chem. Theory Comput.* 2013, *9*, 3581.

55. Lin, C.-K., Li, J., Tu, Z., Li, X., Hayashi, M., Lin, S. H. *RSC Adv.* 2011, *1*, 1228.

56. Taylor, D. J., Paterson, M. J. *J. Chem. Phys.* 2010, *133*, 204302.

57. Taylor, D. J., Paterson, M. J. *Chem. Phys.* 2012, *408*, 1.

58. Berardo, E., Hu, H., Kowalski, K., Zwijnenburg, A. M. *J. Chem. Phys.* 2013, *139*, 64313.

59. Zwijnenburg, M. A. *Nanoscale* 2012, *4*, 3711.

60. Zwijnenburg, M. A. *Phys. Chem. Chem. Phys.* 2013, *15*, 11119.

61. Ping, Y., Rocca, D., Galli, G. *Chem. Soc. Rev.* 2013, *42*, 2437.

62. Ping, Y., Rocca, D., Lu, D., Galli, G. *Phys. Rev. B* 2012, *85*, 035316.

63. Hsueh, H. C., Guo, G. Y., Louie, S. G. In *Silicon-Based Nanomaterials*, Li, H., Wu, J., Wang, Z. M., Eds. 2013, Vol. 187, p. 139, New York: Springer.

64. Ismail-Beigi, S. *Phys. Rev. B* 2008, *77*, 233103-1.

65. Egidi, F., Barone, V., Bloino, J., Cappelli, C. *J. Chem. Theory Comput.* 2012, *8*, 585.

66. Avila Ferrer, F. J., Improta, R., Santoro, F., Barone, V. *Phys. Chem. Chem. Phys.* 2011, *13*, 17007.

67. Pedone, A., Prampolini, G., Monti, S., Barone, V. *Phys. Chem. Chem. Phys.* 2011, *13*, 16689.

68. Pedone, A., Bloino, J., Barone, V. *J. Phys. Chem. C* 2012, *116*, 17807.

69. Pedone, A., Gambuzzi, E., Barone, V., Bonacchi, S., Genovese, D., Rampazzo, E., Prodi, L., Montalti, M. *Phys. Chem. Chem. Phys.* 2013, *15*, 12360.

70. Trevisanutto, P. E., Sushko, P. V., Shluger, A. L., Beck, K. M., Henyk, M., Joly, A. G., Hess, W. P. *Surf. Sci.* 2005, *593*, 210.

71. Beck, K. M., Henyk, M., Wang, C., Trevisanutto, P. E., Sushko, P. V., Hess, W. P., Shluger, A. L. *Phys. Rev. B* 2006, *74*, 045404.

72. Beck, K. M., Joly, A. G., Diwald, O., Stankic, S., Trevisanutto, P. E., Sushko, P. V., Shluger, A. L., Hess, W. P. *Surf. Sci.* 2008, *602*, 1968.

73. Wang, X., Zhang, R. Q., Niehaus, T. A., Frauenheim, T., Lee, S. T. *J. Phys. Chem. C* 2007, *111*, 12588.

74. Wang, X., Zhang, R. Q., Lee, S. T., Frauenheim, T., Niehaus, T. A. *Appl. Phys. Lett.* 2008, *93*, 243120.

75. Zhang, R.-Q., De Sarkar, A., Niehaus, T. A., Frauenheim, T. *Phys. Status Solidi B* 2012, *249*, 401.

76. Sundholm, D. *Phys. Chem. Chem. Phys.* 2004, *6*, 2044.

77. Lehtonen, O., Sundholm, D. *Phys. Chem. Chem. Phys.* 2006, *8*, 4228.

78. Berardo, E., Hu, H.-S., van Dam, H. J. J., Shevlin, S. A., Woodley, S. M., Kowalski, K., Zwijnenburg, M. A. *J. Chem. Theory Comput.* 2014, *10*, 5538.

79. Zwijnenburg, M. A., Illas, F., Bromley, S. T. *Phys. Chem. Chem. Phys.* 2011, *13*, 9311.

80. Tapavicza, E., Tavernelli, I., Rothlisberger, U. *Phys. Rev. Lett.* 2007, *98*, 023001.

81. Guiglion, P., Berardo, E., Butchosa, C., Wobbe, M. C. C., Zwijnenburg, M. A. *J. Phys. Condens. Matter.* 2016, in press.

82. Prezhdo, O. V. *J. Phys. Chem. Lett.* 2012, *3*, 2386.

83. Monti, O. L. A. *J. Phys. Chem. Lett.* 2012, *3*, 2342.

84. Persson, P., Bergstrom, R., Lunell, S. *J. Phys. Chem. B* 2000, *104*, 10348.

85. Jono, R., Fujisawa, J.-i., Segawa, H., Yamashita, K. *J. Phys. Chem. Lett.* 2011, *2*, 1167.

86. Sanchez-de-Armas, R., Oviedo, J., San-Miguel, M. A., Sanz, J. F. *J. Chem. Theory Comput.* 2010, *6*, 2856.

87. Sanchez-de-Armas, R., San-Miguel, M. A., Oviedo, J., Marquez, A., Sanz, J. F. *Phys. Chem. Chem. Phys.* 2011, *13*, 1506.

88. Liu, S.-H., Fu, H., Cheng, Y.-M., Wu, K.-L., Ho, S.-T., Chi, Y., Chou, P.-T. *J. Phys. Chem. C* 2012, *116*, 16338.

89. De Angelis, F., Fantacci, S., Mosconi, E., Nazeeruddin, M. K., Graetzel, M. *J. Phys. Chem. C* 2011, *115*, 8825.

90. De Angelis, F., Fantacci, S., Selloni, A. *Nanotechnology* 2008, *19*, 424002.

91. De Angelis, F., Fantacci, S., Selloni, A., Nazeeruddin, M. K., Gratzel, M. *J. Phys. Chem. C* 2010, *114*, 6054.

92. De Angelis, F., Tilocca, A., Selloni, A. *J. Am. Chem. Soc.* 2004, *126*, 15024.

93. Pastore, M., De Angelis, F. *ACS Nano* 2009, *4*, 556.

94. Pastore, M., Fantacci, S., De Angelis, F. *J. Phys. Chem. C* 2010, *114*.

95. De Angelis, F., Fantacci, S., Gebauer, R. *J. Chem. Phys. Lett.* 2011, *2*, 813.

96. Amat, A., De Angelis, F. *Phys. Chem. Chem. Phys.* 2012, *14*, 10662.

97. Marom, N., Moussa, J. E., Ren, X., Tkatchenko, A., Chelikowsky, J. R. *Phys. Rev. B* 2011, *84*, 245115.

Chapter 9

Case Studies

10. Interfaces in Nanocrystalline Oxide Materials

From Powders Toward Ceramics

Oliver Diwald, Keith P. McKenna, and Alexander L. Shluger

10.1 Introduction to Nanocrystalline Oxides 291
 10.1.1 Nanocrystalline Materials and Applications 291
 10.1.2 From Powders to Ceramics 292
 10.1.3 Experimental Probes of Nanocrystals and Interfaces 293
 10.1.4 Magnesium Oxide Nanopowders. 293

10.2 Theoretical Modeling 295
 10.2.1 Multiscale Models of Nanocrystalline Structure 295
 10.2.2 First Principles Prediction of Properties Using Embedded Cluster
  Methods. 298

10.3 Properties of MgO Nanopowders 299
 10.3.1 Properties Inferred from Studies of Isolated MgO Nanocrystals 300
  10.3.1.1 Charge Trapping and Optical Excitations 300
  10.3.1.2 Atom Desorption from MgO Nanocrystals. 301
 10.3.2 Interfaces between MgO Nanocrystals. 303

10.4 Summary 307

Acknowledgments 307

References 307

10.1   Introduction to Nanocrystalline Oxides

10.1.1   Nanocrystalline Materials and Applications

Nanocrystalline metal-oxide materials, such as nanopowders, polycrystalline films, and ceramics, find diverse technological applications in mechanical engineering [1,2], catalysis, energy technology [3], electronics [4,5], and cosmetics [6]. The synthesis of such materials, and the fabrication of devices with new functional properties, is often achieved by the controlled manipulation of microstructures at the atomic level [7]. New approaches to materials design, however, use particle systems as the primary components for microstructure formation. This interdisciplinary field brings together elements of chemistry, solid-state physics, engineering, and materials science [8–11]. Functional nanocrystalline

Chapter 10

materials with particle densities varying between that of a loose nanoparticle aggregate to that of a dense polycrystalline ceramic can be engineered through powder compression and subsequent sintering procedures. The properties of such materials depend on the size, chemical composition, and structure of constituent grains, as well the nature of interfacial regions between grains and intergranular pores. However, probing and controlling the properties and abundance of interfacial regions in nanocrystalline oxides is extremely difficult, which presents a formidable challenge to the design of high-performance functional materials [12].

The formation and properties of nanoparticle interfaces are important for many applications. For most high surface area materials applications, which are determined by adsorption processes and surface reactivity, it is important to keep the concentration of solid–solid interfaces and grain boundaries low. On the other hand, ceramics produced by compacting loose powders contain many interfaces, which affect their functionality. For example, particle-based semiconductor thin films find applications as components for transparent electrodes [4,5] inside photovoltaic devices and for other optoelectronic parts. Here, particle–particle interfaces modify susceptibility to lattice depletion and to n-type doping [13–15], significantly affecting device performance. Printed electronics is another example based on manufacturing electronic components by standard printing techniques, such as ink jet printing or roll-to-roll printing [16]. Processing matters also for the production of highly efficient solar cells, electrochemical storage devices, and electronic devices that rely on the electronic and/or optical functionality of particle systems. Unintended contamination effects during particle synthesis as well as water adsorption at the nanoparticle surfaces at ambient conditions can lead to significant changes of microstructural arrangements of nanoparticle assemblies and, consequently, of resulting photoelectronic properties [13,14]. The achievement of superior device performance requires significant characterization work in order to address how the properties of materials change as a function of processing parameters.

10.1.2   From Powders to Ceramics

Sintering is a critical part of ceramics processing that facilitates the transformation of a loose nanopowder into a dense polycrystalline material. The starting point is an assembly of contacting particles, as shown in Figure 10.1 [17]. The initial interfaces or interfacial spots range from point contacts to highly deformed interfaces. With sintering, the contacts grow in size, and in the initial stage, there is extensive loss of surface area. In parallel to these transformations, the structure of the residual pores becomes rounded and the originally discrete particles are less evident. This is characterized by a tubular, rounded pore structure that is open to the compact surface. The dramatic effect of such structural changes at the nanometer scale on the macroscopic properties is well illustrated by the development of transparent ceramics [18–20]. Residual porosity in wide band gap materials, such as Al_2O_3, ZrO_2 or MgO, leads to substantial loss of visible light transmission via light scattering at the oxide–air interface. For Al_2O_3 a porosity of less than 1% is required to ensure a suitably transparent material for optical applications with excellent resistance to chemical attack at the same time [18,19].

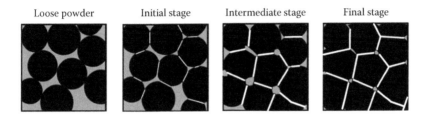

Loose powder   Initial stage   Intermediate stage   Final stage

FIGURE 10.1 From powders to ceramics: Starting with a loose powder, the particle ensemble becomes subsequently sintered in each of the three stages. In the course of this transformation, the initial open pore structure and high porosity become consumed by interparticle neck growth, grain growth, and pore shrinkage, with eventual formation of closed spherical pores in the final stage. (Adapted from German, R.M., *Sintering Theory and Practice*, Wiley-VCH, New York, 1996.)

10.1.3 Experimental Probes of Nanocrystals and Interfaces

Establishing structure–property relationships for nanocrystalline ceramics is particularly challenging [21] and requires the synergy of multiple experimental techniques providing complementary views of a single structure. Transmission electron microscopy (TEM) in conjunction with 3D electron tomography [22–24], high-speed electron backscatter diffraction mapping for the characterization of microstructures in three dimensions [25], and dual-beam focused ion beam scanning electron microscopy (FIB-SEM) has become an established analytical technique for microstructure characterization [26]. It is becoming increasingly possible to study particle systems of variable degrees of consolidation. The 3D interrogation techniques (x-rays, neutrons, FIB-SEM, atom probe) aid in both structure and composition determination at a variety of length scales. Recently developed in situ capabilities [27,28] and the use of complementary characterization methods provide unique insights into the structure formation, functionality, and deformation behavior of complex nanostructures. More sophisticated spectroscopic techniques for the characterization of nanocrystalline materials are positron annihilation spectroscopy [29–31] and EXAFS measurements [32–34]. Related methods are capable of probing the different environments of atoms located inside the crystalline grains and interfacial regions, respectively.

10.1.4 Magnesium Oxide Nanopowders

Synthesis and processing of inorganic materials to create novel microstructures at different length scales is complex [8,35,36]. To test hypotheses about interface-specific functionalities, it is desirable to have a system of nanocrystals with narrow distributions of size, structure, and morphology. This allows one to test the effect of the powder density and, thus, of the concentration of solid–solid interfaces on the overall electronic, chemical, and optical ensemble properties [37]. It has been found that the interfacial regions between MgO nanocrystallines strongly affect the properties of the entire particle ensemble. Elastic properties were found to be markedly affected by the nature of grain boundary networks [38]. Being relevant for the field of heterogeneous catalysis, it is expected that structurally disturbed zones at the grain boundaries areas in densified MgO polycrystals exhibit increased surface basicity. The major enhancement of catalytic activity of densified MgO material caused by significant increase of surface basicity was measured by Knoevenagel condensation or transesterification reactions [39,40].

Particle compression and concomitantly the increase of interfacial area between small particles can either give rise to new optical features or affect the features that appear on uncompressed powders and are related to free particle surfaces [37]. In order to account for such effects, it is necessary to quantitatively assess the optical properties of an entire particle assembly. This is a challenging task since highly dispersed particle powders exhibit enhanced reflectivity. The penetration depths of excitation and emission light strongly depend on the state of aggregation and, thus, on porosity and concentration of particle–particle interfaces in the respective sample volume (Figure 10.2).

An important aspect of solid–solid interfaces and grain boundaries concerns sample preparation for materials characterization. In order to perform spectroscopic measurements, nanoparticle powders are often pressed into pellets, inevitably generating

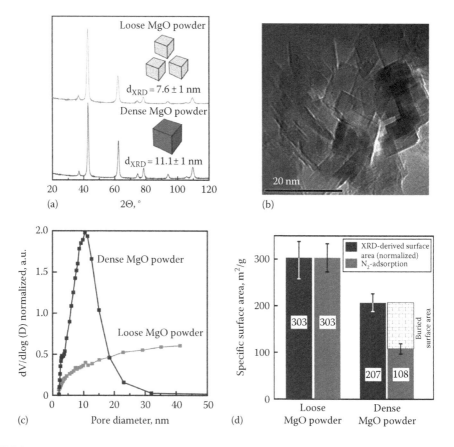

FIGURE 10.2 (a) XRD patterns of loose and dense MgO powder with corresponding average nanocrystal size. (b) Electron microscopy images of chemical vapor synthesis MgO after applying uniaxial pressure of 1.7×108 Pa and subsequent thermal annealing (T = 1170 K, p < 10^{-6} mbar). (c) Distribution in pore sizes for the loose and dense powder as determined from analysis of N_2-adsorption isotherms. (d) Corrected specific surface area either derived from the average nanocrystal sizes or directly determined from the N_2-adsorption isotherms. (Reprinted with permission from McKenna, K.P., Koller, D., Sternig, A., Siedl, N., Govind, N., Sushko, P.V., and Diwald, O., Optical properties of nanocrystal interfaces in compressed MgO nanopowders, *ACS Nano*, 5, 3003. Copyright 2011 American Chemical Society.)

nanocrystal interfaces. Better understanding of how interfaces affect the optical proper-ties may open a way to using optical spectroscopy as a probe of the electronic properties of interfaces between nanocrystals. These interfaces are thought to affect magnetism [41] and the dynamics of electrons and holes under irradiation.

Understanding and predicting these properties requires combining experimental studies with extensive theoretical modeling of the structure, electron and hole trap-ping as well as spectroscopic properties of realistic systems. The MgO crystalline sur-faces and powders are perhaps the best studied systems where the comparison between theory and experiment has been the most extensive. When discussing how the spec-troscopic properties and surface reactivity of nanopowders differ from those of macro-scopic materials, it is common to point to their high specific surface area, the size and shape of the constituent nanocrystals, and the high concentration of surface features, such as steps, kinks, and other low-coordination sites. The external surface of a powder sample has traditionally been thought to be the most exposed to the interaction with gas particles and irradiation. Therefore, the properties of the whole system have long been treated theoretically as a sum of the properties of individual nanocrystals, neglecting their interfaces with other nanocrystals. The development of computational methods recently facilitated more realistic models. In the following text, we briefly review some of the theoretical methods used in these calculations and provide examples of their applications to modeling the properties of MgO powders.

10.2   Theoretical Modeling

10.2.1   Multiscale Models of Nanocrystalline Structure

Computational modeling of the structure and properties of materials ultimately involves computing the interactions between atoms. This can be achieved with various levels of approximation ranging from simple interatomic force fields (e.g., Lennard-Jones) to fully quantum mechanical treatments. The sophistication of the approximation employed determines the quality of the predictions, which can be made and the types of property that can be calculated. However, since more sophisticated approximations are inevitably more expensive computationally, it also determines the size of system that can be con-sidered. The choice of method is usually a trade-off between these two requirements, and in many cases, useful progress can be made through a combination of approaches. For ideal crystals or simple defects, one can often take advantage of periodicity to reduce the number of atoms one needs to consider. Nanocrystalline materials, on the other hand, are inherently nonperiodic, making them more challenging to model.

Broadly, nanocrystalline materials can be defined as materials comprised of a collec-tion of nanocrystals. Therefore, structurally, they can include free surfaces of individual nanocrystals and interfaces between nanocrystals and bulk crystal regions. The relative proportion of these features is an important characteristic of the material. For example, nanopowders consisting of loosely bound aggregates of nanocrystals contain relatively few interfaces and a very high surface area. Interfaces between nanocrystals in such powders may consist of points, lines, or areas of shared contact [42]. On the other hand, dense nanocrystalline ceramics (e.g., produced by sintering a nanopowder) have a very low surface area but many interfaces, which determine many of their properties [43].

Nanocrystals in a ceramic are separated by extended grain boundary defects, which can have very different structures and properties depending on the relative crystallographic orientation of the two grains. To model the structure of such complex materials, it is useful to have a hierarchy of approximations. In the following, we will highlight some of the most useful approaches for modeling nanocrystalline oxides.

- *Modeling mesoscale structure*: Many important properties of nanocrystalline materials are determined by the topological characteristics of the material and its mesoscale structure rather than by its detailed atomic-scale structure. For example, the average distribution of nanocrystal shapes and sizes, the total accessible surface area and interfacial area, and the average coordination number of nanocrystals. Often, useful models of the large-scale structure of nanocrystalline materials can be developed without explicit consideration of interatomic interactions. For example, one can consider average interactions between the nanocrystals comprising the nanocrystalline material in order to simulate the structure. Such approaches have been used to correlate the macroscopic porosity of nanocrystalline TiO_2 with local nanocrystal coordination numbers with relevance to dye-sensitized solar cells [44]. Phase field methods have also proven to be a powerful approach for modeling the solidification of materials resulting in complex polycrystalline structures [45].
- *Classical ionic potentials*: Modeling nanocrystalline materials at the atomic scale requires description of the interactions between the constituent atoms. In metal-oxide materials, ions can often be described adequately as charged point-like particles, with short-range repulsive, long-range attractive, and Coulomb pair-wise interactions [46]. Such models can also be extended to account for the finite polarizability of ions [47]. There are a relatively small number of parameters entering these models, which can be determined by comparison to experimental data or to first principles predictions (e.g., the relative stability of different crystal structures, elastic constants, phonon dispersion relations). Classical ionic potentials are relatively inexpensive computationally, making it possible to consider systems containing millions of atoms and to perform long-time-scale molecular dynamics simulations [48]. Such approaches have been employed to model the structure of nanocrystals [49] and interfaces between nanocrystals [50,51] of various oxide materials. They have also been successfully employed to predict the stable structures of extended grain boundary defects [52,53]. However, a limitation of this type of approach is that it gives no access to electronic properties; only structural properties and the dynamics of ions can be considered. Ionic potentials are also less suitable for more covalent oxide materials where the effective charge on ions may differ significantly in different environments, for example, near a grain boundary or at a surface. A new class of interatomic potentials has been developed over the last decade, which in principle can tackle this problem by allowing the charge on atoms to be determined self-consistently, for example, charge-optimized many-body potentials or reactive force fields (ReaxFF) [54,55]. Providing reliable parameterization is possible: for example, using on-the-fly techniques, such potentials are highly attractive for modeling complex polycrystalline oxide films.
- *Quantum mechanical models*: At the highest level of theory, the interaction between atoms in a nanocrystalline material can be described quantum mechanically by

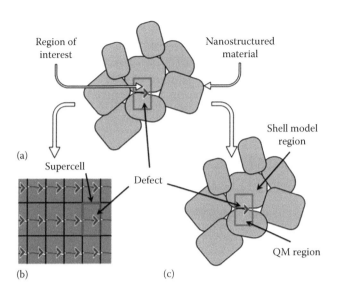

Region of
interest

Nanostructured
material

Shell model
region

(a)

Supercell

Defect

(b)

QM region

(c)

FIGURE 10.3   Schematic of two computational models used for calculating defect properties in nano-structured materials. (a) A nanostructured material as an assembly of nanoparticles. The region of interest is highlighted. The interface or defects in this region can be studied using a periodic model (b), where the region of interest forms a supercell, which is periodically translated, or an embedded cluster method (c). (Reprinted with permission from McKenna, K.P., Sushko, P.V., Kimmel, A.V., Munoz Ramo, D., and Shluger, A.L., Modelling of electron and hole trapping in oxides, *Model. Simul. Mater. Sci. Eng.*, 17, 084004. Copyright 2009 Institute of Physics.)

solving (approximately) the many-electron Schrödinger equation. The most widely used approach is density functional theory (DFT), often with periodic boundary conditions (see Figure 10.3) and a plane wave basis set, and calculations for systems containing hundreds of atoms are now fairly routine [56]. Furthermore, linear scaling DFT approaches are highly suited to metal-oxide materials and extend the size of system that can be considered to thousands of atoms [57–59]. Applications of DFT to model nanocrystalline materials include oxide nanocrystals [60–62], surfaces [63,64], and extended grain boundary defects [65–67]. However, the complexity of system that can be considered is limited by the number of atoms that is computationally feasible. Embedded cluster methods provide an alternative approach (see Figure 10.3) by combining both classical ionic potentials and quantum mechanical descriptions consistently, allowing nonperiodic systems containing hundreds of thousands of atoms to be considered [68–70]. They have proved particularly useful for modeling the optical and chemical properties of nanocrystalline oxides and will be described in more detail in the following section.

The choice of exchange-correlation functional in a DFT-based approach (both periodic DFT and embedded cluster methods) is critical for making reliable predictions. Local or semilocal exchange-correlation functionals can often describe the structure and elastic properties of ideal metal-oxide materials, but fail badly in the description of electronic structure or the properties of charged defects. In particular, band gaps are often underestimated and electrons and holes are often incorrectly predicted to be delocalized rather than trapped at defects [71]. Both of these deficiencies can be corrected by

Chapter 10

employing hybrid density functionals, which mix a proportion of exact Hartree–Fock exchange into the exchange-correlation functional [72]. For example, the B3LYP hybrid density functional combines Becke's hybrid exchange functional with the correlation functional of Lee, Yang, and Parr. The amount of exact exchange that is used can affect the degree of electron and hole localization and predicted excitation energies, so it must be calibrated with respect to experimental data.

10.2.2 First Principles Prediction of Properties Using Embedded Cluster Methods

The key idea behind the embedded cluster approach is the recognition that the properties of complex materials are often determined by localized atomic-scale features contained within them, such as vacancies, impurities, or the presence of specific sites at their surfaces. Modeling the entire system at a quantum mechanical level in order to predict its properties is in most cases computationally prohibitive. The embedded cluster method provides an alternative approach. A complex system is divided into two regions: a region of interest near a particular feature, which is treated quantum mechanically, and the remainder of the system, which is described at some lower level of theory (see Figure 10.3). The entire system is then modeled by allowing atoms in both regions to interact with each other self-consistently. Such a division requires that the electronic states responsible for a given property (e.g., optical, chemical, or electronic) can be adequately described within a finite quantum cluster. This is often the case for metal-oxide materials, where defect-induced electronic states are highly localized. The balance of efficiency and flexibility make the embedded cluster approach ideally suited to modeling complex nanocrystalline materials, which are inherently nonperiodic. The embedded cluster approach has been used successfully to study point defects [73], chemical activity [74], and optical properties [37,75] for a range of oxide materials. While details of the implementation differ for different materials, they share a number of common features. In the following, we describe specific details of the implementation for nanocrystalline MgO, which has been well studied both experimentally and theoretically and which we discuss in detail in the next section.

MgO is a highly ionic material and as such, it can be well described using classical ionic potentials. In particular, polarizable shell models have been shown to be extremely reliable for the description of the properties of the bulk crystal as well as surfaces and defects therein [47]. Therefore, they represent a natural choice for description of the majority of atoms in the embedded cluster approach. A smaller cluster of ions in a particular region of interest (e.g., near a corner or an edge of a nanocrystalline) is treated quantum mechanically using DFT (see Figure 10.4). In principle any basis set could be used for expansion of the wavefunction in this quantum cluster, but Gaussian basis sets are particularly attractive owing to their numerical efficiency. In order to prevent spurious spilling of the wavefunction from the quantum cluster into the classically modeled regions, first and second nearest neighbor Mg ions are modeled using a semilocal effective core pseudopotential possessing either no or a small number of associated basis functions. The total energy of the quantum cluster, in the presence of the electrostatic potential produced by all surrounding classical ions, is calculated by solving

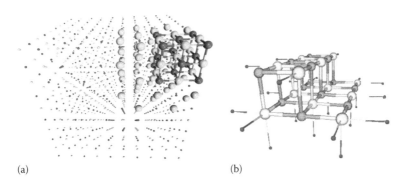

(a)   (b)

FIGURE 10.4 Embedded cluster model for calculating defects at corners and steps of MgO nanocrystallines. (a) A 2 nm side cubic MgO nanocrystal is modeled using classical interatomic potentials; the classical ions are shown as small balls. A corner is described quantum mechanically and the surrounding Mg ions represented by pseudopotentials are shown as medium-sized balls surrounding the quantum mechanical (QM) cluster. (b) The QM cluster used for calculating defects at a step on the MgO (001) surface or on the surface of a nanocrystal shown in (a).

the Kohn–Sham equations. The forces acting on all ions can then be calculated and the geometry of the entire system optimized self-consistently. This type of approach is implemented in a number of quantum mechanical (QM) codes (e.g., GUESS and Chemshell), which interface with different codes for the quantum mechanical calculation.

It is possible to calculate a wide range of properties within the DFT formalism, allowing direct comparison to experiment. For example, optical excitation spectra can be computed using time-dependent DFT, paramagnetic resonance spectra for localized electrons or holes, or vibrational spectra for adsorbed molecules. As discussed earlier, accurate prediction of electronic and optical properties of metal oxides within DFT requires hybrid density functionals. For MgO, the B3LYP hybrid functional has been shown to be reliable for the prediction of a wide range of spectroscopic and optical properties in comparison to experiment [73]. These include the optical excitation energies of low-coordinated sites at surfaces sites, electron paramagnetic resonance spectra of defects, and photostimulated ionic desorption. We describe applications of this approach in more detail in the next section.

10.3 Properties of MgO Nanopowders

MgO nanoparticle systems produced using a chemical vapor synthesis procedure [37] comprise cubic nanocrystals exposing low-index (001) facets with low-coordinated ions at edges, fourfold coordinated (4C), and corners, threefold coordinated (3C). MgO nanopowders of different preparative origin have also been the subject of numerous previous investigations [76–84], and are perhaps some of the best understood ceramic nanomaterials [17,85,86]. Earlier theoretical models assumed that individual MgO nanocrystals can be used to predict the charge trapping and spectroscopic properties of MgO nanopowders; however, the more recent studies recognized the importance of the interactions between nanocrystals and provided a more complete description of these complex systems. In the following, we briefly review the results obtained using the individual nanocrystals model and then discuss the effects of the interaction between crystallites.

Chapter 10

10.3.1 Properties Inferred from Studies of Isolated MgO Nanocrystals

10.3.1.1 Charge Trapping and Optical Excitations

Extensive calculations carried out on relatively simple cubic oxides provided some useful insights into the chemical and spectroscopic properties of surface features, as reviewed in References 81, 87, and 88. In particular, the reduced ion coordination in these crystals leads to significant changes in the crystalline potential, which has been correlated with the optical and chemical properties of the low-coordinated surface sites [76,89]. The presence of high concentration of low-coordinated O^{2-} sites in otherwise stoichiometric samples results in a surface basicity, which is related to the creation of the corresponding energy levels in the gap of the bulk material. On the perfect (001) surface, the reduction of coordination and reduction in the value of the corresponding crystalline potential leads to the appearance of the so-called surface states [90,91], but these surfaces are totally inert. Real cleaved surfaces have a large number of steps, kinks, and cleavage tips [76,92]. Reduced ion coordination at these sites induces occupied and unoccupied states split into the band gap of MgO; however, the relative number of such sites is still negligible. In a powder sample, where these sites represent a significant fraction of the total exposed surface, one observes a significant increase in reactivity, which is attributed to transfer of relatively weakly bound electrons from the electron-rich low-coordinated O^{2-} ions to acceptor molecules [76,88]. Oxygen-deficient samples may contain high concentrations of surface oxygen vacancies with even less bound electrons. On the other hand, surface Mg^{2+} ions have very small electron affinity, that is, they exhibit poor acidic character. However, the empty states of Mg^{2+} cations at steps, edges, corners, and kinks have higher electron affinities and consequently an enhanced acidic (acceptor) character: this is manifested in the enhanced interaction with CO molecules [76,88].

To build a complete model of surface electronic states even for one particular substance is a formidable task. Ideally, such a model should provide a direct correspondence between particular surface features, for example, step edges, kinks, and corners, and their excitation and ionization energies, and electron and hole affinities, as well as their ability to trap electrons and holes. Probing the positions of surfaces states using spectroscopic techniques is challenging due to the dependence of the results on the surface preparation and the surface charging during measurements. The MgO (001) surface demonstrates a great variety of features that depend on its preparation [76,93,94]. Our consideration is therefore inevitably based on simplified models, such as straight steps, normal corners, and simple kinks. Metastable impact electron spectroscopy (MIES) [95] can probe surface electronic states of insulators. The MIES measurements [96] combined with theoretical calculations [69] predicted the position of the top of the valence band of MgO (001) surface terrace at 6.7 ± 0.4 eV with respect to the vacuum level. Further embedded cluster calculations [73] provided vertical ionization energies and electron affinities of other basic low-coordinated surface sites with respect to the vacuum level (see Figure 10.5), confirming the idea of direct correlation between the ion coordination and energies of electronic states at the surface.

The direct relation between excitation and luminescence energies and coordination of surface sites of powdered MgO and CaO has been proposed in a series of works in the 1970s and 1980s [76,97,98] and confirmed theoretically [99,100]. These works demonstrated a dramatic dependence of excitation energies on the number of nearest-neighbor

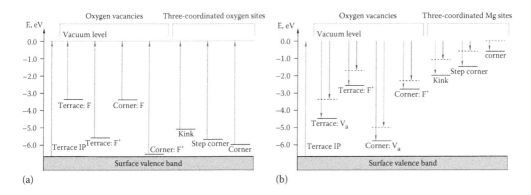

FIGURE 10.5 Energy diagrams of (a) vertical ionization energies and (b) electron affinities of oxygen vacancies at different surface sites as well as O- and Mg-terminated three-coordinated surface sites. (Reprinted with permission from Sushko, P.V., Gavartin, J.L., and Shluger, A.L., Electronic properties of structural defects at the MgO (001) surface, *J. Phys. Chem. B*, 106, 2269. Copyright 2002 American Chemical Society.)

cations surrounding surface oxygen ions. The broad optical absorption peaks with maxima at 4.6 and 5.4 eV have been assigned to three- and four-coordinated surface features, respectively. Extensive EPR studies showed the existence of strongly localized hole and electron states at surfaces [75,101,102]. There have been suggestions that electronically excited states and electrons produced at surface terraces can travel to less coordinated sites, such as kinks and corners [101]. It has been shown that selective excitation or ionization of three-coordinated sites creates stable species, such as O^- ions observed by EPR [101,102]. These results suggest a hierarchy of surface electronic states with different coordination and long-range environment and charge transfer between these states.

The experiments and calculations performed in References 73, 75, and 101 established a direct correlation between common surface features and their spectroscopic and charge trapping properties. They demonstrated that extra electrons and holes are likely to be delocalized at terraces and step edges and localized at three-coordinated terminating sites, such as kinks and corners. The lower electron and hole energies at kink and corner sites suggested that, if produced at terraces and steps, electrons and holes will tend to move to these sites. The results demonstrate that a real surface may possess both deep and shallow electron traps. Deep traps with electron affinities of more than 2 eV include anion vacancies at terraces, steps, and corner sites. However, there are a number of electron traps with electron affinities within the range of 1–2 eV. The demonstrated existence of these traps is important for our understanding of the mechanisms of surface charging and photoinduced surface processes. In particular, it suggests that shallow electron traps may serve as transient states for electron trapping in photoinduced processes, creating electron–hole pairs. One example of such photoinduced process is the desorption of surface atoms discussed in the next section.

10.3.1.2 Atom Desorption from MgO Nanocrystals

It is well established experimentally that both mild electron irradiation and low fluence photoexcitation of MgO nanostructures leads to surface modifications caused by desorption of surface atoms. These nanostructures include nanopowders of small MgO

Chapter 10

nanocrystals [37,83], thick and very rough MgO films produced by reactive ballistic deposition (RBD) on insulating substrates [103], as well as 2–3-layer-thick MgO films deposited on metal substrates [104]. The best documented cases include desorption of O and Mg atoms from MgO nanopowders and RBD films resulting from laser irradiation with weak 4.66 eV photon picosecond pulses [103,105]. These photon energies are much smaller than the bulk band gap and exciton energies and coincide with the energies attributed to the excitation of low-coordinated surface sites, such as corners and kinks. This desorption has been observed experimentally using the resonance-enhanced multiphoton ionization scheme combined with time-of-flight mass spectrometry [103]. Point defects in thin MgO films on metal substrates have been created by very weak electron bombardment and observed using scanning tunneling microscopy [104,106]. These defects have been attributed to oxygen vacancies in different charge states.

Irradiating bulk MgO crystals with single pulses of 248 nm (5 eV photons) laser irradiation with high fluences of 2–12 J/cm^2 leads to much more dramatic consequences. It produces micron-sized holes in the irradiated area [107]. It is assumed that crystal cleavage produces micron-sized sites, which exhibit highly localized absorption, resulting in decomposition, melting, and vaporization of crystal at these sites [107]. Both neutral and charged particles are produced as a result. A large fraction of the ions emitted from MgO have energies well above the energy of the incident photon [108]. The mechanisms of these processes are still not fully understood.

Here we outline the mechanisms of desorption of individual O and Mg atoms caused by weak (~1 mJ/cm^2) 4.66 eV nanosecond laser pulses. The experimental data [103,105] demonstrate that O and Mg atoms desorbed as a result of such irradiation have kinetic energies several times exceeding thermal energies (about 0.15–0.5 eV) and this type of desorption is therefore called hyperthermal. The mechanism of desorption process has been modeled in a series of papers [103,105] using the embedded cluster method. The fact that the 4.66 eV laser irradiation of nanopowders containing many three-coordinated kink and corner surface sites leads to efficient atom desorption indicates that excitation of these surface sites can be key to this process. The main features of hyperthermal O-atom desorption from low-coordinated sites of MgO surfaces can be summarized as follows:

1. Excitation of three-coordinated surface sites leads to creation of singlet and triplet excitons localized at these sites. The long-lived triplet exciton state can be ionized by a 4.66 eV photon, creating a hole. Hole states can be also created by direct ionization of three-coordinated surface sites (see Figure 10.5).

2. Trapping and localization of a hole on a single three-coordinated oxide ion forms an $(O_{3C})^-$ center that serves as a precursor state for the desorption of neutral O atoms. Such centers are readily formed at three-coordinated sites of MgO surfaces using the wide spectrum excitation with a He lamp. They are stable under vacuum conditions and have been observed experimentally using EPR [86,101,102].

3. Photoexcitation of charge-transfer transitions near the $(O_{3C})^-$ centers promotes the formation of neutral oxygen atoms. The potential energy surface of the corresponding electronically excited states has dissociative character: an O atom can spontaneously desorb while simultaneously forming a charged anion vacancy (F^+ center), as shown in Figure 10.6a. The calculated kinetic energies of desorbed O atoms are in good agreement with the experimental data [105].

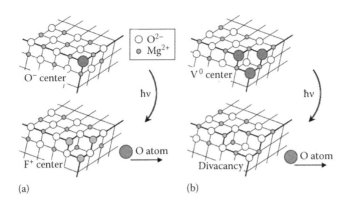

FIGURE 10.6   Schematics of the proposed mechanisms for desorption of hyperthermal O atoms. (a) Hole trapping (h^+) at an O_{3C}^{2-} site followed by charge-transfer photoexcitation leads to desorption of atomic O and formation of an F^+ center. (b) Ionization of the V^- center followed by intra-V^0 charge-transfer photoexcitation induces desorption of an O atom and hence formation of a divacancy. An alternative charge-transfer transition $O^-Mg^{2+} \rightarrow O^0Mg^+$ (not shown) also leads to an O-atom desorption and produces an electron trapped at Mg_{3C} and a single hole shared by two O_{3C}. (Reprinted with permission from Trevisanutto, P.E., Sushko, P.V., Beck, K.M., Joly, A.G., Hess, W.P., and Shluger, A.L., Excitation, ionization, and desorption: How sub-band gap photons modify the structure of oxide nanoparticles, *J. Phys. Chem. C*, 113(4), 1274. Copyright 2009 American Chemical Society.)

4. Desorption from three-coordinated oxygen sites near cation vacancies occurs in a similar way. In this case, the precursor state is generated by trapping electronic holes at the $(O_{3C})^{2-}$ ions. In the case of a neutral Mg vacancy at the corner site (corner V^0 center), irradiation with 4.66 eV or higher energy photons stimulates desorption of an O^0 atom from a 3C site via exciting the $O^-Mg^{2+} \rightarrow O^0Mg^+$ transition while lower energy photons induce intra-V^0 charge-transfer transitions (see Figure 10.6b).

The proposed model rationalizes the linear desorption yield dependence on the laser power observed in Reference 105. According to this model, the precursor for O-atom desorption is a long-lived $(O_{3C})^-$ species that maintains a pseudo-steady-state concentration throughout the desorption measurement. Since the experimental data reflect only the final photoexcitation step leading to desorption, only the single-photon charge-transfer transition associated with $(O_{3C})^-$ centers contributes to this power dependence.

10.3.2   Interfaces between MgO Nanocrystals

As discussed in Section 10.3.1, the spectroscopic properties and surface reactivity of MgO nanocrystals differ significantly from those of bulk crystals. This can often be interpreted in terms of the size and shape of the constituent nanocrystals and the high concentration of surface features, such as steps, kinks, and other low-coordination sites [109]. However, nanocrystals are not isolated in powder samples used in most experiments and the presence of interfaces between nanocrystals is a defining characteristic of real systems. Exposing the role of nanocrystal interfaces on the overall properties of nanopowders requires a combination of experimental characterization and theoretical modeling [33,50,110].

Chapter 10

TEM images of MgO nanopowders [37] demonstrate that loose powders are disordered aggregates of nanocrystals. A TEM investigation into the structure of interfaces between individual MgO nanocrystals performed by C.B. Carter et al. has provided invaluable insight into the types of interfaces that can be present in MgO nanopowders [42]. The MgO nanocrystals used in this study were produced by burning magnesium in air, producing highly cubic nanocrystals with typical dimensions of the order 50–100 nm. The most common type of interfaces seen in the TEM images involves the commensurate contact between two (001) facets. However, there are also a smaller number of interfaces between nanocrystals, which are rotated with respect to each other as well as interfaces involving the contact of nanocrystals along edges or at corners. The preponderance of commensurate (001) interfaces is expected theoretically as these are the most stable thermodynamically. This conclusion is also supported by earlier studies of MgO particles deposited onto a MgO substrate, which found a clear preference for commensurate interfaces, with a smaller fraction of rotated particles (corresponding to high site coincidence twist grain boundaries) [111].

The electronic properties of interfaces in MgO nanopowders have been a source of much speculation. EPR studies of MgO nanopowders indicate that a significant number of electrons and holes can be trapped at various places in the powder, in particular at low-coordinated cations [86,101,102,112]. Sterrer et al. showed that exposing nanopowders to photons of different energy trapped electrons and holes at the surface could be electronically excited, releasing them from their traps [75]. After some time, most of the liberated holes return to the surface, but some remain trapped within the nanopowders in an EPR-invisible configuration. It was suggested that these holes may be delocalized at interfaces between nanocrystallines, an idea that was later supported by theoretical calculations employing the embedded cluster method [50].

Another way to probe the electronic properties of interfaces in nanopowders is optical spectroscopy. Indeed, in order to perform spectroscopic measurements, nanoparticle powders are often pressed into pellets, inevitably generating increased numbers of nanocrystal interfaces. However, optical absorption and luminescence spectra are integrated over the entire nanopowder sample, making it challenging to separate surface, bulk, and interface contributions. Recently, a combined experimental and theoretical approach to directly address this issue has been developed [37,110]. The approach involves mechanical compaction of the nanopowder in order to systematically increase the relative concentration of interface features. Two substantial changes in the UV diffuse reflectance spectra of the nanopowder are observed to be associated with the increase in interface contact area. These are the emergence of new absorption features in the range 4.0–5.5 eV and a depression in intensity at 5.7 eV. To help interpret these experimental observations, theoretical calculations have been performed using the embedded cluster approach as described in Section 10.2.2.

Experimental structural characterization suggests that the main effect of nanopowder compression and annealing is to reduce the size of voids in the powder by forcing the nanocrystals to reorganize (see Figure 10.7). In the process, this forms a more dense nanocrystal network containing more interfaces. Following this transformation, the constituent nanocrystals remain cubic and crystalline, and there is a small increase in their average size, indicating some degree of sintering. However, it is difficult to extract the structure of interfaces present in the nanopowders directly from the experimental measurements. Theoretical modeling can provide some possible configurations.

Atomistic models of a selection of possible metastable configurations involving two adjoined MgO nanocrystals have been obtained on the basis of classical ionic potential simulations (Figure 10.7). The configurations chosen were based on those frequently seen in TEM images and those that are predicted to be most stable energetically. Features formed at interfaces include the edge–terrace feature, which is formed at the contact point between the edge of one nanocrystal and the (001) terrace of another. Another example involves a nanocrystal that is rotated by 45° about an axis parallel to the interface (001) plane, which results in an edge dislocation aligned along the line of contact between the nanocrystals [33]. While in general the structure of nanopowders may be quite complicated, it is expected that they should be predominantly built from these fundamental structural features. An alternative modeling approach would be to

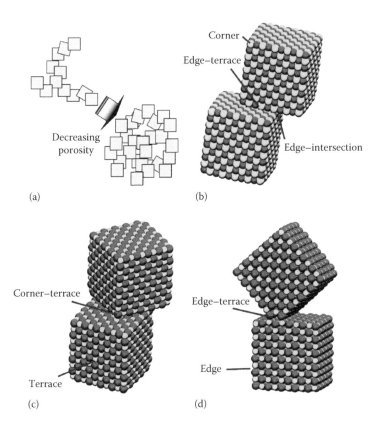

FIGURE 10.7 (a) Model for changes in powder structure on compression. (b) Structural features at the interface between adjoined nanocrystals that are commensurate, (c) rotated about an axis perpendicular to the (001) interface plane (twist), and (d) rotated about an axis parallel to the (001) interface plane (tilt). (Reprinted with permission from Sternig, A., Koller, D., Siedl, N., Diwald, O., and McKenna, K., Exciton formation at solid-solid interfaces: A systematic experimental and ab initio study on compressed MgO nanopowders, *J. Phys. Chem. C*, 116, 10103. Copyright 2012 American Chemical Society.)

employ molecular dynamics to simulate the sintering of nanocrystals directly, as has been done recently for TiO_2 [113].

The atomistic models described were then used in embedded cluster calculations to calculate the optical excitation spectra of a wide variety of features formed at the interfaces between nanocrystals and at their surfaces. In this particular implementation, the quantum cluster is described at the all electron level using a Gaussian 6-311G* basis set and the B3LYP hybrid density functional. Quantum clusters were constructed that were sufficiently large to ensure accurate description of the surface and interface electronic states participating in the electronic transitions (up to 73 atoms). The predicted excitation spectra were able to quantitatively describe the experimentally measured spectra, which is dominated by excitation of corner, edge, and terrace features at the surface of nanocrystals (Figure 10.8). They also predicted that interface features contribute to absorption in the range 4.2–5.4 eV, consistent with the experimental findings. The experimentally observed depression in intensity at 5.7 eV could also be explained due to the reduction of (001) surface area. An important conclusion of this combined experimental and theoretical study is that photons with low energy, for example, 4.6 eV, excite electrons not only at MgO surfaces, as previously believed, but also at interfaces inside the powder. This is the first experimental evidence for the emergence of new electronic states that arise from the contact area between nanocrystals.

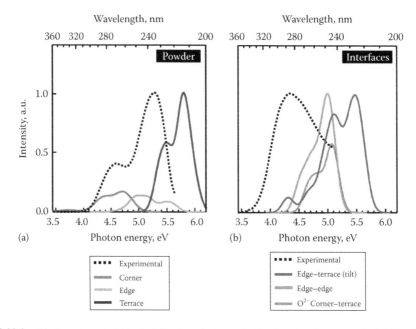

FIGURE 10.8 (a) Experimental (dashed line) and theoretically simulated spectra (solid lines) for low-density MgO powder with dominant contributions from corners, edges, and terraces. (b) Experimental excitation spectrum (dashed line) of the interfaces related emission feature at 2.5 eV in comparison to simulated excitation spectra (solid lines) for various interface features formed between commensurate and rotated nanocrystals (Figure 10.7b through d). (Reprinted with permission from Sternig, A., Koller, D., Siedl, N., Diwald, O., and McKenna, K., Exciton formation at solid-solid interfaces: A systematic experimental and ab initio study on compressed MgO nanopowders, *J. Phys. Chem. C*, 116, 10103. Copyright 2012 American Chemical Society.)

10.4 Summary

In this chapter, we have demonstrated how, by combining experimental studies and theoretical modeling, it has been possible to unravel some of the spectroscopic properties of interfaces between MgO nanocrystals. MgO powders represent a convenient system to study these effects due to the well-defined shape and controllable size distributions of MgO nanocrystals. The possibility to directly address such interfaces by tuning the energy of excitation results from the ionic character of chemical bonding and the strong dependence of the electronic structure on the local coordination of surface ions. This approach may provide new means for functionalization and chemical activation of nanostructures and can help improve performance and reliability for many nanopowder applications.

While we have considered MgO as a model system, the results suggest that understanding exciton generation at interfaces, and the associated electron–hole dynamics, may be important for other materials and applications. By combining theory and experiment, one can identify regions of the spectra that are associated with particular types of interface. The possibility to directly address such interfaces by tuning the energy of excitation is key to understanding exciton generation at interfaces and associated charge separation. Moreover, it may provide new means for functionalization of nanostructures and chemical activation and can help improve performance and reliability for many nanopowder applications, such as photocatalysts, solar cells, and energy-efficient lighting.

Nanocrystalline ceramics are particular in many ways. Between the randomly oriented crystallites, there are grain boundaries or interfacial regions. Reducing the crystallite size to some nanometers and assuming that the average interface thickness is in the subnanometer regime, the volume fraction of interfacial regions can be as high as that of the bulk [12]. Consequently, a significant fraction of the nanocrystalline ceramic material consists of interfaces and these may introduce particular functions to the integral solid. Moreover, the importance of point defects in controlling the grain boundary properties is immense for nonmetallic systems. For all these reasons, the *synthesis* of appropriate networks of particle interfaces and grain boundaries of adjustable density and composition raises an exciting new prospect in the field of functional ceramics. Future materials are expected to evolve from a new synthetic class of nanoheterogeneous but monolithic hybrids that superimpose the functionality of multiple phases in spatial confinement [21].

Acknowledgments

Keith P. McKenna gratefully acknowledges the support from EPSRC (EP/K003151), while Oliver Diwald acknowledges the support from the Austrian Fonds zur Förderung der Wissenschaftlichen Forschung FWF-PI312 (ERA Chemistry). Also Alexander L. Shluger is grateful to the Leverhulme Trust and the FP7 MORDRED project for their financial support.

References

1. Balazs, A. C., Emrick, T., and Russell, T. P. 2006. Nanoparticle polymer composites: Where two small worlds meet. *Science* 314(5802):1107–1110.
2. Shipway, A. N. and Willner, I. 2001. Nanoparticles as structural and functional units in surface-confined architectures. *Chem. Commun.* 20:2035–2045.

Chapter 10

3. Al-Shamery, K., Al-Shemmary, A., Buchwald, R., Hoogestraat, D., Kampling, M., Nickut, P., and Wille, A. 2010. Elementary processes at nanoparticulate photocatalysts. *Eur. Phys. J. B* 75(1):107–114.

4. Long, Y.-Z., Yu, M., Sun, B., Gu, C.-Z., and Fan, Z. 2012. Recent advances in large-scale assembly of semiconducting inorganic nanowires and nanofibers for electronics, sensors and photovoltaics. *Chem. Soc. Rev.* 41(12):4560.

5. Perelaer, J., Smith, P. J., Mager, D., Soltman, D., Volkman, S. K., Subramanian, V., Korvink, J. G., and Schubert, U. S. 2010. Printed electronics: The challenges involved in printing devices, interconnects, and contacts based on inorganic materials. *J. Mater. Chem.* 20(39):8446.

6. Lane, M. E. 2011. Nanoparticles and the skin—Applications and limitations. *J. Microencapsul.* 28(8):709–716.

7. Gleiter, H. 2000. Nanostructured materials: Basic concepts and microstructure. *Acta Mater.* 48(1):1–29.

8. Stoneham, A. M. and Harding, J. H. 2003. Not too big, not too small: The appropriate scale. *Nat. Mater.* 2(2):77–83.

9. Tuller, H. L. and Bishop, S. R. 2011. Point defects in oxides: Tailoring materials through defect engineering. *Annu. Rev. Mater. Res.* 41:369–398.

10. Maier, J. 2005. Nanoionics: Ion transport and electrochemical storage in confined systems. *Nat. Mater.* 4(11):805–815.

11. Preis, W., Kelder, E. M., Merkle, R., Rupp, J. L., and Yildiz, B. 2013. Preface. *Solid State Ionics* 230:1 and references therein.

12. Heitjans, P. and Indris, S. 2003. Diffusion and ionic conduction in nanocrystalline ceramics. *J. Phys. Condens. Matter* 15(30):R1257; Tuller, H. L., Litzelman, S. J., and Jung, W. 2009. Micro-ionics: Next generation power sources. *Phys. Chem. Chem. Phys.* 11(17):3023.

13. Baumann, S. O., Elser, M. J., Auer, M., Bernardi, J., Hüsing, N., and Diwald, O. 2011. Solid-solid interface formation in TiO_2 nanoparticle networks. *Langmuir* 27(5):1946–1953.

14. Siedl, N., Gügel, P., and Diwald, O. 2013. Synthesis and aggregation of In_2O_3 nanoparticles: Impact of process parameters on stoichiometry changes and optical properties. *Langmuir* 29(20):6077–6083.

15. Elser, M. J. and Diwald, O. 2012. Facilitated lattice oxygen depletion in consolidated TiO_2 nanocrystal ensembles: A quantitative spectroscopic O_2 adsorption study. *J. Phys. Chem. C* 116(4):2896–2903.

16. Derby, B. 2010. Inkjet printing of functional and structural materials: Fluid property requirements, feature stability, and resolution. *Annu. Rev. Mater. Res.* 40(1):395–414.

17. German, R. M. 1996. *Sintering Theory and Practice.* Wiley-VCH, New York.

18. Wang, S., Zhang, J., Luo, D., Gu, F., Tang, D., Dong, Z., Tan, G. et al. 2013. Transparent ceramics: Processing, materials and applications. *Prog. Solid State Chem.* 41(1–2):20–54.

19. Hinklin, T. R., Rand, S. C., and Laine, R. M. 2008. Transparent, polycrystalline upconverting nanoceramics: Towards 3-D displays. *Adv. Mater.* 20(7):1270–1273.

20. Krell, A., Klimke, J., and Hutzler, T. 2009. Transparent compact ceramics: Inherent physical issues. *Opt. Mater.* 31(8):1144–1150.

21. Rohrer, G. S., Affatigato, M., Backhaus, M., Bordia, R. K., Chan, H. M., Curtarolo, S., Demkov, A. et al. 2012. Challenges in ceramic science: A report from the workshop on emerging research areas in ceramic science. *J. Am. Ceram. Soc.* 95(12):3699–3712.

22. Kübel, C., Voigt, A., Schoenmakers, R., Otten, M., Su, D., Lee, T.-C., Carlsson, A., and Bradley, J. 2005. Recent advances in electron tomography: TEM and HAADF-STEM tomography for materials science and semiconductor applications. *Microsc. Microanal.* 11(5):378–400.

23. van Tendeloo, G., Bals, S., van Aert, S., Verbeeck, J., and van Dyck, D. 2012. Advanced electron microscopy for advanced materials. *Adv. Mater.* 24(42):5655–5675.

24. Zečević, J., de Jong, K. P., and de Jongh, P. E. 2013. Progress in electron tomography to assess the 3D nanostructure of catalysts. *Curr. Opin. Solid State Mater. Sci.* 17:115–125.

25. Rohrer, G. S. 2011. Measuring and interpreting the structure of grain-boundary networks. *J. Am. Ceram. Soc.* 94(3):633–646.

26. Robertson, I. M., Schuh, C. A., Vetrano, J. S., Browning, N. D., Field, D. P., Jensen, D. J., Miller, M. K. et al. 2011. Towards an integrated materials characterization toolbox. *J. Mater. Res.* 26(11):1341–1383.

27. Ferreira, P., Mitsuishi, K., and Stach, E. 2008. In situ transmission electron microscopy. *MRS Bull.* 33(02):83–90.

28. Espinosa, H. D., Bernal, R. A., and Filleter, T. 2012. In situ TEM electromechanical testing of nanowires and nanotubes. *Small* 8(21):3233–3252.

29. Tuomisto, F. 2013. Open volume defects. Positron annihilation spectroscopy. *Semicond. Semimet.* 88:39–65.

30. Uedono, A., Hattori, N., Ogura, A., Kudo, J., Nishikawa, S., Ohdaira, T., Suzuki, R., and Mikado, T. 2004. Characterizing metal-oxide semiconductor structures consisting of HfSiO$_x$ as gate dielectrics using monoenergetic positron beams. *Jpn. J. Appl. Phys. Part 1* 43(4A):1254–1259.

31. Wang, D., Chen, Z. Q., Wang, D. D., Qi, N., Gong, J., Cao, C. Y., and Tang, Z. 2010. Positron annihilation study of the interfacial defects in ZnO nanocrystals: Correlation with ferromagnetism. *J. Appl. Phys.* 107:023524.

32. Savin, S. L. P., Chadwick, A. V., O'Dell, L. A., and Smith, M. E. 2007. Characterisation of nanocrystalline magnesium oxide by x-ray absorption spectroscopy. *ChemPhysChem* 8(6):882–889.

33. Savin, S. L. P., Chadwick, A. V., O'Dell, L. A., and Smith, M. E. 2006. Structural studies of nanocrystalline oxides. *Solid State Ionics* 177(26–32 Spec. Iss.):2519–2526.

34. Knauth, P., Chadwick, A. V., Lippens, P. E., and Auer, G. 2009. EXAFS study of dopant ions with different charges in nanocrystalline anatase: Evidence for space-charge segregation of acceptor ions. *ChemPhysChem* 10(8):1238–1246.

35. Petkovich, N. D. and Stein, A. 2013. Controlling macro- and mesostructures with hierarchical porosity through combined hard and soft templating. *Chem. Soc. Rev.* 42(9):3721.

36. Lopez-Orozco, S., Inayat, A., Schwab, A., Selvam, T., and Schwieger, W. 2011. Zeolitic materials with hierarchical porous structures. *Adv. Mater.* 23(22–23):2602–2615.

37. Sternig, A., Koller, D., Siedl, N., Diwald, O., and McKenna, K. 2012. Exciton formation at solid-solid interfaces: A systematic experimental and ab initio study on compressed MgO nanopowders. *J. Phys. Chem. C* 116(18):10103–10112.

38. Marquardt, H., Gleason, A., Marquardt, K., Speziale, S., Miyagi, L., Neusser, G., Wenk, H.-R., and Jeanloz, R. 2011. Elastic properties of MgO nanocrystals and grain boundaries at high pressures by Brillouin scattering. *Phys. Rev. B* 84(6):064131.

39. Vidruk, R., Landau, M. V., Herskowitz, M., Talianker, M., Frage, N., Ezersky, V., and Froumin, N. 2009. Grain boundary control in nanocrystalline MgO as a novel means for significantly enhancing surface basicity and catalytic activity. *J. Catal.* 263(1):196–204.

40. Vingurt, D., Fuks, D., Landau, M. V., Vidruk, R., and Herskowitz, M. 2013. Grain boundaries at the surface of consolidated MgO nanocrystals and acid–base functionality. *Phys. Chem. Chem. Phys.* 15:14783–14796.

41. Stoneham, M. 2010. The strange magnetism of oxides and carbons. *J. Phys. Condens. Matter* 22(7):74211.

42. Nowak, J. and Carter, C. 2009. Forming contacts and grain boundaries between MgO nanoparticles. *J. Mater. Sci.* 44:2408–2418.

43. Chen, I.-W. and Wang, X.-H. 2000. Sintering dense nanocrystalline ceramics without final-stage grain growth. *Nature* 404:168–171.

44. van de Lagemaat, J., Benkstein, K. D., and Frank, A. J. 2001. Relation between particle coordination number and porosity in nanoparticle films: Implications to dye-sensitized solar cells. *J. Phys. Chem. B* 105:12433–12436.

45. Warren, J. A., Kobayashi, R., Lobkovsky, A. E., and Carter, W. C. 2003. Extending phase field models of solidification to polycrystalline materials. *Acta Mater.* 51:6035–6058.

46. Catlow, C. R. A. and Stoneham, A. M. 1983. Ionicity in solids. *J. Phys. C Solid State Phys.* 16:4321.

47. Lewis, G. V. and Catlow, C. R. A. 2004. Potential models for ionic oxides. *J. Phys. C Solid State Phys.* 18:1149–1161.

48. Todorov, I. T., Allan, N. L., Purton, J. A., Dove, M. T., and Smith, W. 2007. Use of massively parallel molecular dynamics simulations for radiation damage in pyrochlores. *J. Mater. Sci.* 42:1920.

49. Sayle, T. X. T., Parker, S. C., and Sayle, D. C. 2004. Shape of CeO$_2$ nanoparticles using simulated amorphisation and recrystallisation. *Chem. Commun.* 10:2438–2439.

50. McKenna, K. P., Sushko, P. V., and Shluger, A. L. 2007. Inside powders: A theoretical model of interfaces between MgO nanocrystallites. *J. Am. Chem. Soc.* 129:8600–8608.

51. Sayle, D. C., Mangili, B. C., Price, D. W., and Sayle, T. X. 2010. Nanopolycrystalline materials; a general atomistic model for simulation. *Phys. Chem. Chem. Phys.* 12:8584–8596.

52. Harding, J. H., Harris, D. J., and Parker, S. C. 1999. Computer simulation of general grain boundaries in rocksalt oxides. *Phys. Rev. B* 60:2740–2746.

53. McKenna, K. P. and Shluger, A. L. 2009. First principles calculations of defects near a grain boundary in MgO. *Phys. Rev. B* 79:224116.

54. van Duin, A. C. T., Strachan, A., Stewman, S., Zhang, Q., and Goddard III, W. A. 2003. ReaxFF$_{SiO}$ reactive force field for silicon and silicon oxide systems. *J. Phys. Chem. A* 107:3803–3811.

Chapter 10

55. Shan, T.-R., Devine, B. D., Kemper, T. W., Sinnott, S. B., and Phillpot, S. R. 2010. Charge-optimized many-body potential for the hafnium/hafnium oxide system. *Phys. Rev. B* 81:125328.

56. Janotti, A., Varley, J. B., Rinke, P., Umezawa, N., Kresse, G., and Van de Walle, C. G. 2012. Hybrid functional studies of the oxygen vacancy in TiO_2. *Phys. Rev. B* 81:085212.

57. Bowler, D. R. and Miyazaki, T. 2012. O(N) methods in electronic structure calculations. *Rep. Prog. Phys.* 75:036503.

58. Goedecker, S. 1999. Linear scaling electronic structure methods. *Rev. Mod. Phys.* 71:1085–1123.

59. Skylaris, C.-K., Haynes, P. D., Mostofi, A. A., and Payne, M. C. 2005. Introducing ONETEP: Linear-scaling density functional simulations on parallel computers. *J. Chem. Phys.* 122:084119.

60. Bromley, S. T, de P. R. Moreira, I., Neyman, K. M., and Illas, F. 2009. Approaching nanoscale oxides: Models and theoretical methods. *Chem. Soc. Rev.* 38:2657–2670.

61. Shevlin, S. A. and Woodley, S. M. 2010. Electronic and optical properties of doped and undoped $(TiO_2)_{(n)}$ nanoparticles. *J. Phys. Chem. C* 114:17333–17343.

62. Wolf, M. J., McKenna, K. P., and Shluger, A. L. 2012. Hole trapping at surfaces of m-ZrO_2 and m-HfO_2 nanocrystals. *J. Phys. Chem. C* 116:25888.

63. Reuter, K. and Scheffler, M. 2001. Composition, structure, and stability of RuO_2(110) as a function of oxygen pressure. *Phys. Rev. B* 65:035406.

64. Scanlon, D. O., Galea, N. M., Morgan, B. J., and Watson, G. W. 2009. Reactivity on the (110) surface of ceria: A GGA+U study of surface reduction and the adsorption of CO and NO_2. *J. Phys. Chem. C* 113:11095–11103.

65. Dawson, I., Bristowe, P. D., Lee, M.-H., Payne, M. C., Segall, M. D., and White, J. A. 1996. First-principles study of a tilt grain boundary in rutile. *Phys. Rev. B* 54:13727–13733.

66. McKenna, K. P. and Shluger, A. L. 2008. Electron trapping polycrystalline materials with negative electron affinity. *Nat. Mater.* 7:859–862.

67. Wang, Z., Saito, M., McKenna, K., Gu, L., Tsukimoto, S., Shluger, A. L., and Ikuhara, Y. 2011. Atom-resolved imaging of ordered defect superstructures at individual grain boundaries. *Nature* 479:380–383.

68. Giordano, L., Sushko, P. V., Pacchioni, G., and Shluger, A. L. 2007. Optical and EPR properties of point defects at a crystalline silica surface: Ab initio embedded-cluster calculations. *Phys. Rev. B* 75:024109.

69. Sushko, P. V., Shluger, A. L., and Catlow, C. R. A. 2000. Relative energies of surface and defect states: Ab initio calculations for the MgO (001) surface. *Surf. Sci.* 450:153–170.

70. Keal, T. W., Sherwood, P., Dutta, G., Sokol, A. A., and Catlow, C. R. A. 2011. Characterization of hydrogen dissociation over aluminium-doped zinc oxide using an efficient massively parallel framework for QM/MM calculations. *Proc. R. Soc. A* 467:1900–1924.

71. Perdew, J. P. and Zunger, A. 1981. Self-interaction correction to density-functional approximations for many-electron systems. *Phys. Rev. B* 23:5048–5079.

72. Becke, A. D. 1993. A new mixing of Hartree-Fock and local density-functional theories. *J. Chem. Phys.* 98:1372–1377.

73. Sushko, P. V., Gavartin, J. L., and Shluger, A. L. 2002. Electronic properties of structural defects at the MgO (001) surface. *J. Phys. Chem. B* 106:2269–2276.

74. Vollmer, J. M., Stefanovich, E. V., and Truong, T. N. 1999. Molecular modeling of interactions in zeolites: An ab initio embedded cluster study of NH_3 adsorption in chabazite. *J. Phys. Chem. B* 103:9415–9422.

75. Sterrer, M., Diwald, O., Knözinger, E., Sushko, P. V., and Shluger, A. L. 2002. Energies and dynamics of photoinduced electron and hole processes on MgO powders. *J. Phys. Chem. B* 106(48):12478–12482.

76. Spoto, G., Gribov, E. N., Ricchiardi, G., Damin, A., Scarano, D., Bordiga, S., Lamberti, C., and Zecchina, A. 2004. Carbon monoxide MgO from dispersed solids to single crystals: A review and new advances. *Prog. Surf. Sci.* 76(3–5):71–146.

77. Klabunde, K. J., Stark, J., Koper, O., Mochs, C., Park, D. G., Decker, S., Jiang, Y., Lagadic, I., and Zhang, D. 1996. Nanocrystals as stoichiometric reagents with unique surface chemistry. *J. Phys. Chem.* 100(30):12142–12153.

78. Moon, H. R., Urban, J. J., and Milliron, D. J. 2009. Size-controlled synthesis and optical properties of monodisperse colloidal magnesium oxide nanocrystals. *Angew. Chem. Int. Ed.* 48(34):6278–6281.

79. Chiesa, M., Paganini, M. C., Giamello, E., Di Valentin, C., and Pacchioni, G. 2003. First evidence of a single-ion electron trap at the surface of an ionic oxide. *Angew. Chem. Int. Ed.* 42(15):1759–1761.

80. Chiesa, M., Paganini, M. C., Giamello, E., Di Valentin, C., and Pacchioni, G. 2006. Electron traps on oxide surfaces: $(H^+)(e^-)$ pairs stabilized on the surface of ^{17}O enriched CaO. *ChemPhysChem* 7(3):728–734.

81. Chiesa, M., Paganini, M. C., Giamello, E., Murphy, D. M., Di Valentin, C., and Pacchioni, G. 2006. Excess electrons stabilized on ionic oxide surfaces. *Acc. Chem. Res.* 39(11):861–867.

82. Stankic, S., Bernardi, J., Diwald, O., and Knözinger, E. 2007. Photoexcitation of local surface structures on strontium oxide grains. *J. Phys. Chem. C* 111(22):8069–8074.

83. Sternig, A., Stankic, S., Müller, M., Siedl, N., and Diwald, O. 2012. Surface exciton separation in photoexcited MgO nanocube powders. *Nanoscale* 4(23):7494–7500.

84. Sternig, A., Diwald, O., Gross, S., and Sushko, P. V. 2013. Surface decoration of MgO nanocubes with sulfur oxides: Experiment and theory. *J. Phys. Chem. C* 117(15):7727–7735.

85. Boatner, L. A., Boldu, O. J. L., and Abraham, M. M. 1990. Characterization of textured ceramics by electron paramagnetic resonance spectroscopy. I. Concepts and theory. *J. Am. Ceram. Soc.* 73(8):2333–2344.

86. Boldu, J. L. O., Boatner, L. A., and Abraham, M. M. 1990. Characterization of textured ceramics by electron paramagnetic resonance spectroscopy. II. Formation and properties of textured MgO. *J. Am. Ceram. Soc.* 73(8):2345–2359.

87. Cinquini, F, Di Valentin, C., Finazzi, E., Giordano, L., and Pacchioni, G. 2007. Theory of oxides surfaces, interfaces and supported nano-clusters. *Theor. Chem. Acc.* 117:827–845.

88. Pacchioni, G. and Freund, H.-J. 2012. Electron transfer at oxide surfaces. The MgO paradigm: From defects to ultrathin Films. *Chem. Rev.* 113:4035–4072.

89. Garrone, E., Zecchina, A., and Stone, F. S. 1980. An experimental and theoretical evaluation of surface states in MgO and other alkaline earth oxides. *Philos. Mag. B* 42:683–703.

90. Henrich, V. E., Dresselhaus, G., and Zeiger, H. J. 1976. Observation of excitonic surface states on MgO-Stark-model interpretation. *Phys. Rev. Lett.* 36:158–161.

91. Lee, V.-C. and Wong, H.-S. 1978. Intrinsic surface states of MgO (100) and (110) surfaces. *J. Phys. Soc. Jpn.* 45:895–898.

92. Barth, C. and Henry, C. R. 2003. Atomic resolution imaging of the (001) surface of UHV cleaved MgO by dynamic scanning force microscopy. *Phys. Rev. Lett.* 91:196102.

93. Abriou, D., Creuzet, F., and Jupille, J. 1996. Characterization of cleaved MgO(100) surfaces. *Surf. Sci.* 352–354:499–503.

94. Robach, O., Renaud, G., and Barbier, A. 1998. Very-high-quality MgO(001) surfaces: Roughness, rumpling and relaxation. *Surf. Sci.* 401:227–235.

95. Stracke, P., Krischok, S., and Kempter, V. 2001. Ag-adsorption on MgO: Investigations with MIES and UPS. *Surf. Sci.* 473:86–96.

96. Kantorovich, L. N., Shluger, A. L., Sushko, P. V., Günster, J., Goodman, D. W., Stracke, P., and Kempter, V. 1999. Mg clusters on MgO surfaces: Study of the nucleation mechanism with MIES and ab initio calculations. *Faraday Discuss.* 114:173–194.

97. Anpo, M., Yamada, Y., Kubokawa, Y., Coluccia, S., Zecchina, A., and Che M. 1988. Photoluminescence properties of MgO powders with coordinatively unsaturated surface ions. *J. Chem. Soc., Faraday Trans. I* 84:751–764.

98. Coluccia, S., Deane, A. M., and Tench, A. J. 1978. Photoluminescent spectra of surface states in alkaline earth oxides. *J. Chem. Soc. Faraday Trans. I* 74:2913–2922.

99. Shluger, A. L., Sushko, P. V., and Kantorovich, L. N. 1999. Spectroscopy of low-coordinated surface sites: Theoretical study of MgO. *Phys. Rev. B* 59:2417–2430.

100. Sushko, P. V. and Shluger, A. L. 1999. Electronic structure of excited states at low-coordinated surface sites of MgO. *Surf. Sci.* 421:L157–L165.

101. Diwald, O., Sterrer, M., Knözinger, E., Sushko, P. V., and Shluger, A. L. 2002. Wavelength selective excitation of surface oxygen anions on highly dispersed MgO. *J. Chem. Phys.* 116(4):1707–1712.

102. Pinarello, G., Pisani, C., D'Ercole, A., Chiesa, M., Paganini, M. C., Giamello, E., and Diwald, O. 2001. O-radical ions on MgO as a tool to unravel structure and location of ionic vacancies at the surface of oxides: A coupled experimental and theoretical investigation. *Surf. Sci.* 494(2):95–110.

103. Hess, W. P., Joly, A. G., Beck, K. M., Henyk, M., Sushko, P. V., Trevisanutto, P. E., and Shluger, A. L. 2005. Laser control of desorption through selective surface excitation. *J. Phys. Chem. B* 109:19563.

104. Sterrer, M., Heyde, M., Novicki, M., Nilius, N., Risse, T., Rust, H.-P., Pacchioni, G., and Freund, H.-J. 2006. Identification of color centers on MgO(001) thin films with scanning tunneling microscopy. *J. Phys. Chem. B* 110:46–49.

Chapter 10

105. Trevisanutto, P. E., Sushko, P. V., Beck, K. M., Joly, A. G., Hess, W. P., and Shluger, A. L. 2009. Excitation, ionization, and desorption: How sub-band gap photons modify the structure of oxide nanoparticles. *J. Phys. Chem. C* 113(4):1274–1279.

106. König, T., Simon, G. H., Heinke, L., Lichtenstein, L., and Heyde, M. 2011. Defects in oxide surfaces studied by atomic force and scanning tunneling microscopy. *Beilstein J. Nanotechnol.* 2:1–14.

107. Dickinson, J. T., Jensen, L. C., Webb, R. L., Dawes, M. L., and Langford, S. C. 1993. Interaction of wide band gap single crystals with 248 nm excimer laser radiation. III. The role of cleavage induced defects in MgO. *J. Appl. Phys.* 74(6):3758–3767.

108. Ermer, D. R., Shin, J.-J., Langford, S. C., Hipps, K. W., and Dickinson J. T. 1996. Interaction of wide band gap single crystals with 248 nm excimer laser radiation. IV. Positive ion emission from MgO and $NaNO_3$. *J. Appl. Phys.* 80(11):6452–6466.

109. Stankic, S., Müller, M., Diwald, O., Sterrer, M., Knözinger, E., and Bernardi, J. 2005. Size-dependent optical properties of MgO nanocubes. *Angew. Chem. Int. Ed.* 44:4917–4920.

110. McKenna, K. P., Koller, D., Sternig, A., Siedl, N., Govind, N., Sushko, P. V., and Diwald, O. 2011. Optical properties of nanocrystal interfaces in compressed MgO nanopowders. *ACS Nano* 5:3003–3009.

111. Chaudhari, P. and Matthews, J. W. 1971. Coincidence twist boundaries between crystalline smoke particles. *J. Appl. Phys.* 42:3063–3066.

112. Chiesa, M., Giamello, E., Valentin, C. D., and Pacchioni, G. 2005. The ^{17}O hyperfine structure of trapped holes photo generated at the surface of polycrystalline MgO. *Chem. Phys. Lett.* 403(1–3):124–128.

113. Alimohammadi, M. and Fichthorn, K. A. 2009. Molecular dynamics simulation of the aggregation of titanium dioxide nanocrystals: Preferential alignment. *Nano Lett.* 9:4198–4203.

114. McKenna, K. P., Sushko, P. V., Kimmel, A. V., Munoz Ramo, D., and Shluger, A. L. 2009. Modelling of electron and hole trapping in oxides. *Model. Simul. Mater. Sci. Eng.* 17:084004.

11. Heterogeneous Catalysis

Vanadia-Supported Catalysts for Selective Oxidation Reactions

Monica Calatayud

11.1 Modeling Heterogeneous Catalysis 314
    11.1.1 General Remarks. 314
    11.1.2 The Vanadia/Titania System. 314
        11.1.2.1 DeNO$_x$ 315
        11.1.2.2 Sulfuric Acid Synthesis. 315
        11.1.2.3 Methanol Oxidation to Formaldehyde 315
        11.1.2.4 Alkane Oxide Dehydrogenation (ODH). 315

11.2 Methods and Models. 316
    11.2.1 Methods. 316
        11.2.1.1 From Interatomic Potentials to Quantum Mechanics 316
        11.2.1.2 Exploring Potential Energy Surfaces 317
    11.2.2 Models: Finite versus Periodic Models 318

11.3 The Catalyst: Vanadia Supported on Titania 320
    11.3.1 The Active Phase 320
    11.3.2 The Support 321
    11.3.3 The Additives 321

11.4 Testing the Chemical Reactivity of the Catalysts. 322
    11.4.1 Acid/Base 322
    11.4.2 Redox. 323

11.5 The Reaction: Methanol Oxidation to Formaldehyde 324
    11.5.1 The Reactants, the Products, and the Intermediates 325
    11.5.2 The Mechanisms 325
    11.5.3 Extensions and Limitations of the Model 328
    11.5.4 Impact of Alkali Doping. 328

11.6 Conclusions 330

Acknowledgments. 330

References 330

Chapter 11

11.1   Modeling Heterogeneous Catalysis

11.1.1   General Remarks

Catalysis is the workhorse of chemistry due to the important challenges in environmental, energetic, and industrial applications. Catalysis was born as purely empirical, but it benefits nowadays from basic and applied sciences and has become a multidisciplinary field. It profits from advances in fundamental research, but also motivates the development of novel approaches combining complementary concepts and techniques. Besides the rapid extension of experimental improvements in synthesis and characterization of catalysts and processes, modeling has become a powerful tool to understand the mechanism of catalysis. Interestingly, theoretical works were usually supported/confirmed by experimental results some decades ago, but it is nowadays quite common to include calculations to support experimental data.

Since the discovery of the unique properties of materials at the nanoscale, the catalytic community has found a versatile playground to improve the existing processes and to develop new materials and applications. Two main reasons have driven the exploration of nanostructured catalysts: (1) the reduction of the economic cost when using expensive materials such as noble metals, by decreasing the size of the particles, and (2) the exploration of new chemical paths associated with the specific physicochemical properties of the particles on the nanoscopic scale. An important family of catalysts successfully combines these two mainlines: supported catalysts. In this class of materials, the active phase, usually expensive, is dispersed on a high-area, low-price material, which acts as the support. One of the main objectives when developing supporting catalysts is the control of the size and the nature of the active phase surface sites, leading to a better selectivity of the products obtained in the chemical reaction.

The goal of this chapter is to present a case study of a supported catalyst, highlighting the role of the surface species and the possibilities of the available quantum-mechanical (QM) methods to characterize its properties. The surface structures, the chemical reactivity (acid/base and redox), and selected reaction paths will be described to give an overview of the possibilities of quantum-chemical methods to treat dispersed catalysts at the atomic level.

11.1.2   The Vanadia/Titania System

Vanadium-based materials are widely used in catalysis. The stability of vanadium in different oxidation states (0, II, III, IV, V) makes it very versatile for selective oxidation reactions. Also, it presents important Lewis acidic properties appreciated for certain catalytic reactions. The chemistry of vanadium-containing systems is therefore as rich as its applications. In this chapter, we focus on vanadium oxide where vanadium is in its highest oxidation state, V_2O_5. When supported on titania, TiO_2, it may catalyze a large number of reactions of environmental and industrial interest, some of which will be mentioned in the following texts. The high efficiency of such catalysts is strongly related to the size of the active phase: the presence of vanadia species with less than monolayer coverage is associated with the highest activity, while for more than monolayer coverage bulk V_2O_5 is formed and the catalytic activity drops. The challenge is thus to understand the structure–reactivity relationships for a system containing small particles (nano or below) two-dimensionally dispersed on a support. Let us briefly present some important industrial reactions catalyzed by vanadia/titania dispersed systems.

11.1.2.1 DeNO$_x$

The removal of nitric oxides, NO$_x$, is an important topic for improving the air quality. Among the catalytic reactions used to do so, the selective catalytic reduction (SCR) by NH$_3$ is widely used in industrial plants.[1–4] It has the advantage of producing only N$_2$ and water with no environmental or energetic impact. Despite the exothermicity of the reaction, it needs a catalyst and the mechanism is not completely understood.[5,6] Vanadia/titania systems are found to be very efficient for such process.

$$4NO + 4NH_3 + O_2 \rightarrow 4N_2 + 6H_2O$$

11.1.2.2 Sulfuric Acid Synthesis

Sulfuric acid is among the most important chemicals used for numerous chemical transformations in a variety of applications. Its synthesis is carried out at the industrial level in different steps. The oxidation of SO$_2$–SO$_3$ is catalyzed by vanadia-supported systems,[7] replacing the most expensive and rapidly deactivated Pt catalysts.

$$2SO_2 + O_2 \rightarrow 2SO_3$$

This reaction is, however, undesirable during the SCR of nitrogen oxides found in the flow gas of power plants. Some industrial processes combine the SCR technologies with the production of sulfuric acid.[8,9]

11.1.2.3 Methanol Oxidation to Formaldehyde

This reaction has emerged in the last years as a model system for gas-phase reactions over oxide catalysts.[10] Many experimental and theoretical studies have been conducted to elucidate structure–reactivity relationships, and the steps involved are quite well understood and will be described in more detail in the following. We will use this reaction to explore the different activities of the catalysts, that is, the acidity and the redox character.

$$CH_3OH + 1/2O_2 \rightarrow HCHO + H_2O$$

11.1.2.4 Alkane Oxide Dehydrogenation (ODH)

This reaction involves the formation of olefins: fine chemicals used as platform molecules in a variety of industrial processes. The ethane-to-ethene or propane-to-propene oxidation reactions are catalyzed by vanadia-supported materials.[11–14] The important point here is the selectivity of the desired product, avoiding further oxidation to CO and CO$_2$, for instance.

$$CH_3CH_2CH_3 + 1/2O_2 \rightarrow CH_3CHCH_2 + H_2O$$

Chapter 11

All the catalytic reactions are complex and involve exchange of atoms and electrons with the catalyst and subsequent reoxidation with O_2. Understanding the key factors that stabilize intermediates and transition state structures is therefore crucial to improve the processes. Theoretical and computational methods will be used to describe some of the elementary steps focusing on the structure–reactivity features.

11.2 Methods and Models

11.2.1 Methods

11.2.1.1 From Interatomic Potentials to Quantum Mechanics

There is a variety of methods to treat complex systems such as supported catalysts.[15] The choice will depend on the accuracy and the cost: the most accurate is naturally the most expensive. The cost is related to the size of the system and to the methodological demand. Interatomic potential (IP) is the simplest approach; the atoms are treated as particles that move following Newton equations. Such equations are easily computed at every time step; it is thus possible to consider large systems (hundred thousand atoms) evolving in long timescales (nanoseconds). This approach is of great interest for studying, for instance, phase transitions, complex systems like solid surfaces in contact with a solvent, and a large number of reactants. The interactions between particles must be defined by potential functions beforehand and extensively parameterized. While the energetics and the time are basically well described in IP, the electronic bond forming/breaking processes are not considered unless the so-called reactive force field (FF) is used. IP and FF might thus be used for studying catalytic reactions although the fine electronic structure is in principle absent in the description.

QM methods explicitly consider electrons and are intended to solve the Schrödinger equation. Most often, the electrons and the nuclei are treated separately. Two approaches are common: the wavefunction and the density functional. The former gives a very accurate description of the electronic interactions, which make it possible to use sophisticated formulations such as coupled-cluster or configuration interactions or the simplest Hartree–Fock (HF) method. The former are computationally expensive due to the mathematical formalism and need large localized basis sets to reach their maximum accuracy. The most common approach in solid-state modeling is to use density-functional theory (DFT), which is mathematically simpler to implement and possesses a good accuracy/cost ratio. The electronic density is indeed simpler to obtain since it depends only on three variables, whereas the electronic wavefunction for N particles depends on the 3N variables. The DFT formalism is exact for the ground state; the choice of the functional introduces the inaccuracy and has to be well adapted to the system of study.

The method selected must satisfy the appropriate treatment of the system. It is thus crucial to choose the approximation that is best suited to capture the chemistry and the physics of the object of study. As a general rule, metallic systems are naturally well described by DFT, whereas wavefunction methods are better adapted to insulators. For semiconductors or for processes involving changes in the electronic structure, hybrid methods including some part of HF and some DFT are nowadays very popular.

Note that such methods are not very efficient for periodic plane-wave implementations and are generally very expensive compared to DFT. The accuracy required and the available computational power will also determine the final choice. Note that if the method is critical to describe a physicochemical system, the model employed is also important. The model also has to capture the essential features of the real system. It is thus essential to consider the quality of both the simulation method and the model.

11.2.1.2 Exploring Potential Energy Surfaces

In a chemical reaction, the reactants are transformed into products passing through transition state structures accessible via an energetic barrier. Each of the structures involved has an energy associated. The difference in energy between reactants and products defines the thermodynamics of the reaction: if reactants are more energetic than products, the reaction is exothermic (downhill); otherwise, it is called an endothermic (uphill) process. The energy of the transition state gives the energetic barriers to be overcome and controls the kinetics of the reaction. The catalytic act is related to the ability of a catalyst to decrease the transition state energy so that the kinetics is faster, leaving thermodynamics unaffected. The detailed knowledge of the location and energy of the structures involved in a chemical reaction is known as potential energy surface (PES) and can be determined by computational methods. The approach of computational methods is to assign energy to a given structure so as to build the full PES or important points in the PES.

From a mathematical point of view, stable structures, such as reactants, products, or intermediates, are characterized as minima in the PES, while transition structures are first-order saddle points. This allows the use of mathematical algorithms to find important structures in a multidimensional PES. Stable structures are obtained by optimization algorithms,[16] which evolve from a guess geometry to a better one by minimizing the energy; the optimized structure possesses forces close to zero. Among the most widely used ones, one can find the steepest descendent or the conjugate gradient algorithms. In order to find transition structures, several algorithms are currently being used.[17] The Berny algorithm is based on the energy gradient that minimizes together with the energy. It is important that the guess geometry is close to the transition state. The main limitation is the need to compute the energy gradient at each step of the process. The nudged elastic band constructs a number of intermediate images along the reaction path connecting reactants and products. Each image is energetically minimized with the constraint of keeping equal spacing to neighboring images; otherwise, the optimized images would evolve to reactants or products. This constrained optimization is done by adding spring forces along the band between images and by projecting out the component of the force due to the potential perpendicular to the band. The full reaction path is then determined, the transition structure being close to the image with the highest energy. This method has the limitation of needing the calculation of several images; it is thus more time consuming. Recently, algorithms that involve neither the energy gradient nor a reasonable starting structure have become popular, such as the dimer method.[18-20] Further corrections to the energetics of critical points in a PES include zero-point corrections and entropic terms so the Gibbs free energy PES can be constructed.

Chapter 11

11.2.2 Models: Finite versus Periodic Models

The selection of an appropriate model for the system is crucial. Finite models are generally limited in size to few atoms. The system is therefore represented as a supermolecule, where only the important part of the whole system is explicitly represented. In the case of solid surfaces, the finite models only consider a bit of it and one must be careful with the edge effects; that is, the atoms that have been cut from the surface are undercoordinated and therefore overreactive, and they need to be saturated to avoid artificial enhanced reactivity. These models are suitable for wavefunction sophisticated methods where a fine electronic structure description is needed, for instance, including magnetism or dispersion. In order to get a more accurate description of a local effect, embedded models have been developed, where the region of interest is described accurately and the host lattice is described at a lower level, for instance, point charges. As for QM/MM methods, the choice of the high-level/low-level regions must be done with care paying special attention to the border. For the vanadia/titania systems, several finite systems have been developed; see Figure 11.1. The active phase (V_2O_5) may be modeled as a molecule. Gas-phase vanadium oxide clusters have been obtained experimentally[21] and have been characterized by computational methods.[22,23] These systems are undercoordinated when compared to real catalysts although they can be excellent experimental models for heterogeneous reactions because of their simplicity. To model isolated sites, the smallest model consists of saturating the V^{5+} center by one terminal oxygen and three hydroxyl groups. The support can be included by replacing the H atoms in the small system by TiO_2H units. Note that in all these models, the atoms are in their

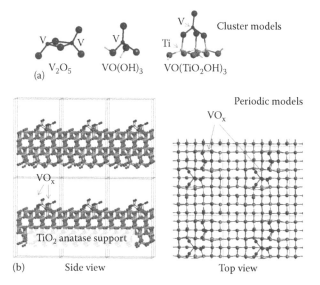

FIGURE 11.1 Models for the vanadia/titania catalyst system. (a) Three cluster models and (b) periodic model in side and top views. The vanadium site and its first neighbors (oxygen sites) are shown in blue. In the cluster models, the system is represented as a molecule, such as V_2O_5, $VO(OH)_3$, or $VO(TiO_2OH)_3$. It is possible to use accurate methods due to their limited size, but one must take care of the dangling bonds. Periodic models correct for the edge termination and allows considering coverage effects, but not all the methods are well suited for them. The unit cell of the periodic models is drawn in the top view.

appropriate oxidation states: that is, V^{5+}, Ti^{4+}, O^{2-} and H^+. In the literature, other cluster models have been used to simulate vanadia-supported catalysts.[24–26]

Periodic boundary conditions repeat a unit cell in one, two, or three dimensions, rendering the system computed infinite. The main advantage is to get rid of the edge effects and allow consideration of the coverage effects. It is usually combined to DFT and the use of plane waves, for which the 3D repetition is required. The definition of a surface is done by building a unit cell with a vacuum, so as to obtain a slab in two dimensions, repeated in the third one at a certain distance to avoid interaction between successive slabs. The slab thickness is a critical point since the slab has to be thick enough to account for the bulk region but limited in size to avoid costly calculations. Also, one has to keep in mind that any modification of the unit cell is replicated, the final system being the result of all the interactions, so that creating a defective site or adding a gas-phase molecule to the unit cell is also done in the neighboring cells. Several periodic models exist for the vanadia–titania system; see, for instance, References 27–32. They model the active phase by V_2O_5 or equivalent units and the titania phase by stable anatase terminations. Figure 11.2 displays some of the models used in this chapter.

In this chapter, we will select the most appropriate method/model to treat different aspects of the vanadia/titania systems. Periodic DFT will be used to obtain models for the heterogeneous catalyst, the coverage effects, and the energetics toward probe molecules. When the description of the fine electronic structure is important, such as in spin-crossing

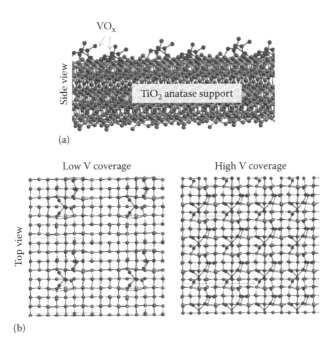

(a)

(b)

FIGURE 11.2 Periodic models for low (0.125 V/Ti$_{surf}$ = 0.84 V nm^{-2}) and high (0.5 V/Ti$_{surf}$ = 3.37 V nm^{-2}) vanadia coverage titania-supported catalysts. (a) Side view of vanadia-supported model and (b) top views of the models for low and high coverage. Vanadia units are displayed in blue. The unit cell is shown in the top view. Low and high coverage is modeled (composition $V_2Ti_{80}O_{165}$ and $V_8Ti_{80}O_{180}$, respectively, for dimensions $15.4 \times 15.4 \times 25$ Å^3).

processes that take place upon reduction of the vanadium site, hybrid functionals and more sophisticated methods like CCSD(T) will be employed using finite models.

11.3   The Catalyst: Vanadia Supported on Titania

The experimental control of the size and morphology of catalysts is a challenge. It is well known that the size of the active phase particles is related to the catalytic activity and selectivity, in particular the presence of polymeric species and hydroxyl groups.[33] Supported catalysts are often obtained by impregnation of the support in an aqueous phase containing a precursor of the active phase.[34] The samples are then dried and calcined to remove residual water and to promote the formation of strong bonds between active phase and support. This method allows the control of the active phase concentration and the polymerization degree via recombination of surface hydroxyl groups. In the last years, deposition techniques have been employed to prepare V_2O_5 thin films on anatase.[35] This method allows controlling the growth of the film in the absence of water (and hydroxyl groups).

Let us present briefly the key aspects of the catalysts and their implication in the catalytic activity.

11.3.1   The Active Phase

It is well known that vanadium oxide polymerizes on TiO_2. The surface structures may thus contain one or several vanadium sites, leading to monomers, dimers, or polymers, or even crystalline structures with specific structural and chemical properties.[36,37] The presence of crystallites on the support deactivates the catalysts; it is thus crucial to stay in the submonolayer range with dispersed surface species. The degree of polymerization determines the activity and selectivity of the catalyst. For instance, the polyvanadates are reported to be more active and selective than monomers in oxidative dehydrogenation reactions[38]; the opposite behavior is observed in the DeNO$_x$ reaction.[39]

The structure of the vanadium sites has been studied from spectroscopic measurements. Raman spectroscopy has confirmed the presence of vanadyl V=O groups, with a peculiar strong bond traditionally related to the catalytic properties. The spectral signature is a band at 1030 cm^{-1} that can be used to monitor the geometric and electronic states of the vanadium centers.[40] The vanadyl groups have been proved to be the reactive sites in vanadia-supported catalysts although in the last years, while other sites may be involved in specific reactivity.[41–43]

The link between the active phase and the support is another important feature of the supported catalysts since the $V-O-M_{support}$ groups only exist in the presence of both vanadia and support.[38,44] Their spectroscopic signature is a Raman vibration around 950 cm^{-1}; this vibration is in the same region as the support vibrations and in practice, it is difficult to unambiguously follow the evolution of the interface groups by spectroscopy. The degree of polymerization can be in principle followed by the $V-O-V$ vibrations, which lie in the same spectroscopic range. The number of polymeric species may be thus measured by the amount of such vibrations, which are absent in isolated species. Scheme 11.1 shows model structures for a monomeric and a dimeric unit of a vanadia-supported catalyst.

It is worth noting that the pyramid arrangement around the vanadium site as in Scheme 11.1 is widely accepted as model but does not exclude other geometries as will be described below.

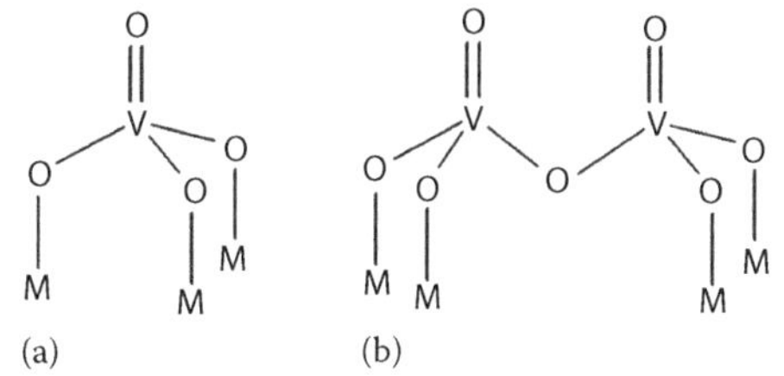

SCHEME 11.1 Surface structures of (a) a vanadate monomer and (b) a dimer. The structures possess a vanadyl V=O group and V−O−M (M, support metal site) and V−O−V bonds.

As regards the oxidation degree of vanadium in supported catalysts, it varies from II, III, IV to V. The most common species contain V^{+5} that are reduced to V^{+4} or V^{+3} during the reaction, then back to V^{+5} by dioxygen in the final reoxidation step.

11.3.2   The Support

While the active phase is associated with the catalytic efficiency, the presence of a support strongly improves the final activity by several orders of magnitude, as evidenced by I. E. Wachs in the 1990s.[44] The so-called support effect is an example of the unique properties of supported catalysts. The role of the support is manifold: besides the physical properties (texture, thermal stability, surface area) and the economic interest (lower price than the active phase), the support participates chemically in the catalytic system. It provides acidic, basic, and redox sites, modifying the overall catalytic efficiency. For instance, the activity of CeO_2-based vanadia catalysts is 1000 times higher than that of SiO_2-supported ones.[44] The polarization of the $V-O-M_{support}$ bonds seems to be intimately related to the outstanding physicochemical properties. Interestingly, the support effect would not affect the kinetic barriers but would increase the number and the performance of the surface reactive sites. For the vanadia/titania catalyst, TiO_2 is in the anatase form. The activity of the titanium sites seems to be the same in the rutile form, but those are less abundant.[45]

11.3.3   The Additives

The catalytic properties can be improved by including additives in the preparation of the catalyst. For instance, the presence of alkali metal oxides strongly affects both the activity and the selectivity of the vanadia/titania system.[46] The physicochemical properties can be tuned by choosing the appropriate dopant and modifying its content. The most thoroughly investigated alkali dopant is potassium; it has proven to influence the activity of the active sites by modifying acidity and redox potential. It is considered a promoter because of its improvement of selectivity in methanol oxidation,[47] o-xylene oxidation,[48] or propane dehydrogenation,[49,50] but it is considered a poison in the DeNO$_x$ process.[51,52] The role of the additives in the structure and the reactivity is therefore complex since it involves intermediate species that are reaction dependent. In this chapter, the impact of alkali on the geometrical and chemical properties of the vanadia/titania models will be analyzed.

Chapter 11

11.4 Testing the Chemical Reactivity of the Catalysts

The importance of the structure–reactivity relationships in the catalytic efficiency has been highlighted in Section 11.2. In this section, we will report on the acidic/basic and redox properties of the vanadia/titania models calculated by means of ab initio methodology. Periodic DFT is used; the chemical properties are calculated in terms of energetic interaction with probe molecules. The results presented are semiquantitative.

11.4.1 Acid/Base

The acid/base properties described hereafter are governed by the electrostatic interactions. The strength of the interaction of the acidic/basic sites can be measured by computing the energetic of the reaction with a probe molecule M:

$$V_2O_5 + M \rightarrow V_2O_5\text{-M} \quad \Delta E = E[V_2O_5\text{-M}] - E[V_2O_5] - E[M]$$

The more negative the value of ΔE, the more exothermic is the reaction. Cations are electron poor and anions are electron rich; this will give them acidic/basic properties (Lewis sense) that can be computed. The surface reactivity can also be explained by acid/base interactions of the surface sites with the gas-phase molecule.[53,54]

The acidity of the vanadia/titania systems is associated with the cations V^{5+} and Ti^{4+}, the former being in principle more acidic due to a higher formal charge. Taking water as an example, the presence of acidic sites will attract the molecule and will bind the vanadium center, with a formal charge of +5, to the oxygen atom in the molecule. If the interaction is strong enough, the water molecule will split into OH^- and H^+, the latter being transferred to a neighboring oxygen site. As a consequence, VOH and OH groups will form. Scheme 11.2 displays the structures of the molecular and dissociated water on a vanadate monomer. Such structure can be obtained for other vanadates.

It is interesting to look at the coverage effects. ΔE values for selected systems are reported in Table 11.1. Low-coverage models possess Ti^{4+} sites available on the surface for water, together with the V^{5+} sites. It is found that upon interaction with one water molecule, TiOH groups are formed by hydrogen bonds with the vanadia unit.[55] The adsorption of a second water molecule to the system results in the formation of VOH groups together with TiO_2 groups. At higher vanadia coverage, only VOH groups are formed due to the fact that the support is not available since it is covered almost

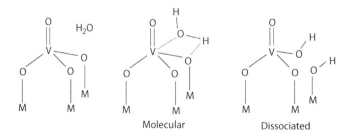

Molecular   Dissociated

SCHEME 11.2 Representative structures obtained for the interaction of water with a monovanadate. M, surface metallic site.

Table 11.1 ΔE Values per Water Molecule for the Hydration Reaction in Selected Systems

	ΔE (eV)	Surface Species
TiO_2 support	−2.15	TiOH
Low coverage	−2.67	VOH and TiOH
$2H_2O$	−2.45	VOH and TiOH
High coverage	−2.22	VOH
$2H_2O$	−1.83	VOH

Note: Water dissociates, forming surface hydroxyl groups.

completely by vanadia. Interestingly, the ΔE values calculated indicate that the low-coverage system has more affinity for water than the high-coverage one (−2.67 vs. −2.22 eV, respectively, for $1H_2O$; see Table 11.1). The adsorption on the support alone is less exothermic than when vanadia is present, indicating a higher acidity of the latter.

The basic sites are the oxygen centers; they can be probed by an acidic species such as H^+ that will form hydroxyl groups. Their role is crucial in stabilizing protons as well as acidic species, and they are involved in hydrogen-transfer processes,[55] exhibiting also Brønsted acidity.

11.4.2 Redox

The redox property of V_2O_5 is associated with the reducibility of V^{+5}. Upon interaction with a reducing species, the vanadium center will be reduced to V^{4+} or V^{3+}. A probe molecule for this reaction is H_2, which will transfer the two electrons to the vanadium center, and the H^+ atoms will bind to oxygen sites as in an acid/base interaction.

$H_2 \rightarrow 2H^+ + 2e^-$

$2V^{5+} + 2e^- \rightarrow 2V^{4+}$ (or $1V^{3+}$)

$2H^+ + 2O^=_{surf} \rightarrow 2OH^-_{surf}$

Global reaction: $H_2 + 2V^{5+} + 2O^=_{surf} \rightarrow 2V^{4+} + 2OH^-_{surf}$

The catalytic activity is intimately related to the ability of V^{5+} to keep the electrons, that is, its reducibility. Experimentally, it is quite common to characterize the redox properties of a catalyst by making it react with H_2 in the so-called temperature-programmed reduction (H_2-TPR): the lower the temperature at which H_2 is taken up, the more reducible is the system.[56] From calculations it is possible to measure the energy of the reaction; the more exothermic corresponds to the more reducible models. Moreover, the localization of the electrons is characterized by the presence of a single electron in the vanadium site for V^{4+}. The computational method should thus include spin polarization to properly consider this feature. The oxygen sites at which the H^+ will bind can also be probed, providing valuable information on the basic sites. In Reference 42, the surface oxygen sites are probed with an H atom and the calculated adsorption energies are V−O−Ti

Chapter 11

(−2.95 eV), V=O (−2.60 eV), V−O−V (−2.17 eV), and Ti−O−Ti (−2.07 eV). This means that the most reactive site toward hydrogenation is the support one. Therefore, the coordination of the oxygen site does not seem to be the most important factor upon hydrogenation: it might be expected that the terminal oxygen V=O, since poorly coordinated, should be the preferred site, but clearly, this is not the case and the support sites are directly involved. Applying reactivity indices based on Pearson's hard and soft acid base HSAB, the different reactivity of the oxygen sites is explained by the hard/soft character.[57] The support V−O−Ti site would be the hardest one and would therefore react with hard electrophiles like H, whereas the softest site is near the vanadyl V=O site. It is important to note at this point that the selectivity of the catalyst depends on the nature of both the reactants and the surface sites. In other words, hard reactants will interact with hard surface sites and soft reactants with soft sites. Hard and soft sites are not located on the same regions in vanadia/titania catalysts,[57] making thus possible to differentiate between them by choosing properly the reactants.

11.5   The Reaction: Methanol Oxidation to Formaldehyde

Methanol is commonly used in catalysis as a *smart* probe.[10] This molecule is sensitive to both acid/base and redox properties. Thus, the presence of acidic sites will lead to its transformation to dimethyl ether, while basic sites will transform it into CO and CO_2. Redox sites will partially oxidize it to formaldehyde. The final proportion of products will depend on the preparation conditions, temperature, additives, etc. and will serve as a measure of the different properties. Recently the selectivity of methanol to different products has been evidenced by experimental and theoretical methods for vanadium-containing zeolites,[58] illustrating the different reactivity one may find in a catalyst.

In order to better understand the reactivity of the catalyst, let us present the mechanism in successive steps:

- *Gas phase*: $CH_3OH \rightarrow [H_2CO + 2H^+ + 2e^-] \rightarrow H_2CO + H_2$
- *Methanol dissociation*: $CH_3OH \rightarrow CH_3O^- + H^+$
- *Adsorption*: $CH_3O^- + V^{5+} \rightarrow CH_3O^--V^{5+}$ and $H^+ + O^=_{surf} \rightarrow HO^-_{surf}$
- *Methoxide oxidation*: $CH_3O^- \rightarrow H_2CO + H^+ + 2e^-$
- *Electron transfer to vanadium*: $V^{5+} + 2e^- \rightarrow V^{3+}$
- *Proton transfer to surface oxygen*: $H^+ + O^=_{surf} \rightarrow HO^-_{surf}$
- *Global mechanism*: $CH_3OH + V^{5+} + 2O^=_{surf} \rightarrow H_2CO + V^{3+} + 2HO^-_{surf}$

We have chosen to analyze the reaction paths obtained from hybrid B3LYP calculations using a cluster model. The transition state structures are determined and so the energetic barriers. Note that the overall mechanism involves the two types of properties described: (1) acid/base, when the methanol dissociates and adsorbs on the surface sites, and (2) redox, when the methoxide group gives two electrons that are transferred either to the H^+ groups, forming H_2 (gas-phase mechanism), or to the vanadium center that becomes reduced. The role of the catalyst is therefore to improve the reaction path by splitting it into several less costly steps.

The methanol oxidation to formaldehyde has been extensively studied both experimentally[10,59,60] and theoretically.[61–65] The rate-limiting step is the oxidation of methanol to formaldehyde, which is reported to occur via an open-shell singlet structure and requires therefore a finer description of the electronic structure. This justifies the selection of a cluster model for which more accurate algorithms and methods are available. The pyramid arrangement around V is also widely accepted in the literature although more coordinated species like pentahedral V may be found and be also reactive. The results presented hereafter have been obtained with the hybrid functional B3LYP/6-311G(2d, p) level; see References 66–68 for more details.

11.5.1   The Reactants, the Products, and the Intermediates

Figure 11.3 illustrates two possible reaction pathways for the methanol oxidation to formaldehyde. Reactants are the methanol molecule and the cluster model. The cluster contains a pyramidal monomeric vanadia unit bonded to a titania support. As the molecule approaches the cluster, a complex with the cluster model is formed and stabilized by hydrogen bonds ~29 kJ mol^{-1}. Next, the molecule dissociates and forms a methoxide group on the vanadium site and a hydroxyl group. These groups are stabilized by electrostatic interactions following an acid/base mechanisms described in Section 11.5: CH_3O^- adsorbs on V^{5+} acidic site and H^+ on the basic O^{2-}_{surf}. Note that the oxygen site was initially bonded to both V and Ti. Then, methanol adsorption induces the break of a V−O bond to form a O−H one. The resulting system is more stable than the reactants by 71 kJ mol^{-1}. Indeed, the methoxide intermediate is the most stable one found in the reaction path.

Formaldehyde is the final product of the reaction. The methoxide group needs to transfer an H atom and to release two electrons to evolve to HCHO. One possible pathway is to combine the H atom and the OH$^-$ surface group with the electrons to obtain CH_2O and H_2 (Path 1 in Figure 11.3). The products that would be obtained in the gas-phase reaction without catalyst are the same. The catalytic cycle is respected. A different pathway, Path 2 in Figure 11.3, is possible if the H atom moves to the vanadyl V=O site, forming a VOH group. The electrons will be transferred to the vanadium site and V^{5+} will become V^{3+}. Note that the V^{3+} species is now in a triplet electronic state (the vanadium center contains two unpaired electrons). The energy of this complex, CH_2O interacting with the reduced cluster, is endothermic by 65 kJ mol^{-1} with respect to *the gas-phase* reactants. Desorption of formaldehyde would destabilize this system by ~17 kJ mol^{-1}. The overall reaction proceeds thus uphill, and the final step to finish the catalytic cycle, not shown, is supposed to occur with low activation barriers. Interestingly, the reaction path involves a spin-crossing process: the reactants are in a singlet electronic state, whereas the products are in a triplet state. In the next section, the energetic barriers of the two paths are presented and the fine electronic structure of Path 2 is discussed.

11.5.2   The Mechanisms

The two reaction pathways presented in Figure 11.3 proceed through very different reaction barriers. The energetic profile for the two paths considered is displayed in Figure 11.4. The formation of the methoxide intermediate, common to both mechanisms, proceeds with a low barrier of 45 kJ mol^{-1}, which means that in real conditions, upon contact of methanol with the catalyst, the surface will be covered by methoxide groups.

Chapter 11

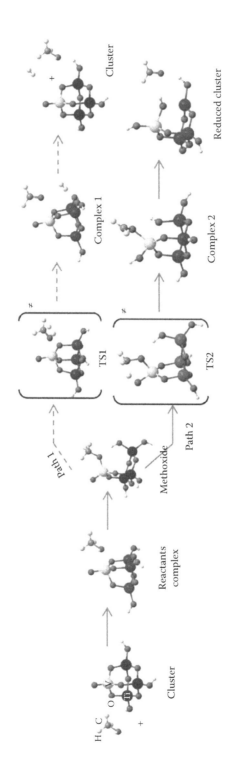

FIGURE 11.3 Reaction pathways for the reaction of methanol oxidation to formaldehyde on a supported catalyst. After the dissociation of methanol forming a methoxide species, the reaction evolves (1) to the formation of H_2 (Path 1) or (2) to the reduction of the vanadium site (Path 2).

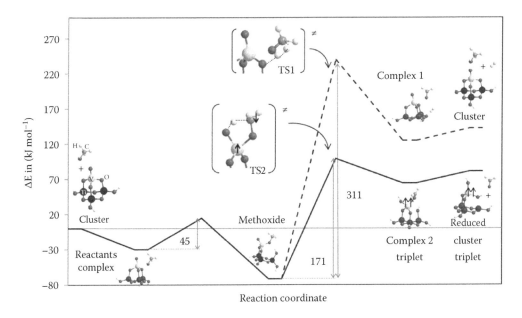

FIGURE 11.4 Energetic profile for the reaction paths shown in Figure 11.3.

This is confirmed experimentally by spectroscopic measurements that determine that roughly 90% of the surface methanol is dissociated, whereas 10% is in its molecular form.

The rate-limiting step in the two paths is the redox process. In Path 1, the transition structure involves the transfer of a hydride H^- species from the methoxide to the surface OH^- group. The electrons and the protons must leave the methanol unit recombine and form an H_2 molecule. The energetic barrier is calculated to be of 311 kJ mol^{-1}. This value is close to that of the pure gas-phase mechanism, 334 kJ mol^{-1}. Both reactants and products are in singlet electronic states and the role of the catalyst is thus to stabilize the methoxide intermediate.

Path 2 is a more interesting alternative. The H atom of the methoxide group is transferred to a neighboring oxygen site to form a hydroxyl group, and at the same time, the electrons are transferred to the vanadium site. Interestingly, the nature of the corresponding transition state is biradicaloid. Such peculiar structure possesses one alpha electron on the vanadium site and one beta electron on the carbon atom of the methoxide group, as schematized in Figure 11.4. The energetic barrier, 171 kJ mol^{-1}, is obtained from a projection of the corresponding energy for the transition structures in the closed shell, the open-shell singlets, and the triplet states by means of the broken-symmetry approach. This technique gives reasonable results as regards the activation barrier for this step. The accuracy of the value is confirmed by more sophisticated methods like CCSD(T).[61] The role of the vanadium site is here to stabilize the transition structure: (1) by withdrawing one electron from the methoxide unit and (2) by stabilizing a reduced cluster. The proximity of several electronic states in the vicinity of the transition structure makes the hopping between them possible. It is important that this fine description of the electronic structure is respected when modeling the catalytic reaction.

Note that the reduced cluster is not the final product of the reaction, since the catalytic cycle is not finished. This intermediate is however more stable than the corresponding one arising from Path 1. A further reoxidation by O_2 is supposed to occur with lower activation barriers so that the catalyst is regenerated.

11.5.3   Extensions and Limitations of the Model

We described in Sections 11.2.2 and 11.5 that the use of a cluster model is well suited for the exploration of complex PESs including fine descriptions of the electronic structure. It has however some drawbacks as regards the choice of the active sites. First, the pyramidal arrangement around the vanadium site is a natural model since such species are characterized experimentally to be very stable.[69] In catalysis, very stable species may be regarded as poorly reactive, the ideal situation being a smooth energetic profile so that the intermediates do not block the active sites once formed. That might be the case for the tetrahedral methoxide intermediate. We have explored a pentahedral coordination around vanadium that leads to almost equivalent energetic profile but where the activation barriers are decreased.[67] Also, the reactivity of vanadyl V=O groups has been many times invoked in the literature[61,63,70] although there is no consensus about their role in reactive-supported catalysts.[41] We have explored the possibility of H transfer to V$-$O$-$Ti or Ti$-$OH groups instead of the V=O group in Path 2, with slightly higher activation barriers.[66] Due to the important role played by the support, the V$-$O$-$M groups may offer alternative paths in other supports and should be considered as potential reactive sites. Besides, the model only considers one isolated vanadium site. High coverage may provide alternative routes for a reactive path by considering the presence of neighboring units, dimers, or polymers possessing V$-$O$-$V species absent in the model described in Figures 11.3 and 11.4.

11.5.4   Impact of Alkali Doping

As introduced in Section 11.3, the presence of dopants on the surface of the catalysts deeply modifies its reactivity. Experimentally, the impact of potassium in the products obtained after methanol exposure is known to depend on the preparation of the catalyst. It is possible to design $K_2O-V_2O_5/Al_2O_3$ systems to create catalysts 100% acidic, 100% basic, 100% redox, mixed redox–acidic, and mixed redox–basic, the nature of which is controlled by the order of impregnation of the precursors, additives, and their content.[47,71] The effect of having potassium is mainly to elongate the V=O bond and to decrease both the acidity and the reducibility of the vanadia surface species.[49,72] In order to account for coverage and support effects, we have chosen a periodic model containing high coverage in vanadia units.[73] We describe here the impact of the presence of alkali atoms in both the geometry and the acid/base and redox reactivity as defined in Section 11.5 for the case of methanol adsorption and H_2.

Alkali are modeled by adding a neutral metal M (M: H, Li, Na, K). The system spontaneously responds with the ionization of the alkali atom to form M^+ alkali cations. Note that the experimental procedure involves the addition of a basic oxide (K_2O is used in Reference 47 on a vanadia/alumina catalyst). The geometric distortions observed experimentally are recovered with the M neutral model and in preliminary calculations with a MOH model. The more stable models are found to be those where the alkali atoms are located between the vanadia units and during interaction with the support,

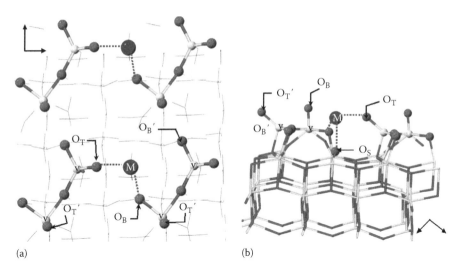

FIGURE 11.5 Top (a) and side view (b) of a high-coverage vanadia/titania periodic model. M, metal; O, oxygen sites (T, terminal; B, bridging; S, support). The metal atom is stabilized between vanadia units and is in contact with the support.

as depicted in Figure 11.5. The systems stabilize by electrostatic interaction between M$^+$ and the oxygen sites. In particular, V=O groups elongate in the presence of alkali cations, as observed by red shift of the V=O Raman band. The affinity of the alkali cations for the support may have important consequences for the vanadia units since it may induce monomerization of polymeric units.

Upon adsorption of methanol, methoxide species are formed as observed in the experiments[47,71] and previous calculations.[73] The interaction energy is exothermic for all cases indicating a favorable process. However, the presence of alkali decreases the energy gain in the series bare > H doped > Li doped > Na doped > K doped. This trend reflects the increase in basicity from Li to K, K being the most basic system in the series. The acidity of the vanadia/titania catalysts therefore decreases upon interaction with alkali.

As regards reducibility, Table 11.2 shows the interaction energy calculated for the H adsorption. As for the methanol interaction, a general decrease in ΔE is observed, the systems becoming less reducible in the series bare > H doped > Li doped > Na doped > K doped. Additionally, the reduction of the vanadium center is checked by the analysis

Table 11.2 Adsorption Energies ΔE in eV for Methanol and Hydrogen Interaction with Alkali-Doped Slabs

ΔE	CH$_3$OH	H (1/2H$_2$)
Bare	−1.46	−0.70
H doped	−1.50	−0.48
Li doped	−1.23	−0.45
Na doped	−1.00	−0.38
K doped	−0.72	−0.29

Source: Calatayud, M. and Minot, C., *J. Phys. Chem. C*, 111, 6411, 2007.

Chapter 11

of the density of states where the vanadium states, unoccupied in the bare slab, become occupied by the electron left by the hydrogen (or by the metal alkali).

We conclude that doping the vanadia/titania catalysis with alkali leads to a decrease in interaction energies. This corresponds to a decrease in the catalytic activity observed in the experiments.[47,71] However, this decrease in activity is positive for the selectivity of the reactions, in particular as regards methanol oxidation: the intermediate is less stabilized on the surface, so its residence time is reduced and this avoids undesired products like CO and CO_2 coming from overoxidation of the methoxide intermediate. The decrease in activity is thus not necessarily a problem but a successful way to control parallel reactions.

11.6 Conclusions

Heterogeneous catalysts are complex systems of high importance in industrial technologies for environmental and economic processes. The success of supported catalysts lies in the fact that the active phase is dispersed on the support-forming surface species at the (sub)nanolevel. It is thus crucial to understand the structure of the surface species, their size and structure, and their implication in chemical reactivity. Controlling the degree of polymerization, the chemical nature and the geometrical structure plays a key role in developing novel materials with selected properties. Recent advances in computational techniques and modeling allow treating with a high level of accuracy the structural and reactivity feature characteristic of these systems, to explain both with experimental observations and in a predictive manner. In the case study of vanadia-/titania-supported catalysts, the careful choice of model and method helps in understanding the nature of the surface species and their role in the catalytic activity and selectivity.

Acknowledgments

The author gratefully thanks all the collaborators for their valuable contribution to the understanding of structure and reactivity in vanadia-supported catalysts: A. Lewandowska, P. Gonzalez-Navarrete, H. Si-Ahmed, L. Gracia, M. A. Bañares, F. Tielens, C. Minot, and B. Mguig.

This work was performed using HPC resources from GENCI-CINES/IDRIS (Grants 2012-x2012082131 and 2013-x2013082131) and the CCRE-DSI of Université P. M. Curie. The COST action CM1104 is also acknowledged.

References

1. Topsøe, N. 1994. Mechanism of selective catalytic reduction of nitric oxide by ammoniac elucidated by situ on-line Fourier transform infrared spectroscopy. *Science*, 265: 1217–1219.
2. Topsøe, N.-Y., Dumesic, J. A., Topsøe, H. 1995. Vanadia/titania catalysts for selective catalytic reduction of nitric oxide by ammonia. II. Studies of active sites and formulation of catalytic cycles. *J Catal*, 151: 241–252.
3. Topsøe, N.-Y., Topsøe, H., Dumesic, J. A. 1995. Vanadia/titania catalysts for selective catalytic reduction (SCR) of nitric oxide by ammonia. *J Catal*, 151: 226–240.
4. Busca, G., Lietti, L., Ramis, G., Berti, F. 1998. Chemical and mechanistic aspects of the selective catalytic reduction of NO_x by ammonia over oxide catalysis: A review. *Appl Catal B Environ*, 18: 1–36.

5. Calatayud, M., Mguig, B., Minot, C. 2004. Modeling catalytic reduction of NO by ammonia over V$_2$O$_5$. *Surf Sci Rep*, 55: 169–236.

6. Jug, K., Homann, T., Bredow, T. 2004. Reaction mechanism of the selective catalytic reduction of NO with NH$_3$ and O$_2$ to N$_2$ and H$_2$O. *J Phys Chem A*, 108: 2966–2971.

7. Dunn, J. P., Stenger, H. G., Wachs, I. E. 1999. Oxidation of sulfur dioxide over supported vanadia catalysts: Molecular structure-reactivity relationships and reaction kinetics. *Catal Today*, 51: 301–318.

8. Armor, J. N. 1992. Environmental catalysis. *Appl Catal B Environ*, 1: 221–256.

9. Ohlms, N. 1993. DESONOX process for flue-gas cleaning. *Catal Today*, 16: 247–261.

10. Badlani, M., Wachs, I. E. 2001. Methanol: A "smart" chemical probe molecule. *Catal Lett*, 75: 137–149.

11. Martinez-Huerta, M. V., Gao, X., Tian, H., Wachs, I. E., Fierro, J. L. G., Banares, M. A. 2006. Oxidative dehydrogenation of ethane to ethylene over alumina-supported vanadium oxide catalysts: Relationship between molecular structures and chemical reactivity. *Catal Today*, 118: 279–287.

12. Mitra, B., Wachs, I. E., Deo, G. 2006. Promotion of the propane ODH reaction over supported V$_2$O$_5$/Al$_2$O$_3$ catalyst with secondary surface metal oxide additives. *J Catal*, 240: 151–159.

13. Beck, B., Harth, M., Hamilton, N. G., Carrero, C., Uhlrich, J. J., Trunschke, A., Shaikhutdinov, S. et al. 2012. Partial oxidation of ethanol on vanadia catalysts on supporting oxides with different redox properties compared to propane. *J Catal*, 296: 120–131.

14. Rao, C. N. R., Raveau, B. 1995. *Transition Metal Oxides*. New York: VCH Publishers.

15. Schleyer, P. v. R. ed. 1998. *Encyclopedia of Computational Chemistry*. U.K.: John Wiley & Sons, Ltd.

16. Farkas, O., Schlegel, H. B. 2003. Geometry optimization methods for modeling large molecules. *J Mol Struct (Theochem)*, 666–667: 31–39.

17. Sheppard, D., Terrell, R., Henkelman, G. 2008. Optimization methods for finding minimum energy paths. *J Chem Phys*, 128: 134106.

18. Heyden, A., Bell, A. T., Keil, F. J. 2005. Efficient methods for finding transition states in chemical reactions: Comparison of improved dimer method and partitioned rational function optimization method. *J Chem Phys*, 123: 224101.

19. Henkelman, G., Jónsson, H. 1999. A dimer method for finding saddle points on high dimensional potential surfaces using only first derivatives. *J Chem Phys*, 111: 7010.

20. Olsen, R. A., Kroes, G. J., Henkelman, G., Arnaldsson, A., Jónsson, H. 2004. Comparison of methods for finding saddle points without knowledge of the final states. *J Chem Phys* 121: 9776.

21. Bell, R. C., Zemski, K. A., Kerns, K. P., Deng, H. T., Castleman, Jr. A. W. 1998. Reactivities and collision-induced dissociation of vanadium oxide cluster cations. *J Phys Chem*, 102: 1733–1742.

22. Calatayud, M., Beltran, A., Andrés, J. 2001. A systematic density functional theory study of V$_x$O$_y^+$ and V$_x$O$_y$ (x = 2–4, y = 2–10) systems. *J Phys Chem A*, 105: 9760–9775.

23. Vyboishchikov, S. F., Sauer, J. 2001. (V$_2$O$_5$)$_n$ Gas-phase clusters (n = 1 – 12) compared to V$_2$O$_5$ Crystal: DFT calculations. *J Phys Chem A*, 105: 8588–8598.

24. Ferreira, M. L., Volpe, M. 2000. On the nature of highly dispersed vanadium oxide catalysts: Effect of the support on the structure of VO$_x$ species. *J Mol Catal A Chem*, 164: 281–290.

25. Kachurovskaya, N., Mikheeva, E. P., Zhidomirov, G. M. 2002. Cluster molecular modeling of strong interaction for VO$_x$/TiO$_2$ supported catalyst. *J Mol Catal A Chem*, 178: 191–198.

26. Bredow, T., Homann, T., Jug, K. 2004. Adsorption of NO, NH$_3$ and H$_2$O on V$_2$O$_5$/TiO$_2$ catalysts. *Res Chem Intermed*, 30: 65–73.

27. Sayle, D. C., Catlow, C. R. A., Perrin, M. A., Nortier, P. 1996. Computer modelling of the V$_2$O$_5$/TiO$_2$ interface. *J Phys Chem*, 100: 8940–8945.

28. Sayle, D. C., Gay, D. H., Rohl, A. L., Catlow, C. R. A., Harding, J. H., Perrin, M. A., Nortier, P. 1996. Computer modelling of V$_2$O$_5$: Surface structures, crystal morphology and ethene sorption. *J Mater Chem*, 6: 653–660.

29. Vittadini, A., Casarin, M., Selloni, A. 2005. First principles studies of vanadia-titania monolayer catalysts: Mechanisms of NO selective reduction. *J Phys Chem B*, 109: 1652–1655.

30. Vittadini, A., Selloni, A. 2004. Periodic density functional theory studies of vanadia-titania catalysts: Structure and stability of the oxidized monolayer. *J Phys Chem B*, 108: 7337–7343.

31. Calatayud, M., Mguig, B., Minot, C. 2003. A periodic model for the V$_2$O$_5$–TiO$_2$ anatase catalyst. Stability of dimeric species. *Surf Sci*, 526: 297–308.

Chapter 11

32. Calatayud, M., Mguig, B., Minot, C. 2005. Hydration of the V_2O_5/TiO_2 catalyst and stability of monomeric species. *Theor Chem Acc*, 114: 29–37.
33. Wachs, I. E. 2005. Recent conceptual advances in the catalysis science of mixed oxide catalytic materials. *Catal Today*, 100: 79–94.
34. Bañares, M. A., Wachs, I. E. 2002. Molecular structures of supported metal oxide catalysts under different environments. *J Raman Spectrosc*, 33: 359–380.
35. Gao, W., Altman, E. I. 2006. Growth and structure of vanadium oxide on anatase (101) terraces. *Surf Sci*, 600: 2570.
36. Kim, T., Wachs, I. E. 2008. CH_3OH oxidation over well-defined supported V_2O_5/Al_2O_3 catalysts: Influence of vanadium oxide loading and surface vanadium-oxygen functionalities. *J Catal*, 255: 197–205.
37. Zhao, C. L., Wachs, I. E. 2008. Selective oxidation of propylene over model supported V_2O_5 catalysts: Influence of surface vanadia coverage and oxide support. *J Catal*, 257: 181–189.
38. Khodakov, A., Olthof, B., Bell, A. T., Iglesia, E. 1999. Structure and catalytic properties of supported vanadium oxides: Support effects on oxidative dehydrogenation reactions. *J Catal*, 181: 205–216.
39. Amiridis, M. D., Wachs, I. E., Deo, G., Jehng, J.-M., Kim, D. S. 1996. Reactivity of V_2O_5 catalysts for the selective catalytic reduction of NO by NH_3: Influence f vanadia loading, H_2O, and SO_2. *J Catal*, 161: 247–253.
40. Wachs, I. E., Briand, L. E., Jehng, J. M., Burcham, L., Gao, X. T. 2000. Molecular structure and reactivity of the group V metal oxides. *Catal Today*, 57: 323–330.
41. Routray, K., Briand, L. E., Wachs, I. E. 2008. Is there a relationship between the M=O bond length (strength) of bulk mixed metal oxides and their catalytic activity? *J Catal*, 256: 145–153.
42. Calatayud, M., Minot, C. 2004. Reactivity of the oxygen sites in the V_2O_5-TiO_2 anatase catalyst. *J Phys Chem B*, 108: 15679–15685.
43. Calatayud, M., Minot, C. 2006. Reactivity of the V_2O_5–TiO_2-anatase catalyst: Role of the oxygen sites. *Top Catal*, 41: 17–26.
44. Deo, G., Wachs, I. E. 1994. Reactivity of supported vanadium-oxide catalysts—The partial oxidation of methanol. *J Catal*, 146: 323–334.
45. Wachs, I. E. 1990. Molecular structures of surface vanadium oxide species on titania supports. *J Catal*, 124: 570–573.
46. Deo, G., Wachs, I. E. 1994. Effect of additives on the structure and reactivity of the surface vanadium-oxide phase in V_2O_5 TiO_2 catalysts. *J Catal*, 146: 335–345.
47. Wang, X., Wachs, I. E. 2004. Designing the activity/selectivity of surface acidic, basic and redox sites in the supported K_2O–V_2O_5/Al_2O_3 catalytic system. *Catal Today*, 96: 211–222.
48. Jiménez-Jiménez, J., Mérida-Robles, J., Rodríguez-Castellón, E., Jiménez-López, A., Granados, M. L., Val, S. d., Cabrera, I. M. et al. 2005. Oxidation of *o*-xylene on mesoporous Ti-phosphate-supported VO_x catalysts and promoter effect of K^+ on selectivity. *Catal Today*, 99: 179–186.
49. Garcia Cortez, G., Fierro, J. L. G., Bañares, M. A. 2003. Role of potassium on the structure and activity of alumina-supported vanadium oxide catalysts for propane oxidative dehydrogenation. *Catal Today*, 78: 219–228.
50. Lemonidou, A. A., Nalbandian, L., Vasalos, I. A. 2000. Oxidative dehydrogenation of propane over vanadium oxide based catalysts: Effect of support and alkali promoter. *Catal Today*, 61: 333–341.
51. Bartholomew, C. H. 2001. Mechanisms of catalyst deactivation. *Appl Catal A Gen*, 212: 17–60.
52. Kamata, H., Takahashi, K., Odenbrand, C. U. I. 1999. The role of K_2O in the selective reduction of NO with NH_3 over a $V_2O_5(WO_3)/TiO_2$ commercial selective catalytic reduction catalyst. *J Mol Catal A Chem*, 139: 189–198.
53. Barteau, M. A. 1996. Organic reactions at well-defined oxide surfaces. *Chem Rev*, 96: 1413–1430.
54. Calatayud, M., Markovits, A., Minot, C. 2004. Electron-count control on adsorption upon reducible and irreducible clean metal-oxide surfaces. *Catal Today*, 89: 269–278.
55. Lewandowska, A. E., Calatayud, M., Tielens, F., Banares, M. A. 2011. Dynamics of hydration in vanadia-titania catalysts at low loading: A theoretical and experimental study. *J Phys Chem C*, 115: 24133–24142.
56. Besselmann, S., Freitag, C., Hinrichsen, O., Muhler, M. 2001. Temperature-programmed reduction and oxidation experiments with V_2O_5/TiO_2 catalysts. *Phys Chem Chem Phys*, 3: 4633–4638.

57. Calatayud, M., Tielens, F., De Proft, F. 2008. Reactivity of gas-phase, crystal and supported V$_2$O$_5$ systems studied using density functional theory based reactivity indices. *Chem Phys Lett*, 456: 59–63.

58. Tranca, D., Keil, F., Tranca, I., Calatayud, M., Dzwigaj, S., Trejda, M., Tielens, F. 2015. Methanol oxidation to formaldehyde on VSiBEA zeolite: A combined DFT/vdW/transition path sampling and experimental study. *J Phys Chem C*, 119: 13619–13631.

59. Burcham, L. J., Briand, L. E., Wachs, I. E. 2001. Quantification of active sites for the determination of methanol oxidation turn-over frequencies using methanol chemisorption and in situ infrared techniques. 1. Supported metal oxide catalysts. *Langmuir*, 17: 6164–6174.

60. Burcham, L. J., Briand, L. E., Wachs, I. E. 2001. Quantification of active sites for the determination of methanol oxidation turn-over frequencies using methanol chemisorption and in situ infrared techniques. 2. Bulk metal oxide catalysts. *Langmuir*, 17: 6175–6184.

61. Dobler, J., Pritzsche, M., Sauer, J. 2005. Oxidation of methanol to formaldehyde on supported vanadium oxide catalysts compared to gas phase molecules. *J Am Chem Soc*, 127: 10861–10868.

62. Romanyshyn, Y., Guimond, S., Kuhlenbeck, H., Kaya, S., Blum, R. P., Niehus, H., Shaikhutdinov, S. et al. 2008. Selectivity in methanol oxidation as studied on model systems involving vanadium oxides. *Top Catal*, 50: 106–115.

63. Khaliullin, R. Z., Bell, A. T. 2002. A density functional theory study of the oxidation of methanol to formaldehyde over vanadia supported on silica, titania, and zirconia. *J Phys Chem B*, 106: 7832–7838.

64. Goodrow, A., Bell, A. T. 2007. A theoretical investigation of the selective oxidation of methanol to formaldehyde on isolated vanadate species supported on silica. *J Phys Chem C*, 111: 14753–14761.

65. Bronkema, J. L., Leo, D. C., Bell, A. T. 2007. Mechanistic studies of methanol oxidation to formaldehyde on isolated vanadate sites supported on high surface area anatase. *J Phys Chem C*, 111: 14530–14540.

66. Gonzalez-Navarrete, P., Gracia, L., Calatayud, M., Andres, J. 2010. Density functional theory study of the oxidation of methanol to formaldehyde on a hydrated vanadia cluster. *J Comput Chem*, 31: 2493–2501.

67. Gonzalez-Navarrete, P., Gracia, L., Calatayud, M., Andres, J. 2010. Unraveling the mechanisms of the selective oxidation of methanol to formaldehyde in vanadia supported on titania catalyst. *J Phys Chem C*, 114: 6039–6046.

68. Gracia, L., Gonzalez-Navarrete, P., Calatayud, M., Andres, J. 2008. A DFT study of methanol dissociation on isolated vanadate groups. *Catal Today*, 139: 214–220.

69. Wachs, I. E., Weckhuysen, B. M. 1997. Structure and reactivity of surface vanadium oxide species on oxide supports. *Appl Catal A Gen*, 157: 57–90.

70. Rozanska, X., Fortrie, R., Sauer, J. 2007. Oxidative dehydrogenation of propane by monomeric vanadium oxide sites on silica support. *J Phys Chem C*, 111: 6041–6050.

71. Zhao, Z., Liu, J., Duan, A., Xu, C., Kobayashi, T., Wachs, I. E. 2006. Effects of alkali metal cations on the structures, physico-chemical properties and catalytic behaviors of silica-supported vanadium oxide catalysts for the selective oxidation of ethane and the complete oxidation of diesel soot. *Top Catal*, 38: 309–325.

72. Si-Ahmed, H., Calatayud, M., Minot, C., Diz, E. L., Lewandowska, A. E., Banares, M. A. 2007. Combining theoretical description with experimental in situ studies on the effect of potassium on the structure and reactivity of titania-supported vanadium oxide catalyst. *Catal Today*, 126: 96–102.

73. Calatayud, M., Minot, C. 2007. Effect of alkali doping on a V$_2$O$_5$/TiO$_2$ catalyst from periodic DFT calculations. *J Phys Chem C*, 111: 6411–6417.

Chapter 11

12. Metal-Supported Oxide Nanofilms

Marek Sierka

12.1 Introduction 335
12.2 General Overview 337
    12.2.1 Preparation Methods. 337
    12.2.2 Characterization. 337
    12.2.3 Applications 339
12.3 Structure and Properties 340
    12.3.1 Experimental Methods 340
    12.3.2 Role of Theory 341
        12.3.2.1 Stability: Ab Initio Thermodynamics 342
        12.3.2.2 Local versus Global Structure Optimization. 344
        12.3.2.3 Model Validation. 346
12.4 Metal-Supported Low-Dimensional Silica. 347
    12.4.1 Mo(112) Substrate 348
    12.4.2 Ru(0001) Substrate 354
12.5 Ordered Water Layers on MgO(100) Ultrathin Films. 359
    12.5.1 Synopsis 363
Acknowledgments. 363
References 363

12.1 Introduction

Metal oxides are a diverse and technologically very important type of materials with versatile applications including ceramics, electronic and optical devices, sensors, energy storage devices, and catalysts, to name a few. They exhibit a wide range of physical properties covering metals, semiconductors, and insulators, and also diverse optical, magnetic, and chemical behaviors, with the latter ranging from very reactive substances to almost completely inert ones. In the past decade, there has been an increasing scientific and technological interest in the exploitation of nanostructured metal oxides as nanoparticles and nanofilms (Fernández-García et al. 2004; Freund and Pacchioni 2008; Pacchioni and Valeri 2012). At the nanoscale, some properties of these systems, including chemical, electronic, optical, and magnetic ones, can be very different from those of their atomic and bulk counterparts (Goniakowski et al. 2008; Netzer et al. 2010; Pacchioni 2012). The extreme confinement due to the proximity of the interfaces stabilizes new structures and phases that otherwise cannot be obtained as bulk materials. This opens new possibilities for the development of advanced materials with controlled and tunable properties. In particular, thin and ultrathin oxide films supported on metallic substrates are of current interest due to their potential use as protective films, insulating layers in integrated circuits, elements of nanodevices, and supports for metal nanoparticles in sensors and catalysis (Chen and Goodman 2008; Freund and Pacchioni 2008; Freund 2010; Pacchioni and Valeri 2012; Shaikhutdinov and Freund 2012). In addition, they provide a convenient solution for charging problems that severely hamper the study surface properties of highly insulating oxides by means of surface science techniques that involve electrically charged probes (Diebold et al. 2010; Nilius et al. 2011; Shaikhutdinov and Freund 2012).

One of the key prerequisites for understanding and control of physical and chemical properties of metal-supported nanofilms is a detailed characterization of their atomic structure. In many cases, the nanofilms are amorphous, offering only a limited control over their structure and properties. Therefore, an increasing attention has been devoted to preparation and characterization of crystalline, highly ordered oxide nanofilms that allow for much better structural control (Freund and Pacchioni 2008; Netzer et al. 2010; Pacchioni 2012). However, even in case of the ordered films, these materials frequently present complex structures to solve. This is because very often not only their atomic structure but also their chemical composition differs substantially from that of the corresponding bulk crystalline phase. In addition, although surface science techniques can provide some information about possible arrangement of atoms, the interpretation of experimental data relies to a large extent on intuition (Sierka 2010). Therefore, it is extremely valuable to combine experimental studies and computational modeling. Harnessing the potential of computer simulations offers not only a unique possibility to explore structure and properties of these materials but also to guide the discovery of new systems with tailored properties. In particular, the advent of modern density functional theory (DFT) based on accurate gradient corrected and hybrid functionals, together with the possibility to perform calculation on systems containing hundreds of atoms, opens the possibility to obtain detailed insight into the atomic structure and properties at the nanoscale. Nowadays, DFT simulations have become an indispensable tool in structure identification of metal-supported oxide nanofilms by assigning experimental signatures to structure models and by predicting stabilities of different models depending on experimental conditions such as temperature or pressure (Freund and Pacchioni 2008; Sierka 2010; Pacchioni 2012).

The purpose of this chapter is to provide some illustrative examples of studies on metal-supported oxide nanofilms in which the combination of theory and experiment is pivotal for structure prediction and characterization. In order to provide the reader with necessary background information, this chapter starts with a general overview in Section 12.2. It includes some of the experimental methods that can be employed in the preparation and characterization of oxide nanofilms followed by selected applications with particular emphasis on model heterogeneous catalysis. Section 12.3 provides a brief discussion of the basic surface science methods and modeling methodology needed for atomic level characterization of the nanofilms. In Sections 12.4 and 12.5, two specific examples of joint experimental and computational studies of oxide nanofilms are given: metal-supported low-dimensional silica and ordered water layers on MgO(100) ultrathin films. Finally, Section 12.5.1 gives a synopsis with conclusions and outlook.

12.2 General Overview

12.2.1 Preparation Methods

A number of methods that can be employed in the preparation and synthesis of ultra-thin metal oxide films on metal surfaces exist, which can generally be classified based on chemical and physical processes (Netzer et al. 2010; Valeri and Benedetti 2012). The main chemical routes are chemical vapor deposition (CVD) and sol-gel techniques. These methods are particularly well suited for commercial preparation of thin films involving large areas or complicated shapes. In CVD, the substrate is exposed to one or a mixture of vapor-phase precursors that undergo chemical reactions or decomposition at the surface (Jones and Hitchman 2009). A particular modification of CVD, which allows atomic-scale deposition control and fabrication of highly crystalline oxide films, is the atomic layer deposition (ALD) in which vapor-phase precursors pulses are provided sequentially, separated by pulses of a neutral purge gas or evacuation of the CVD reactor (Ritala and Niinisto 2009). In sol-gel processes, liquid precursors are used for film preparation, usually leading to oxide films of low crystallinity and relatively high roughness (Valeri and Benedetti 2012).

The main physical methods for the preparation of thin and ultrathin oxide films are based on physical vapor deposition (PVD) or controlled oxidation of metal surfaces or predeposited thin metal or semiconductor films (Valeri and Benedetti 2012). PVD comprises a variety of methods in which the film is deposited on the surface by condensation of a vaporized form of the source material (Mattox 2009). It offers an advantage of utilizing virtually any type of material, on a broad range of surfaces, and in highly controlled conditions. Therefore, in most of the cases, the preparation of metal-supported oxide nanofilms employs PVD-based approaches such as reactive evaporation (RE) or post-oxidation (PO) methods (Netzer et al. 2010; Netzer and Surnev 2012). In RE, material is evaporated from a source with a partial pressure of oxygen or another oxidizing gas present in the chamber and deposited on a clean or preoxidized surface. In order to obtain crystalline oxide films, annealing usually follows the deposition step. The PO method involves deposition of metal atoms on the desired surface in ultrahigh vacuum (UHV) followed by oxidation. Other approaches use direct oxidation of the metal substrate, for example, the preparation of crystalline Al_2O_3 films on NiAl(110) (Jaeger et al. 1991).

12.2.2 Characterization

The small thickness of metal-supported oxide nanofilms allows the application of electron- and ion-mediated surface science methods without charging problems. Consequently, there exist a multitude of experimental techniques that can be employed in their characterization. Table 12.1 provides summary of the most important methods along with the physical effect employed by a particular method and the type of information obtained. For example, information about the topography of a surface can be obtained with an atomic resolution using atomic force microscopy (AFM) or scanning tunneling microscopy (STM). The image generation in AFM is based on the force interaction between the sensor tip and the surface. Since it does rely on sample conductivity, it can be applied to all materials, including isolators. In contrast, STM probes the local

Chapter 12

Table 12.1 Summary of the Most Important Surface Science Methods That Can Be Applied in Characterization of Metal-Supported Oxide Nanofilms along with the Physical Effect Employed and the Type of Information Obtained

Method	Physical Effect	Information
Spectroscopy		
Infrared reflection absorption spectroscopy (IRAS)	Vibrational excitations	IR spectrum, vibrational modes
Raman spectroscopy	Inelastic light scattering	IR spectrum, vibrational modes
Ultraviolet photoelectron spectroscopy (UPS)	Photoelectric effect	Electronic structure
X-ray photoelectron spectroscopy (XPS)	Photoelectric effect	Oxidation state and chemical composition
Near-edge x-ray absorption fine structure (NEXAFS, XANES)	X-ray adsorption	Atomic and electronic structure, oxidation, and coordination state
Auger electron spectroscopy (AES)	Auger effect	Oxidation state and concentration of elements, film thickness
Metastable impact electron spectroscopy (MIES)	Auger effect, deexcitation of metastable He states	Electronic structure
Electron energy loss spectroscopy (EELS)	Electron scattering	Chemical composition, film thickness
High-resolution electron energy loss spectroscopy (HREELS)	Electron scattering	Excitation energies, vibrational modes
Scanning tunneling spectroscopy (STS)	Tunnel effect	Local electronic structure (density of states)
Microscopy		
Scanning tunneling microscopy (STM)	Tunnel effect	Images of local electron density (LDOS)
Atomic force microscopy (AFM)	Force between cantilever and surface	Surface topology
Transmission electron microscopy (TEM)	Electron transmission	Structure
Diffraction		
X-ray diffraction (XRD)	Diffraction of x-ray photons	Diffraction patterns, structure
Low-energy electron diffraction (LEED)	Diffraction of low-energy electrons	Diffraction patterns, structure
Kinetic methods		
Temperature-programmed desorption (TPD)	Surface desorption	Desorption kinetics, surface coverage

density of states (LDOS) of the surface of a conducting material, and therefore, it images its electronic structure. In addition to providing topographical information, STM can also be used as a spectroscopic tool (scanning tunneling spectroscopy [STS]) giving information about local electronic structure of surfaces and materials on surfaces at the atomic scale (Oura et al. 2003; Nilius 2009). Transmission electron microscopy (TEM) is another microscopy method in which a beam of electrons is transmitted through an ultrathin probe, interacting with it as it passes through and finally forming an image (Oura et al. 2003).

The atomic and electronic structures of surfaces can also be probed using a whole family of photoemission spectroscopic techniques involving x-ray and ultraviolet radiation such as x-ray photoelectron (XPS) and ultraviolet photoemission (UPS) spectroscopies. XPS, in particular using higher intensity synchrotron radiation, can be employed to measure the elemental composition, and chemical and electronic state of the elements within the nanofilms. Infrared radiation (IR) spectroscopies such as infrared reflection adsorption spectroscopy (IRAS) are an important source of information about the vibrational excitations of surfaces and adsorbed molecules. Electronic excitations and vibrational modes of the surface and adsorbates can also be investigated using high-resolution electron energy loss spectroscopy (HREELS), which utilizes inelastic scattering of electrons from surfaces.

Scattering and diffraction of radiation as well as charged and neutral particles such as electrons, ions, and atoms can be used to provide structural and chemical data of surfaces. These techniques include surface x-ray diffraction (SXRD), photoelectron diffraction (PD), low-energy electron diffraction (LEED), ion scattering spectroscopy, and low-energy ion scattering spectroscopy. The list of available surface science methods given here is far from being complete and can hardly be summarized within this chapter. A more comprehensive overview can be found in Oura et al. (2003), Fernández-García et al. (2004), and Grinter and Thronton (2012).

12.2.3 Applications

The application areas of thin and ultrathin oxide films are as versatile as the variety of their properties. Depending on their composition, structure, and thickness, they can be soft and very reactive or show great hardness and chemical resistance. The electrical properties of oxide films range from insulators to conductors. Similarly, they can exhibit different ferroelectric, ferromagnetic, and optical properties that can in addition be fine-tuned by using dopants. The biggest and probably the oldest application of oxide thin films has been fabrication of electronic semiconductor devices (Stoneham et al. 2007; Bersuker et al. 2012; Pacchioni 2012), in which an insulating oxide layer performs a number of specific tasks including metal-oxide-semiconductor field-effect transistor gate layer and insulating layers. Until recently, this field of application has been dominated by silica (SiO_2). However, continuous decrease of the size of integrated circuits poses new challenges to the dielectric materials that cannot be supported by SiO_2 (Bersuker and Zeitzoff 2007; Bersuker et al. 2012). For this reason, ultrathin films of new classes of dielectric materials have been introduced, which are based on rare earth or transition metal oxides, for example, HfO_2 (Stoneham et al. 2007). Thin films of metal oxides formed on metals and alloys play also an important role in providing protection against corrosion (Marcus 2012), sensor applications (Brunet et al. 2012), and in a range of new technologies (Pacchioni 2012) including solid oxide fuel cells (Ishihara 2012) and solar energy materials (Granqvist 2012).

One of the important applications of metal-supported oxide nanofilms is their use as model systems in studies of heterogeneous catalysis (Ertl and Freund 1999; Freund and Pacchioni 2008; Freund 2010; Nilius et al. 2011; Shaikhutdinov and Freund 2012). Industrial, high surface area solid catalysts are very complex materials that very often present ill-defined surfaces and wide size distributions of the crystallites of the various phases present. This complicates the task of clearly identifying the influence of the underlying microscopic structure of the surface on the catalytic performance. Moreover, the capabilities of typical surface science techniques can only be applied in a limited way to typical heterogeneous catalysts because of their high surface roughness, presence of surface contaminants, and heterogeneity of the catalyst surface (Henry 1998; Freund et al. 2001; Besenbacher et al. 2009). Therefore, models of solid catalysts are used to investigate complex phenomena that are not accessible when working with real material (Boudart 2000; Freund et al. 2000; Vang et al. 2008). This knowledge leads to a better understanding of a catalytic phenomenon and can be applied to design and synthesize new or improve existing catalysts. The simplest model catalysts are small gas-phase aggregates of catalytically active species, which can give access to energetics and kinetics of catalytic reactions and the factors influencing them (Böhme and Schwarz 2005). However, gas-phase studies on isolated reactants lack the intrinsic heterogeneity present in applied catalysis. Nanoclusters of catalytically active species supported on well-ordered ultrathin oxide films on metallic surfaces allow, unlike isolated gas-phase systems, a study of size and support effects in heterogeneous catalysis under well-defined conditions by a large number of surface science techniques (Freund et al. 2000; Libuda and Freund 2005; Freund 2012).

12.3   Structure and Properties

12.3.1   Experimental Methods

Many of the surface science methods described in Section 12.2.2 can be applied to provide some structural information on metal-supported oxide films down to the atomic scale. The most obvious method is STM since it has the ability to directly image atomic-scale features of the outermost surface layers (Van Hove et al. 1997; Nilius 2009; Grinter and Thronton 2012). However, the interpretation of STM images in terms of atomic position is far from being straightforward. The reason is that STM images electronic structure and not atomic positions directly (Van Hove et al. 1997; Woodruff 2002; Van Hove 2009). As an example, electronegative atoms such as oxygen adsorbed on close-packed metal surfaces generally (but not always) appear as depressions, whereas electropositive or more neutral atoms and molecules, such as carbon, hydrocarbons, and other metal atoms, appear in most of the cases as protrusion in an STM image. Additional complication is that the tunneling current depends on geometric and electronic structure of the tip, which is difficult to control experimentally and in fact often changes during the experiment. Therefore, STM images alone are only able to provide *a range of speculative structural models* (Woodruff 2002). Similar problems are encountered by AFM, since an AFM image does not reflect the true surface topography, but rather represents a convolution of the shape of the tip and the shape of the sample (Eaton and West 2010).

Another method that is commonly used for structure analysis of nanofilms is LEED, which employs bombardment of a surface with a collimated low-energy electron and observation of resulting diffraction pattern (Van Hove 1997, 1999; Van Hove et al. 1997). It provides structural information such the size, symmetry, and rotational alignment of the surface unit cell of the film with respect to the substrate unit cell. Recording intensities of diffracted beams as a function of incident electron beam energy can be employed to generate the so-called I–V curves. Additionally, some other diffraction methods such as PD and x-ray absorption fine structure (XAFS) can be applied in a similar way as LEED. However, due to a complicated nature of scattering phenomenon, these methods cannot be used to determine atomic structures of the films directly. Rather, the atomic structure is determined by a trial-and-error procedure that requires first specifying atomic coordinates and then refining them to fit the experimental data such as I–V curves. This implies a priori knowledge of possible structure candidates that are improved stepwise until a satisfactory agreement between experimental and simulated data is obtained (Van Hove 1999, 2009).

One of the other frequently used methods for structure analysis of oxide nanofilms is SXRD (Van Hove 1999; Grinter and Thronton 2012), in which x-rays are diffracted from the surface under a small angle of incidence, resulting in additional structure along so-called truncation rods that are continuous rods of intensity connecting Bragg peaks along the surface normal direction. Because the volume of the surface region is less compared to the bulk crystal, it results in much less intense diffraction spots, making synchrotron x-ray source necessary in order to acquire the data in a reasonable amount of time. For more complex surface structures, a sufficiently large set of experimental data is necessary to determine the unknown structural parameters. This can make the acquisition of the data a lengthy process, during which time a surface can change, particularly due to adsorption from the always imperfect vacuum. Direct methods based on XRD scattering need a particularly large amount of data, and they are not yet often used. Therefore, similar to LEED, the structure determination of surfaces based on x-ray diffraction (XRD) employs a trial-and-error approach, which requires an initial guess of the atomic positions (Van Hove 1999).

12.3.2 Role of Theory

The past two decades have witnessed a tremendous increase of computing power as well as advances in computational methods and algorithms. Particularly, the development of modern DFT methods based on accurate gradient corrected and hybrid functionals, along with algorithmic improvements that allow simulations of systems containing hundreds of atoms, opens the possibility to obtain detailed insight into the structure and properties at the atomic scale. In principle, computer simulations could be used on their own to predict and characterize nanostructures. However, theory faces its own problems when applied to such complex materials. For small molecules and clusters containing only a few atoms, the accuracy of the approximate methods such as DFT can be assessed by comparison with the results of accurate, high-level quantum chemical calculations.

Chapter 12

In case of nanostructures, the accuracy of the approximate methods remains basically unknown. In addition, the complexity of the nanostructures very often results in materials where not only the atomic structure but also the chemical composition differs substantially from that of the corresponding stoichiometric molecular or bulk phase. Structural characterization of such systems is challenging, because the number of possible structural candidates increases rapidly with the system size. Therefore, due to the difficulty of measuring and simulating structure and processes at the nanoscale, it is extremely valuable to combine experiments and computational modeling (Kresse et al. 2005; Wu et al. 2009; Sierka 2010; Barcaro et al. 2012).

12.3.2.1   Stability: Ab Initio Thermodynamics

One of the issues of atomistic simulations of surfaces in general and metal-supported oxide nanofilms in particular is the problem of comparing the relative stability of different structure models. For systems with identical chemical composition, such as two isomers of a given molecule, the simplest measure of the stability can be obtained by comparing calculated total energies. However, how can we compare the stability of two systems that differ in their chemical composition, such as two metal surfaces with a different number of adsorbed O_2 molecules? Such a comparison can be achieved by combining electronic structure calculations and thermodynamic concepts. This *ab initio thermodynamics* approach not only provides the means of comparing the stability of structures that differ in their composition but also allows determining the equilibrium composition of a system in contact with a chemical environment at a given temperature and pressure. Under conditions of constant temperature T and pressure p, which correspond to most of the experiments performed on surfaces, the appropriate thermodynamic potential that has to be used is the Gibbs energy $G(T,p)$. The following process is introduced to formally describe the formation of a given surface structure

$$[S] + \sum_{i=1}^{n} v_i X_i \rightarrow \frac{[v_i X_i]_n}{[S]}, \tag{12.1}$$

where
   [S] denotes the initial surface configuration
   X_i are the species (e.g., Si, Ti, O_2) forming the surface layer $[v_i X_i]_n/[S]$
   v_i denote stoichiometric coefficients

The Gibbs energy change $\Delta G_{\text{form}}(T,p)$ associated with this process is given by

$$\Delta G_{\text{form}}(T,p) = G_{[v_i X_i]_n/[S]} - G_{[S]} - \sum_{i=1}^{n} v_i \mu_{X_i} = \Delta G_{\text{form}}(0\ \text{K}) + \Delta G_{[v_i X_i]_n/[S]} - \Delta G_{[S]} - \sum_{i=1}^{n} v_i \Delta \mu_{X_i}, \tag{12.2}$$

where
   $G_x = G_x(T,p)$ and $x = [S], [v_i X_i]_n/[S]$ are the Gibbs energies of [S] and $[v_i X_i]_n/[S]$
   $\mu_x = \mu_x(T,p_x)$ and $x = X_i$ are the chemical potentials

The Gibbs energy of formation at 0 K, $\Delta G_{\text{form}}(0 \text{ K})$, is

$$\Delta G_{\text{form}}(0 \text{ K}) = E^0_{[\nu_i X_i]_n/[S]} - E^0_{[S]} - \sum_{i=1}^{n} \nu_i E^0_{X_i},$$ (12.3)

where $E^0_x = E^{\text{el}}_x + E^{\text{ZPV}}_x$ depends only on the total (electronic), E^{el}_x, and zero-point vibrational energies, E^{ZPV}_x. For gas-phase species, the relative chemical potentials $\Delta\mu_{X_i} = \Delta\mu_{X_i}(T, p_i)$ can be expressed within the ideal-gas, rigid-rotor harmonic oscillator approximation as

$$\Delta\mu_{X_i} = \mu_{X_i} - E^0_{X_i} = -RT \ln\left(Q^{\text{vib}}_{X_i} Q^{\text{rot}}_{X_i} Q^{\text{trans}}_{X_i}\right) + RT,$$ (12.4)

where
volume work of the gas phase is replaced by RT (ideal gas)
Q_x are the partition functions for vibrations (vib), rotations (rot), and translations (trans)

This corresponds to referencing the chemical potentials to the chemical potential at 0 K as a standard. Most of the available computer programs for electronic structure calculations provide tools for convenient evaluation of the partition functions and chemical potentials using statistical mechanics. Under the assumption of thermodynamic equilibrium, $\Delta\mu_{X_i}$ can also be related to the relative Gibbs energy per formula unit, Δg_{M_i}, of some bulk material M_i acting as a source of X_i, for example, bulk oxide (Reuter and Scheffler 2002). Neglecting volume changes of solid components, the relative Gibbs energies, ΔG_x, in Equation 12.2 include only vibrational contributions

$$\Delta G_x = G_x - E^0_x = -RT \ln\left(Q^{\text{vib}}_x\right).$$ (12.5)

Since the calculation of the vibrational partition function Q^{vib}_x for solid materials is much more demanding than obtaining the electronic energy, this term is often neglected, that is, $\Delta G_{[\nu_i X_i]_n/[S]} = \Delta G_{[S]} = 0$ in Equation 12.2. In addition, since $Q^{\text{vib}}_{X_i}$ does not contribute to the pressure dependence of the chemical potentials for gas-phase species, its contribution is also often only roughly estimated or neglected (Reuter and Scheffler 2002). For homogeneous surfaces, $\Delta G_{\text{form}}(T,p)$ scales linearly with the surface area A. Therefore, the Gibbs energy change per unit area, $\Delta\gamma$, is introduced:

$$\Delta\gamma = \Delta\gamma(T, p) = \frac{1}{A}\Delta G_{\text{form}}(T, p).$$ (12.6)

Using Equations 12.2 through 12.6, $\Delta\gamma$ can be plotted as a function of the relative chemical potentials $\Delta\mu_{X_i}$, yielding a surface phase diagram and providing means for comparing stability of structures with different chemical compositions—at any given set of $\Delta\mu_{X_i}$ the structure with the lowest value of $\Delta\gamma(T,p)$ is the most stable one. The values of $\Delta\mu_{X_i}$ can be related to experimental conditions, such as temperature, partial pressures, or mole fractions of source bulk materials using statistical mechanics (Zhang et al. 2004).

Chapter 12

12.3.2.2 Local versus Global Structure Optimization

Atomic level characterization of surface structures and nanofilms using computer simulations is often complicated by a large number of possible structural isomers. In addition, their atomic structures may differ significantly from those of their molecular or bulk counterparts. Brute-force determination of the most stable configuration by manual construction of possible structural models and following local structure optimizations is a difficult task, which often leads to erroneous structure assignments. Therefore, several techniques for automatic location of most stable surface structures have been proposed that rely on global structure optimization methods (Ciobanu et al. 2009). One of the promising approaches is genetic algorithm (GA) that finds the global minimum structure by a method that mimics the process of natural evolution (Chuang et al. 2004; Hartke 2011). Due to a high computational cost of global minimization algorithms, most implementations rely on parameterized interatomic potential functions or semiempirical methods. However, in some cases, the application of more accurate DFT or quantum chemical approaches is necessary. Although the global optimization techniques in combinations with DFT are computationally demanding, they start to play an increasing role in the structure determination of surface structures (Sierka 2010).

To illustrate the details of GA application to global optimization of surfaces in some more details, I use the example of our own implementation (Sierka 2010; Sierka and Włodarczyk 2013) with schematic flow chart shown in Figure 12.1. In general, a

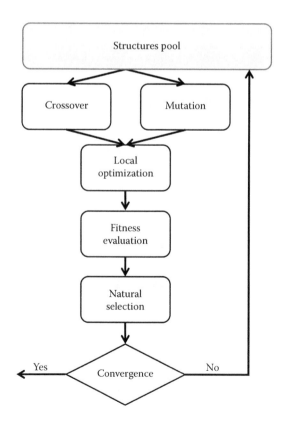

FIGURE 12.1 Schematic flow chart of a global structure optimization using genetic algorithm.

population of candidate surface structures is evolved toward improved solutions and variables to be optimized are atomic coordinates of surface atoms. The algorithm can operate in two modes, either constant composition or constant chemical potential mode. In the former one, the objective function to be minimized is the total (electronic) energy and the global minimum of a system with a known chemical composition is found. In the latter one, the objective function is the Gibbs energy change per unit area $\Delta\gamma$, Equation 12.6, which is globally minimized for given values of the relative chemical potentials $\Delta\mu_{X_i}$. The number of atoms is allowed to vary, permitting simultaneous determination of both the composition and bonding topology of the most stable structure under given experimental conditions, for example, oxygen pressure and temperature. In both modes, the GA optimization starts with a pool of randomly generated surface models that are optimized to the nearest local minimum using a conjugate-gradient technique. Two evolutionary operators—crossover and mutation—are used to exchange structure information between the pool members. In the crossover, pairs of structures are chosen to act as parents that will produce a child structure for the next generation. Within each parent pair, random pieces are exchanged to form a new mixed structure, the child, which is subsequently optimized to the nearest local minimum. This cut and splice crossover operation is illustrated in Figure 12.2. The hope is that the child will combine the good structural features of the two parents and thus will be more stable than either one. Evolutionary pressure toward improved children is added to bias the search in the right direction. This is achieved by selecting parents that have a relatively high fitness f_i defined as

$$f_i = \frac{\exp(-\alpha\varepsilon_i)}{\sum_i \exp(-\alpha\varepsilon_i)}, \tag{12.7}$$

where
  α is a constant
  ε_i is the dynamically scaled relative energy

$$\varepsilon_i = \frac{E_i - E_{min}}{E_{max} - E_{min}}, \tag{12.8}$$

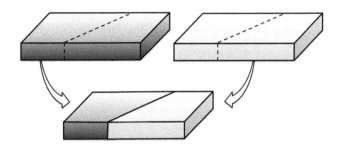

FIGURE 12.2 Schematic representation of the cut and splice crossover operation for surfaces. Two randomly chosen pieces of parent surface models are spliced together to form a child structure.

with E_i as the total energy and $E_{\min}$ and $E_{\max}$ as the lowest and highest total energies in the population, respectively. In the constant chemical potential mode, total energies are replaced by the corresponding values of $\Delta\gamma$. Elitism or natural selection is applied since the child structures do not always present a better fitness than their parents. It is achieved by simply replacing the worse parents by better children in the structures pool. In order to maintain a maximum diversity during the GA runs, similarity recognition is used, allowing only distinct structures to be included in the pool. To prevent trapping of the population in local minima, mutation is added in which random changes are introduced to randomly chosen structures in the pool. Figure 12.3 shows an example of structure evolution in a GA global optimization of water layer adsorbed on MgO(100)/Ag(100) ultrathin film (Włodarczyk et al. 2011).

12.3.2.3 Model Validation

It has to be stressed that even performing global structure optimization does not guarantee that an important structure is not missing. In addition, the description of the system at the DFT level may not be accurate enough to discern structure models that are close in energy. Therefore, the results of simulations have to be validated by a comparison between observable quantities calculated for the most stable structure models and experimental data. One direct method of verification is the comparison of simulated and experimental, atomic resolution STM images. The simplest and the most popular method for generating STM images using the results of electronic structure calculations is the Tersoff–Hamann approach (Tersoff and Hamann 1985). It is based on the assumption that the tunneling current is proportional to LDOS (at the Fermi level) of the surface, at the position of the STM tip. This implies that in the STM experiment, the

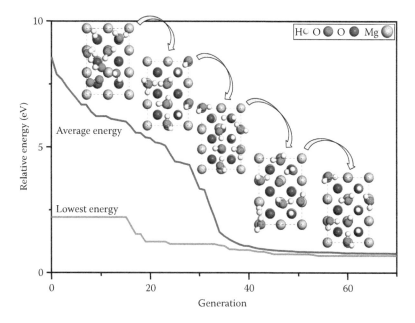

FIGURE 12.3 Evolution of structures in the genetic algorithm (GA) global optimization of a water layer containing seven H₂O molecules adsorbed on a (3 × 2) MgO(100)/Ag(100) ultrathin film (Włodarczyk et al. 2011). Top view of several most stable structures found in the course of the GA run is shown.

tip follows a contour of constant LDOS. Therefore, STM images can be simulated plotting an isosurface of LDOS calculated at the Fermi level.

Vibrational spectroscopies (e.g., IRAS and HREELS) are useful and often used tools for characterization of ultrathin oxide films. IR vibrational frequencies can be routinely calculated by most of the DFT software packages using analytical or numerical second derivatives of energy with respect to nuclear displacements. The corresponding IR intensities can be obtained from the dipole moment derivatives. However, care has to be taken when calculating IR spectra and comparing the results with experimental data. The reason are the selection rules in surface spectroscopies that allow only selected vibrational modes to be present in IRAS or HREELS spectra of thin layers on metal surfaces (Bradshaw and Richardson 1996). For IRAS, the selection rule states that only vibrational modes with a component of the dynamic dipole moment perpendicular to the metal surface will contribute to the IR absorption. These vibrational modes are, in general, only the totally symmetrical modes of the film. In a dipole scattering regime, HREELS obeys the same selection rule that applies to IR spectroscopy—only the completely symmetrical vibrational modes are allowed when the interaction is between the electrical field of the incoming electron and the vibrational dipole moment of the surface layer. Due to these selection rules, vibrational intensities for thin layers on metal surfaces are usually obtained only from the derivatives of the dipole moment component perpendicular to the surface. The agreement between calculated harmonic vibrational frequencies and observed fundamentals can be improved using an empirical scaling factor derived from calculated and observed frequencies of known compounds (Scott and Radom 1996).

In addition to STM images and vibrational spectra, other experimental data such as NEXAFS and XPS can be used to verify calculated structure models. For example, surface core-level shifts (Lizzit et al. 2001) can be calculated and compared to XPS data. The SCLS is defined as the difference in core-level binding energies of an atom on the surface and in bulk solid. The binding energies are obtained from the energy difference between the ground state structure and a structure where one electron is removed from the core of one particular atom and added to the valence or conduction band.

12.4 Metal-Supported Low-Dimensional Silica

Silica (SiO_2) is one of the most abundant materials in the Earth's crust. It has a myriad of applications starting with microelectronics over materials science up to catalysis and photonics, making it one of the technologically most important minerals. In particular, thin SiO_2 films play a very important role in the fabrication of integrated circuits. The continuing miniaturization of microchips makes understanding of the properties of silica layers at the atomic-scale essential for the performance of new generations of electronic devices. Another important application of silica is its use as support in heterogeneous catalysis. Therefore, the use of well-ordered, ultrathin metal-supported SiO_2 films as a model of the support material provides means to capture some of the complexity represented by the real catalysts by allowing the application of a multitude of surface science techniques to study surfaces at the atomistic level. There exist only few examples of ultrathin, highly ordered silica films. Among them, the films grown on surfaces of molybdenum and ruthenium are very instructive examples of systems,

Chapter 12

where full atomic structure determination and characterization could only be achieved by combining experimental studies and computer simulations.

12.4.1   Mo(112) Substrate

The first preparation of well-ordered silica films was reported by Schroeder et al. (2000) on a Mo(112) substrate using repeated cycles of silicon deposition and subsequent oxidation, followed by a final annealing procedure. Experimental data indicated complete coverage of the Mo(112) substrate by a hexagonal, $c(2 \times 2)$-symmetric highly crystalline silica film with a commensurate relationship to the substrate. XPS studies of the films showed silicon solely in the Si^{4+} state and at least two types of oxygen atoms, which were assigned to $Si-O-Si$ and $Si-O-Mo$ species (Schroeder et al. 2001). UPS valence band study demonstrated features similar to known silica compounds with corner-sharing $[SiO_4]$ tetrahedra as building units and high-resolution STM images revealed a honeycomb-like surface structure. IRAS spectrum of the film showed only three absorption bands, a strong one at 1048 cm⁻¹ and two weak ones at 771 and 675 cm⁻¹ (Schroeder et al. 2002). This simple spectrum indicated a high symmetry and/or relatively simple atomic structure of the film. Inconsistent results were obtained concerning the film thickness. Based on Auger electron spectroscopy (AES) and angle-resolved XPS spectra, the film thickness of 5–8 Å was estimated (Schroeder et al. 2000; Kim et al. 2003). However, Chen et al. (2004) using AES concluded that the film is only 3 Å thick.

Despite the wealth of experimental data collected for the crystalline SiO_2/Mo(112) film, its precise atomic structure remained a mystery for over 5 years. The first structure models were suggested based solely on experimental data. Schroeder et al. (2002) proposed two models based on structures of low-index surfaces of known dense silica polymorphs, β-tridimite/β-cristobalite as well as α-/β-quartz, which consist of corner-sharing $[SiO_4]$ tetrahedra attached to the Mo(112) surface. Chen et al. (2004) suggested another structure model based on a comparison of HREEL spectra of the silica film with several Si and Mo oxide compounds available in the literature. The authors argued that the absence of an IR absorption band in the 1150–1200 cm⁻¹ region, which is characteristic for the $Si-O-Si$ asymmetric stretching in bulk silica, was an indication of the absence of $Si-O-Si$ bonds. The intense band at 1048 cm⁻¹ was assigned to the asymmetric stretching vibration of $Si-O-Mo$ species. Based on these considerations, they proposed a model consisting of a layer of isolated $[SiO_4]$ clusters arranged in a $c(2 \times 2)$ structure with all oxygen atoms bonding to Mo atoms of the Mo(112) substrate as shown in Figure 12.4.

In the first computational study of the SiO_2/Mo(112) film, Ricci and Pacchioni (2004) assumed film thickness of about 10 Å corresponding to three layers of $[SiO_4]$ tetrahedra. Using DFT, the authors investigated OH-terminated and fully dehydroxylated silica models cleaved from β-tridimite and β-cristobalite as well as α- and β-quartz bulk structures. They compared the stability and adhesion energies of the models and concluded that the β-cristobalite-derived structure was the most stable and exhibited the strongest adhesion energy to the Mo substrate. The experimental IR absorption band at 1048 cm⁻¹ was assigned to $Si-O$ stretching frequencies, and the formation of two-membered $Si-O-Si$ rings was invoked to explain the weak band at around 795 cm⁻¹. However, other properties calculated for this model did not show satisfactory agreement with experimental data (Ricci and Pacchioni 2004; Todorova et al. 2006).

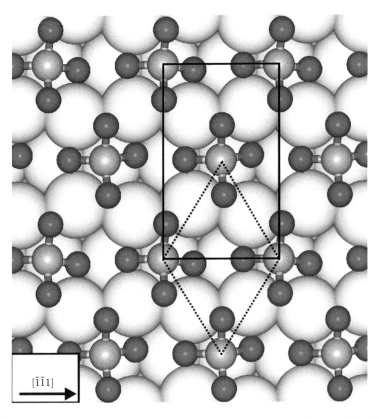

FIGURE 12.4 Isolated [SiO₄] clusters arranged in a $c(2 \times 2)$ structure proposed as model of the crystalline SiO₂/Mo(112) film by Chen et al. (2004) (Si, orange; O, red). Black rectangle indicates the surface unit cell. (Reprinted with permission from Todorova, T.K., Sierka, M., Sauer, J., Kaya, S., Weissenrieder, J., Lu, J.-L., Gao, H.-J., Shaikhutdinov, S., and Freund, H.-J., *Phys. Rev. B*, 73, 165414. Copyright 2006 by the American Physical Society.)

The final elucidation of the atomic structure of the ultrathin SiO₂/Mo(112) film was only possible combining computational end experimental studies (Weissenrieder et al. 2005). DFT calculations were pivotal for selecting the right structure among various structural candidates constructed based on the known experimental data: (1) $c(2 \times 2)$ periodicity derived from the LEED and STM, (2) simple atomic structure and/or high symmetry suggested by XPS and IRAS, (3) hexagonal arrangement of the building blocks suggested by STM results and fully saturated surface indicated by chemical inertness of the film, and (4) an average thickness of the film of about 5–8 Å suggested by AES and angle-resolved XPS spectra. Based on structures found in micas and clay minerals (Liebau 1985), several structures were constructed containing one (1 ML) or two (2 ML) monolayers of [SiO₄] tetrahedra, as shown in Figure 12.5. Since the 1 ML and 2 ML structures differ by their chemical composition, their stability was compared using Gibbs energy change per unit surface, $\Delta\gamma$, of a hypothetical formation process of the films from a clean Mo(112), bulk α-quartz, and molecular oxygen (c.f. Section 12.3.2)

$$\text{Mo}(112) + m(\text{SiO}_2)_{\alpha\text{-quartz}} + \frac{n}{2}\text{O}_2 \rightarrow (\text{SiO}_2)_m \times n\frac{\text{O}}{\text{Mo}(112)}, \tag{12.9}$$

Chapter 12

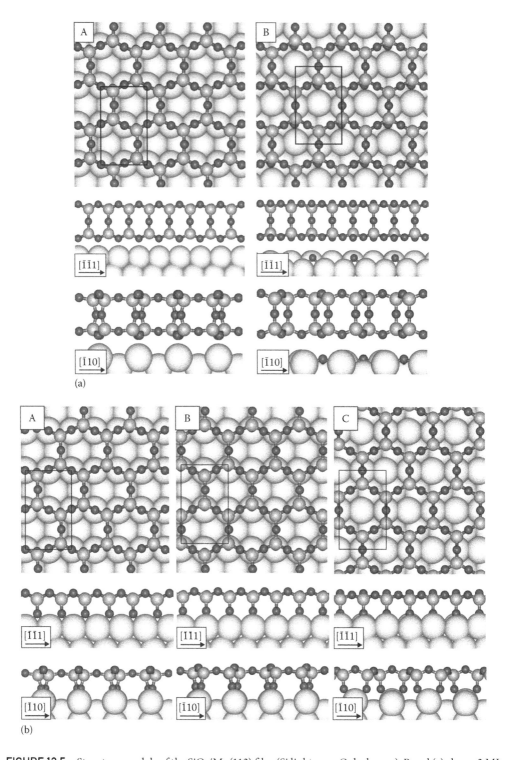

FIGURE 12.5   Structure models of the SiO_2/Mo(112) film (Si light gray, O dark gray). Panel (a) shows 2 ML models A and B. Panel (b) shows 1 ML models A–C. (Reprinted with permission from Todorova, T.K., Sierka, M., Sauer, J., Kaya, S., Weissenrieder, J., Lu, J.-L., Gao, H.-J., Shaikhutdinov, S., and Freund, H.-J., *Phys. Rev. B*, 73, 165414. Copyright 2006 by the American Physical Society.)

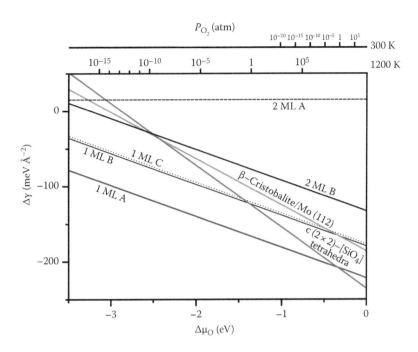

FIGURE 12.6 Gibbs energy change per unit surface $\Delta\gamma(T,p)$ for the most stable models of the SiO$_2$/Mo(112) film as a function of oxygen chemical potential $\Delta\mu_O(T,p)$: 2 ML and 1 ML models, β-cristobalite film (Ricci and Pacchioni 2004), and the isolated $c(2 \times 2)$-[SiO$_4$] tetrahedra model (Chen et al. 2004). In the top horizontal axis, the dependence on $\Delta\gamma(T,p)$ has been cast into a pressure scale at fixed temperatures of $T = 300$ and 1200 K. (Reprinted with permission from Todorova, T.K., Sierka, M., Sauer, J., Kaya, S., Weissenrieder, J., Lu, J.-L., Gao, H.-J., Shaikhutdinov, S., and Freund, H.-J., *Phys. Rev. B*, 73, 165414. Copyright 2006 by the American Physical Society.)

where m and n are the number of SiO$_2$ units and oxygen excess in the films, respectively. As can be seen in Figure 12.6, the surface model denoted 1 ML A is the most stable over the relevant range of relative oxygen chemical potentials $\Delta\mu_O(T,p)$. This model, shown in Figure 12.5, consists of a 2D network of corner-sharing [SiO$_4$] tetrahedra with three oxygen atoms of each tetrahedron forming Si−O−Si bonds. The fourth oxygen atom from each tetrahedron forms Si−O−Mo bond to the protruding rows of the Mo(112) surface. All atoms within the silica layer are fully saturated, which explains its chemical inertness. Comparison between simulated and experimental XPS and IRAS spectra as well as STM images definitely confirmed the 1 ML A structure as the model of the SiO$_2$/Mo(112) film. In particular, the calculated IRAS spectra show an intense signal at 1059 cm^{-1} and two less intense signals at 771 and 675 cm^{-1} that are in excellent agreement with experimental data. A detailed analysis of the calculated IRAS spectrum of the film by Kaya et al. (2007) demonstrated that its simple structure (only three bands are observed) is directly linked to the high symmetry of the film due to the already mentioned surface selection rules (Section 12.3.2). In case of the 1 ML A model, the high symmetry of its surface unit cell allows only three IR active vibrational modes at 1059, 771, and 675 cm^{-1}, which are precisely the three modes observed in the IRAS experiments. These results demonstrate that IR spectra of ultrathin oxide films cannot be simply interpreted by a straightforward comparison with bulk materials and thicker

Chapter 12

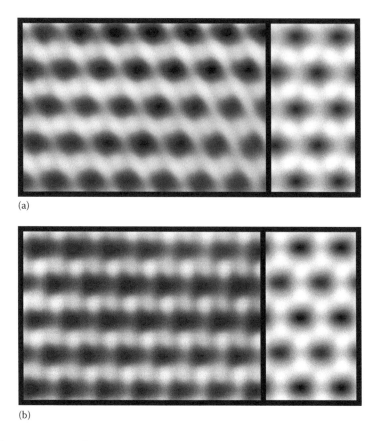

FIGURE 12.7 Comparison of experimental high-resolution (left, color) and simulated (right, b/w) STM images. The following are the image sizes and tunneling parameters for experimental images: (a) 5×2.5 nm^2, $V_S = 0.65$ V, and $I = 0.8$ nA, and (b) 5×2.5 nm^2, $V_S = 1.2$ V, and $I = 0.35$ nA. (Reprinted with permission from Todorova, T.K., Sierka, M., Sauer, J., Kaya, S., Weissenrieder, J., Lu, J.-L., Gao, H.-J., Shaikhutdinov, S., and Freund, H.-J., *Phys. Rev. B*, 73, 165414. Copyright 2006 by the American Physical Society.)

films. Such an assumption was the source of the erroneous structure model of the film proposed by Chen et al. (2004).

Additional confirmation of the 1 ML A model as structure of the SiO$_2$/Mo(112) film came from the comparison of simulated and experimental high-resolution STM images shown in Figure 12.7. The experimental images display a honeycomb-like structure, which is in a very good agreement with the STM images calculated for the 1 ML A model. In contrast, STM image simulated for the isolated [SiO$_4$] cluster model suggested by Chen et al. (2004) differs substantially from the experimental data (Todorova et al. 2006). Furthermore, simulations identified the protrusions imaged by STM as Si atoms at a large tip-surface distance and as O atoms at smaller distance. It has to be noted that the same conclusion about the structure of the film was independently reached by Giordano et al. (2005) based on DFT simulations. Finally, the 1 ML A model of the SiO$_2$/Mo(112) film was also confirmed experimentally by grazing scattering of fast atoms (Seifert et al. 2009) and I–V LEED (Kinoshita and Mizuno 2011).

In an attempt to directly measure the thickness of the SiO$_2$/Mo(112) film using STM by preparing the film at low Si (submonolayer) coverage, Lu et al. (2006) observed

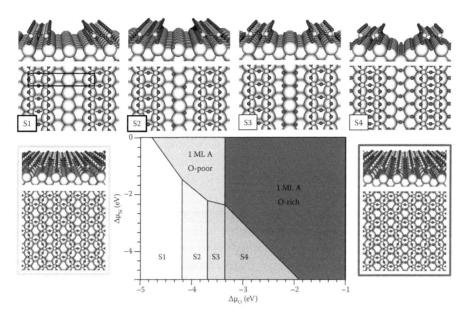

FIGURE 12.8 Calculated phase diagram of silica on the Mo(112) surface showing most stable phases as a function of relative chemical potentials $\Delta\mu_{Si}$ and $\Delta\mu_O$: 1D silica stripes (S1–S4) as well as O-poor and O-rich silica films (Si light gray, O dark gray, Mo white).

formation of 5 Å wide 1D silica stripes. The atomic structure of the stripes, which was determined using combining DFT calculations and surface science techniques, involves corner-sharing rows of $[SiO_4]$ tetrahedra bound to the (1×3) reconstructed Mo(112) surface, as shown in Figure 12.8 (structure S4). The metal surface is fully covered with oxygen atoms including surface area underneath the silica stripes.

The observation of this new low-dimensional silica structures on the Mo(112) surface formed under different experimental conditions was a motivation for more extensive computational study of this system (Sierka et al. 2006). Over 60 models of 1D and 2D silica structures on the Mo(112) substrate were constructed. For all the models, the most stable surface configuration (reconstructed vs. unreconstructed) and distribution of surface oxygen atoms were found by global structure optimizations at the DFT level, employing the GA implementation described in Section 12.3.2. The target function used in the global optimizations was the Gibbs energy change per unit surface associated with a hypothetical formation process from a clean Mo(112) surface, bulk silicon, and molecular oxygen:

$$[Mo_z]_{(112)} + mSi_{(bulk)} + \frac{n}{2}O_2 \rightarrow (SiO_2)_m \frac{O_{n-2m}}{[Mo_{z-u}]_{(112)}} + u[Mo]_{(bulk)}, \tag{12.10}$$

where
  m and n are the number of Si and O atoms in the unit surface cell, respectively
  $(SiO_2)_m \cdot O_{n-2m}/[Mo_{z-u}]_{(112)}$ defines the composition of a particular structure model

The term $[Mo]_{bulk}$ denotes bulk metal Mo atoms and appears in case of a surface reconstruction that results in a change of Mo content in the unit cell. The results of the

Chapter 12

calculations were combined in a 2D phase diagram shown in Figure 12.8 that shows the stability regions of different silica phases as a function of relative chemical potentials $\Delta\mu_{Si}$ (i.e., amount of Si on the surface) and $\Delta\mu_O$ (i.e., oxygen partial pressure). In agreement with known experimental data, the calculations predict various 1D silica stripes to be energetically most favorable at low values of $\Delta\mu_{Si}$. At very low $\Delta\mu_O$ the (1×3) stripes on a clean, unreconstructed Mo(112) surface constitute the most stable model (S1, Figure 12.8). Increasing $\Delta\mu_O$ leads to increased adsorption of oxygen on the Mo surface (S2 and S3, respectively, Figure 12.8). At higher $\Delta\mu_O$ values, the most stable structure was the already known model of 1D silica stripes on the reconstructed Mo(112) surface covered by oxygen atoms (S4, Figure 12.8). At high values of $\Delta\mu_{Si}$ and low values of $\Delta\mu_O$, the 2D 1 ML A structure of the silica film involving corner-sharing $[SiO_4]$ tetrahedra is formed.

However, in addition to the known 1 ML A structure, a new, oxygen-rich phase was predicted to be the most stable at even higher values of $\Delta\mu_O$. It contains four additional oxygen atoms per surface unit cell adsorbed on the metal surface beneath the silica film, as shown in Figure 12.8. Structural parameters for the new, oxygen-rich and the original, oxygen-poor models were virtually identical, demonstrating that the additional oxygen atoms on the Mo surface have little influence on the structure of the silica film. Thus, the properties of these two films were expected to be very similar. These theoretical findings were subsequently confirmed by a successful synthesis of the O-poor and O-rich silica films by Sierka et al. (2006). Both calculated and experimental STM images for the O-rich and O-poor films showed no discernible differences. Also no changes were observed in the LEED patterns, confirming that additional oxygen atoms chemisorbed on the Mo surface do not change the symmetry of the film. However, the analysis of vibrational properties and electronic structures using IRAS and XPS revealed small but detectable changes, which agreed well with theoretical predictions. Due to subtle differences between the properties of the O-poor and O-rich films, the existence of the latter might not have been discovered without the aid of theory. Although the oxygen atoms adsorbed on the Mo surface beneath the silica film have little effect on its network structure, they have to be taken into account in reactions of metal or metal oxide clusters deposited on the film.

12.4.2 Ru(0001) Substrate

Experimental attempts to grow well-ordered multilayer silica films on Mo(112) were not successful and resulted in amorphous structures (Stacchiola et al. 2008; Shaikhutdinov and Freund 2013). Recently, Loffler et al. (2010) reported formation of crystalline silica films on Ru(0001). Similar to $SiO_2/Mo(112)$, the films showed (2×2) LEED pattern and atomically resolved STM and AFM images revealed again a honeycomb-like hexagonal structure. The surface lattice constant of 5.42 Å determined by LEED was very close to that of $SiO_2/Mo(112)$, that is, 5.2–5.5 Å. However, the infrared reflection-absorption (IRA) spectra of the SiO_2 films on Ru(0001) and Mo(112) showed substantial differences. Very sharp and intense bands were observed at 1302 and 692 cm^{-1} for the film on Ru(0001). The band at 1302 cm^{-1}, which was about 250 cm^{-1} higher than that of Si$-$O$-$Mo stretching vibrations in $SiO_2/Mo(112)$, has never been observed on silica films and bulk. Based on the XPS results, IRA spectra of the ^{18}O-labeled films, and symmetry considerations, this band was associated with stretching vibrations of the Si$-$O$-$Si bonds oriented perpendicular to the Ru(0001) surface. Taking into account that silica films formed on Ru possessed approximately twice as much Si as on Mo(112) led to the suggestion that the film is composed of two layers of corner-sharing [SiO$_4$] tetrahedra bonded together by a linking oxygen, which acts as a plane of mirror symmetry, as shown in Figure 12.9 (structure D1).

This bilayer silica model was examined by DFT calculations. Since the silica sheet having two fully saturated surfaces with no dangling bonds interacts only weakly with Ru(0001) surface, it was necessary to include semiempirical dispersion correction (DFT+D) by Grimme (2006) to account for van der Waals interactions missing in DFT. Indeed, the calculated adhesion energy of the silica layer to the Ru(0001) support was only about 2 kJ mol^{-1} A^{-2}, with the main contribution coming from the dispersion term. In agreement with experimental data, the calculated vibrational spectrum revealed only two IR active modes, a very intense one at 1296 cm^{-1}, corresponding to an in-phase combination of asymmetric Si−O−Si stretching vibrations of the vertical Si−O−Si bonds, and a less intense one at 642 cm^{-1} due to a combination of symmetric Si−O−Si stretching vibrations of the horizontal Si−O−Si bonds. DFT simulations of combined experimental studies revealed that in analogy to SiO$_2$/Mo(112), the SiO$_2$/Ru(0001) film may exist in *O-poor* and *O-rich* forms, depending on the amount of O atoms adsorbed onto the Ru(0001) surface underneath the silica sheet (Włodarczyk et al. 2012). The amount of adsorbed oxygen can be reversibly varied, resulting in the change of the distance between the silica sheet and metal support. This results in gradual, reversible variation

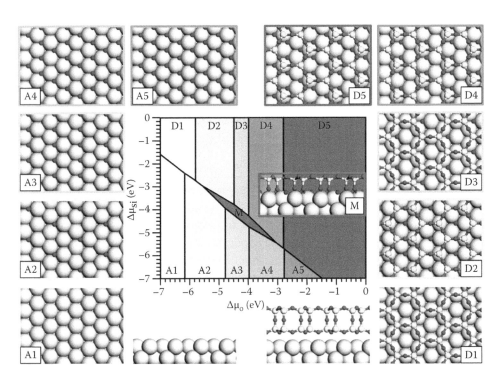

FIGURE 12.9   Calculated phase diagram of silica on Ru(0001) showing most stable phases as a function of relative chemical potentials Δμ$_{Si}$ and Δμ$_O$: chemisorbed oxygen (A1–A5), bilayer (D1–D5), and monolayer (M) silica films. Oxygen atoms adsorbed directly on the Ru(0001) surface are shown in blue. The remaining O and Si atoms are represented by red and yellow spheres, respectively.

of the SiO$_2$/Ru(0001) electronic states that opens the possibility for tuning the electronic properties of oxide/metal systems without altering the structure of an oxide overlayer.

Following the successful atomic structure determination, Yang et al. (2012) investigated dependence of the structure of the SiO$_2$/Ru(0001) film on growth conditions using DFT simulations and a multitude of surface science techniques. They found that at a Si coverage of about one monolayer, the Ru(0001) surface becomes almost fully covered by the silica film, showing again a honeycomb-like structure with a 5.4 Å periodicity. This observation could not be explained assuming that silica films grow exclusively as a bilayer structure since in such a case they would have covered only 50% of the surface. Indeed, the IRA spectra did not show the band at 1302 cm^{-1} as a characteristic of the bilayer silica film. Instead, a sharp and intense signal at 1134 cm^{-1} was observed along with less intense bands at 1074, 790, and 687 cm^{-1}, which clearly resemble those observed for the monolayer structure of SiO$_2$/Mo(112). In order to fully understand the structures of the films, an extensive computational study of silica on Ru(0001) was performed. Different models of 1 ML and 2 ML ordered silica films containing varying amount of O atoms adsorbed directly on the Ru(0001) surface were considered (Figure 12.9). Using the procedure described in Section 12.3.2, the stability of the models was compared, employing Gibbs energy change per unit surface of a hypothetical formation reaction:

$$\text{Ru(0001)} + m\text{Si}_{\text{bulk}} + \frac{n}{2}\text{O}_2 \rightarrow \frac{\text{Si}_m\text{O}_n}{\text{Ru(0001)}}. \tag{12.11}$$

Figure 12.9 displays the resulting 2D phase diagram, which shows the stability regions of different models as a function of $\Delta\mu_{O_2}$ (i.e., oxygen partial pressure) and $\Delta\mu_{Si}$ (i.e., amount of Si on the surface). At high values of $\Delta\mu_{Si}$ the silica bilayer structure model is the most stable phase. Increasing the amount of oxygen (i.e., $\Delta\mu_{O_2}$) leads to oxygen adsorption on the metal surface underneath the silica film. The amount of adsorbed oxygen as a function of $\Delta\mu_{O_2}$ is approximately the same as for clean Ru(0001). This is consistent with experimental data indicating the same oxygen adsorption/desorption behavior of film covered and clean Ru(0001) surface. At very low silicon content, various oxygen-covered Ru(0001) structures are the most stable phases. The monolayer silica film is stable only in a narrow region of relative chemical potentials. It turned out that among all of the monolayer models considered, only one, containing additional oxygen atoms directly adsorbed on Ru(0001), is stable. In agreement with experimental data, IR spectrum calculated for this model showed one intense mode at about 1160 cm^{-1} originating from the in-phase combination of asymmetric stretching vibrations of the Si−O−Ru bonds, a less intense mode at 1076 cm^{-1} involving combinations of symmetric O−Si−O stretching vibrations, and two modes at 820 and 677 cm^{-1} that are the combinations of asymmetric stretching vibrations of Si−O−Ru bonds and O−Si−O bending modes. Also the XPS spectrum simulated for this structure was in good agreement with experiment.

When preparing the bilayer silica films on Ru(0001), the LEED patterns occasionally showed both (2 × 2) spots and a diffraction ring representing the same lattice constant (Shaikhutdinov and Freund 2013). Such diffraction rings are characteristic for glassy (vitreous) silica. It turned out that fast cooling after the high-temperature oxidation and annealing step leads to a formation of well ordered but, at the same time, partially amorphous

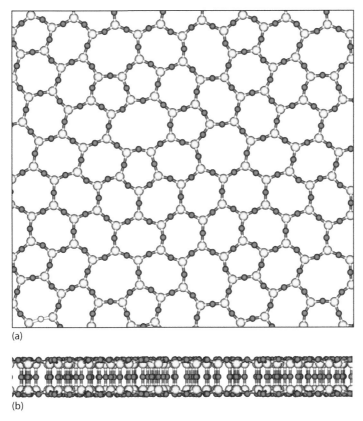

FIGURE 12.10 Model of vitreous silica bilayer film on Ru(0001) (Si light gray, O dark gray). (a) Top and (b) side view; metal surface is not shown.

SiO_2/Ru(0001) films (Lichtenstein et al. 2012). Figure 12.10 shows the corresponding structure model derived based on STM imaging and DFT simulations. The model corresponds to a network of corner-sharing $[SiO_4]$ tetrahedra forming four-membered Si−O−Si rings with ring planes arranged perpendicular to the metal surface and connected randomly, forming the 2D ring network. While the film is amorphous in the substrate plane, it is highly ordered in the direction perpendicular to the metal surface. To estimate stability of the amorphous film, DFT calculations were performed for a model of the ordered film with a defect consisting of two 5- and two 7-membered Si−O−Si rings embedded in a 5 × 3 surface unit cell, as shown in Figure 12.11. This structure was 177 kJ mol^{-1} less stable than the corresponding model of the fully ordered film. Several additional models containing silica rings of different sizes shown in Figure 12.11 were 249–308 kJ mol^{-1} higher in energy than the ordered film model. Based on these results, the estimated energy difference between the ordered and vitreous bilayer silica film is 5.5–9.6 kJ mol^{-1} per SiO_2 unit, suggesting that the amorphous structure is a metastable phase. Indeed, the estimated barrier for its transition into the ordered film by the process shown in Figure 12.11 is high: 338 kJ mol^{-1} per SiO bond involved. Therefore, formation of the vitreous film must be kinetically controlled. The process of the film formation may, for example, start from isolated silica rings and double rings. Such models of crystallization centers of the amorphous film were found to be quite close in energy as shown in Figure 12.11. Indeed, experiments showed that once

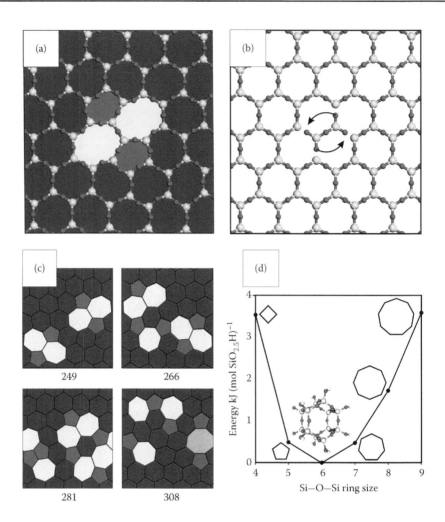

FIGURE 12.11   The simplest model of an amorphous film consisting of two 5- and two 7-membered rings embedded in a 5 × 3 surface unit cell of the ordered film (Si, yellow; O, red; H, gray). (b) The rotation of one of the $(SiO_2)_4$ units, which leads to the model shown in (a). (c) Different models of an amorphous film and their relative energies (kJ mol^{-1}) with respect to the ordered film. (d) Relative energies of the isolated hydroxylated double silica rings with composition $(SiO_{2.5}H)_{2n}$, where n is the Si−O−Si ring size. The atomic structure of the six-membered ring is shown. (From Lichtenstein, L., Buchner, C., Yang, B., Shaikhutdinov, S., Heyde, M., Sierka, M., Włodarczyk, R., Sauer, J., and Freund, H.J.: The atomic structure of a metal-supported vitreous thin silica film. *Angew. Chem. Int. Ed.* 2012. 51. 404–407. Copyright Wiley-VCH Verlag GmbH & Co. KGaA. Adapted with permission.)

formed, the vitreous silica film cannot be transformed back into the crystalline state by reoxidation of the same sample followed by slow cooling, and vice versa. In conclusion, this model of vitreous silica, which can be investigated by well-established surface science tools, provided for the first time the unique possibility to resolve the atomic structure of a silica glass and to study the glass transition with atomic resolution in real space. Recently, Huang et al. (2012) prepared crystalline bilayer silica films grown on graphene, that display close similarities to the structure of $SiO_2/Ru(0001)$ films obtained by Lichtenstein et al. (2012). Both results demonstrate that this new class of 2D glasses can be prepared on different materials.

12.5   Ordered Water Layers on MgO(100) Ultrathin Films

The understanding of water interaction with metal oxides is of fundamental importance in a variety of applied fields, including materials and environmental sciences, geochemistry, and heterogeneous catalysis. For this reason, the study of H_2O adsorption on oxide surfaces continues to attract a great interest, as demonstrated by the number of experimental and theoretical studies as well as several reviews (see, e.g., Brown et al. 1999; Henderson 2002; Verdaguer et al. 2006). Among different oxide surfaces, MgO(100) is considered to be a prototype system for adsorption studies due to its structural and electronic simplicity and an important role of magnesium oxide in many chemical processes. Since MgO is a highly insulating oxide, the use of ultrathin films supported on metal surfaces plays a crucial role in characterization of adsorption process on MgO surfaces by allowing to overcome charging problems associated with many surface science techniques.

Water adsorption on MgO(100) single crystals and supported MgO(100) ultrathin films has attracted particular attention also because it is one of the very few known examples of oxide systems where H_2O forms ordered monolayers. The first indication of the ordered structure formation on a cleaved MgO(100) single crystal was obtained using LEED and FTIR spectroscopy by Heidberg et al. (1995), who found that water forms an ordered $c(4 \times 2)$ phase when adsorbed on MgO(100) at about 150 K. Later, different experiments performed for water and D_2O adsorbed on MgO(100) single crystals and ultrathin films confirmed the existence of the $c(4 \times 2)$ structure and indicated formation of a new, ordered $p(3 \times 2)$ phase stable at higher temperatures (Ferry et al. 1996, 1997, 1998; Xu and Goodman 1997). It has been found that the transition from the $c(4 \times 2)$ to the $p(3 \times 2)$ structure occurs at about 185 K and involves a partial desorption of water from the surface. The $p(3 \times 2)$ phase, which cannot be reversed to the $c(4 \times 2)$ one by cooling without the addition of H_2O, was found to be stable under UHV conditions up to about 235 K. Above this temperature, water desorbs, leaving bare MgO(100) surface. The conclusions drawn from these studies concerning the atomic structure of the water monolayer phases can be summarized as follows: (1) nearly flat arrangement of adsorbed H_2O molecules in both $c(4 \times 2)$ and $p(3 \times 2)$ structures with the molecular plane parallel to the surface and the (2) presence of domains rotated 90° with respect to each other and a glide plane perpendicular to the largest side of the unit cell in the $p(3 \times 2)$ phase. However, different authors reached contradicting conclusions about molecular versus dissociative H_2O adsorption for the two water monolayer structures. Some experimental studies (e.g., Kim et al. 2002a,b; Yu et al. 2003) reported partial H_2O dissociation and formation of surface hydroxyl species, whereas others indicated the presence of only physisorbed water (e.g., Ferry et al. 1996, 1997, 1998; Xu and Goodman 1997; Hawkins et al. 2005).

Computational studies have played the leading role in structure determination and characterization of the ordered water layers on MgO(100). This has, however, come a long way with several structure models suggested over more than a decade. The first models were constructed by either simply drawing the structures (Heidberg et al. 1995; Ferry et al. 1997; Xu and Goodman 1997) or employing simulations at the interatomic potentials level and assuming no water dissociation (Ferry et al. 1998). Most of the early computational ab initio studies using DFT focused on the $p(3 \times 2)$ phase of the $H_2O/$ MgO(001) system. The first two models of this phase including partially dissociated

Chapter 12

water were independently proposed by Odelius (1999) and Giordano et al. (1998). Although the suggested structures differ in the arrangement of water molecules and hydroxyl groups on the surface, they are both consistent with the known experimental data. Both models contain one H_2O per MgO surface unit and show a glide symmetry plane with the water monolayer arranged nearly flat with some of hydrogen atoms twisted toward the surface. One-third of the H_2O molecules in the monolayer are dissociated with protons donated to surface O atoms. About 2 years later, Delle Site et al. (2000) investigated the structure of the $p(3 \times 2)$ phase using DFT molecular dynamics (MD). Starting from the flat physisorbed structure obtained using interatomic potential calculations by Ferry et al. (1998) and performing MD simulations at 300 K, they observed spontaneous transfer of two protons from the water layer to surface oxygen atoms, leading to a stable chemisorbed structure. After quenching and optimizing the dissociated MD configuration, an asymmetric structure was obtained, which, although 18 kJ mol^{-1} more stable than the structure suggested by Giordano et al. (1998), contained no glide plane. Starting from the Giordano et al.'s structure and using short MD followed by a quench, the authors obtained about 33 kJ mol^{-1} more stable *flat dissociated structure* that maintains the glide plane. The new model contained more hydrogen boding within the plane of adsorbed molecules and no hydrogen atoms twisted toward the surface. Two years later, the same authors carried out a systematic search using DFT for stable configurations of the $p(3 \times 2)$ phase (Lynden-Bell et al. 2002). They have found six minima with structures satisfying the observed symmetry constraints, that is, $p(3 \times 2)$ cell with a glide plane, including a new one, which was slightly more stable than the *flat dissociated structure*. The new model also contained one-third of dissociated H_2O and differed from the previously proposed model only by relative orientation of the H_2O molecules. Later, Honkala et al. (2010) confirmed by DFT calculations that the same dissociated structure of the $p(3 \times 2)$ phase is energetically preferred also on thin MgO films supported on Ag(100).

In contrast to the $p(3 \times 2)$ phase of water monolayer on MgO(100), the low-temperature $c(4 \times 2)$ structure attracted less attention. The first structure model suggested by Heidberg et al. (1995) on the basis of experimental data was obtained by drawing one water molecule per MgO surface unit, maintaining the $c(4 \times 2)$ symmetry. Obviously, no water dissociation could have been predicted this way. The first model including partially dissociated water has been suggested by Cho et al. (2000) based on DFT calculations. It was constructed based on the $p(3 \times 2)$ structure suggested by Giordano et al. (1998) and included four H_2O molecules in the primitive $c(4 \times 2)$ surface unit cell, two of which were dissociated. However, the suggested structure contains the same amount of H_2O as the model of the $p(3 \times 2)$ phase. Therefore, it cannot account for the observation that the phase transition from $c(4 \times 2)$ to $p(3 \times 2)$ is accompanied by the desorption of water molecules from the surface.

The water monolayers on MgO(100) are another instructive example of systems where the final structure resolution and characterization could only be achieved by global structure optimizations on DFT potential surface and statistical thermodynamics complemented by experimental studies (Włodarczyk et al. 2011). Using GA implementation described in Section 12.3.2, global minima of different stoichiometric numbers of H and O atoms on (3 × 2) and (4 × 2) models of MgO(100) ultrathin film supported on Ag(001) surface were determined. The stability of the resulting $nH_2O/MgO(100)$ structures was compared using the Gibbs energy change per unit area, $\Delta\gamma$, for adsorption of n water molecules on a clean MgO(100) surface:

$$n\text{H}_2\text{O} + \text{MgO(100)} \rightarrow \frac{n\text{H}_2\text{O}}{\text{MgO(100)}}. \tag{12.12}$$

Figure 12.12 displays $\Delta\gamma$ at a pressure of $p = 10^{-10}$ mbar as a function of temperature T for the most stable $n\text{H}_2\text{O}/\text{MgO(001)}$ structure models found. This phase diagram demonstrates existence of only two stable monolayer structures, one with $p(3 \times 2)$ symmetry and 6 water molecules per unit cell $(6\text{H}_2\text{O}/(3 \times 2))$ and one with $c(4 \times 2)$ symmetry and 10 water molecules per unit cell $(10\text{H}_2\text{O}/(4 \times 2))$. The calculated monolayer water coverage of 1.0 and 1.25 for the $6\text{H}_2\text{O}/(3 \times 2)$ and $10\text{H}_2\text{O}/(4 \times 2)$ structure models, respectively, is in excellent agreement with experimentally determined values of 1 ± 0.1 for the $p(3 \times 2)$ and 1.3 ± 0.1 for the $c(4 \times 2)$ phase (Stirniman et al. 1996; Ferry et al. 1998). In both cases, the most stable models contain two dissociated H_2O molecules per unit cell. The thermodynamic stability of the two phases is in perfect agreement with the results of temperature-programmed desorption (TPD) (Włodarczyk et al. 2011) and HAS experiments (Ferry et al. 1996). At low temperatures, the most stable phase is the $10\text{H}_2\text{O}/(4 \times 2)$ structure, which, upon temperature increase, transforms at about 200 K into the $6\text{H}_2\text{O}/(3 \times 2)$ one that is then stable until complete monolayer desorption. The clean MgO(100) surface becomes the most stable configuration for temperatures higher than 202 K, which explains the single desorption peak observed at 215 K in the TPD experiment. Another small desorption peak observed at 185 K is in line with the

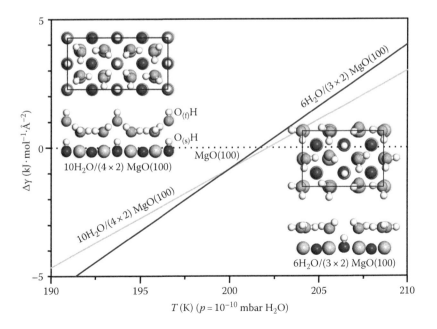

FIGURE 12.12 Gibbs energy change per unit surface $\Delta\gamma(T,p = 10^{-10}$ mbar) for the most stable models of ordered H_2O layers on Mg(100) thin film as a function of temperature. Side and top views of the most stable $6\text{H}_2\text{O}/(3 \times 2)$ and $10\text{H}_2\text{O}/(4 \times 2)$ models (Mg large gray spheres, O small light and dark gray spehres, H white).

transition from the $10H_2O/(4 \times 2)$ to the less dense $6H_2O/(3 \times 2)$ phase, accompanied by desorption of a fraction of the H_2O molecules.

The knowledge of the atomic structure of the two water monolayer phases on MgO(100) allowed to identify and assign vibrational features present in experimental IRAS and sum frequency generation spectra. However, a satisfactory agreement between calculations and experiment could only be achieved when anharmonic effects in the simulated IR spectrum were included. To obtain the anharmonic spectra, NVT MD simulations were performed for the $10H_2O/(4 \times 2)$ and $6H_2O/(3 \times 2)$ D_2O phases at the DFT level. The IR spectra were obtained as the Fourier transform of the autocorrelation function of the dipole moment component perpendicular to the MgO(100) surface. The spectra were then corrected for the lower cutoff for the plane wave basis set used in MD simulations. As shown in Figure 12.13, the main effect of the anharmonicity is a significant

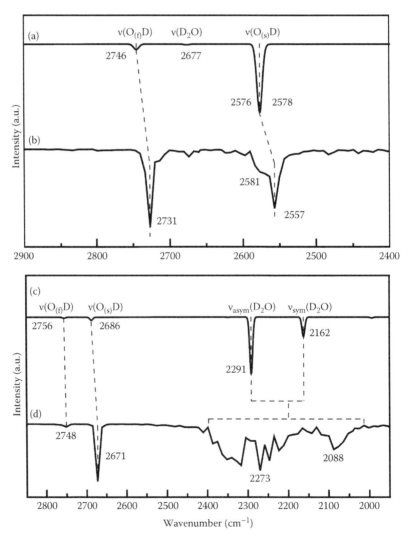

FIGURE 12.13 Simulated IR spectra for the most stable models of ordered H_2O layers on Mg(100) thin film: (a) harmonic and (b) anharmonic spectrums of the $6H_2O/(3 \times 2)$ model; (c) harmonic and (d) anharmonic spectrum of the $10H_2O/(4 \times 2)$ model.

increase of the intensities of high-frequency bands and broadening of the low-frequency ones. In case of the $10H_2O/(4 \times 2)$ structure, the high-frequency signals at 2748 and 2671 cm^{-1} could be assigned to $\nu(O_{(f)}-D)$ and $\nu(O_{(s)}-D)$ stretching modes of O–D species resulting from D_2O dissociation, whereas the broad band at 2020–2400 cm^{-1} corresponds to a combination of stretching modes of remaining D_2O molecules. Analysis of vibrational modes in the $6H_2O/(3 \times 2)$ structure revealed substantial differences in comparison to the $10H_2O/(4 \times 2)$ model. The highest frequency band at 2731 cm^{-1} corresponds to the $\nu(O_{(f)}-D)$ stretching mode of O–D groups within the D_2O monolayer. The broad low-frequency band at 2581 and 2557 cm^{-1} was assigned to different combinations of $\nu(O_{(s)}-D)$ stretching modes of O–D groups located directly on the MgO(100) surface. In contrast to the $10H_2O/(4 \times 2)$ phase, the remaining water molecules do not contribute to the IR spectrum due to their alignment almost parallel to the surface.

12.5.1 Synopsis

Metal-supported oxide nanofilms represent an interesting and promising type of materials. Thanks to their unique properties, they already gained a prominent role in several areas of advanced technologies. The extreme confinement of the nanofilms due to the proximity of the interfaces stabilizes new structures that otherwise cannot be obtained as bulk materials. This opens new possibilities for the development of novel advanced materials with controlled and tunable properties. In comparison to amorphous systems, highly ordered nanofilms allow for much better structural control. However, the examples described in this chapter demonstrate that even in the case of crystalline films, their characterization down to the atomic scale can only be achieved only when computer simulations complement experimental studies. This combination of modern electronic structure theory and advanced surface science techniques has enabled tremendous advances in the understanding of existing and design of new systems. Certainly, the interest in metal-supported oxide nanofilms will continue to grow in the years to come, providing unforeseen opportunities in science and technology.

Acknowledgments

The work described in this chapter would have not been possible without the contribution of many colleagues, in particular Hans-Joachim Freund and Joachim Sauer, as well as their coworkers and students. I gratefully acknowledge financial support from Deutsche Forschungsgemeinschaft (Center of Excellence UniCat and Sonderforschungsbereich 546), Fonds der Chemischen Industrie, Turbomole GmbH and the Thüringer Ministerium für Bildung, and Wissenschaft und Kultur (TMBWK, project NANOSOR).

References

Barcaro, G., E. Cavaliere, L. Artiglia, L. Sementa, L. Gavioli, G. Granozzi, and A. Fortunelli. 2012. Building principles and structural motifs in TiO$_x$ ultrathin films on a (111) substrate. *J Phys Chem C* 116:13302–13306.

Bersuker, G., K. McKenna, and A. Shluger. 2012. Silica and high-*k* dielectric thin films in microelectronics. In *Oxide Ultrathin Films: Science and Technology*, G. Pacchioni and S. Valeri, eds., pp. 47–73. Weinheim, Germany: Wiley-VCH.

Bersuker, G. and P. Zeitzoff. 2007. Requirements of oxides as gate dielectrics for CMOS devices. In *Rare Earth Oxide Thin Films: Growth, Characterization, and Applications*, M. Fanciulli and G. Scarel, eds., pp. 367–377. Berlin, Heidelberg, Germany: Springer.

Besenbacher, F., J.V. Lauritsen, T.R. Linderoth, E. Laegsgaard, R.T. Vang, and S. Wendt. 2009. Atomic-scale surface science phenomena studied by scanning tunneling microscopy. *Surf Sci* 603:1315–1327.

Böhme, D.K. and H. Schwarz. 2005. Gas-phase catalysis by atomic and cluster metal ions: The ultimate single-site catalysts. *Angew Chem Int Ed* 44:2336–2354.

Boudart, M. 2000. Model catalysts: Reductionism for understanding. *Top Catal* 13:147–149.

Bradshaw, A.M. and N.V. Richardson. 1996. Symmetry, selection rules and nomenclature in surface spectroscopies. *Pure Appl Chem* 68:457–467.

Brown, G.E., V.E. Henrich, W.H. Casey, D.L. Clark, C. Eggleston, A. Felmy, D.W. Goodman et al. 1999. Metal oxide surfaces and their interactions with aqueous solutions and microbial organisms. *Chem Rev* 99:77–174.

Brunet, E., G.C. Mutianti, S. Steinhauer, and A. Köck. 2012. Oxide ultrathin films in sensor applications. In *Oxide Ultrathin Films: Science and Technology*, G. Pacchioni and S. Valeri, eds., pp. 239–263. Weinheim, Germany: Wiley-VCH.

Chen, M.S. and D.W. Goodman. 2008. Ultrathin, ordered oxide films on metal surfaces. *J Phys Condens Matter* 20:264013.

Chen, M.S., A.K. Santra, and D.W. Goodman. 2004. Structure of thin SiO_2 films grown on Mo(112). *Phys Rev B* 69:155404.

Cho, J.-H., J.M. Park, and K.S. Kim. 2000. Influence of intermolecular hydrogen bonding on water dissociation at the MgO(001) surface. *Phys Rev B* 62:9981–9984.

Chuang, F.C., C.V. Ciobanu, V.B. Shenoy, C.Z. Wang, and K.-M. Ho. 2004. Finding the reconstructions of semiconductor surfaces via a genetic algorithm. *Surf Sci* 573:L375–L381.

Ciobanu, C.V., C.-Z. Wang, and K.-M. Ho. 2009. Global optimization of 2-dimensional nanoscale structures: A brief review. *Mater Manuf Process* 24:109–118.

Delle Site, L., A. Alavi, and R.M. Lynden-Bell. 2000. The structure and spectroscopy of monolayers of water on MgO: An ab initio study. *J Chem Phys* 113:3344–3350.

Diebold, U., S.-C. Li, and M. Schmid. 2010. Oxide surface science. *Annu Rev Phys Chem* 61:129–148.

Eaton, P.J. and P. West. 2010. *Atomic Force Microscopy*. Oxford, U.K.: Oxford University Press.

Ertl, G. and H.-J. Freund. 1999. Catalysis and surface science. *Phys Today* 52:32–38.

Fernández-García, M., A. Martínez-Arias, J.C. Hanson, and J.A. Rodriguez. 2004. Nanostructured oxides in chemistry: Characterization and properties. *Chem Rev* 104:4063–4104.

Ferry, D., A. Glebov, V. Senz, J. Suzanne, J.P. Toennies, and H. Weiss. 1996. Observation of the second ordered phase of water on the MgO(100) surface: Low energy electron diffraction and helium atom scattering studies. *J Chem Phys* 105:1697–1701.

Ferry, D., A. Glebov, V. Senz, J. Suzanne, J.P. Toennies, and H. Weiss. 1997. The properties of a two-dimensional water layer on MgO(001). *Surf Sci* 377:634–638.

Ferry, D., S. Picaud, P.N.M. Hoang, C. Girardet, L. Giordano, B. Demirdjian, and J. Suzanne. 1998. Water monolayers on MgO(100): Structural investigations by LEED experiments, tensor LEED dynamical analysis and potential calculations. *Surf Sci* 409:101–116.

Freund, H.-J. 2010. Model studies in heterogeneous catalysis. *Chem Eur J* 16:9384–9397.

Freund, H.-J. 2012. Oxide films as catalytic materials and models of real catalysts. In *Oxide Ultrathin Films: Science and Technology*, G. Pacchioni and S. Valeri, eds., pp. 145–179. Weinheim, Germany: Wiley-VCH.

Freund, H.-J., M. Bäumer, and H. Kuhlenbeck. 2000. Catalysis and surface science: What do we learn from studies of oxide-supported cluster model systems? *Adv Catal* 45:333–384.

Freund, H.-J., N. Ernst, T. Risse, H. Hamann, and G. Rupprechter. 2001. Models in heterogeneous catalysis: Surface science quo vadis? *Phys Status Solidi A* 187:257–274.

Freund, H.-J. and G. Pacchioni. 2008. Oxide ultra-thin films on metals: New materials for the design of supported metal catalysts. *Chem Soc Rev* 37:2224–2242.

Giordano, L., J. Goniakowski, and J. Suzanne. 1998. Partial dissociation of water molecules in the (3 × 2) water monolayer deposited on the MgO(100) surface. *Phys Rev Lett* 81:1271–1273.

Giordano, L., D. Ricci, G. Pacchioni, and P. Ugliengo. 2005. Structure and vibrational spectra of crystalline SiO_2 ultra-thin films on Mo(112). *Surf Sci* 584:225–236.

Goniakowski, J., F. Finocchi, and C. Noguera. 2008. Polarity of oxide surfaces and nanostructures. *Rep Prog Phys* 71:016501.

Granqvist, C.-G. 2012. Transparent conducting and chromogenic oxide films as solar energy materials. In *Oxide Ultrathin Films: Science and Technology*, G. Pacchioni and S. Valeri, eds., pp. 221–238. Weinheim, Germany: Wiley-VCH.

Grimme, S. 2006. Semiempirical GGA-type density functional constructed with a long-range dispersion correction. *J Comput Chem* 27:1787–1799.

Grinter, D.C. and G. Thronton. 2012. Characterization tools of ultrathin oxide films. In *Oxide Ultrathin Films: Science and Technology*, G. Pacchioni and S. Valeri, eds., pp. 47–73. Weinheim, Germany: Wiley-VCH.

Hartke, B. 2011. Global optimization. *WIREs Comput Mol Sci* 1:879–887.

Hawkins, S., G. Kumi, S. Malyk, H. Reisler, and C. Wittig. 2005. Temperature programmed desorption and infrared spectroscopic studies of thin water films on MgO(100). *Chem Phys Lett* 404:19–24.

Heidberg, J., B. Redlich, and D. Wetter. 1995. Adsorption of water vapor on the MgO(100) single crystal surface. *Ber Bunsenges Phys Chem* 99:1333–1337.

Henderson, M.A. 2002. The interaction of water with solid surfaces: Fundamental aspects revisited. *Surf Sci Rep* 46:1–308.

Henry, C.R. 1998. Surface studies of supported model catalysts. *Surf Sci Rep* 31:235–325.

Honkala, K., A. Hellman, and H. Gronbeck. 2010. Water dissociation on MgO/Ag(100): Support induced stabilization or electron pairing? *J Phys Chem C* 114:7070–7075.

Huang, P.Y., S. Kurasch, A. Srivastava, V. Skakalova, J. Kotakoski, A.V. Krasheninnikov, R. Hovden et al. 2012. Direct imaging of a two-dimensional silica glass on graphene. *Nano Lett* 12:1081–1086.

Ishihara, T. 2012. Oxide ultrathin films for solid oxide fuel cells. In *Oxide Ultrathin Films: Science and Technology*, G. Pacchioni and S. Valeri, eds., pp. 201–220. Weinheim, Germany: Wiley-VCH.

Jaeger, R.M., H. Kuhlenbeck, H.-J. Freund, M. Wuttig, W. Hoffmann, R. Franchy, and H. Ibach. 1991. Formation of a well-ordered aluminum-oxide overlayer by oxidation of NiAl(110). *Surf Sci* 259:235–252.

Jones, A.C. and M.L. Hitchman. 2009. Overview of chemical vapour deposition. In *Chemical Vapour Deposition: Precursors, Processes and Applications*, A.C. Jones and M.L. Hitchman, eds., pp. 1–36. Cambridge, U.K.: The Royal Society of Chemistry.

Kaya, S., M. Baron, D. Stacchiola, J. Weissenrieder, S. Shaikhutdinov, T.K. Todorova, M. Sierka, J. Sauer, and H.-J. Freund. 2007. On the geometrical and electronic structure of an ultra-thin crystalline silica film grown on Mo(112). *Surf Sci* 601:4849–4861.

Kim, Y.D., R.M. Lynden-Bell, A. Alavi, J. Stulz, and D.W. Goodman. 2002a. Evidence for partial dissociation of water on flat MgO(100) surfaces. *Chem Phys Lett* 352:318–322.

Kim, Y.D., J. Stultz, and D.W. Goodman. 2002b. Dissociation of water on MgO(100). *J Phys Chem B* 106:1515–1517.

Kim, Y.D., T. Wei, J. Stultz, and D.W. Goodman. 2003. Dissociation of water on a flat, ordered silica surface. *Langmuir* 19:1140–1142.

Kinoshita, T. and S. Mizuno. 2011. Surface structure determination of silica single layer on Mo(112) by LEED. *Surf Sci* 605:1209–1213.

Kresse, G., M. Schmid, E. Napetschnig, M. Shishkin, L. Köhler, and P. Varga. 2005. Structure of the ultrathin aluminum oxide film on NiAl(110). *Science* 308:1440–1442.

Libuda, J. and H.-J. Freund. 2005. Molecular beam experiments on model catalysts. *Surf Sci Rep* 57:157–298.

Lichtenstein, L., C. Buchner, B. Yang, S. Shaikhutdinov, M. Heyde, M. Sierka, R. Wlodarczyk, J. Sauer, and H.J. Freund. 2012. The atomic structure of a metal-supported vitreous thin silica film. *Angew Chem Int Ed* 51:404–407.

Liebau, F. 1985. *Structural Chemistry of Silicates: Structure, Bonding and Classification*. Berlin, Germany: Springer.

Lizzit, S., A. Baraldi, A. Groso, K. Reuter, M.V. Ganduglia-Pirovano, C. Stampl, M. Scheffler et al. 2001. Surface core-level shifts of clean and oxygen-covered Ru(0001). *Phys Rev B* 63:205419.

Loffler, D., J.J. Uhlrich, M. Baron, B. Yang, X. Yu, L. Lichtenstein, L. Heinke et al. 2010. Growth and structure of crystalline silica sheet on Ru(0001). *Phys Rev Lett* 105:146104.

Lu, J.-L., S. Kaya, J. Weissenrieder, T.K. Todorova, M. Sierka, J. Sauer, H.J. Gao, S. Shaikhutdinov, and H.-J. Freund. 2006. Formation of one-dimensional crystalline silica on a metal substrate. *Surf Sci* 600:L164–L168.

Lynden-Bell, R.M., L. Delle Site, and A. Alavi. 2002. Structures of adsorbed water layers on MgO: An ab initio study. *Surf Sci* 496:L1–L6.

Marcus, P. 2012. *Corrosion Mechanisms in Theory and Practice*. Boca Raton, FL: CRC.

Mattox, D.M. 2009. *Handbook of Physical Vapor Deposition (PVD) Processing*. Oxford, U.K.: Elsevier Science.

Chapter 12

Netzer, F.P., F. Allegretti, and S. Surnev. 2010. Low-dimensional oxide nanostructures on metals: Hybrid systems with novel properties. *J Vac Sci Technol B* 28:1–16.

Netzer, F.P. and S. Surnev. 2012. Synthesis and preparation of oxide ultrathin films. In *Oxide Ultrathin Films: Science and Technology*, G. Pacchioni and S. Valeri, eds., pp. 47–73. Weinheim, Germany: Wiley-VCH.

Nilius, N. 2009. Properties of oxide thin films and their adsorption behavior studied by scanning tunneling microscopy and conductance spectroscopy. *Surf Sci Rep* 64:595–659.

Nilius, N., T. Risse, S. Schauermann, S. Shaikhutdinov, M. Sterrer, and H.-J. Freund. 2011. Model studies in catalysis. *Top Catal* 54:4–12.

Odelius, M. 1999. Mixed molecular and dissociative water adsorption on MgO[100]. *Phys Rev Lett* 82:3919–3922.

Oura, K., V.G. Lifshits, A.A. Saranin, A.V. Zotov, and M. Katayama. 2003. *Surface Science: An Introduction*. Berlin, Germany: Springer.

Pacchioni, G. 2012. Two-dimensional oxides: Multifunctional materials for advanced technologies. *Chem Eur J* 18:10144–10158.

Pacchioni, G. and S. Valeri, eds. 2012. *Oxide Ultrathin Films: Science and Technology*. Weinheim, Germany: Wiley-VCH.

Reuter, K. and M. Scheffler. 2002. Composition, structure, and stability of $RuO_2(110)$ as a function of oxygen pressure. *Phys Rev B* 65:035406.

Ricci, D. and G. Pacchioni. 2004. Structure of ultrathin crystalline SiO_2 films on Mo(112). *Phys Rev B* 69:161307.

Ritala, M. and J. Niinisto. 2009. Atomic layer deposition. In *Chemical Vapour Deposition: Precursors, Processes and Applications*, A.C. Jones and M.L. Hitchman, eds., pp. 158–206. The Royal Society of Chemistry.

Schroeder, T., M. Adelt, B. Richter, M. Naschitzki, M. Baumer, and H.-J. Freund. 2000. Epitaxial growth of SiO_2 on Mo(112). *Surf Rev Lett* 7:7–14.

Schroeder, T., J.B. Giorgi, M. Bäumer, and H.-J. Freund. 2002. Morphological and electronic properties of ultrathin crystalline silica epilayers on a Mo(112) substrate. *Phys Rev B* 66:165422.

Schroeder, T., A. Hammoudeh, M. Pykavy, N. Magg, M. Adelt, M. Bäumer, and H.-J. Freund. 2001. Single crystalline silicon dioxide films on Mo(112). *Solid State Electron* 45:1471–1478.

Scott, A.P. and L. Radom. 1996. Harmonic vibrational frequencies: An evaluation of Hartree-Fock, Møller-Plesset, quadratic configuration interaction, density functional theory, and semiempirical scale factors. *J Phys Chem* 100:16502–16513.

Seifert, J., D. Blauth, and H. Winter. 2009. Evidence for 2D-network structure of monolayer silica film on Mo(112). *Phys Rev Lett* 103:017601.

Shaikhutdinov, S. and H.-J. Freund. 2012. Ultrathin oxide films on metal supports: Structure-reactivity relations. *Annu Rev Phys Chem* 63:619–633.

Shaikhutdinov, S. and H.J. Freund. 2013. Ultrathin silica films on metals: The long and winding road to understanding the atomic structure. *Adv Mater* 25:49–67.

Sierka, M. 2010. Synergy between theory and experiment in structure resolution of low-dimensional oxides. *Prog Surf Sci* 85:398–434.

Sierka, M., T.K. Todorova, S. Kaya, D. Stacchiola, J. Weissenrieder, J. Lu, H. Gao, S. Shaikhutdinov, H.J. Freund, and J. Sauer. 2006. Interplay between theory and experiment in the quest for silica with reduced dimensionality grown on a Mo(112) surface. *Chem Phys Lett* 424:115–119.

Sierka, M. and R. Włodarczyk. 2013. *Program DoDo*. Humboldt-Universität zu Berlin; since 2012 Firiedrich Schiller Universität Jena, Germany.

Stacchiola, D.J., M. Baron, S. Kaya, J. Weissenrieder, S. Shaikhutdinov, and H.-J. Freund. 2008. Growth of stoichiometric subnanometer silica films. *Appl Phys Lett* 92:011911.

Stirniman, M.J., C. Huang, R.S. Smith, S.A. Joyce, and B.D. Kay. 1996. The adsorption and desorption of water on single crystal MgO(100): The role of surface defects. *J Chem Phys* 105:1295–1298.

Stoneham, M., J. Gavartin, D.M. Ramo, and A. Shluger. 2007. The vulnerable nanoscale dielectric. *Phys Status Solidi A* 204:653–662.

Tersoff, J. and D.R. Hamann. 1985. Theory of the scanning tunneling microscope. *Phys Rev B* 31:805–813.

Todorova, T.K., M. Sierka, J. Sauer, S. Kaya, J. Weissenrieder, J.-L. Lu, H.-J. Gao, S. Shaikhutdinov, and H.-J. Freund. 2006. Atomic structure of a thin silica film on a Mo(112) substrate: A combined experimental and theoretical study. *Phys Rev B* 73:165414.

Valeri, S. and S. Benedetti. 2012. Synthesis and preparation of oxide ultrathin films. In *Oxide Ultrathin Films: Science and Technology*, G. Pacchioni and S. Valeri, eds., pp. 1–26. Weinheim, Germany: Wiley-VCH.

Vang, R.T., J.V. Lauritsen, E. Laegsgaard, and F. Besenbacher. 2008. Scanning tunneling microscopy as a tool to study catalytically relevant model systems. *Chem Soc Rev* 37:2191–2203.

Van Hove, M.A. 1997. Determination of complex surface structures with LEED. *Surf Rev Lett* 4:479–487.

Van Hove, M.A. 1999. Atomic-scale surface structure determination: Comparison of techniques. *Surf Interface Anal* 28:36–43.

Van Hove, M.A. 2009. Atomic-scale structure: From surfaces to nanomaterials. *Surf Sci* 603:1301–1305.

Van Hove, M.A., J. Cerda, P. Sautet, M.-L. Bocquet, and M. Salmeron. 1997. Surface structure determination by STM vs LEED. *Prog Surf Sci* 54:315–329.

Verdaguer, A., G.M. Sacha, H. Bluhm, and M. Salmeron. 2006. Molecular structure of water at interfaces: Wetting at the nanometer scale. *Chem Rev* 106:1478–1510.

Weissenrieder, J., S. Kaya, J.-L. Lu, H.-J. Gao, S. Shaikhutdinov, H.-J. Freund, M. Sierka, T.K. Todorova, and J. Sauer. 2005. Atomic structure of a thin silica film on a Mo(112) substrate: A two-dimensional network of SiO$_4$ tetrahedra. *Phys Rev Lett* 95:076103.

Włodarczyk, R., M. Sierka, K. Kwapien, J. Sauer, E. Carrasco, A. Aumer, J.F. Gomes, M. Sterrer, and H.-J. Freund. 2011. Structures of the ordered water monolayer on MgO(001). *J Phys Chem C* 115:6764–6774.

Włodarczyk, R., M. Sierka, J. Sauer, D. Löffler, J.J. Uhlrich, X. Yu, B. Yang, I.M.N. Groot, S. Shaikhutdinov, and H.-J. Freund. 2012. Tuning the electronic structure of ultrathin crystalline silica films on Ru(0001). *Phys Rev B* 85:085403.

Woodruff, D.P. 2002. Solved and unsolved problems in surface structure determination. *Surf Sci* 500:147–171.

Wu, Q.H., A. Fortunelli, and G. Granozzi. 2009. Preparation, characterisation and structure of Ti and Al ultrathin oxide films on metals. *Int Rev Phys Chem* 28:517–576.

Xu, C. and D.W. Goodman. 1997. Structure and geometry of water adsorbed on the MgO(100) surface. *Chem Phys Lett* 265:341–346.

Yang, B., W.E. Kaden, X. Yu, J.A. Boscoboinik, Y. Martynova, L. Lichtenstein, M. Heyde et al. 2012. Thin silica films on Ru(0001): Monolayer, bilayer and three-dimensional networks of [SiO$_4$] tetrahedra. *Phys Chem Chem Phys* 14:11344–11351.

Yu, Y.H., Q.L. Guo, S. Liu, E.G. Wang, and P.J. Møller. 2003. Partial dissociation of water on a MgO(100) film. *Phys Rev B* 68:115414.

Zhang, W., J.R. Smith, and X.G. Wang. 2004. Thermodynamics from ab initio computations. *Phys Rev B* 70:024103.

Chapter 12

13. Cosmic and Atmospheric Nanosilicates

Stefan T. Bromley and John M. C. Plane

13.1 Introduction 369
    13.1.1 Silicate Occurrence and Composition 369
    13.1.2 Stellar Origin of Nanosilicates and Cosmic Dust 371
    13.1.3 Meteoric Interactions with the Earth's Atmosphere 374

13.2 Nanosilicate Formation 379
    13.2.1 Validity of Classical Nucleation Theory 379
    13.2.2 Kinetic Modeling of Silicate Dust Nucleation 380
    13.2.3 Finding Low Energy Structures of Nanosilicate Clusters 385

13.3 Chemistry on Nanosilicates 394
    13.3.1 Catalyzed H_2 Formation and Dissociation 394
    13.3.2 Nanosilicates as Nuclei for H_2O Ice Condensation 396
    13.3.3 Hydroxylation of Nanosilicates 399

13.4 Summary and Outlook 405

References 405

13.1 Introduction

13.1.1 Silicate Occurrence and Composition

Silicates are inorganic ionic compounds having a silicon-based anion. The most common examples of such an anion is the tetrahedral $[SiO_4]^{4-}$ species. Although these ions can be found as isolated units within solids, by linking up via shared oxygen atoms they can also form a variety of charged rings, chains, and sheets. This topological diversity is augmented by the vast potential choice of cations required to provide charge compensation, leading to thousands of known silicates. Many silicates are well known to geologists and mineralogists as they make up approximately 90% of the Earth's crust. The crust itself is a relatively thin (3–30 km) solidification product of melts derived from the deeper, hotter mantle, where the latter is a silicate rocky shell about 2900 km thick that constitutes about 84% of the Earth's volume. The dominant rock of the upper part of the Earth's mantle is peridotite: a dense, coarse-grained igneous rock consisting mostly of the silicate minerals olivine and pyroxene. Olivine possesses isolated $[SiO_4]^{4-}$ species surrounded by a mixture of Mg^{2+} and Fe^{2+} cations with the general composition $Mg_{2-n}Fe_nSiO_4$ (where $n = 0$–2). Pyroxene is made up of Mg^{2+}/Fe^{2+}-compensated infinite linear chains of oxygen-sharing $[SiO_4]^{4-}$ tetrahedra (often described as $[SiO_3]_n^{2-}$) and has the chemical formula $Mg_{1-n}Fe_nSiO_3$ (where $n = 0$–1). We note that when $[SiO_4]^{4-}$ tetrahedra share all four oxygen centers, we obtain networks of

Chapter 13

neutral SiO_2 units (i.e., silica) with no need for charge-compensating cations, which we also consider as a (nano)silicate herein. Terrestrial olivine and pyroxene are typically very rich in Mg and thus often are close to the $n = 0$ end members, forsterite and enstatite, respectively (see Figure 13.1). Such a situation is not specific to the Earth's intrinsic geology; Mg-rich silicates are a common constituent of meteorites and are found in a wide range of extraterrestrial environments (e.g., comets, circumstellar disks, supernovae, quasars) [1]. Astronomical detection of such silicates typically takes advantage of the fact that, depending on the environment, silicate dust absorbs or emits light in the infrared (IR) part of the spectrum. In some astronomical environments (e.g., circumstellar shells), and accounting for about 10% of all observed cosmic dust, detailed IR spectral characteristics of crystalline forsterite and enstatite have been identified. Typically, however, cosmic silicate dust spectra consist of two significantly broadened characteristic peaks with wavelengths at around 10 and 18 µm (corresponding to Si–O stretching and O–Si–O bending modes respectively), which are judged to correspond to noncrystalline silicate dust, usually with pyroxene or olivine chemical composition [1]. Because of their broad spectral signatures, our knowledge of dust particles in space is more limited than that of specific molecules. Through a combination of lab characterization of pristine material [2,3], astronomical observations [4,5], and analyses of spectral features [6], the general properties of dust particles in various environments have been deduced to a certain extent [2]. Fitting the observed spectra with a combination of lab spectra of materials of different crystallinity, shape, size, and composition gives insight into the possible identity of dust particles in space [2,7]. Such dust is found to be ubiquitous throughout the interstellar medium (ISM) and is the source of all cosmic and terrestrial silicate solids. In fact, silicates are thought to be the most abundant form of solid atomic matter (as opposed to the much higher fraction of gas phase atoms and molecules) in the known Universe.

In this chapter, we highlight the role that computational modeling plays in providing important insights into the detailed structure, properties, and formation of small silicate particles. Specifically, we focus on theoretical efforts to understand nanosilicates in atmospheric and astronomical environments. First, in order to better understand the abundance and chemical composition of silicate dust in the Universe, we consider how it is formed.

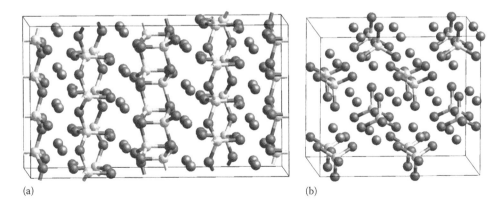

(a) (b)

FIGURE 13.1 Crystal structures of $MgSiO_3$ enstatite (a) and Mg_2SiO_4 forsterite (b). Si atoms: yellow, Mg atoms: blue, O atoms: red.

13.1.2   Stellar Origin of Nanosilicates and Cosmic Dust

As with all stable elements with atomic masses greater than boron, the constituent elements of silicate cosmic dust originate in nucleosynthetic processes in stars. These heavy elements form only about 2% of all atomic matter in our galaxy, with the rest being mainly hydrogen (74%) or helium (24%). Of all the stellar-nucleosynthesized elements, oxygen is the most abundant, constituting about 71 times less of the elemental mass fraction of our galaxy than hydrogen. Constituting approximately half the mass fraction of oxygen, carbon is the next most abundant element. After carbon, the next six most abundant elements in decreasing order are Ne, Fe, N, Si, Mg, and S, each one constituting a mass fraction between 30% and 10% of that of C. Thereafter, the elemental abundance starts to drop off more quickly, with the heavier elements tending to be proportionally less plentiful. In all low- to intermediate-mass stars (0.6–10 solar masses), the thermal and gravitational conditions eventually permit the energetically favorable fusion of the ^{4}He nuclei (or alpha particles) produced by the initial fusion of ^{1}H. Stable ^{12}C and ^{16}O happen to be the most common products of nuclear reactions involving these alpha particles, via the triple-alpha and alpha processes, respectively [8]. Some heavier elements with nuclei corresponding to integer multiples of alpha particles can also be formed by the alpha process (e.g., ^{20}Ne, ^{24}Mg, and ^{28}Si), but with significantly lower efficiencies. These reactions create a situation in which stellar nucleosynthesis produces large amounts of carbon and oxygen, but only a small fraction of these elements is converted into neon and heavier elements. In the latter stages of an intermediate-mass star's life, when the supply of He is ending, the star cools and burns any residual H and He in an erratic fashion, driving convective mixing processes. At this *dredge-up* [9] stage, some deep-lying carbon, oxygen, and other heavier elements can reach the cooling expanding atmosphere. Such stars in this stage of their life (so-called asymptotic giant branch [AGB] stars) are typically found to have oxygen-rich atmospheres. Stellar winds further expel elements in the atmosphere from the star, forming a circumstellar envelope within which chemical reactions between the elements can start. Close to the star's surface, only very strongly bound molecules can form due to the high temperatures. One of the initial reactions to occur is simply that between C and O to form the extremely stable and rather chemically inert CO molecule. In the cooler regions further from the star, any excess oxygen that has not been used up in CO formation can form other molecules. Of the remaining atomic species present, silicon forms the strongest bond with oxygen to form SiO molecules. Although it thus seems clear that this primary population of SiO monomers plays an essential role in the subsequent nucleation of silicates, there is some debate as to: (1) how Mg (and Fe) is incorporated into a SiO-based nucleation process to form silicate dust grain precursors with pyroxene- and olivine-type compositions, and (2) what are the primary seeds that initiate silicate dust nucleation.

The first point highlights the discrepancy between the assumed pure chemical composition of the nucleating SiO monomers and that of the observed Mg-rich silicate dust. In other words, even if we can propose a way by which high temperature SiO nucleation could be initiated, the simple accretion of SiO species (i.e., $(SiO)_N + (SiO)_M \rightarrow (SiO)_{N+M}$) cannot be the only contributing route to the observed circumstellar formation of Mg-rich silicates, typically with pyroxene-type ($Mg_{n-1}Fe_nSiO_3$ with $n \approx 0$) composition. The puzzle of how the O:Si ratio increases and how Mg is incorporated must also be an integral and essential part of any full explanation of silicate dust nucleation.

Through extensive explorations of many possible intermediate nanoclusters with variable $Mg_xSi_yO_z$ compositions and their calculated stabilities, computational modeling has begun to shed some light on this complex problem [10]. Gail and Sedlmayr [11,12] argue that in the silicate dust condensation zone (1200–1000 K, 0.1–0.001 Pa), O is locked up in H_2O and SiO, while Mg is atomic. From this premise, one can begin to theoretically search for clusters composed by adding these components (assuming H_2 release with oxidative H_2O additions), which are stable under typical circumstellar conditions. We note that such a search quickly becomes impracticable with the increase in both cluster size and chemical complexity and global optimization techniques (discussed in Chapter 1 and Section 13.2.3) are required. For condensation processes, the dominating contribution to the entropy of reaction is generally the loss of translation entropy, which is only partially compensated for by an increase in rotational and vibrational entropy. At the high temperatures under consideration, this huge entropy loss weighs heavily on the Gibbs free energy of reaction: $\Delta G_{rxn} = \Delta H_{rxn} - T\Delta S_{rxn}$, with $T\Delta S_{rxn}$ of the order of 200–300 kJ mol^{-1} for any bimolecular addition reaction at 1000 K. Consequently, only very exothermic reactions ($\Delta H_{rxn} < 400$ kJ mol^{-1}) can occur under these conditions. While the homomolecular nucleation of SiO ends effectively at the dimer stage, subsequent oxidation and Mg incorporation steps are sufficiently exothermic to compensate for the enormous entropic costs involved in the condensation processes at these high temperatures. This bottom–up approach strongly suggests that homogeneous homomolecular condensation of SiO is unfavorable in the dust condensation zone of stellar winds, while homogeneous heteromolecular nucleation of magnesium silicates could be feasible. One of many possible routes from the set of monomeric species (SiO, Mg, H_2O) to a small pyroxene cluster is shown in Figure 13.2. We note that here we only consider whether the reactions are thermodynamically favored but possible high reaction barriers can potentially also play an important role in the overall nucleation process.

The second question arises due to the fact that the production of silicate dust around stars is found to begin at temperatures as high as 1200 K, whereas calculated predictions of the onset temperature for SiO condensation, based on laboratory measurements of the vapor pressure of SiO, are at least 200 K lower [13–15]. We note that these latter theoretical predictions are currently based on classical nucleation theory (CNT, see also Chapter 6), which we will evaluate critically with respect to dust nucleation (see Section 13.2.1). Even if we reject the idea that SiO molecules homogeneously nucleate and adopt a richer heteronuclear approach to silicate nucleation, as described above, the very first species to condense at ~1200 K is likely to be more stable than any silicate. The presence of such extremely stable refractory species could also act as nucleation seeds, which could strongly favor the heteronuclear nucleation of magnesium silicates. In Section 13.2.2, we show how computational modeling has provided evidence that the relatively low abundant element Ti might form a range of small but very stable oxidic species, which could help to kick-start silicate dust nucleation.

Clearly, an understanding of nanosized silicate clusters, or nanosilicates, is essential for understanding the initial stages of silicate dust nucleation around stars. We also note that although AGB stars are commonly thought to be by far the main producers of silicate dust, brown dwarf stars and particularly supernovae can also contribute to the total silicate dust quotient. Moving away from a dust-producing AGB star, the average size of typical silicate grains quickly becomes significantly larger than a few nanometers

1. $2SiO \rightarrow Si_2O_2$

2. $Si_2O_2 + H_2O \rightarrow Si_2O_3 + H_2$

3. $Si_2O_3 + \begin{Bmatrix} Mg \rightarrow MgSi_2O_3 \\ \hline H_2O \rightarrow Si_2O_4H_2 \end{Bmatrix}$

4. $\begin{Bmatrix} MgSi_2O_3 + H_2O \\ \hline Si_2O_4H_2 + Mg \end{Bmatrix} \rightarrow MgSi_2O_4 + H_2$

5. $MgSi_2O_4 + \begin{Bmatrix} Mg \rightarrow Mg_2Si_2O_4 \\ \hline H_2O \rightarrow Mg_2Si_2O_5 + H_2 \end{Bmatrix}$

6. $\begin{Bmatrix} Mg_2Si_2O_4 + H_2O \rightarrow Mg_2Si_2O_5 + H_2 \\ \hline MgSi_2O_5 + H_2O \rightarrow MgSi_2O_6H_2 \end{Bmatrix}$

7. $\begin{Bmatrix} Mg_2Si_2O_5 + H_2O \\ \hline MgSi_2O_6H_2 + Mg \end{Bmatrix} \rightarrow Mg_2Si_2O_6 + H_2$

(a)

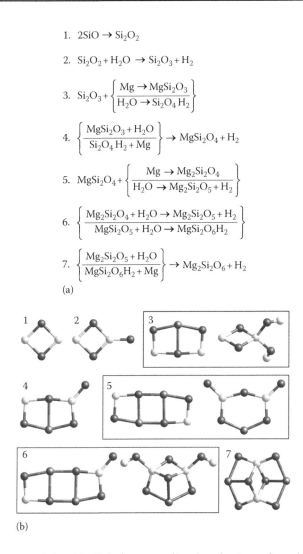

(b)

FIGURE 13.2 A seven-step $2SiO \rightarrow Mg_2Si_2O_6$ (pyroxene dimer) nucleation pathway (a) with corresponding ground state geometries and the cluster species involved (b). Si atoms: yellow, Mg atoms: blue, O atoms: red, and H atoms: white.

and soon the dust grain population becomes dominated by larger grains. These dust grains are increasingly processed by shocks and sputtering in the ISM, where they are amorphized and become mostly olivinic. Throughout these stages of dust condensation, destruction, and coagulation, nanosized silicates could play an important role. Indeed, a substantial (~10%) mass fraction of the silicate grain population in the diffuse ISM could be in the form of very small nanosilicates with <1.5 nm diameters [16].

While dust particles make up only one mass percent of the total matter in the ISM, they play a crucial role in its chemical evolution [17] where they catalyze molecule formation. The formation of molecular H_2, for example, is thought to be primarily produced through the reactions on dust grain surfaces. In Section 13.3.1, we will highlight how we can gain detailed insights into this important dust-catalyzed reaction using *ab initio* computational chemistry modeling. As diffuse clouds of gas and dust contract to

Chapter 13

form denser clouds, silicate dust also scatters the strong interstellar radiation, preventing molecular photodissociation. In such regions, silicate dust grains gradually acquire icy mantles, largely formed from the aggregation of H_2O molecules [18]. The role of nanosilicates as initiators of heterogeneous nucleation of water–ice is also relevant for terrestrial atmospheric processes. Computational modeling allows us to follow in detail the initial steps of H_2O condensation on nanosilicates (see Sections 13.3.2 and 13.3.3).

Contracting dense clouds in the ISM form dark clouds in which, eventually, new generations of stars are born. Subsumption into newly forming stars or their nascent planetary systems marks the end of the cosmic life cycle for much of silicate dust. Some dust, however, remains much unchanged for millennia in the cold bodies of comets or simply between the planets. For such so-called interplanetary dust particles (IDPs), a few may meet a fiery fate through entering the atmosphere of a planet, such as the Earth, where they may initiate a range of chemical processes.

13.1.3   Meteoric Interactions with the Earth's Atmosphere

The main sources of IDPs are the sublimation of comets when they approach the sun while orbiting through the solar system, and collisions between asteroids, which mostly occur in a belt between the orbits of Mars and Jupiter [19,20]. Dust particles from long-decayed cometary trails and the asteroid belt give rise to a continuous input of sporadic meteoroids into the Earth's atmosphere. In addition, the dust-rich trails from comets which recently crossed the Earth's orbit (i.e., within the past century or so) are the source of meteor showers, although these provide a much smaller mass flux on average than the sporadic background.

A recent astronomical model of interplanetary dust starts with the orbital properties of comets and asteroids and then follows the dynamical evolution of dust particles after ejection from these sources [21]. The model predicts that ~90% of the dust in the inner solar system comes from Jupiter family comets. These are comets with short orbital periods of a few decades, and an aphelion close to the orbit of Jupiter. The remaining dust comes from the asteroid belt, and Halley family and Oort cloud comets, which have orbital periods of centuries. IDPs drift into the inner solar system because of Poynting–Robertson drag, where solar photon pressure causes the orbital velocities of IDPs with a radius larger than ~1 μm to decelerate. The model predicts that most of the dust mass is contained in particles with masses in the range 1–10 μg (radius ~50–100 μm), which enter the terrestrial atmosphere from a near-prograde orbit with a mean speed around 14 km s^{-1}.

The magnitude of the total dust input to the Earth's atmosphere is actually very uncertain [22]. Even very recent estimates of the IDP input vary from 5 to 270 t day^{-1} (tonnes per day). In essence, the IDP input rate is so uncertain because no single technique can observe particles over the mass range from about 10^{-12} to 1 g, which make up the bulk of the incoming material [20]. Zodiacal cloud observations [21] and space-borne dust detection [23] indicate a daily input of up to 300 t day^{-1}, which is mostly in agreement with the accumulation rates of cosmic elements in polar ice cores [24,25] and deep-sea sediments [26,27]. In contrast, measurements in the middle atmosphere by radar [28–31], aircraft sampling [32], and satellite remote sensing [33] imply that the input is less than ~50 t day^{-1} [22]. This range indicates that either vertical transport in the middle atmosphere must be considerably faster than what is currently thought to be the case in general circulation models of the atmosphere [34,35], so that meteoritic material is removed more rapidly from the atmosphere in order to sustain a higher rate of injection, or transport mechanisms from the middle atmosphere to the Earth's surface cause *focusing* of meteoric debris at polar latitudes [24].

IDPs enter the atmosphere at speeds between 11 and 72 km s^{-1}, if they originate within the solar system. High-energy inelastic collisions with air molecules lead to sputtering of elements from the particle surface. Most particles with masses in excess of 1 µg (radius > 50 µm, assuming a density around 2 g cm^{-3} for amorphous metal silicate particles) also rapidly heat until they reach a melting point around 1850 K. Evaporation of atoms and oxides from the molten particle—a process termed meteoric ablation—is then very rapid, leading to complete evaporation of the particle [36]. Ablation tends to occur where the atmospheric pressure is around 1 µbar. In the case of the Earth, the peak ablation rate is around 90 km, compared with 80 km on Mars, 115 km on Venus, and 500 km on Titan [37].

The physics of ablation has been treated in detail by several investigators [38–40]. The problem becomes manageable for particles smaller than about 250 µm in radius, because heat conductivity through the particle is then fast enough for the particle to be treated as isothermal. The processes of sputtering and ablation have recently been incorporated into a chemical ablation model (CABMOD), which uses melt thermodynamics from the well-established MAGMA code [41], and calculates the evaporation rates of individual species by assuming Langmuir evaporation, that is, the rate of evaporation into a vacuum is equal to the rate of evaporation needed to balance the rate of uptake of a species in a closed system [36]. Figure 13.3 illustrates the elemental injection profiles calculated by CABMOD for a meteoroid entering the atmosphere with the most

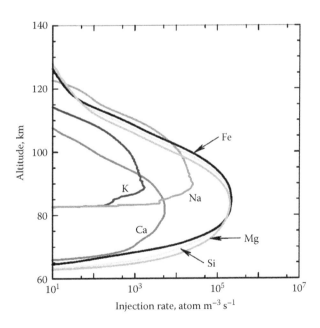

FIGURE 13.3 Injection rates of individual elements, obtained using the CABMOD model (From Vondrak, T. et al., *Atmos. Chem. Phys.*, 8, 7015, 2008) integrated over the meteoroid mass and velocity distributions determined form the *Long Duration Exposure Facility*, an orbital impact detector placed on a spacecraft. (From Love, S.G. and Brownlee, D.E., *Science*, 262, 550, 553, 1993.)

Chapter 13

likely mass and velocity determined from the *Long Duration Exposure Facility* (LDEF), an orbital impact detector placed on a spacecraft [23]. Inspection of Figure 13.3 shows that CABMOD predicts differential ablation, that is, the most volatile elements—Na and K—ablate first, followed by the main constituents Fe, Mg, and Si, and finally the most refractory elements such as Ca.

Meteoric ablation is the source of the layers of neutral metal atoms, which occur between 85 and 95 km in the terrestrial atmosphere [42]. Several of these layers—Na, K, Ca, and Fe—can be observed using ground-based resonance lidars, where the laser in the transmitter is tuned to a strongly allowed optical transition of the metal (see Figure 13.4). Note that atomic Mg, which is a major meteoric metal, cannot be observed by ground-based lidar, because the Mg optical transition at 285 nm is strongly absorbed by the Hartley band of O_3 in the stratospheric ozone layer. However, Mg (and Mg^+) can be observed by resonant scattering of sunlight, using satellite-borne optical spectrometers [43,44].

In the terrestrial atmosphere, atomic Fe is oxidized in a series of reactions involving O_3, O_2, and H_2O to form reservoir species including $OFeO_2$, $Fe(OH)_2$, and FeOH [45]. FeOH may also be oxidized by O_3 to form FeOOH (which is the building block of the mineral goethite). Figure 13.3 illustrates these reaction pathways, and also shows that the reactions, which cycle these species back to Fe, involve atomic O and H. The concentrations of O and H are only significant above about 82 km [42], which is why the Fe layer has a steep underside at this height (Figure 13.4); this is where these molecular reservoir species build up. Figure 13.5 summarizes the analogous chemistry of Mg, whose main reservoir is probably $Mg(OH)_2$ [46], and Si, which is oxidized rapidly to SiO and SiO_2 [47,48]. SiO_2 is then most likely hydrated by the addition of H_2O to form $OSiO_2$ and $Si(OH)_4$ (see Section 13.2.3). This mixture of Fe–, Mg–, and Si–containing species can then polymerize in the mesosphere over several days to form nanometer-sized meteoric smoke particles (MSPs), which provides a permanent sink for gas-phase metallic compounds. Figure 13.5 lists a number of possible types of compounds that can form. Electronic structure calculations show that these polymerization reactions are strongly

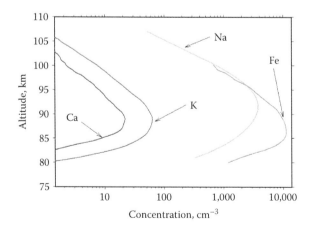

FIGURE 13.4 Vertical profiles of the annual mean concentrations of Fe, Na, K, and Ca measured by lidar at a number of mid-latitude locations in the United States and Europe. (Adapted from Plane, J.M.C., *Chem. Rev.*, 103, 4963, 2003.)

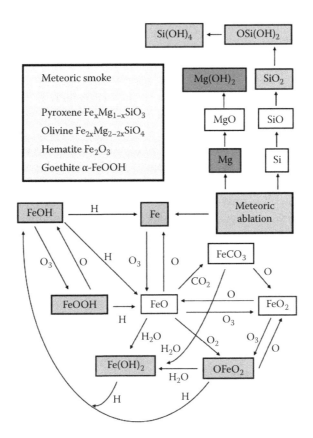

FIGURE 13.5 Formation of meteoric smoke from ablated Fe, Mg, and Si. The pathways for oxidizing Fe are shown in detail. The most stable reservoir species of each element are shown in shaded boxes.

exothermic (Section 13.2.3), so that polymerization is spontaneous even at low temperatures. This has been verified in laboratory experiments to study the formation and growth kinetics of metal oxide/silicate nanoparticles of 1 nm radius and larger [49–51].

MSPs can be measured directly by rocket-borne particle detectors above 80 km [52–55]. These detectors measure the fraction of MSPs which are charged, so the total number has to be obtained by estimating the charged fraction using a dusty plasma model. It appears that the majority of MSPs have a radius between 0.5 and 1 nm at this altitude, in accordance with models [34,56,57]. A new design of particle detector has recently been flown, which contains a pulsed VUV lamp to photodetach electrons from negatively charged particles, and photoionize neutral MSPs [58]. This work has shown that some MSPs have a surprisingly low *work function* of around 4 eV. Electronic structure calculations indicate that these particular particles are likely composed of iron and magnesium hydroxide clusters, rather than silicates [59].

MSPs have also been observed by optical extinction measurements between about 40 and 75 km using a visible/near-IR spectrometer on the *Aeronomy of Ice in the Mesosphere* (AIM) satellite [60]. Measurements at several different wavelengths indicate that the MSP composition is probably olivine ($Mg_{2x}Fe_{2-2x}SiO_4$, $x = 0.4$). However, the refractive indices used for this study were for bulk crystalline *minerals*, whereas MSPs are nanometer-size (and likely amorphous) particles, which may have quite different refractive indices. The particles are also likely to be chemically weathered by H_2O and H_2SO_4 during the months they spend descending into the stratosphere; this may change their optical properties significantly.

An important reason for studying MSPs in the mesosphere is their relation to noctilucent clouds. These clouds were first reported over northern Europe in 1885, and have been growing brighter and spreading to lower latitudes through much of the last century [61]. This has naturally led to noctilucent clouds being regarded as an *early warning* of climate change. The clouds occur between 80 and 86 km, at high latitudes in the summer where the temperature falls below 150 K, which is the approximate frost point (H_2O vapor has a mixing ratio of only a few parts per million) [62]. An important uncertainty in research on noctilucent clouds is the nature of the nuclei on which the ice particles grow. There is increasing evidence that MSPs provide an important source of ice nuclei [63]. This topic is discussed further in Section 13.3.3.

In the mesosphere below 80 km and the stratosphere, ultrafine particles (radius < 50 nm) do not sediment rapidly. Instead, these particles are transported by the bulk atmospheric circulation. Models predict that MSPs should be swept to the winter pole by the mean meridional circulation in the mesosphere before downward transport within the polar vortex to the lower stratosphere [34,35,56]. This has been confirmed by measurements inside the winter Arctic vortex [64]. During the months that MSPs spend in the mesosphere and upper stratosphere, the particles are likely to grow by agglomerative coagulation, which can be very rapid because of the long-range magnetic dipole forces between the Fe-containing particles [50]. Models predict that the particles could grow to around 40 nm in radius by the time they reach the middle stratosphere around 30 km [34,56,57]. Metal-rich MSPs should readily remove gas-phase acids such as H_2SO_4 and HNO_3. This may explain the marked decrease of gas-phase H_2SO_4 measured by mass spectrometry on balloons in the upper stratosphere [35,65–67].

In the lower stratosphere, air-borne aerosol mass spectrometer measurements have shown that the sulfuric acid particles, which make up the so-called Junge layer around 20 km, contain ~0.75 and ~0.2 wt% of meteoric Fe and Mg, respectively [32]. This finding is consistent with a recent laboratory study showing that amorphous Fe-Mg-silicate nanoparticles dissolve in concentrated H_2SO_4 solutions at stratospheric temperatures in less than a week [35]. These high levels of metal sulfates may elevate the freezing point of H_2SO_4–H_2O solutions to form solid sulfuric acid tetrahydrate (SAT) [68]. MSPs may also be involved in the heterogeneous nucleation of nitric acid trihydrate (NAT) from the tertiary HNO_3–H_2SO_4–H_2O system in the winter polar stratosphere [69]. MSPs could therefore modify the degree of O_3 depletion in the lower stratosphere caused by chlorine activation and denitrification.

Finally, MSPs probably enter the troposphere at midlatitudes, where stratosphere–troposphere exchange is driven by waves generated by mountain ranges and storm tracks over the North Atlantic, North Pacific, and the Southern Ocean. Within the troposphere, wet deposition (by rain and snow) of these particles dominates over dry deposition [25]. A recent study [70] using a general circulation model predicts strong deposition over the Southern ocean (50°S–60°S), where the supply of bioavailable iron to phytoplankton is limited [71]. The resulting Fe fertilization could have important climate feedbacks [22].

13.2 Nanosilicate Formation

13.2.1 Validity of Classical Nucleation Theory

CNT was first developed to explain the formation of water droplets in the Earth's atmosphere under equilibrium conditions of constant supersaturation [72]. It is a theory that applies to *homogeneous* nucleation, where gas-phase molecules of the same species nucleate to form a small particle. The free energy needed to form a spherical cluster of radius r from a population of monomer molecules is given by

$$\Delta G = \frac{4}{3}\pi r^3 \Delta G_v + 4\pi r^2 \sigma$$

The first term represents the free energy gained by creating a new droplet volume (ΔG_v is negative if formation of droplets is favorable). The second term shows that this gain is offset by the surface energy required to create the new interface with the surroundings, where σ is the surface tension of the droplet. Once an initial cluster forms, there is a free energy cost to add further molecules, because $dG/dr > 0$. However, at a critical radius r^* where $dG/dr = 0$ and $r^* = -2\sigma/\Delta G_v$, addition of further monomers leads to a release of free energy. This favorable situation then leads to *runaway* nucleation, where there is no thermodynamic constraint on droplet growth. The free energy needed to form a droplet of radius r^* is then given by

$$\Delta G^\star = \frac{16\pi\sigma^3}{3\Delta G_v^2}$$

In CNT, the rate of particle nucleation is then given by

$$J^\star = \tau^{-1} N^\star$$

where

> τ is the time constant for addition of a molecule to increase the cluster size beyond r^*, that is, the capture rate constant multiplied by the equilibrium concentration (or number density) of monomer
> N^* is the number density of clusters of critical radius, which can be evaluated using ΔG^*, that is, assuming equilibrium between the critical cluster and monomer

For nucleation to be efficient, a high supersaturation ratio is necessary (i.e., $S \gg 1$) so that N^* is nonnegligible. However, once nuclei (or *seeds*) have formed, subsequent condensation of monomers onto these nuclei to produce macroscopic dust grains can take place in only slight supersaturated conditions. (See also Chapter 6 for a more detailed general description of CNT.)

   CNT was first used to model dust formation in the ISM by Draine [73], who considered the formation of very small clusters of about 10 monomers. Much of the subsequent work on dust grain formation in stellar outflows using CNT has been carried out by Gail,

Sedlmayr, and colleagues (e.g., [74–76]). Taking into account the abundance of molecular species and the temperatures at which they exist in the solid phase, possible nucleation seed candidates that have been investigated include SiO, SiO_2, MgO, Fe, Al_2O_3, TiO, and TiO_2. For studies of cluster formation, data on the Gibbs free energy of formation $\Delta_f G(N)$ of clusters with N monomers are required. In most cases, relevant experimental data are not available, so theoretical calculations using semi-empirical formalisms have been used (e.g., Kohler and others [74], Jeong and others [77]). The nucleation rates, J^*, are then determined for these high temperature condensates using CNT, assuming chemical equilibrium in the gas phase. In many cases, the critical cluster size was found to be ~20 monomers [75].

However, there are a number of difficulties with CNT. These were first pointed out by Donn and Nuth [78], and have also been discussed more recently by Cherchneff and Dwek [79]. First is the fact that low-density environments such as stellar outflows and supernova ejecta are not in local chemical equilibrium, since the concentrations of monomers may be changing more rapidly—due to gas-phase production and/or loss, and transport dynamics—than the formation of clusters. Second, the bulk thermodynamic properties and surface tension of the condensed phase, which are required to calculate J^*, may well not be applicable at the cluster size scale, which can sometimes involve only tens of atoms. Third, while the concept of surface tension giving rise to a surface free energy has some significance to a droplet, it is not obviously applicable to a small amorphous cluster where the surface *roughness* and chemical inhomogeneity is significant compared to the cluster dimension. Fourth, formation of dust often proceeds via the formation of small primary particles (a few nanometers in radius), followed by rapid growth through agglomeration, producing a smaller number of large fractal-like particles. This process is well known in soot formation and has also been observed in metal oxide and silicate particles [50,51,80]. In contrast, classical CNT predicts that once a cluster grows larger than r^* and becomes thermodynamically stable, further growth occurs only through surface condensation of the monomer.

Although there have been attempts to account for the effects of nonlocal equilibrium in CNT (e.g., Patzer and others [81], Paquette and others [13]), and the lack of chemical equilibrium may not always be a serious problem [82], the other inherent difficulties with the theory still remain. This is why the kinetic approach, described in the next section, is a more rigorous—though much more computationally demanding—treatment of dust nucleation.

13.2.2   Kinetic Modeling of Silicate Dust Nucleation

As an alternative to CNT, a kinetic modeling approach to silicate dust formation can be used (e.g., Cherchneff and Dwek [79]; Plane [83]). This approach describes explicitly the formation of the molecular precursors in the gas phase, which polymerize to form embryonic dust particles. The full kinetic approach involves three stages. First, electronic structure calculations are used to determine energetically possible reaction pathways to form the molecular precursors (i.e., the monomer species). Second, the molecular properties (rotational constants and vibrational frequencies) of stable intermediates and transition states along these pathways are used to calculate reaction rates over typically complex potential energy surfaces (PESs). Third, the resulting rate coefficients are then employed in a model, which couples gas-phase chemistry with particle growth kinetics.

A recent study of nanoparticle formation in stellar outflows exemplifies this approach [83]. During the late stages of stellar evolution, AGB stars lose a significant fraction of their mass in a short period, during which they are surrounded by an optically thick shell of dust [84]. Spectroscopic observations indicate that a major component of the dust are amorphous magnesium-iron-silicates, which raises the important question of the nature of the condensation nuclei (CN) on which the relatively abundant species Mg, Fe and SiO condense from the gas phase to form the metal silicates [12]. The CN must be highly refractory nanoparticles, since they appear to form in a region of the outflow where the temperature ranges from ~1000 to 1200 K [77].

This is a challenging environment in which nanoparticles have to be formed from gas-phase constituents, which are present in the outflow. There are several factors to consider. First, the molecules (building blocks) from which nanoparticles form need to be relatively stable to survive thermal dissociation in the 1000–1200 K temperature range, which requires that their bond energies are at least ~400 kJ mol^{-1}. Second, the major constituent of the outflow is H_2, so there must also be significant quantities of H atoms; both H and H_2 are chemically reducing species, which prevent the oxidation of metals and silicon oxides.

Third, the total pressure in a stellar outflow is very low, typically less than 0.01 Pa in the region where nanoparticles form. Recombination reactions, which involve two species associating in the presence of a *third body* M (i.e., $A + B + M \rightarrow AB + M$), are pressure dependent, because M is required to stabilize the product. Furthermore, the formation of these compounds involves reactions between at least two constituents in the outflow, whose concentrations will be roughly proportional to the total density (or pressure) in the outflow. Hence, the rates of these reactions will have at least a second-order dependence on the total density in the outflow, and thus decrease rapidly as the flow expands away from the AGB star.

Due to the difficulties in reconciling the estimated upper temperature limit for pure SiO condensation with the highest temperatures observed for silicate dust formation around stars, other candidate CN constituents have been sought. Although Ti is approximately 300 times less abundant than Si in stellar outflows, solid Ti_xO_y oxides are typically very stable and are potentially good candidates for forming the primary CNs. Following this line of thinking, the nucleation of pure $(TiO_2)_N$ has been theoretically studied [12,77]. Considering the high abundance of SiO molecules, we may also consider reactions with TiO_2 and SiO. The smallest possible species of this type is the $SiTiO_3$ molecule, which has been predicted to be a possible kick-starter for subsequent Mg-rich silicate dust condensation [85]. Perhaps the most stable of all solid titanium-based compounds that could possibly be present in a stellar outflow is $CaTiO_3$. In its bulk crystalline form, $CaTiO_3$ is a stable refractory material and is the prototypical member of the perovskite structured family of compounds; thus it is often referred to simply as *perovskite*. From calculations of condensation sequences of solids from high temperatures, $CaTiO_3$ usually appears first [12] and thus before the pure Ti-oxides. In the following, the possibility of molecules of $CaTiO_3$ forming one of the first CNs in stellar outflows is considered.

Although Ca and TiO are relatively minor constituents in stellar outflows, they can form $CaTiO_3$, which has several attractive features as a building block compared to more abundant constituents such as Fe, Mg, and SiO [83]. First, $CaTiO_3$ is extremely stable with respect to thermal dissociation: $CaTiO_3 \rightarrow CaO + TiO_2$ is endothermic by 585 kJ mol^{-1}. Second, reaction with atomic H must be slow, because it is quite endothermic: $CaTiO_3 + H \rightarrow CaOH + TiO_2$, $\Delta H = 148$ kJ mol^{-1}. Third, $CaTiO_3$ has a very large dipole moment of 10.2 Debye, so there will be strong dipole–dipole forces and hence an unusually high polymerization rate for $CaTiO_3$ cluster formation. Fourth, $CaTiO_3$ binds to $CaTiO_3$ polymers with energies well in excess of the lower limit of 400 kJ mol^{-1} required for reasonable thermal stability. The net result is that $CaTiO_3$ can survive long enough, at high temperatures and in the presence of significant H atom concentrations, to polymerize rapidly and form clusters.

Having established that $CaTiO_3$ should be an effective building block, the problem is to quantify its rate of formation from Ca and TiO under the conditions of high temperatures and low pressures in a typical outflow. One feasible pathway involves the following reactions [83]:

$$TiO_2 + H_2O \, (+M) \rightarrow OTi(OH)_2 \tag{13.1}$$

$$Ca + OTi(OH)_2 \rightarrow CaTiO_3 + H_2 \tag{13.2}$$

where TiO_2 is produced from TiO by oxidation with OH. In order to calculate the rate coefficients of these reactions for inclusion in a kinetic model, electronic structure calculations must first be used to map the stationary points—reactants, products, intermediates, and transition states—on the electronic PESs. Molecular geometries are first optimized, and then rotational constants and vibrational frequencies are determined for the master equation calculations described in the following. Since a molecule such as $CaTiO_3$ contains some heavy atoms (even more the case for the dimer and trimer), a computationally efficient but reasonably accurate theoretical model needs to be used. A well-established general purpose method is the hybrid density functional B3LYP, which includes some exact Hartree–Fock exchange. This needs to be combined with a reasonably large, flexible basis set such as the 6–311+G(2d,p) triple zeta basis set, which has both polarization and diffuse functions added to the atoms. Previous theoretical benchmarking studies indicate an expected uncertainty in the calculated reaction enthalpies at this level of theory on the order of ±25 kJ mol^{-1} [86]. More accurate energies can then be determined using higher level methods such as the *Complete basis set* (CBS-Q) method of Petersson and coworkers [87].

Figure 13.6 illustrates the PES for reaction given as Equation 13.1. This shows that after initially forming a TiO_2–H_2O complex, rearrangement over a barrier submerged below the energy of the entrance channel leads to the product $OTi(OH)_2$ where three O atoms are now bound to the Ti, that is, it is effectively oxidized to the +4 state of the titanate ion. For reaction given as Equation 13.2, Figure 13.7 shows that addition of the Ca atom leads to a $CaTiO_3H_2$ intermediate. This intermediate can then rearrange by migration of one of the H atoms onto the Ti, forming the more stable $CaTi(H)O_3H$ intermediate via transition state TS1. $CaTi(H)O_3H$ can in turn rearrange by twisting the hydroxyl H so that H_2 can form, leading via transition state TS2 to the products $CaTiO_3 + H_2$. As in reaction given as Equation 13.1, TS1 and TS2 are submerged with respect to the entrance channel. The calculations shown here were performed using the Gaussian 09 suite of programs [88].

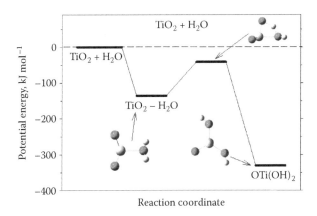

FIGURE 13.6 PES calculated at the CBS-Q level of theory for the recombination of TiO_2 and H_2O. Atom key: O, dark gray; H, small light gray; Ti, light gray. (Adapted from Plane, J.M.C., *Philos. Trans. R. Soc. A*, 371, art. no. 20120335, 2013.)

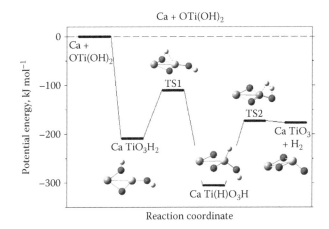

FIGURE 13.7 PES for the reaction of Ca with $OTi(OH)_2$, calculated at the CBS-Q level of theory. Atom key: O, dark gray; H, small light gray; Ti, light gray; Ca, medium gray.

Reaction rate coefficients can then be estimated using Rice–Ramsperger–Kassel–Markus (RRKM) theory [89]. For reactions where there is a small or nonexistent barrier in the entrance channel (such as reactions in Equations 13.1 and 13.2), RRKM theory can be utilized by a master equation solution based on the inverse Laplace transform method [90]. This method has been used to model successfully the measured rate coefficients of reactions involving metal-containing species where a stable intermediate is present on the PES [91]. These reactions proceed via the formation of an excited adduct from the two reactants. This adduct can then dissociate back to reactants, rearrange to other intermediates connected by transition states, or dissociate to bimolecular products. Any of the intermediates can also be stabilized by collision with a third body. The time evolution of all these possible outcomes is modeled using the master equation.

For complex PESs involving several intermediates (e.g., Figure 13.7), a multiwell energy-grained master equation has to be used [92]. One example of such a master equation solver is the open source master equation program, MESMER (Master Equation

Chapter 13

Solver for Multiwell Energy Reactions) [93]. The internal energies of the intermediates on the PES are divided into a contiguous set of grains (typical width = 200 cm^{-1}), each containing a bundle of rovibrational states. Each grain is then assigned a set of microcanonical rate coefficients linking it to other intermediates; these rates are calculated by RRKM theory [89]. For dissociation to products or reactants, microcanonical rate coefficients are determined using inverse Laplace transformation to link them directly to the capture rate coefficient, $k_{capture}$. For these reactions involving neutral species with no barrier, $k_{capture}$ can be set to a typical capture rate coefficient of 3×10^{-10} $(T/1000$ K$)^{1/6}$ cm^3 molecule^{-1} s^{-1} [94]. The small positive temperature dependence is characteristic of a long-range potential governed by dispersion and dipole–dipole forces.

The probability of transfer between grains when the adduct collides with the third body, M, is usually estimated using the exponential down model, where the average energy for downward transitions is typically $\langle \Delta E \rangle_{down}$ = 200–300 cm^{-1} for H$_2$ [89]. The probabilities for upward transitions are calculated by detailed balance. MESMER determines the temperature- and pressure-dependent rate coefficient from the full microcanonical description of the system time evolution by performing an eigenvector/eigenvalue analysis similar to that described by Bartis and Widom [95].

Having calculated the appropriate rate coefficients, the rate of production of particle building blocks (e.g., individual CaTiO$_3$ molecules) can be estimated for the conditions in the outflow. The gas-phase part of the outflow model must then be coupled to a particle agglomeration model, in order to follow the growth and evolution of the size distribution of nanoparticles. One standard approach is the volume-conserving model of Jacobson [96], which has been applied previously to model the growth of metal silicate particles [50]. Particle sizes are separated into radius space in a number of fixed center bins, where the first bin size, r_1, is set at a molecular radius corresponding to the equivalent size of a spherical monomer. Subsequent bin sizes are increased geometrically by a fixed volume ratio of α (typically α = 1.2–1.7), so that the particle radius of the ith bin is $(\alpha^{1/3})^{i-1} r_1$. For modeling CaTiO$_3$ particle growth [83], r_1 was set to 0.28 nm, which is estimated by assuming an amorphous particle density of 2500 kg m^{-3}. With 45 bins and α = 1.5, the particles in the largest bin have r_{45} = 107.1 nm. The concentration of monomer in the first bin is updated with production of gas-phase CaTiO$_3$ at each time step of the gas-phase module.

The model then assumes that particle growth is dominated by Brownian diffusion/coagulation, where collisions between pairs of particles cause coalescence, that is, the spherical morphology and compact structure of the particles are preserved. The collision rate coefficients (or kernels) for Brownian coagulation of the small particles (<10 nm) are calculated using the expression for the free molecular regime (Knudsen number, $K_n \gg 1$), interpolated into the transition regime for larger particles [97]. It is worth noting that the volume-conserving particle growth model can be adapted to treat the agglomerative growth of fractal-like particles, in particular where the rapid growth of Fe-containing particles is driven by magnetic dipoles once they reach a primary particle radius around 4 nm [50]. A final point is that coagulation is the same process as recombination (see earlier text) and so, in principle, is also pressure dependent. Indeed, at the exceptionally low pressures of a stellar outflow, the rate for dimerization of a monomer such as CaTiO$_3$ is slightly pressure dependent [83]. However, polymerization to larger polymers (beyond the dimer) becomes essentially pressure independent because of the increasing number of atoms involved, which gives rise to a large number of vibrational modes and a correspondingly large density of rovibrational states. The condensation of Fe, Mg, SiO, and H_2O on these $CaTiO_3$ nanoparticles, once the outflow has cooled below 1000 K, can then produce the olivine/pyroxene particles of over 100 nm radius observed around AGB stars.

In the Earth's upper atmosphere, the formation of silicate nanoparticles occurs through a different route. Meteor-ablated Si is oxidized rapidly to SiO_2 [48], which then undergoes hydrolysis [83]:

$$SiO_2 + H_2O \, (+M) \rightarrow OSi(OH)_2 \qquad (13.3)$$

This reaction has a similar PES to the analogous reaction of TiO_2, and hence is rapid. Further hydration can then occur to produce silicylic acid:

$$OSi(OH)_2 + H_2O \rightarrow Si(OH)_4 \qquad (13.4)$$

which is exothermic by 260 kJ mol^{-1} at the B3LYP/6–311+g(2d,p) level of theory. Unlike in the environment of a stellar outflow where large concentrations of H atoms destroy metal oxides and hydroxides [83], in the MLT region of the atmosphere, Mg and Fe form quite stable compounds like $Mg(OH)_2$ [46] and FeOH [45]. According to theory, these hydroxides should condense with $Si(OH)_4$:

$$Mg(OH)_2 + Si(OH)_4 \rightarrow HOMgOSi(OH)_3 + H_2O \qquad (13.5)$$

$$FeOH + Si(OH)_4 \rightarrow FeOSi(OH)_3 + H_2O \qquad (13.6)$$

where these reactions are, respectively, 61 and 21 kJ mol^{-1} exothermic, with no barriers involved in the rearrangement required to expel a H_2O. The resulting hydrated silicates can then lose H_2O in the dry mesosphere to produce $MgSiO_3$ and $FeSi(OH)O_2$, which then polymerize to make MSPs.

13.2.3   Finding Low Energy Structures of Nanosilicate Clusters

Often simply by virtue of their size, nanosilicates are extremely difficult to analyze by experimental characterization methods alone. In some recent studies, it has been demonstrated that detailed nanoscale information for well-defined laboratory-based nanosilicate systems can be accurately ascertained from a synergistic combination of computational modeling and experimental data [98]. For less-well-defined nanosilicate systems, which are not prepared in the lab, such as atmospheric or cosmic silicates, computational modeling can play an essential role in providing realistic nanoscale structural models, which would be otherwise unobtainable. These candidate nanomodels can subsequently be used to calculate properties (e.g., atomic and electronic structure, chemical reactivity, spectroscopy), which can be compared and used together with any relevant observational data to help explain the phenomenon under study. As highlighted in the previous section, for example, quantum chemical calculations are particularly useful for providing (free) energies for nanocluster/molecular species and the barrier heights for their reactions for use in understanding nucleation. Even when no experimental/observational data are available, computational modeling can also provide detailed predictions and helpful insights that may help guide future experimental research.

Although the various methods of computational modeling constitute a powerful and, often indispensable, toolkit, the theoretical treatment of nanoscale materials is still not a routine task. The first obstacle is that of scale. Although, almost by definition, the nanoscale is small from a typical experimental materials perspective, even a nanocluster of only ~10 nm diameter possesses thousands of atoms. Although this is, in principle, of a size that is readily treated by computational modeling methods using rapidly evaluated classical interatomic potentials, the aim in many theoretical investigations of nanosystems is to achieve an accurate quantum mechanical based description. For many accurate and efficient *ab initio* electronic structure modeling approaches (e.g., density functional theory [DFT] implemented in a well-parallelized code), nanosystems consisting of a thousand or more atoms are at a size and complexity that is still significant for routine calculations.

The second problem to be tackled is the high configurational complexity of many nanosystems. For small- to moderate-sized molecules and bulk crystals, one often has sufficient structural information from experiment or chemical databases to start directly with *ab initio* quantum chemical calculations. Nanosystems lie in a size regime between these two extremes for which, currently, there is a dearth of detailed structural information. Although knowledge of the chemical composition of a small molecule or a crystal with a small number of atoms in its unit cell can be sufficient for a rapid exhaustive search through all chemically sensible structures, this is typically not the case for nanosystems. For silicate nanoclusters, for example, the number of possible structural configurations for even modest sizes (e.g., 20–30 atoms) becomes very large and increases combinatorially with increasing cluster size. The space of isomeric structures with respect to their energetic stability is often termed the potential energy surface (PES). Typically, one wants to find the most stable nanocluster structures on the PES. The complexity of the PES, and thus in turn the ease of with which one can find low-energy clusters, depends on many factors. A strong tendency to form atomically ordered closed packed clusters, for example, can simplify the PES. On the other hand, less dense systems that permit a high degree of structural freedom (e.g., bond angles, coordination number) and/or have multiple atom types, which can permute positions, tend to have PESs that are more difficult to search efficiently. In the latter case in particular, searches for low-energy structures by hand, guided by chemical intuition, or simply randomly guessing, rapidly become untenable approaches and one must employ efficient global optimization methods. Such methods and their application to inorganic nanocluster systems are described in Chapter 1. Here, as a simple example of applying such an approach for nanosilicates, we examine the case of pure silica $(SiO_2)_N$ nanoclusters from $N = 2$ to 30.

Pure anhydrous bulk silica (SiO_2) exhibits a wide range of both crystalline and amorphous bulk polymorphs. The vast majority of these materials are constructed from frameworks of silicon atoms joined together by bridging oxygen atoms in such a way that every silicon atom is attached to four others. Although the oxygen atoms around each silicon center have a strong tendency to maintain O–Si–O angles close to 109.47° (i.e., to be near-ideally tetrahedral), the Si–O–Si linkages between silicon centers are approximately two orders of magnitude are more flexible. This structural property allows the formation of highly complex three-dimensional networks spanning a wide range of densities and topologies. A number of mathematical-based searches have so far accumulated approximately 100,000 periodic tetrahedral nets [99], thousands of which have been shown by calculations of their energies to be stable as silica materials [100]. The complexity of silica has also been compared with that of water, another tetrahedrally ordered material. Calculations have shown that water clusters exhibit a particularly complex PES characteristic of the so-called *strong* liquid [101]. From comparative investigations of the structures and stabilities of low-energy silica and water clusters, topological analogies between water and silica are found to persist down to the nanocluster size range [102], indicating a similar level of complexity. One important difference between water and silica, however, is that the former is a discrete molecular-based system, whereas silica is a continuously bonded material. Although water nanoclusters have an inherently closed-shell electronic structure by virtue of being constituted by discrete water molecules, those of silica must invoke structural and electronic reconstructions to avoid energetically costly terminating open-shell *dangling* bonds. The structure and properties of nanoclusters in general are largely governed by their inherently high proportion of *surface* atoms. Knowledge of how silica deals with surface terminations is thus key to understanding silica at the nanoscale.

Stoichiometric silica nanoclusters are found to maintain closed-shell electronic structures at their surfaces via the formation of (a) three coordinated silicon centers terminated by singly coordinated silanone oxygen centers, (b) closed Si_2O_2 rings, or *two-rings*, and (c) pairs of terminating singly-coordinated oxygen centers and three-coordinated oxygen centers. We note that although silanones are often written as $Si=O$ implying a double bond, there is strong evidence that a more ionic Si^+-O^- description is more appropriate [103]. Silanones, along with the topological two-ring defect, tend to be highly reactive with water to spontaneously give silanol $Si-OH$ groups. From theoretical studies, it is also known that if a silica network is not unduly strained, the presence of two silanones is thermodynamically unstable with respect to two-ring formation [104]. Type (c) defects involve a single electron charge transfer from a three-coordinated oxygen center to a singly coordinated oxygen center (i.e., $\equiv O^+\cdots-O^-$). This type of defect was first proposed to be present in bulk amorphous silica [105,106] in the 1970s based on Mott's valence alternation pair (VAP) model [107]. In low-energy silica nanoclusters, both the singly terminated O^- species and the donating $\equiv O^+$ center are present exclusively as a surface species [108,109]. Such surface-terminating VAP defects (or compensated nonbridging oxygen [CNBO] defects [108]) have also been observed in modeling studies of metastable surfaces of alpha-quartz [110,111]. Furthermore, it is notable that when the $\equiv O^+$ center is in a subsurface site, the remaining terminating O^- species is unreactive with water [108] and, much like the related but more reactive terminating silanolate group $Si-O^-$, may even have the propensity to order water molecules [112] or act as a catalyst [113]. In Figure 13.8, we show the structures of the three closed-shell surface defect types (a–c).

The candidate ground state $(SiO_2)_N$ clusters in the range $N = 2-27$ [109,114–116] shown in the following were all found by extensive global optimization searches using Monte Carlo basin hopping (MCBH) [117] (see also Chapter 1). Although the MCBH algorithm is thought to be one of the least hindered global optimization methods with respect to the specific topology of the PES [118], its success with respect to clusters of a real material, as with all global optimization methods, relies further on a sufficiently accurate yet efficient

Chapter 13

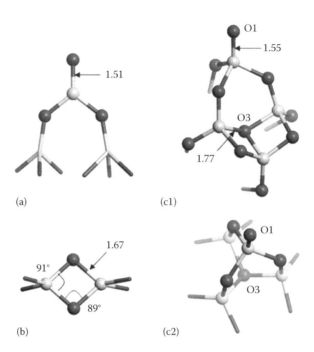

FIGURE 13.8　Surface defects exhibited by low-energy silica clusters: (a) silanone, (b) Si_2O_2 two-ring (c1 and c2), two examples of a surface-VAP defect. Distances are in Angstroms. O1 and O3 correspond to singly and triply coordinated oxygen atoms. Si atoms: gray, O atoms: red.

representation of the PES of that material at the nanoscale. For the smaller $(SiO_2)_N$ clusters (approx. $N < 16$), an interatomic potential set was used that has been specifically parameterized to accurately predict the energies and structures of silica nanoclusters [114]. For the larger clusters (approx. $N \geq 16$), the TTAM [119] silica potential was also employed, which although parameterized for bulk silica has also been shown to be of some use in global optimization studies of silica clusters [115,120]. From the resulting large number of isomers from the global optimizations (typically a few hundred for each $(SiO_2)_N$ cluster size), the 20–30 lowest-energy structures and selected higher-energy isomers with high symmetry were taken for energetic and structural refinement with DFT employing no symmetry constraints using the GAMESS-UK code [121]. The postoptimizations employed the B3LYP functional [122] and a 6–31G(d,p) basis set, which has been shown in numerous previous studies to be a suitable level of theory for calculating reliable structures and relative energetics of silica nanoclusters [123]. The resulting set of $(SiO_2)_n$ ground state clusters are shown in Figures 13.9 and 13.10. In the following, we briefly describe the evolution of the structures of the $(SiO_2)_N$ nanosilica ground states with increasing size, noting the occurrence of the three defect types.

For $(SiO_2)_N$ clusters with small N values, the low-energy isomer spectrum contains relatively few isomers and can be explored accurately by manual construction of isomers, which means that global optimization is largely unnecessary. Early studies following this intuitive approach were the first to suggest that the ground states for $(SiO_2)_N$ clusters $N = 2–5$ are silanone-terminated two-ring chains [124,125] (see Figure 13.9; note that the relaxed linear SiO_2 monomer is not shown). This theoretical prediction was later confirmed by global optimization studies [109] and also supported by cluster beam

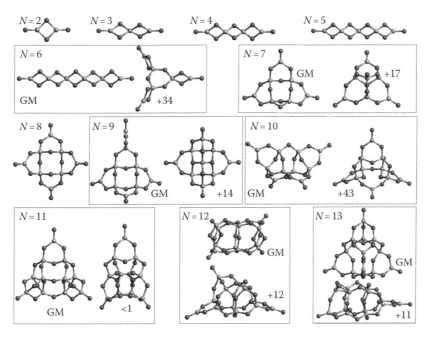

FIGURE 13.9 Low energy $(SiO_2)_N$ clusters for $N = 2$–13. Global minima are labeled GM and the energetically next nearest isomer is shown in kJ mol^{-1} (total energy with respect to the corresponding GM) when <50 kJ mol^{-1} higher in energy. Si atoms: light gray, O atoms: dark gray.

experiments on anionic $(SiO_2)_N$ species for $N = 2$–4 [126]. For $(SiO_2)_6$, the ground state is still theoretically predicted to marginally be the two-ring chain but other nonlinear isomers incorporating (Si_3O_3) three rings become energetically competing structures. This cluster size marks a crossover point in the evolution of silica cluster structure with increasing size where we pass from a preference for one-dimensional growth to more complex growth trends for $(SiO_2)_N$, $N > 6$. It is noted that this increase in structural complexity is mirrored in the increased difficulty in searching the low-energy $(SiO_2)_6$ PES.

The $(SiO_2)_7$ ground state has a C_{3v} symmetric trigonal structure and is energetically degenerate with a C_s symmetric isomer. The former cluster isomer is the first ground state candidate to contain a surface VAP defect in addition to its three silanone terminations. The lowest energy $(SiO_2)_8$ cluster is a D_{2d} symmetric cross-like structure containing four silanone terminations and no other defect centers. The cluster's highly symmetric form seems to be strongly energetically favorable with respect to other $(SiO_2)_8$ cluster isomers, with the next lowest energy $(SiO_2)_8$ isomer being 99 kJ mol^{-1} higher in energy. For $(SiO_2)_9$, the two lowest energy structures have the advantage of the low energy symmetric core of the $(SiO_2)_8$ ground state cluster. The lowest energy cluster adds one SiO_2 unit to one of the terminating silanone centers of the $(SiO_2)_8$ ground state forming a pendant two-ring and lowering the symmetry to C_s. The next lowest energy $(SiO_2)_9$ cluster results from adding a SiO_2 unit onto the center of the $(SiO_2)_8$ ground state, creating two compensating-pair VAP defects and a structure having C_{2v} symmetry. Both these $(SiO_2)_9$ structures have almost the same energy.

In the two lowest energy $(SiO_2)_{10}$ clusters, there is a tendency to move away from cluster structures built upon smaller low-energy forms. Both $(SiO_2)_{10}$ clusters shown

Chapter 13

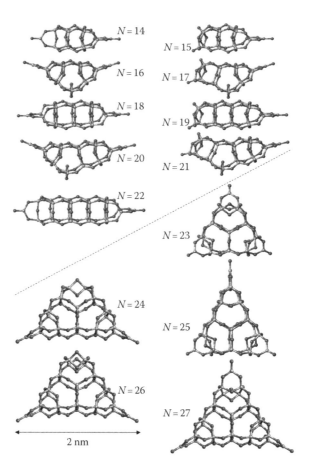

FIGURE 13.10 Candidate global minima $(SiO_2)_N$ clusters for $N = 14–27$. The dashed line indicated a transition between columnar and disk-like growth. Si atoms: light gray, O atoms: dark gray.

have complex three-dimensional structures predominately formed from interlocking (Si_3O_3) three rings. Both these clusters display four silanone centers and have C_2 and C_s symmetries, respectively. The two lowest energy $(SiO_2)_{11}$ clusters have a similar three-dimensional triangular structural form with two VAP defects and one silanone, and are essentially degenerate in energy, differing by only 4 kJ mol^{-1}. Both $(SiO_2)_{12}$ lowest energy isomers shown in Figure 13.9 have a compact elongated form containing numerous four rings. The ground state isomer contains a compensating-pair VAP termination at each end, joined together at a single four ring. The second lowest energy isomer, in contrast, has three oxygen terminations: two silanones and one compensating-pair VAP defect.

With regard to $(SiO_2)_N$ cluster structure the size $N = 13$, as for $N = 6$, marks a transitional size for cluster structural preference. For $(SiO_2)_N$ where $N = 6–13$, the clusters seem to have no well-defined structural type but can be roughly characterized as possessing a large proportion of three rings, typically three or four defective oxygen terminations, and often having pyramidal/trigonal-like structures. Only for $(SiO_2)_{12}$ does this trend appear to be opposed, with the lowest energy cluster isomers having more compact elongated forms with two or three terminations and more four rings. For $(SiO_2)_{13}$, we see structures of both types in the two lowest energy isomers.

The structural growth trend for $(SiO_2)_N$ $N = 14–22$ clusters is found to be remarkably simple and proceeds via the addition of Si_4O_4 four rings to make progressively longer compact elongated clusters (hereafter referred to as columnar clusters) of a similar form that is seen for the low energy isomers for $(SiO_2)_{12}$ and $(SiO_2)_{13}$. This four-ring addition growth pattern can be seen through $N = 14, 18, 22$ and $N = 15, 19$ in Figure 13.10. For nanoclusters between these values of N, the number of SiO_2 units is not sufficient to make a column of complete four-rings and instead a Si_2O_2 two-ring is inserted into the side of the cluster (e.g., $N = 16, 17, 20, 21$). For all columnar clusters, the dominant defect termination is via a silanone, either at both ends of the column for even N, or only at one end for odd N. For clusters $N = 15–21$, the odd number of SiO_2 units does not allow for a twofold symmetric termination with only silanone groups, and the odd-N columnar clusters are instead terminated by one silanone defect and a less energetically favorable compensating-pair defect. In all these cases, however, the connected four-ring columnar skeleton is maintained, which still appears to be the energetically favored structural route to obtain the lowest energy for this size range.

Although persistent, the energetic preference for one-dimensional columnar growth is overcome after a $(SiO_2)_N$ cluster size of $N = 22$ by the emergence of a more compact two-dimensional disk-like structure based on the centrally symmetric sharing of three double (Si_5O_5) five-ring cages. For the N-odd $(SiO_2)_N$ clusters for $N \geq 23$, it appears that having one silanone termination and one less energetically favored compensating-pair termination in a columnar cluster is outweighed by having three silanone terminations and a more compact structure. For even-N for $N > 23$, the energetic benefit of having a disk-like form as opposed to a columnar form appears to solely be structural with both forms having the same double silanone defect termination.

Considering the range of candidate ground state $(SiO_2)_N$ clusters between $n = 2$ and 27, some trends with increasing size can be identified. A common way to analyze models of annealed amorphous bulk silica glasses is in terms of the average $(SiO)_R$ ring size, which tends to have a distribution centered around $\langle R \rangle \approx 6$. Rings with $R < 6$ are intrinsically more strained and thus their presence is energetically destabilizing. The dominant ring size in silica cluster ground states gradually increases from $R = 2$ for the smallest two-ring chain-based clusters to $R = 5$ for the largest. Evidently, the clusters in this size range are still some way from being simple cuts from annealed silica glass.

Cluster evolution can also be analyzed by the change in relative total energy per SiO_2 unit (E_{rel}) with increasing size. In Figure 13.11, we show the gradual decrease in E_{rel} (with respect to the energy per unit of alpha-quartz, which is set to zero) of the $(SiO_2)_N$ cluster ground states for $N = 1–27$. For dense spherical clusters, we expect this energy difference to decrease proportionally to $N^{-1/3}$ (see Johnston [127]). Although the silica clusters reported herein can hardly be considered as corresponding to this idealized model, it is interesting that for $N > 13$, the cluster binding energies follow such a trend, possibly reflecting the near regular columnar and disk-like growth pattern in this size range (see Figure 13.10). We note that by extrapolating this trend we expect to reach typical bulk like energies per SiO_2 unit for silica glass at a cluster size of approximately $N = 7000$.

Another measure of cluster stability, also shown in Figure 13.11, is the first-order energy difference, ΔE, which measures the energetics of monomeric SiO_2 addition: $\Delta E = [E_N - (E_{N-1} + E_{SiO_2})]$. The initial $(SiO_2)_N$ ground states for $N = 2–6$ show evident regular linear chain growth based on two rings (see Figure 13.9), which is reflected in the high but stable values of ΔE. The ground state clusters $(SiO_2)_N$ $N = 7–9$ are based on a common structural

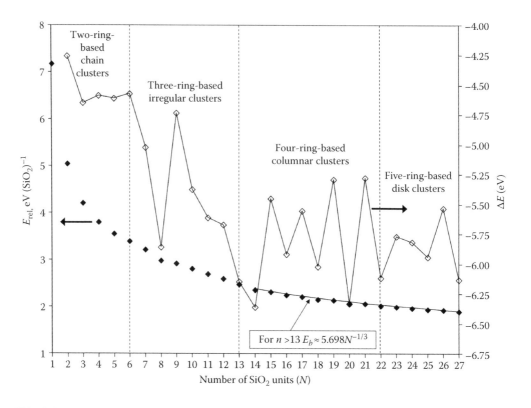

FIGURE 13.11 Left axis: decrease in total energy difference of $(SiO_2)_N$ cluster ground states with respect to alpha-quartz with increasing size (filled diamond data points). Right axis: first-order energy difference, i.e., $\Delta E = E[(SiO_2)_N] - E[(SiO_2)_{N-1}] - E(SiO_2)$, open diamond data points.

motif built upon four connected three rings, which is most clearly and symmetrically exhibited for the $N = 8$ cluster. The pronounced dip in ΔE for growth of, and then away from, the $N = 8$ cluster is indicative of its particular energetic stability or "magicness." Due to this property, it has been speculated that it could act as building block for new bulk polymorphs [128]. For $N > 9$ until $N = 13$, the clusters continue to be based mainly on three rings with a monotonic decrease in ΔE. For $N > 13$ till $N = 22$, the clusters form columnar structures based upon four rings (see Figure 13.10) with odd–even fluctuations in ΔE caused by the types of defect terminations allowed for even-n (two silanones) versus odd-n (silanone and VAP defect). In the last region considered, the clusters for $22 > N > 28$ are based upon five-ring disk-line clusters (see Figure 13.10) for which the ΔE values are relatively smaller and stable indicating a gradual trend toward bulk-like behavior.

For $(SiO_2)_N$ clusters with $N > 27$, it is increasingly more difficult for global optimization methods to find low-energy ground state minima due, mainly, to the exponentially increasing complexity of the PES with increasing cluster size. In an attempt to predict the likely structure of larger silica clusters, one can instead expand upon an idea planted in a study of the size-dependent transition from two silanone defects to a two-ring in a model nanochain-to-nanoring system [104]. In this study, fully coordinated (FC) rings of two rings were found to be energetically favored over defective chains after a certain size was reached. Other DFT calculations of larger nonglobally optimized compressed nanoslabs of 36–48 SiO_2 units have also observed a competition between silanones and two rings as surface-energy-reducing mechanisms [129]. As the reaction between two separated silanones to make an FC two-ring is an energetically favorable and barrierless process [130], depending on the constraints of the bonding topology of the cluster, it is expected that all silanone terminations will eventually lose out to FC two-rings. Using an algorithm specifically designed to explore the PES of low-energy fully reconstructed directionally bonded nanosystems (e.g., FC nanoclusters), it has been shown that as silica nanoclusters grow in size, complex non-cage-like FC structures become increasingly energetically stabilized [131]. Using this approach, it has been predicted that FC clusters to become the most stable form of nanosilica beyond a system size of approximately 100 atoms (i.e., $(SiO_2)_N$ of approx. $N > 27$) and before the eventual emergence of bulk crystalline structures. An example of a selection of FC $(SiO_2)_{24}$ nanoclusters, with respect to the silanone-terminated ground state, is shown in Figure 13.12.

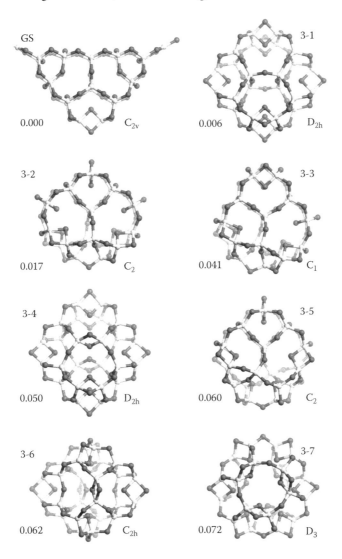

FIGURE 13.12 Structures, symmetries and total energies (eV $(SiO_2)^{-1}$) of lowest energy $(SiO_2)_{24}$ FC nanoclusters (3-1 to 3-7) relative to the ground state (GS). Si atoms: light gray, O atoms: dark gray.

In summary, low-energy $(SiO_2)_N$ isomers of pure silica containing up to at least ~100 atoms are found to be structurally rich and different compared to the bulk. These nanosilica ground states form a stability baseline that is fundamental for understanding the chemical, physical, and structural properties of nanosilicates in general.

13.3 Chemistry on Nanosilicates

13.3.1 Catalyzed H_2 Formation and Dissociation

In addition to its essential role in star formation, the abundance of molecular hydrogen in the diffuse clouds is important in the evolution of molecular complexity that occurs in these regions. In particular, the amount of energy deposited in the external and internal degrees of freedom of H_2 molecules influences both the IR emission in interstellar clouds [132,133] and their gas phase reactivity. For example, many reactions, such as $O(^3P) + H_2$, become efficient when H_2 is vibrationally excited [134]. Due to the low gas density and the very low temperatures in the ISM, H_2 cannot form in the gas phase efficiently enough via two-body radiative association or three-body collision processes [135,136]. Hence, it is generally accepted that H atom recombination takes place on the surface of dust grains that act as a catalyst for the stepwise release of the 434 kJ mol^{-1} excess energy in a time comparable to the vibration period of the highly vibrationally excited state in which it is formed [135,136]. The fundamentals of astrochemistry in the gas phase are relatively well established, in contrast to the special relevance attributed to processes involving cosmic dust grains. Silicate dust grains are thought to be good candidates for catalyzing the $H + H \rightarrow H_2$ reaction in the ISM. Theoretically, formation of H_2 on the crystalline forsterite silicate surface (010) has been studied with DFT using an embedded cluster approach [137] and with periodic plane wave codes [138,139]. Calculations on a small pyroxene silicate nanoparticle with respect to its reactivity toward H_2 have also been reported [140]. Experimentally, the formation of H_2 has been studied on the surface amorphous silicates [141]. In the following, we outline some results from some DFT calculations in light of competing mechanisms of catalytic H_2 formation and dissociation on a stable nanosilicate with a forsterite-type $(MgO)_6(SiO_2)_3$ composition (further details can be found in Kerkeni and Bromley [142]).

The nanosilicate dust grain cluster employed, although not crystalline, corresponds to the most energetically stable structure found through an MCBH global optimization search employing suitable interionic potentials [114,143] (see Figure 13.13). All structures and energies in this section were derived from DFT calculations as implemented in the Gaussian 09 code [88]. The MPWB1K functional was used [144], which is found to reproduce well experimental reaction barriers and thus, ultimately, reaction rates, and also captures nonbonded interactions that are important for physisorption energies. For Mg and Si atoms, a 6–31G(d) basis set was used, and for O and H atoms, a 6–31+G(d,p) set of basis functions was employed. In all optimizations, all atom positions in the cluster together with the reactants/products were fully relaxed without any symmetry constraint. The structures resulting from the optimizations were found to have all positive frequencies and thus have true local energy minima. Transition states were obtained using the Berny algorithm [145] and were all verified to have a single imaginary vibrational frequency.

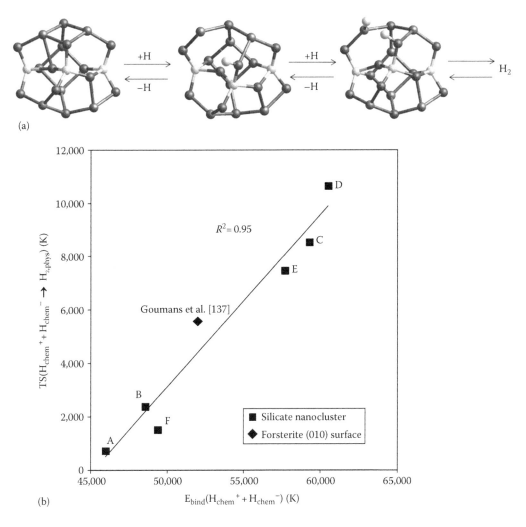

(a)

(b)

FIGURE 13.13 Typical intermediate structural configurations for H atom chemisorption on the $Mg_6Si_3O_{12}$ nanosilicate cluster (a). Si atoms: yellow, Mg atoms: blue, O atoms: red, H atoms: gray. Plot of $TS(H_{chem}^+ + H_{chem}^-)$ recombination barrier height versus $E_{bind}(H_{chem}^+ + H_{chem}^-)$ for six reaction pathways (A through F) on the nanosilicate cluster and that calculated for the extended forsterite surface (b). The good linear fit is also indicated by the coefficient of determination (R^2).

Considering first the adsorption of one H atom, a number of energetically favorable chemisorption sites on top of oxygen anions on the nanosilicate cluster were found. These reactions induce subsequent structural relaxations with respect to the positions of the oxygen centers that the H atom binds to, and the immediate neighboring atoms of this O anion. Chemically the H protonates an oxygen atom, forming an O–H group with the excess electron localized at an adjacent Mg cation. After testing numerous pathways for an incoming gas phase H atom to approach the cluster along the reaction path, no barrier to H chemisorption was apparent.

The quantum chemical calculations reveal that the nanosilicate cluster can also adsorb a second gas phase H atom without a barrier, which could, in principle, catalyze H_2 formation. It is found that the second H atom approaching the cluster always binds most strongly to the Mg center, which gained charge from the first protonation event, forming a hydride-like Mg–H bond (1.7 Å) without any barrier. Due to the difference in the chemisorption state of each of the two separated H atoms with the nanosilicate cluster, this state is referred to as $H_{chem.}^{\cdot+} + H_{chem}^-$; here the binding energy is taken as the total energy of the chemisorbed state relative to the sum of the two separate H atoms and the nanosilicate. The distance between the two chemisorbed H atoms varies between 1.49 and 4.44 Å. Typical configurations of the intermediate energy minima found in the calculations are shown in Figure 13.13. In order to see if these two separated chemisorbed H atoms could recombine and desorb in the cold ISM to form gas phase H_2, a search for possible relevant transition states was conducted. From these calculations several reaction paths of two separated chemisorbed H atoms to recombine and desorb into the gas phase were found. The most energetically favorable path corresponds to a reaction path, which starts from the least strongly chemisorbed $H_{chem.}^{\cdot+} + H_{chem}^-$ configuration. This reaction path has a small barrier (6 kJ mol^{-1}) and a high exothermicity (−86 kJ mol^{-1}) and would efficiently catalyze H_2 formation. The activation barrier for H + H recombination following this reaction path could easily be overcome by the high binding energy gained by the two chemisorbed H atoms and would even lead to the formation of vibrationally excited H_2 molecules. Two other similar exothermic pathways to H_2 formation were also found. Three endothermic reaction paths were also found corresponding to the largest $H_{chem.}^{\cdot+} + H_{chem}^-$ binding energies with the nanocluster and the largest recombination barriers. These energy profiles are exothermic toward H_2 dissociation and may play a role in attenuating H_2 formation. However, each of these pathways has quite large barriers (50–55 kJ mol^{-1}), which would tend to inhibit the dissociation process. The calculated reaction energy profile corresponding to H + H recombination on the forsterite (010) surface [137] lies midway between the exothermic and endothermic pathways we find on the nanosilicate cluster, with respect to both $H_{chem}^+ + H_{chem}^-$ binding energy and recombination barrier height, and is itself exothermic with respect to H_2 formation.

In Figure 13.13, the recombination barrier height versus the $H_{chem.}^{\cdot+} + H_{chem}^-$ binding energy is plotted for six different reaction pathways on the nanosilicate and for that calculated for the forsterite (010) surface [137]. The plot clearly shows a linear relation between the two quantities: the ease of H_2 formation is largely determined by how strongly both H atoms chemisorb with a silicate dust grain surface. More generally, this suggests that there may be a general Brønsted–Evans–Polanyi (BEP) relation for H_2 dissociation on bare silicate grains independent of dust grain size or crystallinity. Interestingly, these results provide evidence that ultrasmall noncrystalline nanosilicate dust grains do not simply tend to be more, or less, effective for H_2 formation relative to larger, more crystalline grains but, rather, that nanosilicates provide a greater range and variety of chemical pathways for both H_2 formation and dissociation. In addition, this work provides a lower size limit benchmark for the astrochemical relevance of such nanosilicate species with respect to that known for the upper size limit of large grains with bulk crystalline facets [137].

13.3.2 Nanosilicates as Nuclei for H$_2$O Ice Condensation

Noctilucent clouds occur at high latitudes in the summer mesosphere when the temperature falls below 150 K [62,146]. Such a low temperature is required for water–ice formation because H_2O vapor is present at mixing ratios of only a few parts per million [147]. Heterogeneous condensation occurs on available nuclei—see the following—although at very low temperatures (<120 K), which sometimes occur in the polar mesosphere [148], homogeneous nucleation of H_2O vapor may be possible [149]. Ice particles larger than 20 nm in radius can be observed by lidar (e.g., von Cossart and others [150]) and satellite-borne spectrometers [33,63]. Particles larger than about 50 nm can be seen by ground-based observers after sunset when the sun is about 6° below the horizon (this gives rise to the name *noctilucent*, which means *night-shining*). The ice particles also become charged by uptake of electrons. This produces sharp gradients in the electron density profile above and below a cloud, which causes strong radar backscatter signals known as polar mesospheric summer echoes (PMSE) (e.g., Rapp and Lübken [151]).

Meteoric ablation has been proposed as a major source of CN. Two distinct types of nuclei have been considered. First are metallic ions such as Fe^+, Mg^+, and Na^+, which then undergo successive hydration with falling temperature [152]. One problem with this idea is that laboratory studies show that metallic ions should be rapidly neutralized below 95 km by forming molecular ions, which then undergo dissociative electron recombination (e.g., Cox and Plane [153], Woodcock and others [154]). This has been confirmed by rocket-borne mass spectrometry [155], so that it is unlikely that metallic ions are a major source of CN. Although proton hydrates of the form $H^+(H_2O)_n$ sometimes occur at sufficiently high concentrations below 90 km and play a potential role as ice nuclei [156], they are also likely to undergo dissociative electron recombination before they grow to a size where a stable fraction of the cluster would survive neutralization [62,157].

The second type of condensation nucleus that has been proposed is MSPs (see Section 13.1.3). According to CNT, smoke particles need a radius of at least 2 nm radius to overcome the classical nucleation barrier to ice particle growth at typical polar summer mesopause conditions [62]. Recent global circulation models [34,56] describing the growth and transport of MSPs in the upper mesosphere have revealed a potential problem with the hypothesis of MSPs as nuclei. There is a strong meridional circulation from the summer to the winter pole, so there is insufficient time for MSPs to build up to a radius of 2 nm before they are transported from the summer polar mesosphere, where ice nucleation to form noctilucent clouds occurs, to lower latitudes [157]. Two solutions have been proposed to explain how nucleation can occur on particles significantly smaller than 2 nm. The first is that nucleation occurs on small particles that are charged by electron attachment, since there is a weak plasma in this part of the atmosphere [42]. In principle, a charged particle should bind H_2O molecules more strongly, thereby overcoming the classical nucleation barrier [157,158]. However, in practice only a small fraction of MSPs is likely to be charged in a weak plasma: around 5%–10% at NLC altitudes [55,157]. Furthermore, electron attachment to small particles with large dipole moments does not in fact make H_2O bind to them more strongly, as discussed in the following.

Another explanation for how ice can nucleate on very small (sub-nm) particles is that the particles have exceptionally large dipole moments. This possibility was first explored by Plane [159], who showed that the thermodynamics for the attachment of 2 H_2Os to sodium bicarbonate ($NaHCO_3$) is quite favorable because $NaHCO_3$ has a relatively large dipole moment (6.8 Debye). However, although $NaHCO_3$ is the major reservoir for meteor-ablated sodium in the MLT, sodium is a minor meteoric component and $NaHCO_3$ is depleted in the daytime upper mesosphere by photolysis [160] and undergoes reaction with atomic H [161]. Much more promising candidates are the silicates formed from the three most abundant meteoric elements Fe, Mg, and Si, which should ablate at similar rates and height ranges in the upper mesosphere [36]. As discussed in Section 13.2.3, the smallest MSPs that should form from the recombination of these constituents are the metal silicates $FeSiO_3$ (known as pyroxene in the solid phase) and $MgSiO_3$ (perovskite).

Electronic structure calculations have been used to show that these molecules have dipole moments of 9.5 and 12.2 Debye, respectively, and bind H_2O molecules even more favorably than Mg^+ and Fe^+ ions [162]. The geometries of $MgSiO_3$ and $FeSiO_3$ and their clusters with up to eight H_2Os were first optimized using DFT calculations employing the B3LYP functional with a flexible 6–311+g(2d,p) triple zeta basis set. This combination of theory and basis set was chosen to provide a compromise between tolerable accuracy and the demanding computational resources required to perform calculations (including vibrational frequencies) on systems with up to 29 atoms. Figure 13.14 shows that $MgSiO_3$ has a planar kite-shaped structure of $MgSiO_3$ with C_{2v} symmetry (as does $FeSiO_3$). The resulting structures for $MgSiO_3 \cdot (H_2O)_n$ ($n = 1 - 8$) are also illustrated in Figure 13.14. Even the largest of these clusters, $MgSiO_3 \cdot (H_2O)_8$, is still less than 1 nm in radius.

Following geometry optimization, the calculated vibrational frequencies are used to make zero-point energy corrections to the cluster binding energies and to compute binding entropies and hence Gibbs free energies. The free energies for addition of successive H_2Os to $MgSiO_3$ (i.e., $\Delta G_{n-1,n}$ n is the number of H_2O) are also listed in Figure 13.14, at a representative upper mesospheric summer temperature of 135 K. This shows that the free energy change is favorable (i.e., negative), even up to the addition of an eighth H_2O, when ΔG approaches that for sublimation of H_2O to a bulk ice surface [163]. Note that hydroxylation of the SiO_3 group spontaneously occurs on addition of the fifth H_2O to $MgSiO_3$, resulting in two pendant –OH groups. This reaction step is significantly more exothermic than simple electrostatic binding of H_2O as a cluster ligand. Hydroxylation of nanosilicates is the theme of Section 13.3.3.

Interestingly, charging of these metal silicates by electron attachment results in metal silicate anions that do not attach H_2O as strongly as the neutral species. This arises because the negative charge is mostly located on the metal atom, reducing the (Mulliken) charge on Mg from +1.0 to +0.2. The binding energy for the first H_2O to $MgSiO_3^-$ is only -65 kJ mol^{-1}, compared with -132 kJ mol^{-1} to $MgSiO_3$.

Extrapolating the binding free energies of the $MgSiO_3 \cdot (H_2O)_n$ and $FeSiO_3 \cdot (H_2O)_n$ clusters from $n = 8$ to the bulk can then be used to explore the H_2O concentration during which runaway nucleation will occur at a particular temperature in the MLT [162]. In this case, nucleation can be (arbitrarily) defined to have occurred once the temperature is low enough for the most likely cluster size n to exceed 50 H_2O molecules. For H_2O mixing ratios around 3–5 ppm, typical of the high latitude summer mesopause [147], nucleation of ice particles on metal silicates should occur around 140 K [162]. This is in accordance with atmospheric observations [62].

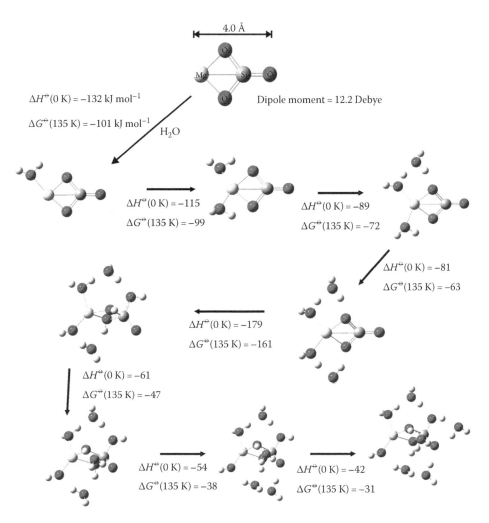

FIGURE 13.14 Optimized geometries of MgSiO$_3$ and the first eight H$_2$O clusters, at the B3LYP/6–311+g(2d,p) level of theory. The standard enthalpy change at 0 K ($\Delta H^{\ominus}(0$ K)) and the Gibbs fee energy change at 135 K ($\Delta G^{\ominus}$ (135 K)) for the addition of sequential H$_2$O molecules are in units of kJ mol^{-1}. Note that hydration of the SiO$_3$ group occurs on addition of the fifth H$_2$O. (Adapted from Plane, J.M.C., *J. Atmos. Solar Terr. Phys.*, 73, 2192, 2011.)

13.3.3 Hydroxylation of Nanosilicates

From their essential role in dissolution and nucleation in terrestrial geologic and atmospheric processes, to the formation and growth of icy dust grains in astronomical environments, hydroxylated silicate nanoclusters are of ubiquitous fundamental importance. From a technological perspective, such species are also inherently involved in the synthesis of widely used nanoporous silicate materials such as zeolites. Despite their widespread significance, the experimental determination of the structures and properties of silicate nanoclusters is hindered by their structural and dynamic complexity. Typically, however, modeling studies have focused upon the energetics and structures of very small hydroxylated pure silica species, $(SiO_2)_M(H_2O)_N$, such as dimers ($M = 2$), up to tetramers ($M = 4$) [164–166] and their oligomerization reactions [165,167,168]. In this section, computational modeling is employed to follow the stepwise hydroxylation of both a set of pure silica nanoclusters of different sizes and a small magnesium silicate cluster of pyroxene-type composition.

Silica is increasingly being employed in a wide array of functionalized nanostructures [169] and bio-inspired (nano)materials [170] where hydroxylation reactions at its surfaces are key. Computational modeling methods have contributed greatly to the understanding of the complex silica-water system with uniquely detailed microscopic insights into the mechanisms of silica hydroxylation reactions [171]. From such studies, water molecules are generally predicted to preferentially attack the surfaces of bulk silica at terminating point defects and strained Si–O–Si sites (see Section 13.2.3) with modest reaction barriers [172–175]. On the contrary, regular nondefective crystalline silica surfaces are known from experiment to be often very resistant to reactions with water (e.g., thin ordered films [176], zeolite interiors [177]). As established in Section 13.2.3, even the most stable silica nanoclusters with up to at least ~100 atoms inherently exhibit terminating surface defects and strained $(SiO)_n$ rings and thus they are expected to be easily and energetically favorably hydroxylated. In the following, we briefly summarize some of the main results from a number of computational modeling studies [115,178–180], and explore and compare the structures and stabilities of silicate clusters: $(SiO_2)_M(H_2O)_N$, $M = 4, 8, 16, 24$ through a systematic stepwise molecular hydroxylation from their anhydrous state until an $N{:}M$ ratio ($R_{N/M}$) of ≥ 0.5.

The candidate global minima for each of the silica clusters reported were derived from a two-step approach. First, low energy clusters of composition $(SiO_2)_M(H_2O)_N$, $M = 4, 8, 16, 24$ (in each case from $N = 0$ until $R_{N/M} \geq 0.5$) were comprehensively searched for using the MCBH global optimization algorithm [117] and specifically parameterized interatomic potentials [114,181]. Second, the energies and geometries of the 10–15 lowest energy clusters were reoptimized using DFT with the GAMESS-UK code [121], employing the B3LYP hybrid functional [122] and a 6–31G(d,p) basis set with no symmetry constraints. As noted in previous sections, this approach has been shown to provide a combination of accuracy and computational efficiency comparable to very high level methods [167]. In particular this approach has previously been successfully used to find the candidate ground states of bare and partially hydroxylated silica clusters [109,115,116,178,179,182]. Throughout each series, the energy of a suitable number of water molecules is added as required so that all quoted energies for a fixed M correspond to the same chemical composition. When comparing energies of clusters in different series, total energies are normalized by dividing them by the number of SiO_2 units in the respective cluster. The average deviation of the O–Si–O angle with respect to the optimal unstrained value of 109.47° is employed as a measure of tetrahedral distortion. This measure is only calculated for clusters for which all Si atoms have four oxygen neighbors and is thus not given for some clusters with low $R_{N/M}$ values, which have three-coordinated Si centers.

First, the energetic stability and structure is compared for each series for three representative degrees of hydroxylation corresponding to $R_{N/M} = 0.0, 0.25, 0.5$. In Figure 13.15, the normalized energetic stability of the lowest energy cluster isomer for each series and the three values of $R_{N/M}$ are plotted. For $R_{N/M} = 0$, the increase in cluster size leads to an

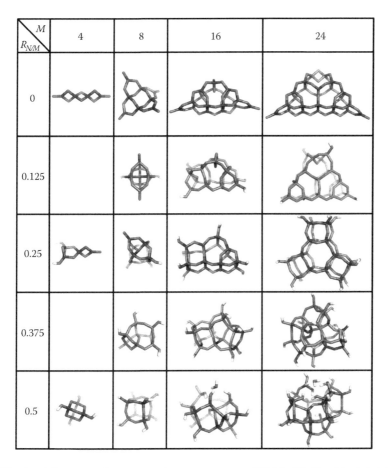

FIGURE 13.15   Candidate ground state nanocluster structures of $(SiO_2)_M(H_2O)_N$, $M = 4, 8, 16, 24$ clusters for $R_{N/M} = 0$–0.5. Atom key: O, red; Si, dark gray; H, light gray.

expected drop in energy per unit following an inverse power law dependence on the cluster size [115,127]. With increasing $R_{N/M}$, however, this decreasing energetic trend becomes less pronounced except for $R_{N/M} = 0.5$; the normalized energies for cluster sizes $M = 8, 16, 24$ are almost identical, with only the $M = 4$ cluster being higher in energy. For $R_{N/M} = 0.5$, the clusters have two Si atoms per hydroxylating water molecule and thus, potentially, one OH group per Si atom. This latter situation naturally provides a means for the formation of cages with N Si–OH vertices perhaps suggesting an underlying structural rational for the increased similarity in normalized energetic stability for $R_{N/M} = 0.5$. From extensive global searches, however, it is found that at $R_{N/M} = 0.5$, such cages are only energetically favored for clusters of size $M = 4, 8$, where a tetrahedron and a cube are obtained, respectively. For $M = 16, 24$ and for $R_{N/M} = 0.5$, instead of cages, it is found that free space clusters energetically prefer to form a dense amorphous cluster core with a discrete water molecule hydrogen-bonded to the surface Si–O–H groups of the cluster. For $M = 16$, this structural phenomenon first occurs for $R_{N/M} = 0.5$, whereas for $M = 24$, it first happens at a slightly lower hydroxylation level of $R_{N/M} = 0.458$. In Figure 13.15, the gradual change in structure of the lowest energy cluster isomers found for a selection of $R_{N/M}$ values for $R_{N/M} \leq 0.5$ is shown. Clearly the water-induced structural evolution of

Chapter 13

the smaller clusters with $N = 4, 8$ is quite distinct from that of the larger clusters with $N = 16, 24$. Although in the former, each addition of a water molecule leads to an energetically favorable hydroxylation and appears to be proceeding to the full dissolution natural limit of N Si(OH)$_4$ monomeric units, the latter clusters are inherently resistant to reaching this limit.

In Figure 13.16, a more general overview of the energetics of hydroxylation is shown for each (SiO$_2$)$_M$(H$_2$O)$_N$, $M = 4, 8, 16, 24$ cluster series through a plot of normalized cluster energy versus $R_{N/M}$ for all clusters found in our investigation. From the anhydrous $N = 0$ clusters in each series, we see an initial steep drop in energy with increasing stepwise hydroxylation, which then becomes less pronounced and eventually reaches a point where each successive energy drop is very similar and at its lowest value. The transition from large energy decreases at lower hydroxylations to a linear minimal energy decreasing regime with increasing hydroxylation is suggestive of a threshold being reached at which subsequent hydroxylation is no longer so effective. The first few hydroxylation steps effectively convert surface defects (e.g., terminal silanone groups, strained small (SiO)$_2$ two-rings), which are common on anhydrous clusters, into more relaxed four-coordinated centers. The effects of such structural reconstructions are not localized to the surface-most atoms of a structure but also permit greater structural relaxation throughout a cluster. Once a cluster has been sufficiently hydroxylated to optimize the structural relaxation, any further hydroxylation is ineffective at reducing strain and, for smaller clusters, may even cause the structure to start to be broken open. Thus, in the latter stages of hydroxylation, each energy drop is by a fairly fixed amount mainly due to adsorption of H$_2$O without any extra energetic stabilization due to structural relaxation. As the size of the cluster increases, proportionally fewer atoms are involved in strained

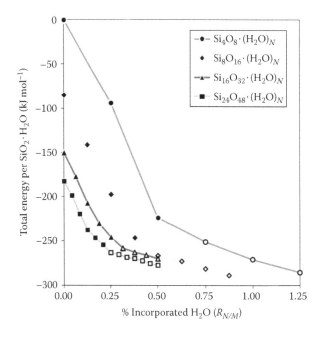

FIGURE 13.16 Relative total energies of (SiO$_2$)$_M$(H$_2$O)$_N$, $M = 4, 8, 16, 24$ clusters with respect to $R_{N/M}$. Open symbols indicate the linearly decreasing energy regime.

(surface) defects and so the linear hydroxylation regime is encountered at progressively smaller $R_{N/M}$ values. This effect is analyzed in detail by Jelfs et al. [180].

In order to examine whether the interaction of water molecules with magnesium silicate nanoclusters follows that predicted for pure silica nanoclusters, the stepwise addition of H_2O molecules to a $Mg_4Si_4O_{12}$ nanocluster of pyroxene-type composition has also been followed using the same computational methodology [140]. In order to compare the results with those reported for pure silica nanoclusters, the number of oxygen atoms in the magnesium silicate cluster is used to determine an effective number of SiO_2 units. In the case of the $Mg_4Si_4O_{12}$ cluster, we take this to be 6 (i.e., 12/4) effective Si cations. In this way, after normalization with respect to the number of effective and/or real Si centers, we can compare the impact of having two Mg^{2+} cations for each *replaced* Si^{4+} cation on the energetics of hydroxylation.

In Figure 13.17, the normalized stabilization energies of the $Mg_4Si_4O_{12}$ cluster due to interaction with $(H_2O)_N$, for $N = 1-5$, is compared with that calculated for the pure silica clusters: Si_8O_{16} and $Si_{16}O_{32}$. The most stable product of the interaction of a single water molecule with the $Mg_4Si_4O_{12}$ nanocluster is also a hydroxylated species, with a stabilization energy per effective SiO_2 unit of 52 kJ mol^{-1}. This is very similar to the values of 56 and 54 kJ mol^{-1} calculated for Si_8O_{16} and $Si_{16}O_{32}$, respectively, indicating that a similar strain release/defect healing may also be occurring. Looking at the structure of the $Mg_4Si_4O_{12}$ nanosilicate cluster before and after this initial water addition step (see Figure 13.17), we, indeed, see that two of the Si centers become hydroxylated and a strained $(SiO)_3$ ring in the anhydrous cluster is no longer present after hydroxylation. As the hydroxylated nanosilicate structure was found by global optimization, however, it does not provide any information with respect to possible reaction pathways from the original anhydrous state. From the previous section, we know that adding discrete water molecules to a small magnesium silicate species does not (initially) spontaneously lead to hydroxylation and thus a barrier from the physisorbed state to the chemisorbed state

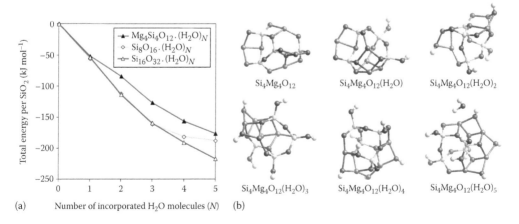

(a)   Number of incorporated H_2O molecules (N)   (b)

FIGURE 13.17   (a) Total relative energies of $Si_8O_{16}(H_2O)_N$, $Si_{16}O_{32}(H_2O)_N$ and $Si_4Mg_4(H_2O)_N$ for $N = 0-5$. All energies are with respect to the corresponding anhydrous cluster and five discrete water molecules, which is set to zero in all three cases. All energies are also normalized with respect to the effective number of SiO_2 units (where each MgO counts as 0.5). (b) The most energetically stable structural configurations for the $Si_4Mg_4(H_2O)_N$ nanoclusters for $N = 0-5$. Atom key: O, red; Si, yellow; Mg, blue; H, light gray.

Chapter 13

exists. Whether this barrier is *submerged* (i.e., lower that the energy gained from physisorption), which would facilitate hydroxylation, is currently unknown.

The addition of a second water molecule to the $Mg_4Si_4O_{12}$ nanocluster further hydroxylates the cluster but stabilizes it by 20 kJ mol^{-1} per SiO_2 unit less than the first hydroxylation step. Conversely, the respective stabilization energies for the second H_2O addition for Si_8O_{16} and $Si_{16}O_{32}$ stay constant. Unlike pure silica, only half the cations in a pyroxyene-type nanocluster are associated with network-forming SiO_2 centers, with Mg cations helping to bind these centers together and saturate potential Si–O terminating defects. Considering this, it may be expected that the observed strong initial stabilization by water addition is limited to relatively fewer water molecule additions than for pure silica. The lowest energy structure found for two water molecules hydroxylating the $Mg_4Si_4O_{12}$ nanocluster has two hydroxyls on two Si centers and two OH groups each bound to two Mg centers (see Figure 13.17). This isomer is found to be 18 kJ mol^{-1} more stable than the most stable isomer with four Si–OH groups. Although, in general, Mg-hydroxylation is likely to be less stabilizing than Si-hydroxylation due to the smaller ionic charge of the Mg centers, one must also consider the destabilizing structural distortion induced by increasing Si–OH formation (see earlier text), which may help explain the pattern of hydroxylation at this stage. The third and forth water molecules further hydroxylate the $Mg_4Si_4O_{12}$ nanocluster with a gradual reduction in the stabilization energy for each successive H_2O molecule addition. For the subsequent addition of a fifth water molecule, the lowest energy nanosilicate cluster isomer is found to exhibit a hydrogen-bonded H_2O molecule. This isomer is found to be only ~6 kJ mol^{-1} more stable than the lowest energy fully hydroxylated isomer found. Although the stabilization energy associated with the addition of the hydrogen bonded fifth water molecule is only 20 kJ mol^{-1} per SiO_2, this follows the gradual decrease in stabilization with an increasing number of added water molecules and does not appear to mark a change to a linear stabilization regime as seen for pure silica nanoclusters (see earlier text). For the Si_8O_{16} cluster, for example, a leveling off of the stabilization curve starting at four added water molecules (see Figures 13.16 and 13.17) is observed. The smoothly decaying stabilization trend of the $Mg_4Si_4O_{12}$ nanocluster for two to five added water molecules is more akin to that displayed by larger clusters such as $Si_{16}O_{32}$. Further, the tendency of the $Mg_4Si_4O_{12}$ nanocluster to energetically prefer to be only hydroxylated up to a certain number of added water molecules (and beyond which hydrogen bond discrete water molecules) is also a characteristic of larger silica clusters.

The fact that relatively small magnesium silicate nanoclusters appear to act more like relatively larger silica clusters with respect to their interactions with water may make them (1) relatively stable to hydrolysis into hydroxylated monomers and (2) able to absorb relatively high numbers of water molecules while maintaining reasonable absorption energies. These properties are what make small magnesium silicate species significant for water ice nucleation in the upper atmosphere (see Section 13.3.2) and, in an astronomical context, in dense molecular clouds. Ice formation in interstellar clouds only becomes possible once they are opaque enough to prevent water photodesorption from the intense UV field in the ISM. Prior to ice formation, highly hydroxylated species such as $Mg_4Si_4O_{12}(H_2O)_4$ could be interesting due to their relatively high oxygen to metal ratio (in this case, using our definition of M above, O/M equals 2.66). This ratio is interesting in view of the recent observations that oxygen depletes more strongly than other elements in high total depletion regions [183,184]. In the denser regions, more oxygen ($\sim$2.4 $10^{-4} \times n_H$) seems to be tied up in solids than can be incorporated in silicates with a bulk O/M ratio of 2 for both olivines and pyroxenes. If indeed, as is estimated, 10% of the interstellar silicate mass consists of nanosized particles [16] and these could incorporate extra O atoms up to a O/M ratio of 2.66, substantially more O could be tied up in these superoxygenated nanosilicates. Nanosilicates [185], thus, by virtue of their higher surface to volume ratio, can adsorb relatively more molecular water than bulk silicates. Although it is likely that there are other unidentified solid particulates or molecules that also contribute to the strong depletion of oxygen, nanosilicates could be a significantly contributing unidentified depleted oxygen carrier [184].

13.4 Summary and Outlook

In this chapter, we have attempted to provide a brief overview of the relevance of nanosilicates with respect to the terrestrial atmosphere and interstellar space. We have particularly focused on how computational modeling is beginning to provide necessary insights into the formation and structure of nanosilicates in these environments and their subsequent physicochemical properties. This chapter primarily deals with studies of rather small nanosilicate species largely due to the dearth of theoretical and experimental data regarding the structures and properties of such species in the size regime between clusters possessing $\sim$100 atoms and significantly larger bulk-like particles. Nevertheless, even within this limited size range, we have shown the importance of computational modeling for obtaining unprecedentedly detailed insights into the low-energy structures of nanosilicates, their formation through nucleation, and their role as seeds for further nucleation of water ice. We have also shown that even such small nanosilicates could be vital catalysts for fundamental chemical reactions in astronomical environments. Whether they have a similarly key function in the Earth's atmosphere is still unknown. Further work in the burgeoning field of cosmic and atmospheric nanosilicates from laboratory experiment, observation, and computational modeling will be key to further understanding of such species. Specifically, from a computational viewpoint, we see many opportunities for further work and a particular need for modeling studies on (1) spectroscopic properties (e.g., IR and UV adsorption), (2) bridging the gap between the properties and structure of small clusters and bulk-like nanosilicates, and (3) investigations into more complex chemical processes on nanosilicates. We hope that this chapter provides some inspiration for further studies in these areas.

References

1. Henning, T. 2010. *Astromineralogy*, Lecture Notes in Physics 815. Springer-Verlag, Berlin, Germany.
2. Messenger, S., L. P. Keller, and D. S. Lauretta. 2005. Supernova olivine from cometary dust. *Science* 309:737–741.
3. Nittler, L. R. 2010. Cometary dust in the laboratory. *Science* 328:698–699.
4. Abraham, P., A. Juhasz, C. P. Dullemond, A. Kospal, R. Van Boekel, J. Bouwman, T. Henning et al. 2009. Episodic formation of cometary material in the outburst of a young Sun-like star. *Nature* 459:224–226.
5. Sloan, G. C., M. Matsuura, A. A. Zijlstra, E. Lagadec, M. A. T. Groenewegen, P. R. Wood, C. Szyszka, J. Bernard-Salas, and J. T. van Loon. 2009. Dust formation in a galaxy with primitive abundances. *Science* 323:353–355.

Chapter 13

6. Bradley, J. P., L. P. Keller, T. P. Snow, M. S. Hanner, G. J. Flynn, J. C. Gezo, S. J. Clemett, D. E. Brownlee, and J. E. Bowey. 1999. An infrared spectral match between GEMS and interstellar grains. *Science* 285:1716–1718.

7. Min, M., L. B. F. M. Waters, A. de Koter, J. W. Hovenier, L. P. Keller, and F. Markwick-Kemper. 2007. The shape and composition of interstellar silicate grains. *Astronom. Astrophys.* 462:667–676.

8. Ostlie, D. A. and B. W. Carroll. 2007. *An Introduction to Modern Stellar Astrophysics*. Addison Wesley, San Francisco, CA.

9. Kwok, S. 2000. *The Origin and Evolution of Planetary Nebulae*. Cambridge University Press, Cambridge, U.K.

10. Goumans, T. P. M. and S. T. Bromley. 2012. Efficient nucleation of stardust silicates via heteromolecular homogeneous condensation. *Mon. Not. Roy. Astronom. Soc.* 420:3344–3349.

11. Gail, H. P. and E. Sedlmayr. 1986. The primary condensation process for dust around late M-type stars. *Astronom. Astrophys.* 166:225–236.

12. Gail, H. P. and E. Sedlmayr. 1998. Inorganic dust formation in astrophysical environments. *Faraday Discuss.* 109:303–319.

13. Paquette, J. A., F. T. Ferguson, and J. A. Nuth III. 2011. A model of silicate grain nucleation and growth in circumstellar outflows. *Astrophys. J.* 732:art. no. 62.

14. Gail, H. P., S. Wetzel, A. Pucci, and A. Tamanai. 2013. Seed particle formation for silicate dust condensation by SiO nucleation. *Astronom. Astrophys.* 555:art. no. A119.

15. Nuth III, J. A. and F. T. Ferguson. 2006. Silicates do nucleate in oxygen-rich circumstellar outflows: New vapor pressure data for SiO. *Astrophys. J.* 649:1178–1183.

16. Li, A. and B. T. Draine. 2001. On ultrasmall silicate grains in the diffuse interstellar medium. *Astrophys. J.* 550:L213–L217.

17. Williams, D. A. and E. Herbst. 2002. It's a dusty Universe: Surface science in space. *Surf. Sci.* 500:823–837.

18. Ferriere, K. M. 2001. The interstellar environment of our galaxy. *Rev. Mod. Phys.* 73:1031–1066.

19. Williams, I. P. 2002. The evolution of meteoroid streams. In *Meteors in the Earth's Atmosphere*. E. Murad, and I. P. WIlliams, eds. Cambridge University Press, Cambridge, U.K., pp. 2–32.

20. Ceplecha, Z., J. Borovicka, W. G. Elford, D. O. Revelle, R. L. Hawkes, V. Porubcan, and M. Simek. 1998. Meteor phenomena and bodies. *Space Sci. Rev.* 84:327–471.

21. Nesvorny, D., D. Janches, D. Vokrouhlicky, P. Pokorny, W. F. Bottke, and P. Jenniskens. 2011. Dynamical model for the zodiacal cloud and sporadic meteors. *Astrophys. J.* 743:art. no. 129.

22. Plane, J. M. C. 2012. Cosmic dust in the earth's atmosphere. *Chem. Soc. Rev.* 41:6507–6518.

23. Love, S. G. and D. E. Brownlee. 1993. A direct measurement of the terrestrial mass accretion rate of cosmic dust. *Science* 262:550–553.

24. Gabrielli, P., C. Barbante, J. M. C. Plane, A. Varga, S. Hong, G. Cozzi, V. Gaspari et al. 2004. Meteoric smoke fallout over the Holocene epoch revealed by iridium and platinum in Greenland ice. *Nature* 432:1011–1014.

25. Lanci, L., D. V. Kent, and P. E. Biscaye. 2007. Meteoric smoke concentration in the Vostok ice core estimated from superparamagnetic relaxation and some consequences for estimates of Earth accretion rate. *Geophys. Res. Lett.* 34:art. no. L10803.

26. Wasson, J. T. and F. T. Kyte. 1987. On the influx of small comets into the Earth's atmosphere II: Interpretation. *Geophys. Res. Lett.* 14:779–780.

27. Peucker-Ehrenbrink, B. 1996. Accretion of extraterrestrial matter during the last 80 million years and its effect on the marine osmium isotope record. *Geochim. Cosmochim. Acta* 60:3187–3196.

28. Janches, D., S. Close, and J. T. Fentzke. 2008. A comparison of detection sensitivity between ALTAIR and Arecibo meteor observations: Can high power and large aperture radars detect low velocity meteor head echoes? *Icarus* 193:105–111.

29. Hughes, D. W. 1978. Meteors. In *Cosmic Dust*. J. A. M. McDonnell, ed. Wiley, London, U.K., pp. 123–185.

30. Baggaley, W. J. 2002. Radar observations. In *Meteors in the Earth's Atmosphere*. E. Murad and I. P. Williams, eds. Cambridge University Press, Cambridge, U.K., pp. 123–148.

31. Mathews, J. D., D. Janches, D. D. Meisel, and Q. H. Zhou. 2001. The micrometeoroid mass flux into the upper atmosphere: Arecibo results and a comparison with prior estimates. *Geophys. Res. Lett.* 28:1929–1932.

32. Cziczo, D. J., D. S. Thomson, and D. M. Murphy. 2001. Ablation, flux, and atmospheric implications of meteors inferred from stratospheric aerosol. *Science* 291:1772–1775.

33. Hervig, M. E., R. E. Thompson, M. McHugh, L. L. Gordley, J. M. Russell, and M. E. Summers. 2001. First confirmation that water ice is the primary component of polar mesospheric clouds. *Geophys. Res. Lett.* 28:971–974.

34. Bardeen, C. G., O. B. Toon, E. J. Jensen, D. R. Marsh, and V. L. Harvey. 2008. Numerical simulations of the three-dimensional distribution of meteoric dust in the mesosphere and upper stratosphere. *J. Geophys. Res.* 113:art. no. D17202.

35. Saunders, R. W., S. Dhomse, W. S. Tian, M. P. Chipperfield, and J. M. C. Plane. 2012. Interactions of meteoric smoke particles with sulphuric acid in the Earth's stratosphere. *Atmos. Chem. Phys.* 12:4387–4398.

36. Vondrak, T., J. M. C. Plane, S. Broadley, and D. Janches. 2008. A chemical model of meteoric ablation. *Atmos. Chem. Phys.* 8:7015–7031.

37. Molina-Cuberos, J. G., J. J. Lopez-Moreno, and F. Arnold. 2008. Meteoric layers in planetary atmospheres. *Space Sci. Rev.* 137:175–191.

38. Love, S. G. and D. E. Brownlee. 1991. Heating and thermal transformation of micrometeoroids entering he Earth's atmosphere. *Icarus* 89:26–43.

39. Kalashnikova, O., M. Horanyi, G. E. Thomas, and O. B. Toon. 2000. Meteoric smoke production in the atmosphere. *Geophys. Res. Lett.* 27:3293–3296.

40. Hunten, D. M., R. P. Turco, and O. B. Toon. 1980. Smoke and dust particles of meteoric origin in the mesosphere and stratosphere. *J. Atmos. Sci.* 37:1342–1357.

41. Fegley, J. B. and A. G. W. Cameron. 1987. A vaporization model for iron/silicate fractionation in the Mercry protoplanet. *Earth Planet. Sci. Lett.* 82:207–222.

42. Plane, J. M. C. 2003. Atmospheric chemistry of meteoric metals. *Chem. Rev.* 103:4963–4984.

43. Correira, J., A. C. Aikin, J. M. Grebowsky, W. D. Pesnell, and J. P. Burrows. 2008. Seasonal variations of magnesium atoms in the mesosphere-thermosphere. *Geophys. Res. Lett.* 35:art. no. L06103.

44. Scharringhausen, M., A. C. Aikin, J. P. Burrows, and M. Sinnhuber. 2008. Global column density retrievals of mesospheric and thermospheric MgI and MgII from SCIAMACHY limb and nadir radiance data. *J. Geophys. Res.* 113:art. no. D13303.

45. Self, D. E. and J. M. C. Plane. 2003. A kinetic study of the reactions of iron oxides and hydroxides relevant to the chemistry of iron in the upper atmosphere. *Phys. Chem. Chem. Phys.* 5:1407–1418.

46. Plane, J. M. C. and C. L. Whalley. 2012. A new model for magnesium chemistry in the upper atmosphere. *J. Phys. Chem. A* 116:6240–6252.

47. Gomez Martin, J. C., M. A. Blitz, and J. M. C. Plane. 2009. Kinetic studies of atmospherically relevant silicon chemistry part I: Silicon atom reactions. *Phys. Chem. Chem. Phys.* 11:671–678.

48. Gomez Martin, J. C., M. A. Blitz, and J. M. C. Plane. 2009. Kinetic studies of atmospherically relevant silicon chemistry. Part II: Silicon monoxide reactions. *Phys. Chem. Chem. Phys.* 11:10945–10954.

49. Saunders, R. W. and J. M. C. Plane. 2011. A photo-chemical method for the production of olivine nanoparticles as cosmic dust analogues. *Icarus* 212:373–382.

50. Saunders, R. W. and J. M. C. Plane. 2006. A laboratory study of meteor smoke analogues: Composition, optical properties and growth kinetics. *J. Atmos. Solar Terr. Phys.* 68:2182–2202.

51. Saunders, R. W. and J. M. C. Plane. 2010. The formation and growth of Fe_2O_3 nanoparticles from the photo-oxidation of iron pentacarbohyl. *J. Aerosol. Sci.* 41:475–489.

52. Rapp, M., I. Strelnikova, and J. Gumbel. 2007. Meteoric smoke particles: Evidence from rocket and radar techniques. *Adv. Space Res.* 40:809–817.

53. Lynch, K. A., L. J. Gelinas, M. C. Kelley, R. L. Collins, M. Widholm, D. Rau, E. MacDonald, Y. Liu, J. Ulwick, and P. Mace. 2005. Multiple sounding rocket observations of charged dust in the polar winter mesosphere. *J. Geophys. Res.* 110:art. no. A03302.

54. Gelinas, L. J., K. A. Lynch, M. C. Kelley, R. L. Collins, M. Widholm, E. MacDonald, J. Ulwick, and P. Mace. 2005. Mesospheric charged dust layer: Implications for neutral chemistry. *J. Geophys. Res.* 110:art. no. A01310.

55. Robertson, S., S. Dickson, M. Horanyi, Z. Sternovsky, M. Friedrich, D. Janches, L. Megner, and B. P. Williams. 2014. Detection of meteoric smoke particles in the mesosphere by a rocket-borne mass spectrometer. *J. Atmos. Solar Terr. Phys.* 118:161–179.

56. Megner, L., D. E. Siskind, M. Rapp, and J. Gumbel. 2008. Global and temporal distribution of meteoric smoke: A two-dimensional simulation study. *J. Geophys. Res.* 113:art. no. D03202.

57. Saunders, R. W., P. M. Forster, and J. M. C. Plane. 2007. Potential climatic effects of meteoric smoke in the Earth's paleo-atmosphere. *Geophys. Res. Lett.* 34.

Chapter 13

58. Rapp, M., I. Strelnikova, B. Strelnikov, P. Hoffmann, M. Friedrich, J. Gumbel, L. Megner et al. 2010. Rocket-borne in situ measurements of meteor smoke: Charging properties and implications for seasonal variation. *J. Geophys. Res.* 115:art. no. D00I16.

59. Rapp, M., J. M. C. Plane, B. Strelnikov, G. Stober, S. Ernst, J. Hedin, M. Friedrich, and U. P. Hoppe. 2012. In situ observations of meteor smoke particles (MSP) during the Geminids 2010: Constraints on MSP size, work function and composition. *Ann. Geophys.* 30:1661–1673.

60. Hervig, M. E., L. L. Gordley, L. E. Deaver, D. E. Siskind, M. H. Stevens, J. M. Russell, S. M. Bailey, L. Megner, and C. G. Bardeen. 2009. First satellite observations of meteoric smoke in the middle atmosphere. *Geophys. Res. Lett.* 36:art. no. L18805.

61. Shettle, E. P., M. T. DeLand, G. E. Thomas, and J. J. Olivero. 2009. Long term variations in the frequency of polar mesospheric clouds in the Northern Hemisphere from SBUV. *Geophys. Res. Lett.* 36:art. no. L02803.

62. Rapp, M. and G. E. Thomas. 2006. Modeling the microphysics of mesospheric ice particles: Assessment of current capabilities and basic sensitivities. *J. Atmos. Solar Terr. Phys.* 68:715–744.

63. Hervig, M. E., L. E. Deaver, C. G. Bardeen, J. M. Russell III, S. M. Bailey, and L. L. Gordley. 2012. The content and composition of meteoric smoke in mesospheric ice particles from SOFIE observations. *J. Atmos. Solar Terr. Phys.* 84–85:1–6.

64. Curtius, J., R. Weigel, H. J. Vossing, H. Wernli, A. Werner, C. M. Volk, P. Konopka et al. 2005. Observations of meteoric material and implications for aerosol nucleation in the winter Arctic lower stratosphere derived from in situ particle measurements. *Atmos. Chem. Phys.* 5:3053–3069.

65. Mills, M. J., O. B. Toon, V. Vaida, P. E. Hintze, H. G. Kjaergaard, D. P. Schofield, and T. W. Robinson. 2005. Photolysis of sulfurid acid vapor by visible light as a source of the polar stratospheric CN layer. *J. Geophys. Res.* 110:art. no. D08201.

66. Arijs, E., D. Nevejans, J. Ingels, and P. Frederick. 1985. Recent stratospheric negative-ion composition measurements between 22 km and 45 km Altitude. *J. Geophys. Res.* 90:5891–5896.

67. Arnold, F., R. Fabian, and W. Joos. 1981. Measurements of the height variation of sulfuric-acid vapor concentrations in the stratosphere. *Geophys. Res. Lett.* 8:293–296.

68. Wise, M. E., S. D. Brooks, R. M. Garland, D. J. Cziczo, S. T. Martin, and M. A. Tolbert. 2003. Solubility and freezing effects of Fe^{2+} and Mg^{2+} in H_2SO_4 solutions representative of upper tropospheric and lower stratospheric sulfate particles. *J. Geophys. Res.* 108:art. no. 4434.

69. Voigt, C., H. Schlager, B. P. Luo, A. D. Dornbrack, A. Roiger, P. Stock, J. Curtius et al. 2005. Nitric acid trihydrate (NAT) formation at low NAT supersaturation in polar stratospheric clouds (PSCs). *Atmos. Chem. Phys.* 5:1371–1380.

70. Dhomse, S. S., R. W. Saunders, W. Tian, M. P. Chipperfield, and J. M. C. Plane. 2013. Plutonium-238 observations as a test of modelled transport and surface deposition of meteoric smoke particles. *Geophys. Res. Lett.* 40:4454–4458.

71. Johnson, K. S. 2001. Iron supply and demand in the upper ocean: Is extraterrestrial dust a significant source of bioavailable iron? *Global Biogeochem. Cycle* 15:61–63.

72. Feder, J., K. C. Russell, J. Lothe, and G. M. Pound. 1966. Homogeneous nucleation and growth of droplets in vapours. *Adv. Phys.* 15:111–178.

73. Draine, B. T. 1979. Time-dependent nucleation theory and the formation of interstellar grains. *Astrophy. Space Sci.* 65:313–335.

74. Kohler, T. M., H.-P. Gail, and E. Sedlmayr. 1997. MgO dust nucleation in M-Stars: Calculation of cluster properties and nucleation rates. *Astronom. Astrophys.* 320:553–567.

75. Gail, H. P. and E. Sedlmayr. 1998. Dust Formation in M Stars. In *The Molecular Astrophysics of Stars and Galaxies.* T. W. Hartquist and D. A. Williams, eds. Clarendon Press, Oxford, U.K., pp. 285–312.

76. Gail, H. P. and E. Sedlmayr. 1999. Mineral formation in stellar winds—I. Condensation sequence of silicate and iron grains in stationary oxygen rich outflows. *Astronom. Astrophys.* 347:594–616.

77. Jeong, K. S., J. M. Winters, T. Le Bertre, and E. Sedlmayr. 2003. Self-consistent modeling of the outflow from the O-rich Mira IRC-20197. *Astronom. Astrophys.* 407:191–206.

78. Donn, B. and J. A. Nuth. 1985. Does nucleation theory apply to the formation of refractory circumstellar grains? *Astrophys. J.* 288:187–190.

79. Cherchneff, I. and E. Dwek. 2010. The Chemistry of population III supervnova ejecta. II. The nuecleation of molecular clusters as a diagnostic for dust in the early universe. *Astrophys. J.* 713:1–24.

80. Wooldridge, M. S. 1998. Gas-phase combustion synthesis of particles. *Prog. Energy Combust. Sci.* 24:63–87.

81. Patzer, A. B. C., A. Gauger, and E. Sedlmayr. 1998. Dust formation in stellar winds—VII. Kinetic nucleation theory for chemical non-equilibrium in the gas phase. *Astronom. Astrophys.* 337:847–858.

82. Paquette, J. A. and J. A. Nuth III. 2011. The lack of chemical equilibrium does not preclude the use of classical nucleation theory in circumstellar outflows. *Astrophys. J. Lett.* 737:art. no. L6.

83. Plane, J. M. C. 2013. On the nucleation of dust in oxygen-rich stellar outflows. *Phil. Trans. R. Soc. A* 371:art. no. 20120335.

84. Gail, H. P., S. V. Zhukovska, P. Hoppe, and M. Trieloff. 2009. Stardust from asymptotic giant branch stars. *Astrophys. J.* 698:1136–1154.

85. Goumans, T. P. M. and S. T. Bromley. 2013. Stardust silicate nucleation kick-started by $SiO + TiO_2$. *Philos. Trans. R. Soc. A* 371:art. no. 20110580.

86. Foresman, J. B. and A. Frisch. 1996. *Exploring Chemistry with Electronic Structure Methods.* Gaussian, Inc., Pittsburgh, PA.

87. Montgomery, J. A., M. J. Frisch, J. W. Ochterski, and G. A. Petersson. 2000. A complete basis set model chemistry. VII. Use of the minimum population localization method. *J. Chem. Phys.* 112:6532–6542.

88. Frisch, M. J., G. W. Trucks, H. B. Schlegel, G. E. Scuseria, M. A. Robb, J. R. Cheeseman, G. Scalmani et al. 2009. *Gaussian 09, Revision A.1.* Gaussian Inc., Wallingford, CT.

89. Gilbert, R. G. and S. C. Smith. 1990. *Theory of Unimolecular and Recombination Reactions.* Blackwell, Oxford, U.K.

90. De Avillez Pereira, R., D. L. Baulch, M. J. Pilling, S. H. Robertson, and G. Zeng. 1997. Pressure and temperature dependence of the multichannel rate coefficients for the $CH_3 + OH$ System. *J. Phys. Chem.* 101:9681.

91. Rollason, R. J. and J. M. C. Plane. 2001. A kinetic study of the reactions of MgO with H_2O, CO_2 and O_2: Implications for magnesium chemistry in the mesosphere. *Phys. Chem. Chem. Phys.* 3:4733–4740.

92. Miller, J. A. and S. J. Klippenstein. 2006. Master equation methods in gas phase chemical kinetics. *J. Phys. Chem. A* 110:10528–10544.

93. Robertson, S. H., D. R. Glowacki, C.-H. Liang, C. Morley, R. Shannon, M. Blitz, and M. J. Pilling. 2013. MESMER (Master Equation Solver for Multi-Energy Well Reactions), 2008–2012; an object oriented C++ program for carrying out ME calculations and eigenvalue-eigenvector analysis on arbitrary multiple well systems, http: //sourceforge.net/projects/mesmer, accessed February 20, 2014.

94. Georgievskii, Y. and S. J. Klippenstein. 2005. Long-range transition state theory. *J. Chem. Phys.* 122:art. no. 194103.

95. Bartis, J. T. and B. Widom. 1974. Stochastic models of the interconversion of three or more chemical species. *J. Chem. Phys.* 60:3474–3482.

96. Jacobson, M. Z. 2005. *Fundamentals of Atmospheric Modeling.* Cambridge University Press, New York.

97. Fuchs, N. A. 1964. *Mechanics of Aerosols.* Macmillan, New York.

98. Weissenrieder, J., S. Kaya, J. L. Lu, H. J. Gao, S. Shaikhutdinov, H. J. Freund, M. Sierka, T. K. Todorova, and J. Sauer. 2005. Atomic structure of a thin silica film on a Mo(112) substrate: A two-dimensional network of SiO_4 tetrahedra. *Phys. Rev. Lett.* 95:076103.

99. Treacy, M. M. J., I. Rivin, E. Balkovsky, K. H. Randall, and M. D. Foster. 2004. Enumeration of periodic tetrahedral frameworks. II. Polynodal graphs. *Micropor. Mesopor. Mat.* 74:121–132.

100. Foster, M. D. and M. M. J. Treacy. http://www.hypotheticalzeolites.net.

101. Wales, D. J., M. A. Miller, and T. R. Walsh. 1998. Archetypal energy landscapes. *Nature* 394:758–760.

102. Bromley, S. T., B. Bandow, and B. Hartke. 2008. Structural correspondences between the low-energy nanoclusters of silica and water. *J. Phys. Chem. C* 112:18417–18425.

103. Zwijnenburg, M. A., A. A. Sokol, C. Sousa, and S. T. Bromley. 2009. The effect of local environment on photoluminescence: A time-dependent density functional theory study of silanone groups on the surface of silica nanostructures. *J. Chem. Phys.* 131:art. no. 034705.

104. Bromley, S. T., M. A. Zwijnenburg, and T. Maschmeyer. 2003. Fully coordinated silica nanoclusters: $(SiO_2)_N$ molecular rings. *Phys. Rev. Lett.* 90:art. no. 035502.

105. Greaves, G. N. 1978. Color-centers in vitreous silica. *Philos. Mag. B* 37:447–466.

106. Lucovsky, G. 1979. Spectroscopic evidence for valence-alternation-pair defect states in vitreous SiO_2. *Philos. Mag. B* 39:513–530.

107. Street, R. A. and N. F. Mott. 1975. States in gap in glassy semiconductors. *Phys. Rev. Lett.* 35:1293–1296.

Chapter 13

108. Hamad, S. and S. T. Bromley. 2008. Low reactivity of non-bridging oxygen defects on stoichiometric silica surfaces. *Chem. Commun.* 44:4156–4158.

109. Flikkema, E. and S. T. Bromley. 2004. Dedicated global optimization search for ground state silica nanoclusters: $(SiO_2)_N$ (N=6–12). *J. Phys. Chem. B* 108:9638–9645.

110. Rignanese, G. M., A. De Vita, J. C. Charlier, X. Gonze, and R. Car. 2000. First-principles molecular-dynamics study of the (0001) alpha-quartz surface. *Phys. Rev. B* 61:13250–13255.

111. Murashov, V. V. 2005. Reconstruction of pristine and hydrolyzed quartz surfaces. *J. Phys. Chem. B* 109:4144–4151.

112. Du, Q., E. Freysz, and Y. R. Shen. 1994. Vibrational spectra of water molecules at quartz water interfaces. *Phys. Rev. Lett.* 72:238–241.

113. Goumans, T. P. M., C. R. A. Catlow, and W. A. Brown. 2008. Catalysis of addition reactions by a negatively charged silica surface site on a dust grain. *J. Phys. Chem. C* 112:15419–15422.

114. Flikkema, E. and S. T. Bromley. 2003. A new interatomic potential for nanoscale silica. *Chem. Phys. Lett.* 378:622–629.

115. Bromley, S. T. and E. Flikkema. 2005. Columnar-to-disk structural transition in nanoscale $(SiO_2)_N$ clusters. *Phys. Rev. Lett.* 95:art. no. 185505.

116. Bromley, S. T. and F. Illas. 2007. Energetics and structures of the initial stages of nucleation of $(SiO_2)_N$ species: Possible routes to highly symmetrical tetrahedral clustersw. *Phys. Chem. Chem. Phys.* 9:1078–1086.

117. Wales, D. J. and J. P. K. Doye. 1997. Global optimization by basin-hopping and the lowest energy structures of Lennard-Jones clusters containing up to 110 atoms. *J. Phys. Chem. A* 101:5111–5116.

118. Doye, J. P. K. and D. J. Wales. 1998. Thermodynamics of global optimization. *Phys. Rev. Lett.* 80:1357–1360.

119. Tsuneyuki, S., M. Tsukada, H. Aoki, and Y. Matsui. 1988. 1st-principles interactomic potential of silica applied to molecular dynamics. *Phys. Rev. Lett.* 61:869–872.

120. Bromley, S. T. 2006. Predicting the low energy landscape of nanoscale silica using interatomic potentials. *Phys. Stat. Sol. A* 203:1319–1323.

121. Guest, M. F., I. J. Bush, H. J. J. Van Dam, P. Sherwood, J. M. H. Thomas, J. H. Van Lenthe, R. W. A. Havenith, and J. Kendrick. 2005. The GAMESS-UK electronic structure package: Algorithms, developments and applications. *Mol. Phys.* 103:719–747.

122. Stephens, P. J., F. J. Devlin, C. F. Chabalowski, and M. J. Frisch. 1994. Ab initio calculation of vibrational absorption and circular dichroism spectra using density functional force fields. *J. Phys. Chem.* 98:11623–11627.

123. Chu, T. S., R. Q. Zhang, and H. F. Cheung. 2001. Geometric and electronic structures of silicon oxide clusters. *J. Phys. Chem. B* 105:1705–1709.

124. Harkless, J. A. W., D. K. Stillinger, and F. H. Stillinger. 1996. Structures and energies of SiO_2 clusters. *J. Phys. Chem.* 100:1098–1103.

125. Nayak, S. K., B. K. Rao, S. N. Khanna, and P. Jena. 1998. Atomic and electronic structure of neutral and charged Si_nO_m clusters. *J. Chem. Phys.* 109:1245–1250.

126. Wang, L. S., S. R. Desai, H. Wu, and J. B. Nichloas. 1997. Small silicon oxide clusters: Chains and rings. *Zeit. Phys. D* 40:36–39.

127. Johnston, R. L. 2002. *Atomic and Molecular Clusters.* Taylor & Francis, London, U.K.

128. Zwijnenburg, M. A., S. T. Bromley, J. C. Jansen, and T. Maschmeyer. 2004. Toward understanding extra-large-pore zeolite energetics and topology: A polyhedral approach. *Chem. Mater.* 16:12–20.

129. Kuo, C.-L., S. Lee, and G. S. Hwang. 2008. Strain-induced formation of surface defects in amorphous silica: A theoretical prediction. *Phys. Rev. Lett.* 100:art. no. 076104.

130. Avramov, P. V., I. Adamovic, K. M. Ho, C. Z. Wang, W. C. Lu, and M. S. Gordon. 2005. Potential energy surfaces of Si_mO_n cluster formation and isomerization. *J. Phys. Chem. A* 109:6294–6302.

131. Flikkema, E. and S. T. Bromley. 2009. Defective to fully coordinated crossover in complex directionally bonded nanoclusters. *Phys. Rev. B* 80:art. no. 035402.

132. Burton, M. G., M. Bulmer, A. Moorhouse, T. R. Geballe, and P. W. J. L. Brand. 1992. Fluorescent molecular-hydrogen line emission in the far red. *Mon. Not. R. Astron. Soc.* 257:P1–P6.

133. Le Bourlot, J., G. P. D. Forets, E. Roueff, A. Dalgarno, and R. Gredel. 1995. Infrared diagnostics of the formation of H_2 on interstellar dust. *Astrophys. J.* 449:178–183.

134. Jaquet, R., V. Staemmler, M. D. Smith, and D. R. Flower. 1992. Excitation of the fine structure transitions of $O(^3P_J)$ in collision with ortho-H_2 and para-H_2. *J. Phys. B* 25:285–297.

135. Gould, R. J. and E. E. Salpeter. 1963. Interstellar abundance of hydrogen molecule. 1. Basic processes. *Astrophys. J.* 138:393–407.

136. Williams, D. A. 2003. In *Solid State Astrochemistry*. V. Pirronello, J. Krelowski, and G. Manico, eds. Kluwer, Dordrecht, the Netherlands.

137. Goumans, T. P. M., C. Richard, A. Catlow, and W. A. Brown. 2009. Formation of H_2 on an olivine surface: A computational study. *Mon. Not. Roy. Astron. Soc.* 393:1403–1407.

138. Garcia-Gil, S., D. Teillet-Billy, N. Rougeau, and V. Sidis. 2013. H atom adsorption on a silicate surface: The (010) surface of forsterite. *J. Phys. Chem. C* 117:12612–12621.

139. Downing, C. A., B. Ahmady, C. R. A. Catlow, and N. H. de Leeuw. 2013. The interaction of hydrogen with the {010} surfaces of Mg and Fe olivine as models for interstellar dust grains: A density functional theory study. *Philos. Trans. R. Soc. A* 371:art. no. 20110592.

140. Goumans, T. P. M. and S. T. Bromley. 2011. Hydrogen and oxygen adsorption on a nanosilicate— A quantum chemical study. *Mon. Not. Roy. Astron. Soc.* 414:1285–1291.

141. Vidali, G., V. Pirronello, L. Li, J. Roser, G. Manico, E. Congiu, H. Mehl et al. 2007. Analysis of molecular hydrogen formation on low-temperature surfaces in temperature programmed desorption experiments. *J. Phys. Chem. A* 111:12611–12619.

142. Kerkeni, B. and S. T. Bromley. 2013. Competing mechanisms of catalytic H_2 formation and dissociation on ultrasmall silicate nanocluster dust grains. *Mon. Not. Roy. Astron. Soc.* 435:1486–1492.

143. Roberts, C. and R. L. Johnston. 2001. Investigation of the structures of MgO clusters using a genetic algorithm. *Phys. Chem. Chem. Phys.* 3:5024–5034.

144. Zhao, Y. and D. G. Truhlar. 2004. Hybrid meta density functional theory methods for thermochemistry, thermochemical kinetics, and noncovalent interactions: The MPW1B95 and MPWB1K models and comparative assessments for hydrogen bonding and van der Waals interactions. *J. Phys. Chem. A* 108:6908–6918.

145. Schlegel, H. B. 1982. Optimization of equilibrium geometries and transition structures. *J. Comp. Chem.* 3:214–218.

146. Lübken, F. J. 1999. Thermal structure of the Arctic summer mesosphere. *J. Geophys. Res.* 104:9135–9149.

147. Seele, C. and P. Hartogh. 1999. Water vapor of the polar middle atmosphere: Annual variation and summer mesosphere conditions as observed by ground-based microwave spectroscopy. *Geophys. Res. Lett.* 26:1517–1520.

148. Lübken, F. J., J. Lautenbach, J. Höffner, M. Rapp, and M. Zecha. 2009. First continuous temperature measurements within polar mesosphere summer echoes. *J. Atmos. Solar Terr. Phys.* 71:453–463.

149. Murray, B. J. and E. J. Jensen. 2010. Homogeneous nucleation of amorphous solid water particles in the upper mesosphere. *J. Atmos. Solar Terr. Phys.* 72:51–61.

150. von Cossart, G., J. Fiedler, and U. von Zahn. 1999. Size distributions of NLC particles as determined from 3-color observations of NLC by ground-based lidar. *Geophys. Res. Lett.* 26:1513–1516.

151. Rapp, M. and F. J. Lübken. 2004. Polar mesosphere summer echoes (PMSE): Review of observations and current understanding. *Atmos. Chem. Phys.* 4:2601–2633.

152. Witt, G. 1969. The Nature of noctilucent clouds. *Space Res.* 9:157–169.

153. Cox, R. M. and J. M. C. Plane. 1998. An ion-molecule mechanism for the formation of neutral sporadic Na layers. *J. Geophys. Res.* 103:6349–6359.

154. Woodcock, K. R. S., T. Vondrak, S. R. Meech, and J. M. C. Plane. 2006. A kinetic study of the reactions $FeO^+ + O$, $Fe^+N_2 + O$, $Fe^+O_2 + O$ and $FeO^+ + CO$: Implications for sporadic E layers in the upper atmosphere. *Phys. Chem. Chem. Phys.* 8:1812–1821.

155. Kopp, E. 1997. On the abundance of metal ions in the lower ionosphere. *J. Geophys. Res.* 102:9667–9674.

156. Balsiger, F., E. Kopp, M. Friedrich, K. M. Torkar, U. Wälchli, and G. Witt. 1996. Positive ion depletion in a noctilucent cloud. *Geophys. Res. Lett.* 23:93–96.

157. Megner, L. and J. Gumbel. 2009. Charged meteoric particles as ice nuclei in the mesosphere: Part 2. A feasibility study. *J. Atmos. Solar Terr. Phys.* 71:1236–1244.

158. Gumbel, J. and L. Megner. 2009. Charged meteoric smoke as ice nuclei in the mesosphere: Part 1—A review of basic concepts. *J. Atmos. Solar Terr. Phys.* 71:1225–1235.

159. Plane, J. M. C. 2000. The role of sodium bicarbonate in the nucleation of noctilucent clouds. *Ann. Geophys. Atmos. Hydrospheres Space Sci.* 18:807–814.

160. Self, D. E. and J. M. C. Plane. 2002. Absolute photolysis cross-sections for $NaHCO_3$, $NaOH$, NaO, NaO_2 and NaO_3: Implications for sodium chemistry in the upper mesosphere. *Phys. Chem. Chem. Phys.* 4:16–23.

Chapter 13

161. Cox, R. M., D. E. Self, and J. M. C. Plane. 2001. A study of the reaction between $NaHCO_3$ and H: Apparent closure on the chemistry of mesospheric Na. *J. Geophys. Res.* 106:1733–1739.

162. Plane, J. M. C. 2011. On the role of metal silicate molecules as ice nuclei. *J. Atmos. Solar Terr. Phys.* 73:2192–2200.

163. Murphy, D. M. and T. Koop. 2005. Review of the vapour pressures of ice and supercooled water for atmospheric applications. *Q. J. R. Meteorol. Soc.* 131:1539–1565.

164. Mora-Fonz, M. J., C. R. A. Catlow, and D. W. Lewis. 2007. Modeling aqueous silica chemistry in alkali media. *J. Phys. Chem. C* 111:18155–18158.

165. Trinh, T. T., A. P. J. Jansen, R. A. van Santen, and E. J. Meijer. 2009. Role of water in silica oligomerization. *J. Phys. Chem. C* 113:2647–2652.

166. White, C. E., J. L. Provis, G. J. Kearley, D. P. Riley, and J. S. J. van Deventer. 2011. Density functional modelling of silicate and aluminosilicate dimerisation solution chemistry. *Dalton Trans.* 40:1348–1355.

167. Schaffer, C. L. and K. T. Thomson. 2008. Density functional theory investigation into structure and reactivity of prenucleation silica species. *J. Phys. Chem. C* 112:12653–12662.

168. Trinh, T. T., X. Rozanska, F. Delbecq, and P. Sautet. 2012. The initial step of silicate versus aluminosilicate formation in zeolite synthesis: A reaction mechanism in water with a tetrapropylammonium template. *Phys. Chem. Chem. Phys.* 14:3369–3380.

169. Mochizuki, D., A. Shimojima, T. Imagawa, and K. Kuroda. 2005. Molecular manipulation of two- and three-dimensional silica nanostructures by alkoxysilylation of a layered silicate octosilicate and subsequent hydrolysis of alkoxy groups. *J. Am. Chem. Soc.* 127:7183–7191.

170. Sanchez, C., H. Arribart, and M. M. G. Guille. 2005. Biomimetism and bioinspiration as tools for the design of innovative materials and systems. *Nature Mat.* 4:277–288.

171. Yang, J. and E. G. Wang. 2006. Reaction of water on silica surfaces. *Curr. Opin. Solid State Mater. Sci.* 10:33–39.

172. Walsh, T. R., M. Wilson, and A. P. Sutton. 2000. Hydrolysis of the amorphous silica surface. II. Calculation of activation barriers and mechanisms. *J. Chem. Phys.* 113:9191–9201.

173. Masini, P. and M. Bernasconi. 2002. Ab initio simulations of hydroxylation and dehydroxylation reactions at surfaces: Amorphous silica and brucite. *J. Phys. Condens. Matter* 14:4133–4144.

174. Du, M. H., A. Kolchin, and H. P. Cheng. 2003. Water-silica surface interactions: A combined quantum-classical molecular dynamic study of energetics and reaction pathways. *J. Chem. Phys.* 119:6418–6422.

175. Ma, Y. C., A. S. Foster, and R. M. Nieminen. 2005. Reactions and clustering of water with silica surface. *J. Chem. Phys.* 122:art. no. 144709.

176. Wendt, S., M. Frerichs, T. Wei, M. S. Chen, V. Kempter, and D. W. Goodman. 2004. The interaction of water with silica thin films grown on Mo(112). *Surf. Sci.* 565:107–120.

177. Demontis, P., G. Stara, and G. B. Suffritti. 2003. Behavior of water in the hydrophobic zeolite silicalite at different temperatures. A molecular dynamics study. *J. Phys. Chem. B* 107:4426–4436.

178. Jelfs, K. E., E. Flikkema, and S. T. Bromley. 2012. Evidence for atomic mixing via multiple intermediates during the dynamic interconversion of silicate oligomers in solution. *Chem. Commun.* 48:46–48.

179. Flikkema, E., K. E. Jelfs, and S. T. Bromley. 2012. Structure and energetics of hydroxylated silica clusters, $(SiO_2)_M(H_2O)_N$, M=8, 16 and N=1–4: A global optimisation study. *Chem. Phys. Lett.* 554:117–122.

180. Jelfs, K. E., E. Flikkema, and S. T. Bromley. 2013. Hydroxylation of silica nanoclusters $(SiO_2)_M(H_2O)_N$, M = 4, 8, 16, 24: Stability and structural trends. *Phys. Chem. Chem. Phys.* 15:20438–20443.

181. Hassanali, A. A. and S. J. Singer. 2007. Model for the water-amorphous silica interface: The undissociated surface. *J. Phys. Chem. B* 111:11181–11193.

182. Bromley, S. T. and E. Flikkema. 2005. Novel structures and energy spectra of hydroxylated $(SiO_2)_8$-based clusters: Searching for the magic $(SiO)_8O_2H_3$ cluster. *J. Chem. Phys.* 122:art. no. 114303.

183. Jenkins, E. B. 2009. A unified representation of gas-phase element depletions in the interstellar medium. *Astrophys. J.* 700:1299–1348.

184. Whittet, D. C. B. 2010. Oxygen depletion in the interstellar medium: Implications for grain models and the distribution of elemental oxygen. *Astrophys. J.* 710:1009–1016.

185. Carrez, P., K. Demyk, P. Cordier, L. Gengembre, J. Grimblot, L. D'Hendecourt, A. P. Jones, and H. Leroux. 2002. Low-energy helium ion irradiation-induced amorphization and chemical changes in olivine: Insights for silicate dust evolution in the interstellar medium. *Meteorit. Planet. Sci.* 37:1599–1614.

Index

A

Acid/base catalyst properties
hydrogen-transfer processes, 323
vanadia/titania systems, 322
water interaction, 322–323
Active phase vanadia supported
titania catalyst
crystallites, 319
DeNO$_x$ reaction, 320
polymerization degree, 319
Raman spectroscopy, 320
surface structures, 320–321
Aeronomy of Ice in the Mesosphere
(AIM) satellite, 377
AFM, *see* Atomic force microscopy
Alkali doping methanol oxidation
adsorption energies, 329
methanol exposure, 328
vanadia/titania periodic
model, 329
Alkali-halide clusters, 35
Alkane oxide dehydrogenation
(ODH), 315–316
All-shell homogeneous conduction
process, 255
Antisymmetric many-electron wave
function, 56
Atom desorption, MgO
nanocrystals
hyperthermal O atoms, 302–303
point defects, 302
surface modifications, 301
Atomic force microscopy (AFM)
atomic resolution, 337
convolution shape, 341
electrostatic approach, 230
graphene stretched image, 231
Ru(0001) substrate, 354
Atomic layer deposition (ALD), 337
Atomic structure; *see also*
Nanoclusters and
nanoparticles
cubic/hexagonal-based
structure, 35
dendritic tree map, 8
energy-based cost function, 7
GM, 32–34, 36
H–O–H bond angle, 31
Li$_2$MnO$_3$ nanoparticle, 9
LM configurations, 7–8, 30
locally ergodic regions, 7
local minimum structure, 31
nanoparticles properties, 32
PBE solution energies, 38
saddle point, 8

stoichiometry, 30, 32
substitution effect, 262
Auger electron spectroscopy
(AES), 348

B

Ballistic electron transport, 249
Ballistic phonon transport, 248, 264
Band structure/quasiparticles
spectra, 277
Becke's hybrid exchange
functional, 298
Bending modulus distortion, 230,
232, 235
Berny algorithm, 317, 394
Bethe–Salpeter equation (BSE),
57, 277
Binary and ternary nanocompounds
cluster-assembled materials
Al$_{13}$-nanocluster, 140
DFT calculations, 142–143
global minimum
structure, 141
icosahedral structure, 140–141
ionic assemblies, 140
MetCars, 142
nanocluster approach,
141–142
quantum chemical
methods, 141
XO$_2$ oxides
DFT periodic
calculations, 137
mechanical properties, 136
nanocluster approach, 138
potential-based
calculations, 137
top to bottom method,
138–139
Bismuth telluride nanowires
lattice thermal conductivity, 260
Umklapp scattering, 259
Bloch wave packets, 53–54, 62, 73
B3LYP hybrid density functional,
298–299, 306
BNNRs, *see* Boron nitride
nanoribbons
Bohr–Oppenheimer MD
(BO-MD), 283
Boltzmann factor, 160, 251
Bond-order potentials, 217
Born–Oppenheimer approximation
electron-nuclear system, 61
PES, 122
polar thin films, 100

Boron nitride nanoribbons
(BNNRs)
GNR, thermal conductance of,
255, 257
hexagonal boron nitride, 255
schematic image of, 256
zigzag, thermal conductance
of, 257
Boron nitride nanotubes
(BNNTs), 251
Bragg peaks, 156, 341
Brillouin zone (BZ), 54, 62, 124, 133
Brittle fracture, 227, 230
Brønsted–Evans–Polanyi (BEP)
relation, 323, 396
Brown dwarf stars, 372
Brownian diffusion/
coagulation, 384
Brute-force determination, 344
Buckingham potential, 12, 237
Building solid blocks, 114–115

C

Canonical ensemble, 157–158
Capillary melting theory
chemical/mechanical type
equilibrium, 153
melting point, 154
nanoparticles, 153
Carbon nanotubes (CNTs), 224,
228, 251
Carnot efficiency, 257
Cartesian coordinates, 5, 20, 122
Casida equation, 275
Catalysts chemical reactivity
acid/base properties
hydrogen-transfer
processes, 323
vanadia/titania systems, 322
water interaction, 322–323
periodic DFT, 322
redox
H$_2$-TPR, 323
reactivity indices, 324
Charge density, 56–57
Charge/heat transport, ideal
systems
Bloch wave packets, 73
group velocity, 70
heat transfer, 72
intrinsic quantum properties, 73
Matthiessen's rule, 72
mobility calculation, 75
molecular wire, 76
Newton's laws, 69, 71, 77

Schrödinger equation, 69
tunneling effects, 75
Chemical ablation model
(CABMOD), 375–376
Chemical vapor deposition
(CVD), 337
Clapeyron–Clausius formula, 187
Classical ionic potentials,
296–298, 305
Classical nucleation theory (CNT)
Clapeyron–Clausius
formula, 187
cluster dimension *vs.* surface
roughness/chemical
inhomogeneity, 380
dust nucleation, 372
Earth's atmosphere, 379
free energy, 197
Gibbs–Duhem relation, 189
nucleus size, 186
supersaturations, 196
thermodynamic potential,
181, 188
two-step mechanism, 197–198
Cluster approach (CA), 122, 124
Cluster-assembled nanomaterials
Al_{13}-nanocluster, 140
assembly cluster conditions, 117
DFT calculations, 142–143
global minimum structure, 141
icosahedral structure, 140–141
ionic assemblies, 140
metastability
global/local minimum, 119
triaxial tensile strength, 120
MetCars, 142
nanocluster approach, 141–142
1:1 stoichiometry
nanoclusters, 127
quantum chemical methods, 141
solids nanocluster
properties, 118
3D modeling
approaches type, 122
GMOs, 123
Cluster assembly conditions
$C@Al_{12}$ dimer, 116–117
HOMO-LUMO gap, 116–117
matrices embed cluster, 115
superatom building blocks, 116
Complete basis set (CBS-Q)
method, 382
Computational melting methods
energy landscapes, 159–160
numerical simulations,
160–162
observables simulations,
162–163
statistical ensembles
canonical, 157–158
grand-canonical, 158–159
microcanonical, 157

Computational modeling methods
Coulombic interactions, 2D/3D
systems, 93–96
free-standing thin films
bulk phase choice,
96–97
nonpolar, parity/in-plane
symmetry, 98–99
polar, 99–101
surface reconstruction,
101–103
Tasker classification,
surface orientation and
termination, 97–98
Condensation nuclei (CN), 381
Core–shell nanowires
ballistic hole transport, 264
Ge core, 264
pristine, thermal conductance
of, 265
VLS method, 263
Correlated wavefunction-based
methods, 276
Cosmic and atmospheric
nanosilicates
cosmic dust, stellar origin of
dredge-up stage, 371
ground state
geometries, 373
IDPs, 374
ISM, 373
nucleosynthetic
processes, 371
SiO homomolecular
nucleation, 372
Earth's atmosphere meteoric
interactions
annual mean concentrations,
vertical profiles of, 376
astronomical model, 374
CABMOD, injection rates
of, 375
early warning, climate
change, 378
meteoric ablation, 376
meteoric debris, 375
MSPs, 377
smoke formation, 377
visible/near-IR
spectrometer, 377
silicates occurrence and
composition
crystal structures, 370
silicon-based anion, 369
Cosmic dust stellar origin
dredge-up stage, 371
ground state geometries, 373
IDPs, 374
ISM, 373
nucleosynthetic processes, 371
SiO homomolecular
nucleation, 372

Coulomb–Hartree–Fock LMs, 35
Coulombic interactions, 2D/3D
systems
Ewald sum, 94
excited electron/hole,
270–271
metal-oxide materials, 296
periodic boundary conditions,
93, 95
pseudo-2D calculation, 95
repulsion potential, 52, 56
Crystal-field effect, 140
Crystal growth
additives role, 208
kinetics/thermodynamics
attachment
equilibrium Wulff
morphology, 205
kinetic control, 206
urea crystals, 206–207
plane orientation, 199
solid nuclei, 198
surfaces
Einstein crystal, 202
equilibrium morphology,
200–201
free energy, 205
Kirkwood–Buff
equation, 202
metadynamics
simulation, 204
slab geometry, 201–202
thermodynamic integration,
202–203
Wulff theorem, 200
Crystalline minerals, 377
Crystal surfaces
Einstein crystal, 202
equilibrium morphology,
200–201
free energy, 205
Kirkwood–Buff equation, 202
metadynamics simulation, 204
slab geometry, 201–202
thermodynamic integration,
202–203
Cubic/hexagonal-based
structure, 35

D

Darwinian/Lamarckian evolution,
29–30
Defect states, one-dimensional
nanosystem
crystallinity, 76
modeling of, 77
Delta SCF (ΔSCF)
DFT, 274
photoluminescence,
281–282
vertical absorption spectra, 279

DeNO$_x$ vanadia/titania system, 315, 320

Density-functional-based tight-binding (DFTB), 220

Density functional theory (DFT)
 catalysts chemical reactivity, 322
 charge density, 57
 cluster-assembled materials, 142–143
 ΔSCF, 274
 electronic structure modeling approaches, 386
 energy/electronic structure theory, 57
 exchange-correlation functional, 297
 formalism properties, 299
 global minima/metastability, 126
 global optimization techniques, 344, 360
 ground-state theory, 274
 GW calculations, 277
 HFLE, 274
 IP, 315–316
 IR spectroscopy, 347
 Kohn–Sham energies, 275
 local minima, 123
 low energy silica clusters, 388
 MD simulations, 222
 MgO(100) ultrathin films, 360
 Mo(112) substrate, 349
 orbital energy differences, 274
 semi-empirical electronic structure, 214
 Si nanowires, 60
 spin-polarized, 142
 triple zeta basis set, 398
 van der Waals interactions, 355
 vertical absorption spectra, 278
 XO$_2$ oxides, 137

Density of states (DOS), 54, 58

DFT, see Density functional theory

Diffusive transport, 255

Dye-sensitized solar cells (DSSCs), 284; see also Photocatalysis

Dynamical simulation properties; see also One-dimensional nanosystems
 band structure calculation, 62
 electronic energy, 61
 incoherent interface, 107
 mean field theory, 56
 PBC, 62
 phonon dispersion, 63–66
 quantum-mechanical level, 62
 Seebeck coefficient, Si nanowires, 67
 wagging motion, 63, 65

E

Earth's atmosphere, meteoric interactions
 annual mean concentrations, vertical profiles of, 376
 astronomical model, 374
 CABMOD, injection rates of, 375
 crust, 369
 early warning, climate change, 378
 meteoric ablation, 376
 meteoric debris focus, 375
 meteoric smoke from, 377
 MSPs, 377
 visible/near-IR spectrometer, 377

Effective core potentials (ECP), 124

Einstein crystal, 202

Elastic constants, 222, 224, 226

Electroless etching (EE) method, 261

Electron affinities (EA), 115, 133, 274

Electron density, 217, 226

Electronic polarizability, 156

Electrostatic slab–slab interaction, 95

Embedded atom model (EAM), 217–218

Embedded cluster methods
 atomic-scale features, 298
 MgO nanocrystallines, 299
 nonperiodic systems, 297
 prediction properties, 298

Empirical potentials, 216, 220, 243

Endothermic (uphill) process, 317

Energy/electronic structure theory
 anion–cation repulsion, 58
 ballistic transport, 60
 band structure, 54, 59–61
 crystalline solid theories, 52
 electron–electron interactions, 56
 Peierl's distortion, 58
 periodic array, 54
 Schrödinger equation, 53–54
 theory levels
  DFT, 57
  Hartree–Fock theory, 56
  many-electron, 57
  semiempirical methods, 56–57

Energy functions vs. nanoclusters
 energy-based cost function, 10–11
 GM cluster, 14
 IPs, 11
 Pauli repulsion, 12
 potential-based methods, 11
 shell model, 13

Energy hypersurface, 5, 7, 41

Energy landscape, 5, 7, 159–160

Equation-of-motion-coupled cluster single doubles and triples (EOM-CCSDT) methods, 276, 279

Equation-of-motion-coupled cluster singles and doubles (EOM-CCSD) methods, 276, 279

Equilibrium morphology, 9–10

Euler–MacLaurin summation formula, 52

Evaporation/low-energy regime transition
 Arrhenius formula, 172
 caloric curves, 173
 harmonic approximation, 172
 thermal dissociation, 171

Ewald's approach, 52, 93–94, 108

Excited-state nano properties
 cartoon-based description, 270
 EBE, 270–272
 fluorescence, 271
 inorganic nanostructures, 270
 phosphorescence, 272
 potential energy surface, 270

Excited-state oxidation potential (ESOP), 285

Excited state relaxation/photoluminescence, 281

Exciton binding energy (EBE)
 bound state, 270
 optical/quasiparticle gap, 271

Exothermic (downhill) process, 317

F

Fermi's golden rule, 72, 279

Finite-temperature relaxation, 238

Finite vs. periodic catalyst models
 periodic models, 319–320
 vanadia/titania catalyst system, 318

Focused ion beam scanning electron microscopy (FIB-SEM), 293

Fourier transform, 52, 248, 362

Free-standing thin films
 bottom-up construction
  monolayer 2D sheets, 88
  stacking fault polytypism, 89
 bulk phase choice, 96–97
 conceptual approaches, 86
 nonpolar, parity/in-plane symmetry, 98–99
 polar, 99–101
 surface reconstruction, 101–103
 tasker classification, surface orientation and termination, 97–98

top-down model
    bulk and surface region, 87
    crystallographic planes, 88
    vacuum-exposed parallel
        surfaces, 85–86
Full configuration interaction (FCI)
    method, 276, 280

G

Gas-phase nanoparticles, 49, 156
Gaussian approximation potential
    method, 219–220
Gaussian 6-311G' basis set, 298, 306
Gaussian penalizing functions,
    52, 162
Generalized gradient approximation
    (GGA), 57, 123
Genetic algorithm (GA), 27,
    344–346
Geometric stability assessment,
    108–109
Geometry optimization approaches
    global minima
        DFT-based energy
            calculations, 126
        energy landscape, 124–125
        and metastability, 125–126
        volume diagrams vs.
            energy, 125
    local minima
        DFT, 123
        Monkhorst–Pack grid, 124
        quantum theory, 123
Gibbs–Duhem relation, 189
Gibbs energy change
    free energy, 317
    hypothetical formation
        process, 353
    MgO(100) ultrathin films, 361
    of reaction, 372
    stability vs. Mo(112) substrate,
        349, 351
    thermodynamics approach, 342
Gibbs–Thomson equation, 153
Global minima (GM) geometry
        optimizations
    atomic structure, 36
    ball-and-stick models, 32, 34, 36
    DFT-based energy
        calculations, 126
    energy functions, 14
    energy landscape, 124–125
    less stable configurations, 33
    and metastability, 125–126
    stable point, 6
    volume diagrams vs. energy, 125
Global optimization algorithms, 8
    independent walkers, 17
    interacting multiple walkers/
        population based, 27–30
    local gradient information, 22–27

non-GM LMs, 16
    single Monte Carlo walker,
        17–22
Grand-canonical ensemble
    isothermal–isobaric, 158
    Lennard–Jones cluster, 159
Graphene nanoribbon (GNR), 255
Graphitic layers, 88–89
Green's functions (GW) formalism,
    57, 277, 280
Ground-state oxidation potential
    (GSOP), 285

H

Haberland's experiment,
    156–157, 166
Hamiltonian methods, 55, 74
Hartree–Fock-like exchange (HFLE)
    charge-transfer excitations, 281
    correlated wavefunction
        methods, 276
    DFT, 274
    hybrid density functional, 382
    mean field, 56–58
Hartree–Fock (HF) method, 316
Helical replication, 221
Heterogeneous catalysis, 293
    catalysts chemical reactivity
        acid/base, 322–323
        redox, 323–324
    finite vs. periodic model,
        318–319
    methanol oxidation,
            formaldehyde reaction
        alkali doping, 328–330
        extensions/limitations, 328
        mechanisms, 325–328
        reactants/products, 325
    methods
        interatomic potentials,
            quantum mechanics,
            316–317
        potential energy
            surfaces, 317
    modeling, 314
        bulk materials, 90
        ionic systems, 108
    vanadia-supported titania
            catalyst, 321
        active phase, 319–321
        additives and support, 321
    vanadia/titania system
        alkane oxide
            dehydrogenation (ODH),
            315–316
        DeNOₓ, 315
        formaldehyde methanol
            oxidation, 315
        sulfuric acid synthesis, 315
HFLE, see Hartree–Fock-like
    exchange

High-resolution electron energy loss
        spectroscopy (HREELS)
    electronic/vibrational
        excitations, 339
    IR spectroscopy, 347
    vs. Mo(112) substrate, 348
High-thermal-conductivity
    ab initio study, 254
    BNNRs
        GNR, thermal conductance
            of, 255, 257
        hexagonal boron nitride, 255
        schematic image of, 256
        zigzag, thermal conductance
            of, 257
    BNNTs, 251
    coherent phonon transport, 254
    diffusive transport, 255
    heat conduits, 250
    microscopic mechanisms, 254
    semiconducting feature, 250
    theoretical and experimental
            studies
        chemical bonds, 252
        intrinsic phonon dispersion
            relationship, 254
        lattice thermal conductivity,
            251
        molecular dynamics
            simulations, 252
        phonon–phonon scattering,
            252–253
        transport characteristic
            lengths, 254
        unstable isotopes effects, 253
    thermal conduction path, 255
Hohenberg–Kohn theorem, 275
H_2O ice condensation nanosilicates
    Gibbs free energies, 398
    homogeneous nucleation, 397
    $MgSiO_3$ optimized
        geometries, 399
    MSP nuclei, 397
    PMSE, 397
Homogeneous nucleation, 379
H_2-temperature-programmed
        reduction (H_2-TPR), 323
Hydroxylation, nanosilicates
    ice formation, 404
    MCBH global optimization
        algorithm, 400
    nanocluster structures, 401
    pyroxene-type composition, 400
    relative total energies, 402
    stabilization energies, 403
Hypothetical structures, 85

I

Ideal excited state method
        properties, 273–274
Independent walkers, 17

Infrared radiation (IR) spectroscopy
  DFT software packages, 347
  HREELS, 347
  silicate dust, 370
  vibrational excitations, 339
Infrared reflection-absorption
        spectra (IRAS)
  absorption bands, 348
  electronic structures, 354
  MgO(100) ultrathin films, 362
  vibrational spectroscopies, 347
Inherent structures, 163
Inorganic clusters/nanoparticles
        melting
  ionic, 169–170
  mechanical properties
    nanoscale, characterization
        of, 222–227
    nanoscaled objects,
        experimental
        approaches, 227–237
    ZnO nanowires, atomistic
        simulations of, 237–242
  metal/nanoalloys, 166–169
  piezoelectricity, 213
  semiconducting and covalent,
        170–171
  simulation techniques and
        models, 214–222
  thermal properties of
    Fourier's law, 248
    nanotubes, 247
    phonons, 248
  thermal transport
    high-thermal, 250–256
    low-thermal, 256–265
    quantized thermal
        conductance, 249–250
    thermal conductivity and
        thermal conductance,
        248–249
  van der Waals method, 164–165
Interacting multiple walkers/
        population based
  atomic structure prediction, 27
  diversity, 28
  phenotype crossover, 29
  population-based algorithms, 27
Interatomic potentials (IP)
  DFT, 315–316
  reactive force field (FF), 316
Internal conversion (IC), 271
Interplanetary dust particles
        (IDPs), 374
Intersystem crossing (ISC), 271
Inverse Laplace transformation, 384
Ionic clusters
  Coulomb force, 169
  Monte Carlo simulations, 170
  and nanoparticles, 169
Ionization potentials (IE), 133
Ion-mobility measurements, 156

IR/Raman spectra, 142
Isolated MgO nanocrystals
  atom desorption, 301–303
  charge trapping, 300
  energy diagrams of, 301
  MIES optical excitations, 300
  properties of, 299
Isotope abundance ratio, 251
Interstellar medium (ISM)
  CNT, 379
  cosmic dust, stellar origin
        of, 371
  dust grains, 373

J

Jahn–Teller-driven distortions, 13
Junge layer, 378

K

Kasha's rule, 271, 281
Kinetic density, 57
Kinetics/thermodynamics crystal
        attachment
  equilibrium Wulff
        morphology, 205
  kinetic control, 206
  urea crystals, 206–207
Kirkwood–Buff equation, 202
Knoevenagel condensation/
        transesterification
        reactions, 294
Kohn–Sham energy equations
  DFT, 57, 275
  exchange-correlation
        functionals, 275
  local geometry
        optimizations, 123
  quantum cluster, 299
Kronig–Penny model, 54–56, 68
Kubo–Greenwood formulism, 74

L

Landscape model properties
  atomic structures, 4
  LM/GM, 6
  locally ergodic regions,
        5–7
  nanoparticles, structure of, 4
  walker, energy landscape,
        5–6
Langmuir evaporation, 375
Lateral distortion, see Bending
        modulus distortion
Lattice thermal conductivity, 251
LDOS, see Local density of states
LEED, see Low-energy electron
        diffraction
Legendre transformations, 158

Lennard–Jones cluster potential
  atomic configurations, 35
  classical-nucleation-style
        mechanism, 196
  dynamical coexistence, 169
  microcanonical ensemble, 159
  van der Waals forces, 164
Lewis acidic properties, 314
Lindemann melting theory, 153
Linear-response approach, 275
Liquid–vapor transition
  evaporation/low-energy regime
    Arrhenius formula, 172
    caloric curves, 173
    harmonic
        approximation, 172
    thermal dissociation, 171
  multifragmentation
    computational modeling, 174
    evaporation, 173
    Fisher model, 174
Local density approximation
        (LDA), 123
Local density of states (LDOS)
  electronic structure, 339
  Fermi level, 347
  model validation, 346–347
Local gradient information, single
        walkers
  catchment area, 24
  classical harmonic springs, 27
  Hessian/inverted Hessian, 24
  MCBH, 25–26
  optimization schemes, 22
  quasi-Newtonian approaches,
        22–23
  take-home messages, 24
Locally ergodic regions, 5–7
Local minimum (LM) geometry
        optimizations
  ball-and-stick model, 36–37
  catchment area, 24
  configurations, 30
  DFT, 123
  energy landscapes, 25
  low-energy atomic structures, 24
  metastable stationary point, 6
  Monkhorst–Pack grid, 124
  quantum theory, 123
Local vs. global structure
        optimization
  atomic structures, 344
  cut and splice crossover
        operation, 345
  GA, 344–346
  structure evolution, 346
Long duration exposure facility
        (LDEF), 376
Longitudinal acoustic (LA)
        mode, 63
Longitudinal optical (LO)
        mode, 63

Long-term interactions; *see also*
 One-dimensional
  nanosystems
 atomistic and electronic
  structure codes, 52
 electrostatic potential, 50
 infinite 1D chain model, 51
 macroscopic local field, 52
 Taylor expansion, 51
 2D lattice, 52
Low-energy electron diffraction
 (LEED)
 diffraction methods, 341
 I–V characteristics, 352
 MgO(100) single crystal, 359
 nanofilm structure analysis, 341
 Ru(0001) substrate, 354, 356
 x-ray diffraction (XRD), 341
Low-thermal-conductivity materials
 bismuth telluride nanowires
  lattice thermal
   conductivity, 260
  temperature dependence, 259
  Umklapp scattering, 259
 Carnot efficiency, 257–258
 core–shell nanowires
  ballistic hole transport, 264
  pristine, thermal
   conductance of, 265
  VLS method, 263
 PGEC, 258
 quantum-confinement effects, 258
 Seebeck coefficient, 258
 silicon nanowires
  low frequency phonons, 261
  phonon-boundary
   scattering, 262
  surface roughness, 261–262
  temperature
   dependence, 261
  ubiquitous elements, 261
 thermoelectrics, 256

M

Magic clusters, 113; *see also*
  Nanocluster-assembled
  materials
Magnesium oxide (MgO)
  nanopowders
 atom desorption
  hyperthermal O atoms,
   302–303
  point defects, 302
  surface modifications, 301
 charge trapping/optical
  excitations
  energy diagrams of, 301
  MIES, 300
 elastic/isolated properties,
  293, 300
 heterogeneous catalysis, 293

nanocrystal interfaces
 atomistic models, 305–306
 electronic properties, 304
 TEM images, 304
 particle–particle interfaces, 294
 spectroscopic properties, 295
Makov–Payne correction, 96
Many-body perturbation theory
 (MBPT), 277, 285
Many-electron Schrödinger
 equation, 57, 276
Master equation solver for multiwell
 energy reactions
 (MESMER), 383–384
Matthiessen's rule, 72
MCBH, *see* Monte Carlo basin
 hopping
Mechanical deformation, 255
Melting and phase transitions
 computational methods
  energy landscapes, 159–160
  numerical simulations,
   160–162
  observables simulations,
   162–163
  statistical ensembles,
   157–159
 inorganic clusters/nanoparticles
  ionic, 169–170
  metal/nanoalloys, 166–169
  semiconducting and
   covalent, 170–171
  van der Waals method,
   164–165
 liquid–vapor transition
  evaporation/low-energy
   regime, 171–173
  multifragmentation, 173–174
 macroscopic systems, 151
 phenomenology
  capillary theory,
   nanoparticles, 153–154
  experimental methods,
   155–157
  fluctuations, 154–155
  Lindemann theory, 153
 scalable regime, 152
Metal clusters and nanoalloys
 covalent bonds, 167
 melting temperatures, 166
 silver-rich nanoclusters, 168
 sodium/aluminum, 166
Metallocarbohedrenes
 (MetCars), 142
Metal-supported low-dimensional
 silica
 Mo(112) substrate
  DFT calculations, 349
  high-resolution *vs.* simulated
   images, 352
  vs. HREEL silica spectra, 348
  isolated clusters, 348–349

  stability *vs.* Gibbs energy
   change, 349, 351
  structure models of, 350
  2D phase diagram of,
   353–354
 Ru(0001) substrate
  amorphous film model,
   357–358
  honeycomb structure, 355–356
  IRA spectra, 354
  LEED, 354, 356
  phase diagram of, 355–356
  vitreous silica bilayer film
   model, 357
 silica (SiO_2), 347
Metal supported-oxide nanofilms
 applications, 339–340
 atomic structure
  characterization, 336
 characterization, 337–339
 local *vs.* global structure
  optimization, 344–346
 low-dimensional silica
  Mo(112) substrate, 348–354
  Ru(0001) substrate, 354–358
 model validation, 346–347
 preparation methods, 337
 stability, thermodynamics,
  342–343
 structure and properties
  experimental methods,
   340–341
  theory, role of, 341–342
  water layers, MgO(100) ultrathin
   films, 359–363
Metastability cluster, 119–120, 124
Metastable impact electron
 spectroscopy (MIES), 300
Metastable stationary point, 6
Meteoric smoke particles (MSPs)
 Earth's atmosphere meteoric
  interactions, 377
 H_2O ice condensation
  nanosilicates, 398
 midlatitudes, mesosphere, 378
Methanol oxidation, formaldehyde
 reaction, 315; *see also*
  Catalysts chemical
  reactivity
 alkali doping
  adsorption energies, 329
  methanol exposure, 328
  vanadia/titania periodic
   model, 329
 energetic profile, 327
 extensions and limitations, 328
 gas-phase reactants, 325
 mechanisms, 325–328
 reaction pathways, 326
 transition structure
  stabilization, 327
Metropolis acceptance criterion, 20

MgO(100) ultrathin films
  DFT molecular dynamics, 360
  flat dissociated structure, 360
  Gibbs energy change, 361
  global structure optimizations, 360
  IRAS, 362
  ordered water layers, 359
  prototype system, 359
  simulated IR spectra, 362–363
Microcanonical ensemble, 157
Microelectromechanical system
      (MEMS), 233
Modeling mesoscale structure, 296
Molecular dynamics (MD)
  full-excited-state, 273
  ground-state, 283
  Monte CarloMC, 190
  nanoscale materials, 240
  simulation strategies, 9–10,
      214–215
  thermal transport, 253, 259, 263
Molecular photodissociation, 373
Møller–Plesset (MP*n*), 57
Monkhorst–Pack scheme, 124, 137
Monte Carlo (MC)-based global
      optimization, 10, 190
Monte Carlo basin hopping
      (MCBH) algorithm
  global optimization, 387, 394
  hydroxylation of, 400
  molecular dynamics, 190
  numerical simulations, 160
  random configuration, 25–26
Morse potential, 12
Mo(112) substrate silica
  DFT calculations, 349
  high-resolution *vs.* simulated
      images, 352
  vs. HREEL silica spectra, 348
  isolated clusters, 348–349
  stability *vs.* Gibbs energy
      change, 349, 351
  structure models of, 350
  2D phase diagram of, 353–354
MSPs, *see* Meteoric smoke particles
Multicanonical ensemble
      sampling, 162
Multidimensional landscape, 6
Multireference wavefunction
      method, 276
Multiscale models, nanocrystalline
      structure
  classical ionic potentials, 296
  computational modeling, 295
  mesoscale structure, 296
  quantum mechanical
      B3LYP hybrid density
          functional, 298
      many-electron Schrodinger
          equation, 297
Multiwalled carbon nanotubes
      (MWCNTs), 229

N

Nanocalorimetry experiments, 156
Nanocluster-assembled materials
  assembly conditions, 115–117
  building blocks, 114–115
  compounds
      binary and ternary, 136–143
      1:1 stoichiometry, 127–136
  geometry optimizations
      global minima/metastability,
          124–126
      local minima, 123–124
      3D cluster modeling, 122–123
  metastability, 119–120
  solid properties, 117–119
Nanoclusters and nanoparticles
  atomic structure of, 30–39
  challenges, 6–10
  definition of, 4
  energy functions, 10–16
  global optimization algorithms
      independent walkers, 17
      interacting multiple walkers/
          population based, 27–30
      local gradient information,
          22–27
      non-GM LMs, 16
      single Monte Carlo walker,
          17–22
  landscape model, properties, 4–6
Nanocrystalline oxide material
      interfaces
  applications, 291–292
  crystal growth, 198–200
      additives role, 208
      kinetics/thermodynamics,
          205–208
      surfaces, 200–205
  experimental probes, 293
  MgO nanopowders properties,
      293–295
      atom desorption, 301–303
      charge trapping/optical
          excitations, 300–301
      nanocrystal interfaces,
          303–306
  nucleation, 184–186
      classical theory, 186–189,
          195–198
      simulation, 189–195
  powders to ceramics, 292–293
  surface tension/pressure, 182
  theoretical modeling
      embedded cluster approach,
          298–299
      multiscale models, 295–298
Nanoparticles/adsorbed molecules
  DSSCs, 284
  mechanical characterization of
      active volume, 226
      auxetic behavior, 224

  CNTs, 224
      elastic regime, 226
      graphite, 225
      isotropic system, 222–223
      Poisson's ratio, 223–224
      single-walled nanotube,
          224–225
      Young's modulus, 223–225
  phenomenology
      capillary theory, 153–154
      experimental methods,
          155–157
      ionic clusters, 169
      Lindemann melting
          theory, 153
      semiconducting and
          covalent, 170
      statistical fluctuations,
          154–155
  TD-DFT, 285
Nanoscaled objects
  AFM, 230–231
  atomistic simulations, 236
  cantilevered boron
      nanowire, 233
  CNTs, 228
  counterelectrode, 232
  elasticity, 228
  image-processing
      procedure, 229
  mechanical
      characterization, 228
  mechanical testing, 231
  MEMS, 233
  microfabricated mechanical
      test, 236
  MWCNTs, 229
  numerical experiment, 236
  platinum deposition, 234
  sword-in-sheath fashion, 233
  ultrafiltration alumina
      membrane, 230
  vibrational amplitude, 229
  wave-like distortion, 232
  Young's modulus, 229,
      234–235
  ZnO nanobelts, 235
Nanosilicates
  chemistry
      catalyzed H_2 formation and
          dissociation, 394–396
      H_2O ice condensation,
          nuclei, 396–399
      hydroxylation of,
          399–405
  cosmic and atmospheric
      cosmic dust, stellar origin of,
          371–374
      Earth's atmosphere meteoric
          interactions, 374–379
      silicates occurrence and
          composition, 369–370

formation
    classical nucleation theory,
        379–380
    low energy cluster
        structures, 385–394
    silicate dust formation,
        kinetic modeling,
        380–385
low energy cluster structures
    alpha-quartz ground
        states, 392
    cluster stability, 391
    computational modeling, 385
    DFT, 388
    global minima, 389
    open-shell dangling
        bonds, 387
    PESs, 386
    strong liquid, 387
    structures, symmetries, 393
    surface defects, 388
    transition clusters, 390
    VAP defects, 387
material properties, 84
silicate dust formation, kinetic
        modeling
    amorphous magnesium-
        iron-silicates, 381
    CBS-Q level, 382–383
    hydrolysis, 385
    perovskite, 381
    PESs, 380, 383
    RRKM theory, 383
Neural network method, 219–220
Newton's laws, 69, 71, 77
Nitric acid trihydrate (NAT), 378
Nonpolar thin films, 98–99
Nucleation
    additive molecules, 186
    classical theory
        Clapeyron–Clausius
            formula, 187
        free energy, 197
        Gibbs–Duhem relation, 189
        nucleus size, 186
        supersaturations, 196
        thermodynamic
            potential, 188
        two-step mechanism, 197–198
    crystallization, 184
    simulation
        carbonate/bicarbonate
            ratios, 191
        collective variable, 192–193
        gradient density, 194
        growing nucleus, 195
        MC/MD, 190
        mechanical
            Hamiltonian, 191
        Steinhardt order parameters,
            192–194
    thermodynamics laws, 185

Numerical simulations, phase
        transitions
    Markov chain, 161
    superposition approach, 160
    thermal equilibrium, 161
    Wang–Landau approach, 162

O

Ohm's law, 248
One-dimensional nanosystems
    applications of, 48
    atomic structure, 47
    charge/heat transport, ideal
        systems, 70–76
    defect states, 76–77
    dynamical properties, 61–67
    energy/electronic structure
        theory, 52–61
    energy landscape, 161–162
    infinite 1D systems, 50
    long-term interactions, 50–52
    structure/phase transitions, 67–70
1D inorganic nanomaterials, 72
1D nanowires, 49
1D silica stripes, 353–354
1:1 stoichiometry nanoclusters
    cluster-assembled solids, 127
    EA, 133
    ground state, 128
    hexagonal faces, 134
    IE, 133
    octahedral cluster, 129, 132
    SOD-$(MX)_{12}$, 128
    top to bottom method, 130–132
    van der Waals solid, 135
    volume diagrams vs. total
        energy, 133, 135
Open-shell transition metal
        cations, 13
Optical absorption spectra, 278
Optical modeling, excited system
    applications
        band structure, 277–278
        excited state relaxation,
            281–284
        nanoparticles/adsorbed
            molecules, charge
            transfer, 284–285
        photoluminescence, 281–284
        quasiparticles spectra,
            277–278
        vertical absorption spectra,
            278–280
    excited-state properties
        modeling, role of, 272–273
        potential energy surface,
            270–272
    methods
        BSE, 277
        correlated wavefunction-
            based, 276

DFT, 274–275
        ideal excited state, 273–274
        MBPT, 277
        time-dependent DFT,
            275–276
Optical spectroscopy interface, 304
Optoelectronics, 4

P

Parity/in-plane symmetry, 98–99
Pauli exclusion principle, 12, 56
Peierl's distortion, 58, 64
Perdew–Burke–Ernzerhof
        exchange-correlation, 58
Periodic boundary conditions
        (PBCs), 49, 62, 122, 124
PESs, see Potential energy surfaces
Phonon-boundary scattering, 252
Phonon frequencies, 63
Phonon-glass electron crystal
        (PGEC), 258
Phonon–phonon scattering,
        252–253, 264
Photocatalysis, 284
Photoelectron diffraction (PD), 339
Photoluminescence
    DSCF, 281–282
    and excited-state electron
        density, 281–282
    TD-DFT, 281
Physical vapor deposition (PVD), 337
Polar mesospheric summer echoes
        (PMSE), 397
Polar thin films; see also Top-down
        process, thin films
    BCT structure, 101
    Born–Oppenheimer
        approximation, 100
    electrostatic (Madelung) energy,
        99–100
    uncompensated polarity, 100
Population-based algorithms, 27
Positron annihilation
        spectroscopy, 293
Potential energy surfaces (PESs)
    contiguous grains, 384
    energy-grained master
        equation, 383
    Gibbs free energy, 317
    low energy cluster structures,
        386, 392
    packed clusters, 386
    strong liquid, 387
Poynting–Robertson drag, 374
Printed electronics, 292
Pyroxene
    oxygen-sharing $[SiO_4]^{4-}$
        tetrahedra, 369
    silicate nanoparticle, 394
    small cluster, 372
    water molecules, 403

Q

Quantized thermal conductance
   ballistic electron transport, 249
   Boltzmann's constant, 249
   CNTs, 250
   Plank's constant, 249
   quantum of, 249–250
   temperature dependence of,
      249–250
Quantum chemistry method, 222
Quantum cluster, 306
Quantum-confinement effects, 258
QUANTUM ESPRESSO program
   package, 142
Quantum mechanical (QM)
   models
   B3LYP hybrid density
      functional, 298
   many-electron Schrödinger
      equation, 297, 316
Quantum Monte Carlo
   technique, 222
Quasi-Newtonian approaches,
   22–23
Quasiparticles spectra
   and band structure, 277
   Green's function–based GW
      method, 277–278

R

Raman spectroscopy, 320
Random phase approximation
   (RPA), 57
Reactive ballistic deposition
   (RBD), 302
Recrystallization, 107
Redox catalysts
   H_2-TPR, 323
   reactivity indices, 324
   semiconductor band edges
      species, 284
Resonant Raman spectroscopy,
   60, 63
Rice–Ramsperger–Kassel–Markus
   (RRKM) theory, 383
Rigid-ion model, 217
Rigid-rotor harmonic oscillator
   approximation, 343
Roly-poly analogy, 21
Runaway nucleation, 379
Runge–Gross theorem, 275
Ru(0001) substrate silica
   amorphous film model,
      357–358
   honeycomb structure, 355–356
   IRA spectra, 354
   LEED, 354, 356
   phase diagram of, 355–356
   vitreous silica bilayer film
      model, 357

S

Satellite-borne optical
   spectrometers, 376
Scanning tunneling microscopy
   (STM)
   model validation, 346
   point defects, 302
   spectroscopic tool, 339
   structure/properties, metal
      oxide films, 340
Schrödinger equation, 54, 70, 74,
   273, 316
Seebeck coefficient, 257–258
Selective catalytic reduction
   (SCR), 315
Self-consistent field (SCF), 56–57
Semiconducting/covalent
   nanomaterials
   miniaturized electronics, 170
   rms bond length fluctuation
      index, 171
Semiempirical Hamiltonian
   method, 56–57, 284
Silicon nanowires
   low frequency phonons, 261
   phonon-boundary
      scattering, 262
   surface roughness, 261–262
   ubiquitous elements, 261
Silver nanoparticles, antibacterial
   properties, 3
Simulating nucleation
   annealing, 20
   carbonate/bicarbonate
      ratios, 191
   collective variable, 192–193
   gradient density, 194
   growing nucleus, 195
   MC/MD, 190
   mechanical Hamiltonian, 191
   Steinhardt order parameters,
      192–194
Simulation techniques,
   nanostructures
   atomistic simulation, 214
   cohesive energy, 217
   conjugate gradient method,
      215, 218
   covalent systems, 217
   DFT, 214, 222
   EAM model, 218
   empirical potential
      approach, 216
   Gaussian process regression, 220
   Hamiltonian operator, 220
   helical replication, 221
   interactions, 216
   MD, 215–216
   microcanonical, 215
   neural network method,
      219–220
   phase transformation, 219
   quantum mechanical
      treatment, 216
   rigid-ion model, 217
   semiempirical electronic
      structure models, 214
   strain energy *vs.* chiral
      angle, 221
   strain tensor, 215
   surface effects, 218
   TB models, 220
   Tersoff many-body
      potential, 215
   unit cell, 220
   wires, relaxed configurations in,
      218–219
Si nanowire
   DFT, 60
   electron mobility, 74
   Schrödinger and Poisson
      equations, 74
   Seebeck coefficient, 67
Single Monte Carlo walker
   energy landscapes, 19
   hold points, 21
   interatomic distance, 17
   LM nanocluster, 18
   local gradient operators, 2
   lowest-energy configuration, 19
   moveclass operator, 20
   pseudorandom path, 19
Single reference wavefunction
   method, 276
Single-walled nanotube (SWNT),
   224, 226
Sol-gel techniques, 337
Solids properties, nanoclusters
   cluster-assembled materials, 118
   physics techniques, 120
   prototypical porous materials,
      119
Spin-polarized DFT
   calculations, 142
Stacking fault polytypism, 88
Stationary points, 5–6
Statistical ensembles, melting
   properties
   canonical, 157–158
   grand-canonical, 158–159
   microcanonical, 157
Statistical fluctuation phase
   transition, 154–155
Steinhardt order parameters,
   192–194
Stillinger–Weber potential,
   170, 217
Stoichiometric silica nanoclusters,
   30, 32, 387
Stress–strain relation, 233
Structure and properties, metal
   oxide films
   I–V curves, 341

local *vs.* global structure optimization
atomic structures, 344
cut and splice crossover operation, 345
GA, 344–346
structure evolution, 346
model validation
Tersoff–Hamann approach, 346
vibrational spectroscopies, 347
stability, thermodynamics equilibrium
composition, 342
Gibbs energy, 342–343
vs. of structures, 343
STM images, 340
SXRD, 341
theory, role of, 341
Structure/phase transitions; *see also* One-dimensional nanosystems
MD simulations, 67
optical eigenmodes, 68
phonon dispersion, 69
Young's modulus, 67
Sulfuric acid synthesis, 315
Sulfuric acid tetrahydrate (SAT), 378
Superatom cluster, 115–116
Supported thin films; *see also* Two-dimensional nanosystems
charge transfer, 103
epitaxial relationships
coherent interface, 104
commensurate interface, 106
cubic lattices schematic representation, 105
electronic redistribution, 103
incoherent interfaces, 107
lattice mismatch, 107
2D square lattices, 104
structureless dielectric continuum, 90
substrate polarization and charge transfer, 103
Surface-disordered Ge–Si core–shell NWs, 262, 265
Surface-hopping molecular dynamics (SH-MD), 283
Surface orientation and termination
bulk crystal, 97
reconstruction
genetic algorithms, 103
microfaceting, 101–102
rock salt surface, 102
symmetric repeat unit, 98
Tasker classification, 97
Surface x-ray diffraction (SXRD), 339

T

Tasker classification surfaces, 97
TEM, *see* Transmission electron microscopy
Temperature gradient, 248–249
Temperature-programmed desorption (TPD), 361
Tensile strength, 233, 238
Ternary/quad–ternary compounds, 38
Terrestrial olivine, 370
Tersoff–Hamann approach, 346
Tersoff manybody potential, 217
Thermal conductivity, 248–249
Thermodynamics stability
equilibrium composition, 342
functions, 163
Gibbs energy, 342–343
vs. structures, 343
Thermoelectrics, 256
3D electron tomography, 293
3D Ewald sum, 94–95, 108
3D interrogation techniques, 293
3D periodicity and convergence, 9, 108
$3N$-dimensional gradient, 5
Time-dependent DFT (TD-DFT)
adsorbed molecules, 285
correlated wavefunction-based methods, 279
groundstate/static, 275
inorganic materials *vs.* organic molecules, 275
nanoparticles, 285
photoluminescence, 281
vertical absorption spectra *vs.* DSCF method, 279
Titanium dioxide-based particles, 3
Top-down process, thin films
bulk and surface region, 87
crystallographic planes, 88
Transition structure stabilization, 327
Transmission electron microscopy (TEM)
anchored NTs, 229
electron beam, 339
nanocrystal interfaces, 304
3D electron tomography, 293
Transparent ceramics development, 292
Transverse acoustic (TA) mode, 63
Transverse optical (TO) mode, 63
Two-dimensional nanosystems
computational modeling structures
bottom-up construction, 88–90
Coulombic interactions, 2D/3D systems, 93–96
free-standing thin films, 96–103
top-down model, 87–88
geometric stability assessment, 108–109

heterostructures, 90–91, 108
nanoscopic, 84
supported thin films, 90
epitaxial relationships, 103–107
substrate polarization and charge transfer, 103
2D nanofilm, 86
2D periodic boundary conditions, 9
2D ring network, 357
2D silica structures, 353

U

Ultrahigh vacuum (UHV), 337
Ultrathin metal oxide films, 337
characterization
HREELS, 339
surface science methods, 337, 339
electrical properties, 339
heterogeneous catalysts, model systems, 340
preparation methods, 337
Umklapp processes, 252
UV–visible (UV–Vis) absorption spectrum, 271–272

V

Valence alternation pair (VAP) model, 387
Valence-forcefield analysis, 142
Vanadia supported titania catalyst
active phase
crystallites, 319
DeNO$_x$ reaction, 320
polymerization degree, 319
Raman spectroscopy, 320
surface structures, 320–321
additives, 321
physicochemical properties, 321
support effect, 321
Vanadia/titania systems
acid/base properties, 322
alkane oxide dehydrogenation (ODH), 315–316
catalysis, 314
DeNO$_x$, 315
Lewis acidic properties, 314
methanol oxidation, formaldehyde, 315
olefins formation, 315
sulfuric acid synthesis, 315
Vanadium dioxide particles, 3
van der Waals clusters
argon clusters, heat capacities of, 165
DFT, 355
forces, 88, 118, 135
inorganic clusters/nanoparticles melting, 164–165
interactions, 12

Lennard–Jones cluster
potential, 164
1:1 stoichiometry nanoclusters, 135
Vapor–liquid–solid (VLS), 261
Vertical absorption spectra
DFT orbital energy
approximation, 278
optical absorption spectrum,
278–279
TD-DFT *vs.* DSCF method, 279
TiO_2 monomer excitations, 280

W

Walker, energy landscape, 5
Wang–Landau approach, 162
Wide-gap semiconductor, 251
Wulff construction theorem, 9–10,
200, 205

X

XO_2 cluster-assembled solids
DFT periodic calculations, 137
mechanical properties, 136
nanocluster approach, 138
potential-based calculations, 137
top to bottom method,
138–139

Y

Yang–Lee theorem, 152
Young's modulus
bulk value, 238
core-shell model, 240
diameter dependence of, 241
size dependence of, 238
transformation, 240–241

Z

ZnO nanowires
atomistic simulations
Binks, 237
bulk crystal, 238
elastic constants, 239
MD simulations, 237
quasi-static fashion, 238
short-range exponential
repulsion, 237
surface relaxation effects, 238
Young's modulus
bulk value, 238
core-shell model, 240
diameter dependence
of, 241
size dependence of, 238
transformation, 240–241

T - #0481 - 071024 - C42 - 254/178/19 - PB - 9780367783044 - Gloss Lamination